INTRODUCTION TO PROGRAMMING LANGUAGES

Arvind Kumar Bansal

Kent State University
Ohio, USA

CRC Press
Taylor & Francis Group
Boca Raton London New York

CRC Press is an imprint of the
Taylor & Francis Group, an **informa** business
A CHAPMAN & HALL BOOK

CRC Press
Taylor & Francis Group
6000 Broken Sound Parkway NW, Suite 300
Boca Raton, FL 33487-2742

Printed on acid-free paper
Version Date: 20140520

International Standard Book Number-13: 978-1-4665-6514-2 (Paperback)

Visit the Taylor & Francis Web site at
http://www.taylorandfrancis.com

and the CRC Press Web site at
http://www.crcpress.com

This book is dedicated to my parents for all their love and care, who, despite their limitations and financial hardships, taught their children to dream.

Contents

Preface

Programming language is a core topic in the undergraduate curricula of computer science. It integrates abstract concepts in computation, programming paradigms—styles to express formal solutions to problems; compilers; low-level execution behavior of programs; and operating systems. As students grow their understanding of computer science, it becomes clear to them that instructions and data representations in various programming languages have a common purpose and features that can be abstracted—identified using their common properties. Once provided with a deeper understanding of abstractions, students can superimpose the syntax on top of these abstractions to learn quickly new programming languages in the fast-changing world of computer science.

There are multitudes of powerful programming languages. Educational institutions teach the first course in programming, and data structures in multiple languages such as C++, Java, PHP, Python, and C. The use of different syntax for the same abstraction tends to confuse students who are fresh from core courses in computer science. The main purpose of this book is to free students from the shackles of syntax variations in languages and biases of specific programming paradigm(s).

This classroom-tested material introduces programming language concepts at an abstract level, freeing them from the restraints of multiple language syntax. The text is designed for computer science/IT courses focusing on the principles or concepts of programming languages. The book is suitable as a textbook for a semester-long course at the sophomore/junior levels to teach concepts of programming language design and implementation. It can also be used as a textbook for an introductory graduate-level course in programming language or as a reference book for other graduate-level courses in programming languages.

The book provides background on programming language concepts and discusses the features of various paradigms. The book teaches: (1) the common features of the programming languages at the abstract level rather than at a comparative level; (2) the implementation model and behavior of programming paradigms at an abstract levels so that students understand the power and limitations of programming paradigms; (3) language constructs at a paradigm level; and (4) a holistic view of the programming language design and behavior.

In addition to the discussion of classical topics such as syntax and semantics, imperative programming, program structures, mechanisms for information exchange between subprograms, object-oriented programming, logic programming, and functional programming, this book adds new topics such as dependency analysis, communicating sequential processes, concurrent programming constructs, web and multimedia programming, event-based programming, agent-based programming, synchronous languages, high-productivity programming on massive parallel computers, and implementation models and abstract machines used in different programming paradigms. Effort has been made to include distributed computing topics such as models for mobile computing, remote procedure calls, remote method invocation, and the corresponding parameter passing mechanisms. With multicore computers, distributed networks of computers and the pervasiveness of the Internet, many of these topics have become relevant. Chapter 2 explains many background concepts at an abstract level. My experience in teaching this course is that these background concepts are taught to students at the program-writing level, when they lack the required abstract understanding. Besides, various schools have differing syllabi. Students will find it useful to refresh their understanding of the concepts through an abstract-level description.

This book illustrates programing constructs with intuitive reserved words instead of dwelling on the specific syntax of some languages. However, examples from newer representative languages such as Ada 2012, C++, C#, Haskell, Java, Lisp, Modula-3, Ruby, Scala, Scheme, Perl, Prolog, and Python have been included to illustrate the concepts.

Chapter Outlines

The scope of this book has been limited to the material covered in a semester-long course that teaches principles and theory of programming languages. However, the book has sufficient material to be used as a reference for the first course in programming languages at the entry level in a graduate program.

It is assumed that students who read this book will have a background of two semesters of programming, introductory courses in discrete structures and data structures, and some intuition about operating systems. The book assumes that students will have knowledge of at least one programming language, and capability of writing around 200 lines of code in a program involving blocks, functions, procedures, parameter passing, and various control structures such as if-then-else, case-statements, while-loop, do-while-loop, function, and procedure calls.

The book is divided into 15 chapters, including a concluding chapter. Concepts have been explained in a simple, intuitive language at an abstract level. Examples and case studies using popular and newer languages have been included and explained enough to take the sting out of understanding syntax. Owing to a multitude of languages being developed, a representative set has been chosen that in no way reflects any preference over other languages.

Chapter 1 discusses the need for a course in programming languages and the learning outcomes of this course. It also introduces the notion of a program, explains the difference between data and control abstractions, and explains the need for abstractions and high-level modeling for specifying solutions to real-world problems. It describes various problem domains and their differing requirements. It also describes briefly the software-development cycle to emphasize the properties of good programming languages. The conflicting natures of many of these properties are explained. A brief history of the evolution of programming languages at the programming paradigm level is explained. Finally, the chapter describes various classifications used to categorize programming languages.

Chapter 2 describes the mathematics and abstract concepts in computation needed to understand design and implementation models of programming languages. The chapter describes briefly properties of the von Neumann machine and its effect on programming style, related discrete structure concepts, related data structure concepts, related operating system concepts, and some required abstract concepts in computation. Discrete structure concepts described are sets and set operations, relations and functions, Boolean and predicate logic, functions, and types of recursion. Related data structure concepts are stacks and queues, depth-first and breadth-first search, trees and graphs, pointers and recursive data structures. Concepts related to abstract computation are the notions of variable, expression and command, environment and store, computational state, and transitions between computational states. Background concepts for concurrent programming such as concepts of processes, threads, and buffers are discussed.

There is some overlap between the abstract model of computation discussed in Chapters 2 and 4. However, due to mutual dependency between syntax and semantics, explained in Chapter 3, and abstractions in program structure and execution, part of the abstraction needed for understanding syntax and semantics is introduced in Chapter 2.

Chapter 3 describes the syntax—techniques to validate programming constructs and semantics—associating meanings to abstract programming constructs in programming language. It introduces the definitions of syntax and semantics, explains how the grammar for a programming language can be expressed and developed in the textual form needed for parsing and the visual form needed for human comprehension. The chapter illustrates the conversions between textual and visual forms. It describes different types of grammars, ambiguity in grammars, their powers and limitations, and their usage in the compilation process. It explains parsing—the process of validating sentence structure, its abstract algorithm, and parse trees—the trees formed during the sentence structure validation. The chapter also explains different types of semantics: operational semantics needed to translate high-level language constructs to low-level abstract instruction sets, axiomatic semantics (semantics based upon modeling a computational state as Boolean expressions), denotational semantics (semantics based upon mapping syntax rules into corresponding semantic rules in a domain), action semantics—(a practical integration of operational, axiomatic, and denotational semantics used by the compilation process), and behavior semantics (a semantics based upon high-level, state transition–based behavior of underlying programming constructs).

Chapter 4 describes abstractions and information exchange mechanisms between program units. The abstractions have been grouped as data abstractions and control abstractions. The chapter describes how this information can be transferred using scopes of the variables, import/export mechanisms across modules, and parameter-passing mechanisms between a calling and a called subprogram. The chapter also describes the mechanism of exception-handling in programming languages for graceful handling of error conditions. It describes the notion of nondeterministic programming style and languages. The chapter introduces the notion of meta-programs—programs that can use other programs as data, programs as first-class objects where a program can be developed as data at run time, and the concept of side effect—changes in a computational state that can cause incorrectness during program execution. The need and various techniques for interoperability between programming languages are also discussed.

Chapter 5 describes the implementation models and an abstract machine for executing imperative programs. The chapter describes two major models of implementation: static implementation and an integrated model that combines static, stack-based, and heap-based memory allocation. It discusses the limitations and advantages of the three memory allocation models: static, stacks based, and heap based; and describes how an integrated model reduces limitations of individual memory-allocation schemes. The chapter also describes how procedure calls and parameter passing can be modeled in von Neumann–style abstract machines, and compares the implementation of different parameter-passing mechanisms including parameter-passing mechanisms in functional programming and distributed computing.

Chapter 6 describes the structure of a heap—a common memory space that is used to allocate recursive and dynamic data objects. It also describes memory allocation and garbage collection—the process of recycling released memory space for memory reallocation. The chapter describes different types of garbage collection techniques grouped under four major categories: stop-and-start garbage collection schemes that suspend programs once garbage collection starts, incremental garbage collection schemes that allow program execution during garage collection, continuous, and concurrent garbage collection techniques. It describes the advantages, abstract algorithms, and limitation of various approaches.

Chapter 7 describes the theory of types. It explains the relationship of data abstraction described in Chapter 4 and set operations, and explains the advantages of type declarations and limitations of type theory to handle run-time errors. It describes various run-time errors that can be caused despite type declaration in compiled languages. The chapter also describes polymorphism—the capability to handle multiple (possible indefinite) data types by a generic function or operator. The chapter discusses the issue of type equivalence—when two variables of different types can be equated. The chapter describes the internal representation of type information during compilation for statically compiled languages. The chapter also describes the implementation models and compile-time type-inference mechanisms for polymorphic languages. It describes a case study of type-rich languages such as Ada 2012, Hope, and C#.

Chapter 8 describes the theory of nondeterministic programming and models of concurrent programming. It describes the notion of automated parallelization of program; program dependency graphs—a graph that shows dependency among program statements based upon data flow and control flow; sequentiality introduced due to writing information into shared data structures; the notion of coarse granularity—grouping the set of statements to be executed on the same processor to reduce data transfer overhead across processors and/or memory storage; and various programming constructs such as threads and monitors needed for concurrent execution of programs. The chapter also discusses the need for sequential consistency to regulate the run-time unpredictable behavior of concurrent programs caused by racing condition—a condition where the order of termination of concurrent process is not predictable. The chapter discusses various synchronization and mutual exclusion techniques including locks and transactional memory. Concepts are illustrated using a case study of four representative languages: Ada, CSP, Emerald, and Java.

Chapter 9 describes functional programming paradigm; lambda expressions and their evaluations; and programming constructs, implementation models for functional programming languages, and integration of functional programming with concurrent programming paradigm. The concept of λ-calculus—an underlying mathematical model for the evaluation of mathematical expressions—is discussed. The concept of a generic functional programming system (FPS) using generic kernel functions and generic function-forming operators to form complex functions is also introduced. Various functional programs are presented to illustrate the advantages and limitations of functional programming style that uses immutable data entities, compared to imperative programming style that supports destructive update of the variable-values. The data abstractions and control abstractions in many functional programming languages such as Lisp, Haskell, Hope, Ruby, and Scala are discussed. The chapter discusses SECD machine—a classical abstract machine for executing functional programs, and a graph-based abstract implementation models. Finally, the chapter describes concurrent functional programming, and the concurrency present in the Lisp family, Haskell, and Ruby.

Chapter 10 describes logic programming paradigm, abstractions in logic programs, and the implementation model of logic programs. It introduces the concept of forward chaining, backward chaining, the concept of unification to equate two logical terms and a parameter-passing mechanism in logic programs. It describes AND-OR tree as a means for the abstract implementation of logic programs, and shows how a logical-query can be mapped to an AND-OR tree. It describes backtracking as means to traverse the search space to generate multiple (possibly all) solutions to a problem. It introduces Prolog as an example of backward chaining backtracking-based language and describes simple features of Prolog. It describes briefly the WAM (Warren abstract machine)—the primary abstract implementation model for the efficient implementation of the AND-OR tree. It describes briefly various extensions of the logic programming paradigm such as temporal logic programming, constraint logic programming paradigm, higher-order logic programming. Finally it describes various attempts to integrate logic programming with functional programming and concurrent programming.

Chapter 11 describes the abstract concepts in the object-oriented programming paradigm, and briefly describes schematics to implement object-oriented programming languages. The chapter describes class hierarchy, inheritance, encapsulation, multiple inheritances, virtual methods, overrides, and dynamic binding. It describes a schematic of an abstract model to implement object-oriented languages that can handle dynamic binding, virtual methods, and multiple inheritance. It discusses inherent limitations of multiple inheritances. The chapter takes examples from Java, C++, Ruby, and Scala to illustrate the object-oriented programming concepts.

Chapter 12 discusses the web and multimedia programming paradigm. It discusses theoretical aspects of code and data mobility, the related security-related issues, compilation issues with the mobile code, and the need for middle-level software such as XML, JVM—Java Virtual Machine—and CLI: Common Language Interface. It describes XML and .NET based specification mechanism. The chapter describes how dynamic XML can be used to model dynamically changing graphs, 3D animated objects, and databases. It discusses multimedia formats to represent multimedia objects such as images, texts, audio, audio visuals, and streams. The issue of perception distortion due to the lack of synchronization is discussed. Various multimedia synchronization mechanisms and constructs are discussed. The last section provides case studies from three representative high-level languages: Synchronous Multimedia Integration Language (SMIL); Alice—a language that integrates visual programming paradigm and object-oriented programming paradigm to model animated objects; and C#—a popular multiparadigm language that integrates web computing, object-oriented programming, and event-based programming.

Chapter 13 discusses the concepts in the remaining programming paradigms: event-based programming, agent-based programming, synchronous programming, and object-based languages currently being developed for massive parallel computers. A detailed example is taken from C# to illustrate the program development and internal working of event-based programming. Agent-based programming is explained, and security issues and fault tolerance issues for agent-based programming are discussed. Partitioned Global Address space is discussed as a memory model for massive-parallel, high-productivity computing. Parallel constructs in X10 and Chapel—two languages integrating object-oriented programming and massive-parallel computing—are discussed. The concept of synchronous computation based upon signal emissions and logical clock is explained using abstract syntax for constructs taken from Estrel—a synchronous language.

Chapter 14 discusses the concepts of scripting languages. As application systems evolved, there has been a need to integrate shell-based programming, control and data abstractions, database programming, and web-based programming. Scripting languages provide this glue and have incorporated multiple features from different programming paradigms to provide a user-friendly programming style. The chapter discusses control and data abstractions in scripting languages, their string- and text processing capability, pattern-matching capabilities, integration with shell-based programming and web-based scripting. The examples have been taken from Perl, Prolog Python, and Ruby. The final section discusses various scripting features of Perl, PHP, Prolog, Python, and Ruby.

Chapter 15 summarizes all the discussions and describes the evolution of programming languages with respect to evolution in paradigms and evolution in implementation of languages. Possible future directions in the language development are also discussed. The last section explains that as the cluster of massive parallel computers becomes available to masses, there will be a need for programming languages that can integrate high-productivity massive parallel programming and user friendliness for robust software maintenance.

Classroom Use of This Book

The suggested details of timing efforts for various chapters are given in the table below. The teaching material takes approximately 28 lectures of 75 minutes each or 42 lectures of 45 minutes each. Chapter 2 discusses the background material and is quite helpful, as different schools have different course descriptions for data structure, discrete structures, and operating systems. Except for Section 2.4, the remaining material in Chapter 2 can be given as self-study or can be covered as needed. At the beginning of each chapter, the required background concepts are mentioned. Section 2.4 can be taught in sequence with other chapters.

Chapter	Time in Minutes	75-Minute Lectures	45-Minute Lectures	Suggested Minimum Coverage for Semester-Long Course
Chapter 1	105	1.5	2.5	Full
Chapter 2	40	0.5	1.0	Section 2.4; remaining self-study
Chapter 3	150	2.0	3.0	Full
Chapter 4	180	2.5	4.0	Full
Chapter 5	150	2.0	3.0	Full
Chapter 6	150	2.0	3.0	Full
Chapter 7	180	2.5	4.0	Full
Chapter 8	180	3.0	4.5	Sections 8.1 through 8.4, 8.7, and 8.8
Chapter 9	150	2.0	3.0	Sections 9.1 through 9.5.1 and 9.7
Chapter 10	150	2.0	3.0	10.1 through 10.3, 10.6
Chapter 11	100	2.0	3.0	11.1 through 11.4, 11.6; languages could be self-study
Chapter 12	100	1.5	2.0	12.1 through 12.5; languages could be self-study
Chapter 13	100	1.5	2.0	13.1 through 13.3
Chapter 14	100	1.0	1.5	14.1, 14.2; languages could be self-study
Chapter 15	50	0.5	1.0	15.1, 15.2
Total	**App. 1900**	**27 lectures**	**41 lectures**	

Acknowledgments

I offer my prayers to Goddess Saraswati for the inspiration to write this book. I acknowledge my parents who infused the love of reading in me and sacrificed a lot to help me achieve my dreams. I acknowledge my elder brother, Professor Arun Agrawal, whom I tried to emulate early in my life. I acknowledge all my students who have pointed out patiently various errors in classroom slides for the last two decades. Even after teaching and correcting the slides for the past two decades, I find errors and knowledge gaps every time I teach in this ever-changing field. I acknowledge various researchers, instructors, and professors who have improved, and will improve in subsequent editions, my knowledge of programming language theory by their book reviews, insightful articles, presentations, and their books. The artwork on the cover page is a generous gift from Prabha Singh—my senior from high school days. I am especially grateful to all those people who have helped me by editing, reviewing, proofreading, and printing this book, which was long overdue and lived only in the form of overheads and my voice in my lectures. I thank Angela Guercio who read a draft version page by page to correct many overlooked mistakes and offered many useful insights. I acknowledge my children, Ambika and Ambuj, who put up with my long working hours during the summer and fall of 2012. Finally, I acknowledge Randi Cohen, the acquiring editor at CRC Press, who made this endeavor possible. Part of this book was completed during a sabbatical from Kent State University.

About the Author

Arvind Bansal is a professor of computer science at Kent State University. He earned a BTech (1979) in electrical engineering, an MTech (1983) in computer science from Indian Institute of Technology at Kanpur, India, and a PhD (1988) in computer science from Case Western Reserve University, Cleveland, Ohio. He has been a faculty member of computer science at Kent State University, Kent, Ohio since 1988. Dr. Bansal has taught undergraduate-level and graduate-level courses regularly in the areas of artificial intelligence, multimedia systems, and programming languages. He has been teaching undergraduate programming languages regularly for the past 23 years.

Professor Bansal's research contributions are in the areas of concurrent logic programming, fault-tolerant agent-based systems, knowledge bases, program analysis, and XML-based multimedia languages and systems. His other research contributions are in bioinformatics, biological computing, and proteomics. He is a member of IEEE and ACM. He has served in more than 50 international conferences as a program committee member in the areas of artificial intelligence, logic programming, multimedia, parallel programming, and programming languages, and has contributed as an area editor to the international journal *Tools with Artificial Intelligence* for many years.

Glossary of Symbols

Symbol	Meaning
$\vee$	Logical-OR of two truth values
$\wedge$	Logical-AND of two truth values
...	More similar information omitted for better readability
$\mapsto$	Maps to. In defining store, the symbol has been used for memory location mapping to r-value and identifier mapping to memory location. In type theory, it has been used to map a domain element to a range element.
$\cup$	Union of two sets
$\prec_s$	Left-side entity is a subclass of the right-side entity
$\subseteq$	Left-side entity is a proper subset of right-side entity
$\bigcap$	Intersection of two sets
$\in$	Left-side entity is a member of right-hand side set
$\oplus$	Insertion of an entity in a bag of entities
$\uplus$	Disjoint union
$\succ$	Left-side entity succeeds right-side entity
$\prec$	Left-side entity precedes right-side entity
$\perp$	Bottom symbol used to show mapping to error condition in functions
$\rightarrow$	Denotes (1) implication, (2) delimiter separating guards from commands in guarded commands, and (3) transition between states from the current state to the next state.
$\times$	It has been used for (1) multiplication of numbers in definitions of grammars, and (2) Cartesian product in type theory.
{...}	Set of data-items
<...>	(1) Nonterminal symbols in syntax grammars, (2) sequence of data items, and (3) generic abstract symbol
[...]	(1) Dimension in multidimensional arrays and (2) list in functional programming languages Haskell and ML
(...)	(1) Enumeration set in program examples and (2) dimension of an array in Ada or Scala syntax

CHAPTER 1

Introduction

Computer science has become an exciting field since its inception in the late 1950s. What was conceived as a tool for large-scale scientific computing during its birth is now being used in almost every aspect of life, including medical science, space exploration, telecommunication, information exchange, remote collaboration, modeling, computer-aided design, automated navigation, automated manufacturing, automated surgery, planning, designing, productivity tools for presentation, electronic transactions and commerce, transportation, and managing utility distributions. Now we cannot think of our life without the use of an embedded or personal computer. Computers are embedded in many modern-day gadgets such as automobiles, cell phones, airplanes, spacecrafts, high-end washers and dryers, ovens, and home security systems. Now we talk of smart homes that will have computers to process multiple sensor information and take care of many routine aspects of the house.

Behind all these computer activities are the smart brains of programmers who have solved many complex problems and presented the solutions using high-level instructions. These instructions are translated to low-level machine instructions by automated language translators. These low-level instructions drive the execution of programs in computers. The machine-level instructions are limited in their expressive power and will not facilitate modeling and modification of the high-level solutions with the evolution of requirements. *Clearly there is a need for high-level languages and language constructs to express solutions to the complex problems and change these solutions as the requirements evolve.* These requirements come into existence due to restructuring of society caused by technological advancement. This calls for incremental improvement and well-regulated modification of software, so that modifying one part of the software does not adversely affect the other part. To localize the effects, the software, should be made modular: different modules with well-defined, non-overlapping functionality. These modules are loosely coupled with limited sharing of internal data structures and operations.

Modern-day problems are complex, and hundreds of thousands of lines of instructions are required to solve these problems. Developing such solutions is not easy and involves many man-years: the number of programmers multiplied with time in years. Such efforts mean large-scale commitment of organizational and financial resources. Such efforts cannot be duplicated and need to be augmented with evolutionary modification in the software. Otherwise, the delay in the development and the cost of the development of software would be tremendous and wasteful.

1.1 MULTITUDE OF PROBLEM DOMAINS

In the modern era, the problems to be solved by computers lie in different problem domains such as scientific computing, text processing, database programming, business applications, system programming, process automation, intelligent systems, web-based applications, and real-time processing. All these domains are quite different from each other with different requirements as follows.

An example of scientific computing is reasoning regarding the universe, where there are billions of stars. In order to reason about their interaction with each other and with our solar system, we have to process and analyze the data from computerized telescopes that get trillions of bytes of data. Another example of scientific computing is the simulation and tracking of atmospheric conditions such as tornadoes that cause devastation in coastal areas. Another example of scientific computing is processing, analyzing, and tracking seismic activities on earth. Processing scientific data and developing useful models to solve scientific problems require the capability to allocate and process large matrices in the computer. The numbers have to be processed accurately. For example, computing the trajectory of a spacecraft or a space shuttle requires precise calculations and high accuracy.

Examples of text processing are word processors and productivity tools that we use every day to write letters and prepare presentations. These kinds of problems require the capability to represent and process large amount of strings, pictures, tables, video clips, and other media objects in efficient ways. While large amount of computation is needed, the computational need is not as large as that needed in the scientific computing domain. However, it requires more interaction with humans and user friendliness.

Database programming requires organizing, processing, and searching huge amounts of data in a logical manner so that they can be accessed immediately without unnecessary duplications. Examples of database programming in real-time world are (1) processing student data in a registrar's office and (2) swiping a credit card at a gas pump. When students walk in the registrar's office and provide their *student IDs*, an accountant would know, in a split second, detailed information about the student's payments, apartment address, and courses being taken.

Business applications require extensive generation of reports that can be presented to the customers and upper division executives in a user-friendly manner. Business applications require capabilities to integrate database programming with user-friendly report generation. However, a part of business application processing can be done in a *batch mode*; real-time processing is not always needed.

System programming requires handling of multiple processes running concurrently, providing virtual environments to improve programmers' productivity, and interfacing with low-level programming to improve the execution efficiency. System programming requires stepping through computer memory, raising exceptions, giving out warnings if a process misbehaves, and interfacing user-level calls to low-level system calls.

Real-time processing requires capturing and processing data in real time. Any account-keeping job should not slow down the capture or processing of data, otherwise the corresponding real-time event may be missed with catastrophic outcomes. For example, if a computer in a nuclear plant responds slowly to an overheating condition, then the nuclear power plant may be damaged. If an onboard computer on a fighter aircraft is slow to respond, the aircraft may be wiped out by an enemy missile. These examples require providing high level of priority to real-time tasks, and taking quick, real-time decisions to facilitate capturing and processing real-time events.

In recent years, intelligent systems have been placed in multiple walks of life such as industrial robots, shop floor planning, airport scheduling, natural language understanding system, game playing such as chess that can beat grandmasters, and analyzing images. These systems have to work in a very large problem space and would need capabilities to logically analyze and intelligently guess the solutions. Generally, these problems are guided by heuristic programming—a sound intelligent guess based on mathematical modeling to quickly move toward the solution state. The problem is complicated due to inherent uncertainty in the behavior of the real-world phenomenon and lack of complete information.

In recent years, web-based programming combined with multimedia systems has put a different requirement on programming languages. The programs should be retrievable from remote websites (URLs) and executed on local machines. While retrieval from the remote machines requires conversion of data structures to strings and vice versa, efficient execution requires efficient translation to low-level instructions. Unfortunately, the translation and execution of web-based programs occur concurrently, slowing down the execution of programs, and needs just-in-time compilation techniques to speed up the processing. In future, we will see domain-specific languages to handle different domains.

1.2 MOTIVATION

Given a set of requirements in a problem domain, programs can be developed for automating a process. However, communication of the solutions to the computers is difficult, because computers lack intelligence to comprehend our intentions, cannot correct our mundane unintended errors, and lack capabilities to understand implications hidden behind our communications. Everything has to be modeled and communicated explicitly to the computers. The entities have to be modeled abstractly and precisely, and the solution needs to be constructed stepwise, without incorporating any ambiguity.

A programming language is an organized way of communicating with a computer, such that the computer behaves faithfully, according to the instructions given by the programmer. The instructions could use any media such as textual, visual, sign, gesture, audio, or their combinations to specify the solutions. However, the important criteria are as follows:

1. The solution of a problem can be easily and completely expressed.
2. There has to be a potential for the evolution of the specification of the solution.

3. There has to be one to one unambiguous translation between the programmer's intention and the action taken by the computer. Since the solutions are specified at a high level, there are indefinite possibilities of miscommunication between the programmer and the computer, unless there are enough unambiguous constructs to express the solutions.

As the level of automation increases, the society restructures itself by absorbing the level of automation, and new problem domains are developed. For example, when the computers were invented in 1950s, the perceived requirement was scientific computing. The requirements quickly evolved to text processing, consumer productivity tools, graphic design, business automation, intelligent systems, web-based transactions, and web-based collaborations.

Evolution of technology, societal restructuring, and the evolution of problem domains are interleaved. As the technology and requirements grow, the solutions to the problems become more intricate. Communicating these solutions need more human comprehensible, yet computer translatable, programming languages that can be automatically translated to low-level instructions for better software maintenance. Internet-based languages and web programming are barely 20 years old and are still evolving. High-level languages for massive parallel computers are still evolving. New complex domains are evolving that need integration with different styles of programming.

The design and the development of programming languages are guided by many aspects such as the evolution of technology, the evolution of computer architecture, the evolution of operating systems, the need to develop large-scale modular software, and the need to maintain the software for a long period. With the development of new problem domains, there will be new requirements, and thus the need for new programming languages.

1.3 LEARNING OUTCOMES

The learning outcomes of this course are as follows:

1. *Reducing the learning curve for new languages*: The requirements are created by the advancement of social infrastructure provided by the automation. The programming language of the future will be more high level, will integrate multiple programming paradigms, and will be used to develop complex software. The programmers would have to learn new multiparadigm programming languages. It is not possible to retool, unless one has a deeper abstract level of understanding of the programming paradigms, abstractions, programming constructs, and their pitfalls. A deeper abstract level of understanding of various programming paradigms will allow the programmers to superimpose the syntax of new languages on the abstractions for programming in new languages.

2. *Programmer will become aware of low-level execution behavior*: The course describes the effect of the high-level constructs by translating them to the low-level instructions that execute on low-level abstract machines. Understanding low-level behavior will help avoid the pitfalls of many programming errors. It will also improve students'

programming style by making them aware of efficient and side-effect free programming. Side-effects are undesirable computational effects caused by abstract computation models, and can cause the program to behave incorrectly.

3. *Programmer will be able to relate to compiler development*: As new, domain-specific languages are being developed, there is a need for the development of new compilers. Understanding the low-level behavior of programming languages is the basis of code generation for efficient execution.

4. *Improvement in programming style*: The students will learn many constructs in different classes of programming languages that will broaden their scope of programming style. They will also be able to express their solution efficiently by choosing appropriate data and control abstractions. Generally, after knowing a limited number of programming languages such as C++, Java, PHP, or C#, programmers get biased by specific programming styles. The knowledge of other programming constructs, used in different programming paradigms, improves their programming.

5. *Programmer will be able to select appropriate programming languages and programming paradigms*: Students will be able to map the problem domains to specific programming paradigms, and choose the appropriate languages for program development.

1.4 PROGRAM AND COMPONENTS

When asked to automate a process or to solve a problem, a system analyst has to make a model of the system, parameterize the input and output behavior, and connect various modules using a flow chart. These modules abstract the real-world process. The programs are specifications of the solutions that are developed to process data and handle dataflow between these modules.

The specification of a solution to a real-world problem is described at a high level for the ease of human understanding, and the computer's action is based on low-level machine instructions. There is a need for a translation process that uniquely maps (without any ambiguity) high-level instructions to a sequence of equivalent low-level instructions. Instructions should have a clear unique meaning to avoid ambiguity.

A program is a sequence of meaningful symbols to formally specify the solution to a complex problem (see Figure 1.1). A program has three major components: *logic* + *abstraction* + *control*. *Logic* means coming up with the high-level specification of a solution to a problem. This requires repeated breaking up of a complex problem into a structured combination of simpler problems, and combining the solutions of these simpler problems using well-defined operations to compute the final solution.

Abstraction means modeling an entity by the desired attributes needed to solve the problem at hand. The entity may have many more attributes. However, all the attributes may not be needed to model the solution of the problem. The advantage of abstraction is that programs are easily comprehended and easily modified resulting into ease of program maintenance.

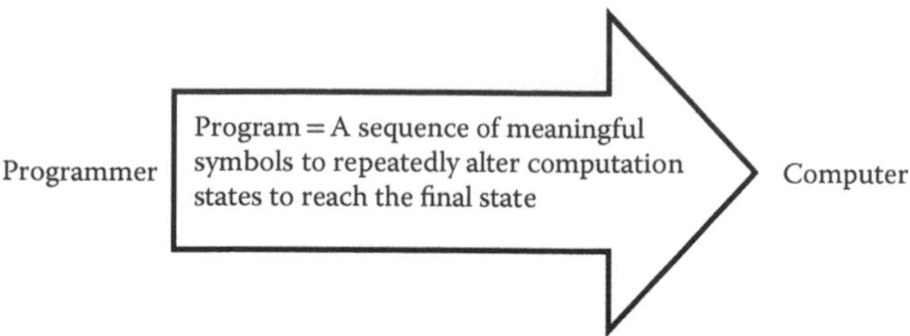

FIGURE 1.1 An abstract definition of programs.

Control means mapping the solution of the problem based on the von Neumann machine, where the memory of the computer is continuously altered to derive a final state of computation that contains a solution. Every instruction alters the state of computation to a new state. The use of explicit control in a program gives a programmer the power to modify the computer memory explicitly. The modification of computer memory to realize the logic varies from programmer to programmer, and makes a program difficult to understand.

Example 1.1

This example illustrates the three components through a simple, bubble-sort program that sorts a sequence of unsorted numbers. Bubble sort utilizes "repeatedly find the next maximum" strategy on progressively smaller subsequences. A program for bubble sort uses three components as follows:

- **Abstraction:** This represents the set of numbers as an indexible sequence.
- **Logic:** This compares adjacent numbers and swaps the position of numbers if the following number is smaller until the end of the sequence is reached. Comparisons of numbers in the current sequence gives the maximum of the current sequence. This element is excluded in future comparisons, and the comparison process is repeated with the remaining elements until one element is left.
- **Control:** This uses the code to repeatedly exchange and update the values in the different memory locations associated with variables.

A program is organized using multiple units: (1) program name, (2) imported routines from the software libraries or modules developed in the past, (3) declaration of the type information and variables needed for expressing the logic, (4) parameters to exchange the information between various program modules, and (5) sequence of commands to manipulate the declared variables.

1.4.1 Abstractions in Programs

There are two types of abstractions a program uses: *data abstractions* and *control abstraction*. Data abstraction is used to model real-world entities by identifying specific attributes needed to solve the problem at hand. For example, to model a class for the grading purpose, the class will be modeled as a set of students, where each student is modeled as a triple of the form (*student ID, student name, letter grade*). Many other attributes of students—such as gender, height, weight, address, and sport interests, and so on—are not used, because they are not needed to solve the problem at hand. Data abstractions can be implemented by various data representation techniques. For example, sequence of students can be represented as an array of students, and a triple can be represented as a "struct" with three fields.

Example 1.2

For example, one such representation of class is

```
const class_size = 20;
struct student {string student-id;
                string student-name;
                char letter-grade;
               }
student class[class_size];
```

Control abstraction classifies various types of program statements into different groups based on their common properties. For example, *assertion* (assignment statement), *block of statements, selection constructs* such as *if-then-else* construct or *case* construct, *indefinite iteration*—conditional repetition of a sequence of statements, *definite iteration*—fixed number of predetermined iterations, *procedural calls*, and *functional calls* are various control abstractions. The major advantage of abstractions is the expression of high-level program constructs that are easily modified and maintained.

Control abstractions are also related to control flow diagrams. For example, *if (<predicate>) then <then-statement> else <else-statement>* is a control abstraction that selects *<then-statement>* or *<else-statement>* based on the truth value of the *<predicate>*. If we represent evaluation of the predicate by a *diamond block* and statements by a *rectangular block*, then an if-then-else statement can be represented, as shown in Figure 1.2a. Similarly,

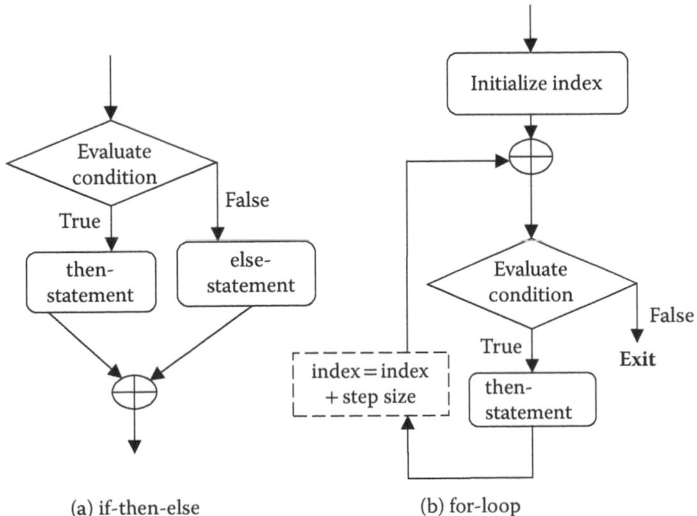

FIGURE 1.2 Control flow diagrams of if-then-else and definite iteration.

we can represent a repeat of a block of statements as *definite iteration, indefinite iteration,* or *data-driven iteration*—iteration constructs where the number of iterations are determined by the number of data elements in a collection.

A *definite iteration* repeats a block of statements for a fixed number of times by using a combination of an index variable, lower bound, upper bound, and a fixed step size. The index variable takes on an initial value, either lower bound (or upper bound), and moves progressively in each step size toward the other end, until it is no more in the range *lower-bound* ≤ *index variable* ≤ *upper-bound.* It is called *definite iteration,* as the number of times the block of statements is executed is fixed by looking at three values: lower bound, upper bound, and step size, and these values cannot be altered within the block of statements inside the definite iteration. A generic control flow diagram for the for-loop is given in Figure 1.2b. The evaluation of the condition is given by the diamond. The block of statements is enclosed in the rectangle, and the dashed box shows the embedded increment of an index.

Indefinite iteration constructs evaluate *Boolean conditions* to execute the next cycle of iteration, and the components that constitute the predicate can be altered in the body of the block of statements within the construct. Owing to this property, there is a probability of indefinite looping in an indefinite iteration.

There can be many types of loops that have been studied that contain multiple conditional statements followed by blocks of statements. However, it has been shown that the functional power of all these different types of loops can be captured by the use of two types of indefinite iteration constructs: while-loop and do-while loop. Do-while loop and while-do loop have a single entry point and single exit point, and no jump from outside the loop is allowed. In addition, they have one conditional statement that decides whether to go through the next iteration cycle or exit out of the loop. The control flow diagrams for while-loop and do-while loop are shown in Figure 1.3.

While-loop checks for a *Boolean condition* before executing the block. *Do-while* loop, alternately called *repeat-until* loop in some languages such as Pascal, executes the block

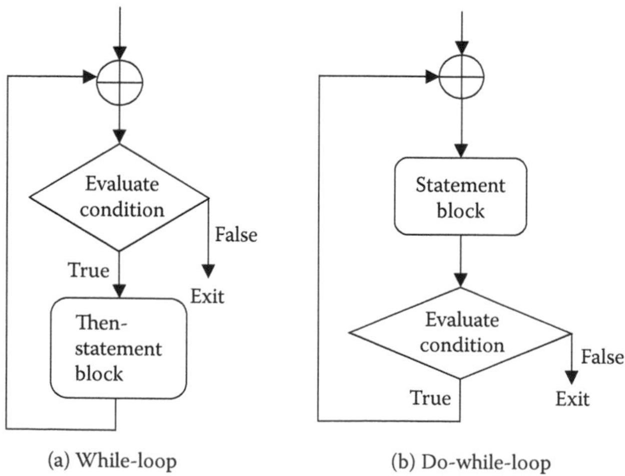

FIGURE 1.3 Control flow diagrams for indefinite iteration.

of statements first and then checks for the condition. The difference between the two constructs is that the do-while loop executes the block of statement at least once, while the while-loop may not execute the block of statement even once.

Data-driven iterations, also called iterators, are generally used in data abstractions that aggregate the data elements such as multisets and linked lists modeling multisets. Unlike classical iteration schemes using arrays, the internal representation of the data structure is hidden from the programmer. These iterators execute a block of statements for every element in the aggregate. Iterators are interesting abstractions, because the number of iterations is dependent upon the data size, and they automatically step through the elements in the list. One such construct is the *foreach-loop* in Lisp and Java that iterates through every element of a list. The general construct for iterators is as follows:

```
foreach element in <multiset> {
        <block of statements> ;}
```

Let us understand the drawing of control flow diagrams and nested control flow diagrams using simple examples. Example 1.3 shows the control flow diagrams for a nested conditional statement. Example 1.4 shows the control flow diagram for a definite iteration nested inside an indefinite iteration. The technique to draw control flow diagrams is to translate the top-level construct and then translate incrementally the next levels of constructs.

Example 1.3

In this example, first we will develop the control flow diagram for the outer if-then-else statement; while developing the control flow diagram for the outer control flow diagram, the complete inner if-then-else statement will be treated as one block of statement. In the second step, we will expand the nested inner if-then-else block. The resulting control flow diagram is given in Figure 1.4.

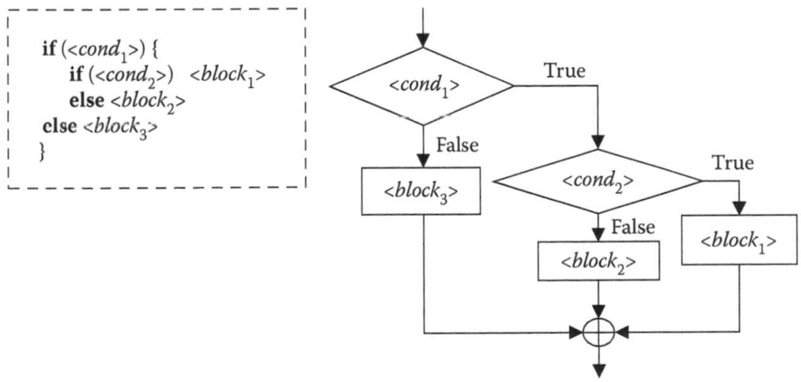

FIGURE 1.4 A control flow diagram for a nested if-then-else statement.

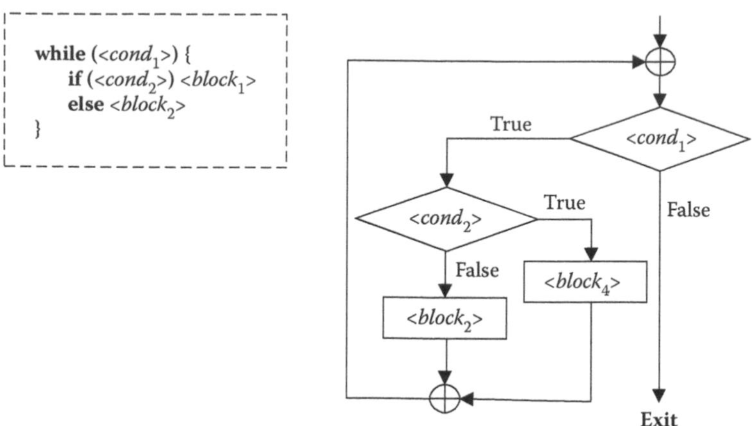

FIGURE 1.5 A control flow diagram for a nested statement.

Example 1.4

This example illustrates the development of the control flow diagram for a nested while-loop. First a control flow diagram for the outer while-loop is drawn by treating the inner if-then-else as a statement within the outer while-loop. Next, the control flow diagram of the embedded if-then-else statement is constructed. The resulting control flow diagram is given in Figure 1.5.

1.4.2 Program Comprehension and Jumps

One of the major issues in the life cycle of a software is the *software maintenance,* since (1) the need evolves, (2) the programmers move on, (3) programmers forget their approach over a period of time, and (4) the architecture or technology changes significantly. Software maintenance is directly related to program comprehension: if the control is complicated, then it is difficult to comprehend, and any attempt to modify the software will lead to bugs in the software. In the absence of proper language features that support comprehension and programming standards, the old software will either become useless, or an organization has to spend significant amounts of time and resources to maintain the software.

Programmers understands a piece of code by (1) understanding the meanings of known abstract constructs, (2) understanding the control flow in an abstract domain, (3) modeling the information flow between various program subunits, and (4) predicting the final condition by progressively predicting the conditions after individual statements in their minds, without the actual execution of the program.

There are many factors that help the human comprehension of programs. Having background knowledge and training in programming is a necessary but not sufficient condition to comprehend code written by others. Some of the factors that affect the human comprehension of a developed code are as follows: (1) the level of abstraction provided in the programming language, (2) simplicity of abstractions needed for translating logic to code, (3) removing excessive control from the code, and (4) writing enough standardized comments at the variable level, block level, module level, and algorithm level that can be understood by others. In addition, many organizations also impose programming styles and the use of a preferred language.

Humans are very good at understanding a program that is structured, easily categorized, modular in functionality, and follows a forward direction of control flow. As the number of lines of code in a module increases, it becomes difficult for humans to understand and categorize the overall functionality of the module. Similarly, too many jumps increase the unstructured nature of programs, and humans cannot mentally visualize the run-time behavior of programs.

It has been shown that the functional power of unconditional jumps (*goto* statements) is maximum. Unconditional jumps have been utilized to implement many control abstractions such as (1) *selection*: if-then-else statement or case statement and (2) *iteration*: while-do loop, do-while loop, and for-loop, and functional and procedure calls. Within a module, jumps are used: (1) to pass the control locally to few instructions away, and (2) to exit out of a block of statements. In the 1970s, a healthy debate took place among the computer scientists about the program comprehension and the use of jumps. The use of unconditional jumps and program comprehension are inversely related to each other. Humans can handle a jump to few instructions away specially in the forward direction. However, jumping backwards, jumping long distances away, and too many jumps significantly reduce program comprehension. Structuring the program into blocks, subprograms, objects, and control abstractions with well-defined functionalities such as while-do loop and do-while loop enhances program comprehension.

Figure 1.6a and b shows the control flow diagram of two equivalent programs. The program in Figure 1.6a moves one instruction at a time in the forward direction, while the program in Figure 1.6b uses multiple jumps in a combination of backward and forward jumps, and the statements are shuffled up with embedded go-to statements.

Let us assume that S_1 is equivalent to S'_1 followed by an unconditional jump to S'_{300}, S_2 is equivalent to S'_{300} followed by an unconditional jump in backward direction to S'_2, S_3 is equivalent to S'_2, S_{300} is equivalent to S'_{299} followed by unconditional forward jump to S'_{301}, S_{301} is equivalent to S'_{301}, and all the statements from S_4 to S_{299} are equivalent to the corresponding statements S'_3 to S'_{298}. If we look at the *functional equivalency*—capability to do the same task—both programs are functionally equivalent. However, it is easy to understand the structured program in Figure 1.6a.

Owing to this lack of comprehension caused by the free use of unconditional jump statements, various control abstractions have been developed. The use of jump statements is limited to the implementation of control abstractions and to exit out of nested blocks. It

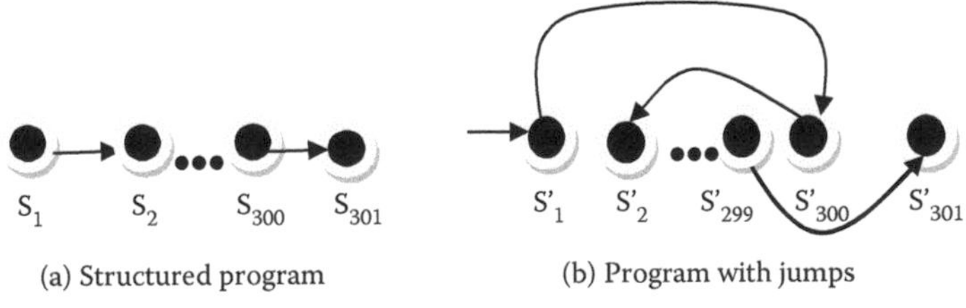

FIGURE 1.6 Block structured programming versus spaghetti code with jumps.

can be shown that programs using jump statements can be translated to a functionally equivalent structured program using a combination of Boolean variables and control abstractions such as if-then-else, while-loop, or do-while loop.

1.4.3 Execution of Programs

Major criteria for the sound execution of a program are as follows:

1. The language constructs of a programming language should be well defined.
2. There should be a unique meaning for every language construct as computers do not handle ambiguities.
3. Each high-level instruction should be translated to a sequence of low-level instructions that perform consistently the same action on a computer every time.
4. A computer should execute consistently the same sequence of low-level instructions producing the same final result.

There are multiple layers of software before the programming language layer, as shown in Figure 1.7. At the lowest level, it is the bare machine and machine code. The next level is the operating system, system utilities, and language translators. The outermost layer is the high-level programming and application layer. A high-level program is translated to low-level instructions and uses multiple intermediate layer interfaces to use the computer system resources and utilities needed for program execution.

There are three ways in which a high-level instruction can be translated to a set of low-level instructions: (1) compile the high-level instructions before the execution of the program; (2) interpret the high-level program, and execute it using an implemented abstract machine; and (3) just-in-time compilation of the high-level language that gives an effect of partially compiled code and partially interpreted code.

As illustrated in Figure 1.8, the process of compilation translates the high-level instructions to low-level instructions before execution. On the other hand, interpreters translate and execute one instruction at a time. In contrast to compiled code, interpreted code goes through the translation and execution cycle for very statement.

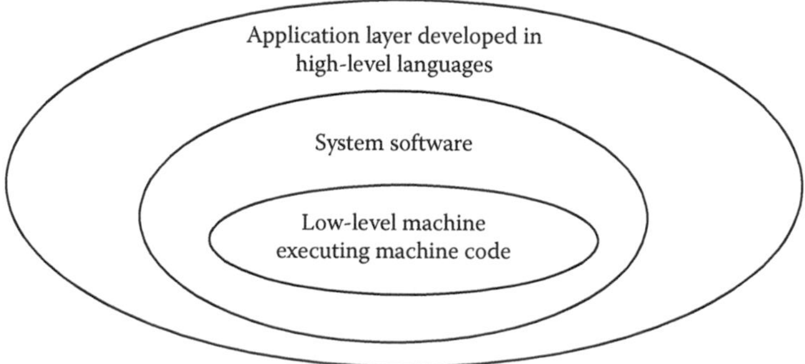

FIGURE 1.7 Different layers of software in a computer.

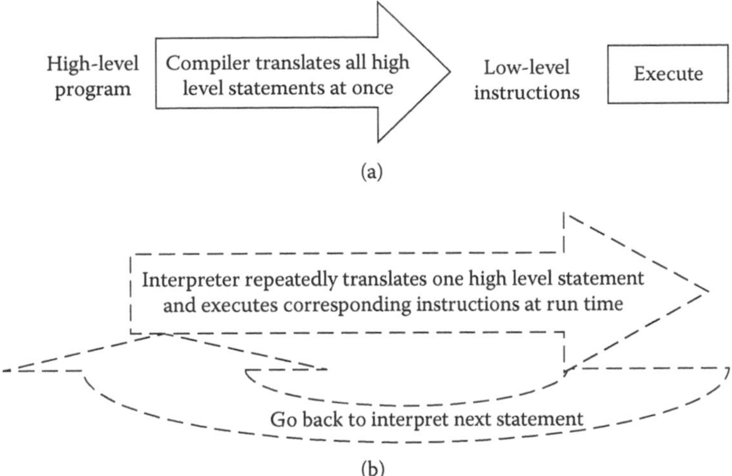

FIGURE 1.8 Compilation versus interpretation. (a) A schematics of compilation (b) a schematics of interpretation.

An advantage of compiled code is that it has no overhead of translation at run time. Another advantage of using compilers is that a large percentage of the programming bugs are detected, and memory allocations are optimized before execution.

The execution efficiency of interpreted code is an order of magnitude slower than the execution efficiency of the compiled code, because the translation process is interleaved with execution at run time. Another drawback of interpreters is that all the errors cannot be detected before program execution, and interpreted programs may crash after executing many instructions, making it unsafe for mission-critical programs. However, interpreters are easy to develop and have been used in early days for the languages for which compiler technology was not well developed.

In declarative programming languages that have the capability of developing programs as data and then transforming them to program at run time, interpreters were a choice. Interpreters facilitate better interactive debugging capability at the instruction level by displaying all the values of variables. Interpreters support execution of dynamically typed languages where a variable can be associated with any type of data object at run time. Interpreters also support better machine independence, since programs are executed within an interpreter, and programs need not be ported across different machines.

In web-based languages, code is sent as a stream to the client computers that have a version of an abstract machine installed for the native operating system. An *abstract machine* is a program that runs on the native machine and executes a standard instruction set that is independent of the host operating system and computer architecture. A Java Virtual Machine (JVM) is one such abstract machine that executes Java programs on multiple platforms. The execution of instructions on JVM is much slower than the execution of machine instructions on a native machine, because JVM uses *zero-addressing* for universal compatibility, while current-day computers use two-address or three-address instruction sets. Execution of code on a higher-address machine is faster than that on zero-address machines, as explained in Section 2.1.

To speed up the execution on higher-address machines, *just-in-time compilers* keep a library of executable methods, using the instruction set of the native operating system that is much faster than executing equivalent instructions in JVM. After checking for security, the fragments of the intermediate code are translated to the native machine code upon the first occurrence of intermediate code and cached in a library for future lookup. On subsequent occurrence of the fragments, the executable machine code is looked up and executed. Similarly, if a high-level Java method already has a compiled code in the native library, then the compiled code is retrieved and executed. If the binary code is unavailable or the fragment cannot be compiled in realistic time, then the high-level code is translated to instructions in JVM and executed. The overall scheme for just-in-time compilation is shown in Figure 1.9.

There is an overhead of translation when the code-fragments are compiled for the first time. However, on the subsequent occurrence of the fragments, there is no overhead of translation. The execution efficiency of a code using just-in-time compilation lies between the execution efficiency of an interpreted code and the compiled code.

Some vendors such as Microsoft have their own intermediate language called *Common Intermediate Language* (CIL). The advantage of a common intermediate language is in providing interoperability across different high-level languages. A high-level language is first translated to CIL, and this CIL is transmitted across the Internet and compiled on various architectures using just-in-time compilation, as shown in Figure 1.10. Languages such as C#, visual C++, F#, and Visual Basic are translated to CIL.

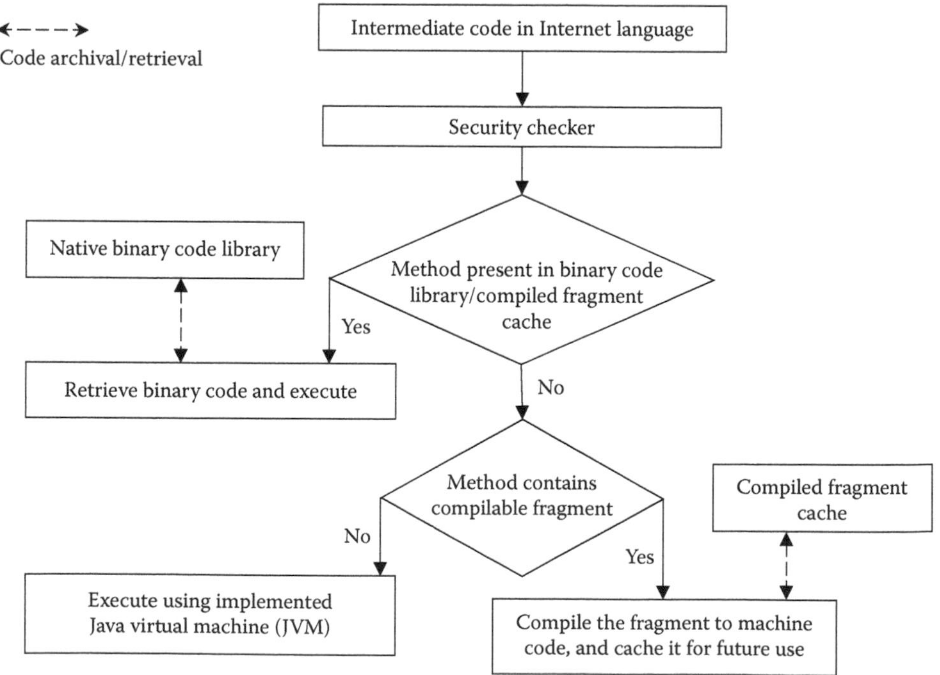

FIGURE 1.9 Schematics of just-in-time compilation.

Modern-day compilers use a two-stage translation for execution of a high-level program on a variety of architectures. Different architectures have different assembly languages, and it is not possible to write compiler for multiple architectures. Instead, an intermediate code has been developed for the programming languages. The first stage of compiler translates high-level programs to an intermediate-level code that is independent of computer architectures. The second stage translates intermediate code to low-level machine code.

The first stage of translation consists of (1) *lexical analysis*, (2) *parsing*, and (3) *semantic analysis* and *code generation* to generate intermediate code, as shown in Figure 1.11.

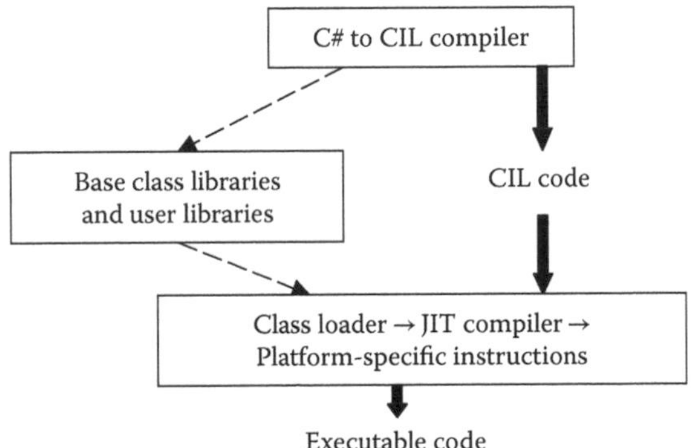

FIGURE 1.10 Just-in-time compilation of C# program.

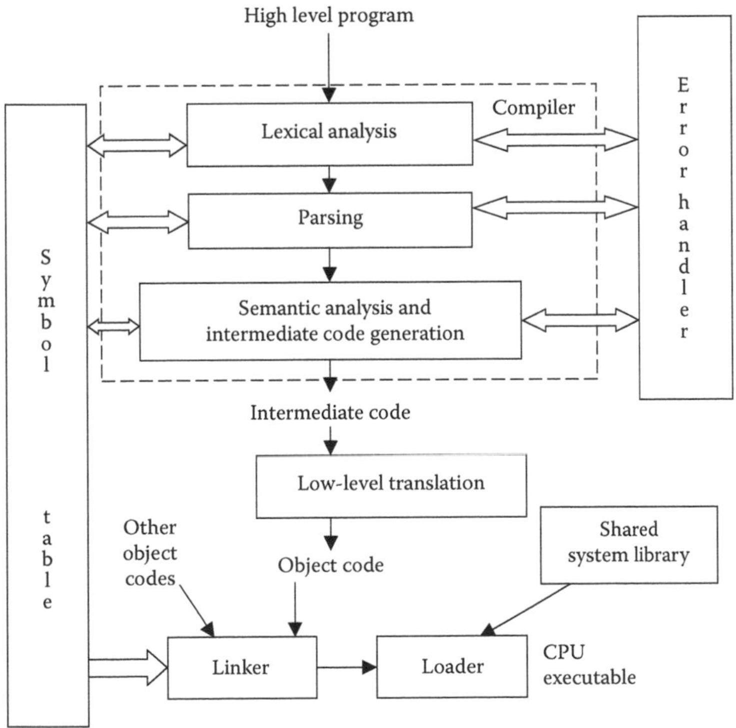

FIGURE 1.11 Translation of high-level program for execution on the computer.

A *lexical analyzer* checks for the reserved words—meaningful words that are part of a language, identifiers, variables and numbers—and converts them into an internal representation called *tokens* for the ease of parsing. The output of lexical analysis is a tokenized stream that becomes input to a parser. A parser validates the structure of a program sentence according to the grammar of a programming language. Its output is in the form of a tree for the ease of efficient internal handling during code generation. The semantic analyzer validates that parsed sentences are meaningful, and the code generator linearizes the tree-based representation to the corresponding *intermediate code*—a sequence of low-level abstract instructions. The optimization level removes the redundant code fragments and enhances the use of processor registers to improve the execution efficiency of the executable code. A *symbol table* is used to store the information from previous stages that may be needed or may need resolution in the following stages. The type of information that a symbol table contains are variable names and their locations, nesting level of procedures, information about nonlocal variables, where a procedure has been called, what procedure has been called, and so on.

In addition to two-stage compilation, a *linker* links multiple object files (compiled code) into one big executable code. Linking is done in the order specified by the programmer in the link command and is independent of the calling pattern of the procedures. The compiled code of a subprogram occurs only once in the executable code and is independent of the frequency of the calling of the subprogram. After linking, the information about the relative location of procedures is fixed. A *loader* loads the linked code to be executed in a memory segment—the memory area given to a user process corresponding to the executable code. The loader relocates the logical address derived after linking to the physical address in the RAM. The loaded executable program is called a *process* and is executed using a combination of operating system software and hardware techniques you will study in a course on operating systems. Dynamic link libraries are shared system libraries to share the executable code by multiple processes. You will learn about the mechanisms of dynamic link libraries in a course on operating systems.

1.5 INTEROPERABILITY OF PROGRAMMING LANGUAGES

Language interoperability is the ability of a code fragment in one language to interact with code fragments written in other programming languages. Modern-day programming languages provide language interoperability to support (1) mixing of problem domains in a complex problem to be automated; (2) utilizing abstractions specific to a different language; (3) efficiency by interfacing with libraries of low-level languages; and (4) to maximize code reuse of an already developed software library, improving software reuse and maintenance.

To provide interoperability, one must have capabilities to include the compiled module developed in other languages, to exchange information with compiled modules written in other languages, and have shared *metadata*—a separately declared information about the module and exported data types that can be used by codes in other languages. In addition, the exchanged information has to be transformed back to proper data abstractions in the host language.

There has been a trend to provide a common language specification and standard specification of common data types for uniform exchange of information between the host language and the interfaced languages. Common language specifications include (1) defining and using types consistent across languages and (2) defining a uniform mechanism for storing and retrieving information about type correspondence in different languages. Some vendors—more specifically Microsoft—have developed a common language specification under .NET project to provide standardized common features and rules needed for interoperability.

1.6 SOFTWARE DEVELOPMENT CYCLE

Now we turn our attention to the relationship of programming languages to software development. After all, programming languages are used to develop large software to automate some real-world process. Software can be hundreds of thousands of lines of codes. A successful development of large software needs teams of programmers, system analysts, and field programmers.

There are multiple stages in the life cycle of software: *requirement analysis, system analysis and design, development, implementation, initial verification, field testing, deployment,* and *evolution.* After every phase, feedbacks from the following processes are used to refine the previous phases before progressing to the next phase. At the end of the deployment, the system capabilities are well understood by the clients, and the clients' needs evolve. There is a need for refinement in the system analysis and the design to satisfy the new needs. The evolution is incremental, learns from the previous analysis and design errors, and benefits from the technology improvements during the next software development cycle.

In the *requirement analysis* phase, a system analyst listens to the client's requests for the automation, incomplete and often ill-defined information, and the type of queries the client intends to ask. The scope of automation is fixed, and the goals of the project are fixed. In the *system analysis* phase, a team of system analysts studies the scope and the system to be automated then develops an information flow model of the process being automated. In the *design phase*, the information flow model is split into various interconnected modules with the help of lead programmers and system designers. These modules have independent functionality, so that evolution of one module does not affect other modules significantly. Input–output of information from various modules is clearly fixed. The designers design the modules with a potential for evolution as the clients' needs evolve. This part of visualizing and predicting the future evolution of the client needs is quite difficult and needs experienced analysts and system designers.

In the *development phase*, after the modules and their linkages (input–output to the modules) have been developed, a team of lead programmers designs the data abstractions, the interfaces, and the algorithms for each module, mindful of technical difficulties and restrictions imposed by the constructs in the available programming languages in the problem domain. If the automation involves more than one problem domain, then modules may be developed using multiple programming languages utilizing interoperability.

In the *implementation phase*, after developing the major high-level algorithms, the program code is developed. This part is done by a team of programmers, and the data flow interface developed in the previous stages is translated into programming module interface between various programming modules. Care is taken to provide enough information hiding to prevent programmers from using local features of other modules to minimize the numbers of unintended bugs. Development of the code is only one-third of the total process of the software implementation. The major effort lies in the verification stage, testing stage, and deployment stage. The programmers should heed the future needs and life cycle of the software and provide ease of expandability of the scope of the software. Otherwise, even small programming or design errors may make the software unusable and cripple the daily essential functions. An example of such errors was the Y2K problem, which cost billions of dollars to fix. The software should also have concessions for future enhancements in low-level technology such as faster computer architectures and improved hardware.

In the *initial verification stage*, programmers try their program on a sample data provided by the client. After its successful completion, the developed software is floated as an alpha version and given to the field programmers for testing on the client side. The field programmers make changes and collect feedbacks from the clients to correct the model and the program. In the field-testing phase, the new corrected version called the *beta version* is tested extensively by the field programmers for any bug on the actual client-supplied data, until the final version is released.

In the deployment stage, the refined version is released for public use (or client use), and field testing and collection of feedback starts. The client sees the potential of the software and suggests improvement after the software is integrated with the overall system in the workplace. The suggested modifications are quickly incorporated in the software at the field sites (workplace) or through *software patches*—small pieces of software to fix the minor improvements, bugs, and security concerns.

Over a period of time, the strengths and the weaknesses of the software are identified, the client identifies new needs to be incorporated in the software, and a new iterative cycle of software evolution starts. During this period, the hardware technology may also change, and the software has to be maintained to make it portable across various architectures. A schematic of software development cycle is given in Figure 1.12.

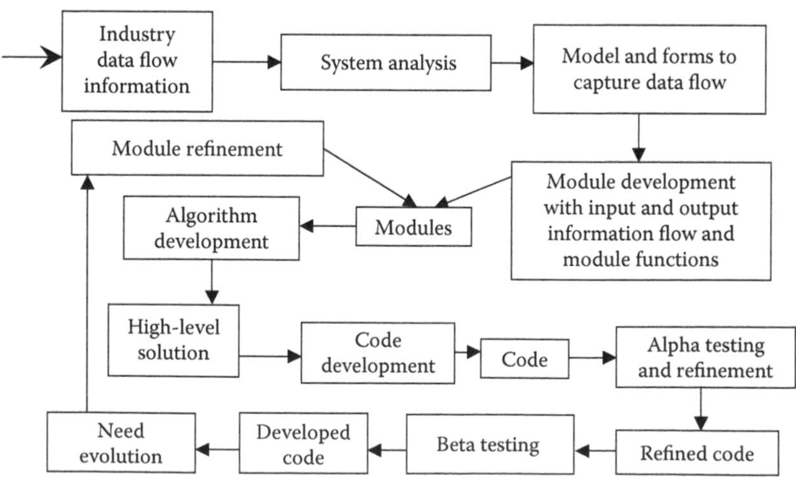

FIGURE 1.12 Software development cycle.

There are many software cycle models such as the "Waterfall model" and "Spiral model" for software development. These models have been implicitly included in the above discussion. However, comparison and detailed study of these models is within the scope of a course in software engineering.

The software keeps evolving with time as the needs evolve, until a new technology makes old technology obsolete. In that case, new software needs to be developed for better automation. There have been four major changes in the history of industrial automation: (1) the development of high-level, object-oriented languages; (2) the development of visual and multimedia techniques for better visualization; (3) the development of high-level database languages; and (4) the development of web-based languages such as XML to facilitate transfer and resource sharing over the Internet.

As is obvious, software development requires teamwork, software maintenance, portability, and continuous evolution. Any task that requires a large team over a long period of time requires software standardization, modularity, user friendliness, and ease of comprehension.

1.7 CRITERIA FOR A GOOD PROGRAMMING LANGUAGE

Criteria for a good programming language are guided by the domain needs; software development, maintenance, and evolution; and interaction and efficiency of execution. Some of these criteria are as (1) *abstraction*, (2) *modularity*, (3) *orthogonality*, (4) *exception handling*, (5) *user friendliness*, (6) *readability*, (7) *ease of comprehension* and *maintenance*, (8) *overall simplicity*, and (9) *portability*. We have already discussed abstraction, modularity, portability, and ease of comprehension. *Orthogonality* means that the constructs in programming languages should be independent of each other and should not be redundant. *Exception handling* capability is necessary to capture the run-time error conditions by invoking a user-defined routine for graceful correction of the error condition or exit to avoid program crashes. In large programs—that is, time-critical programs, such as a program monitoring a nuclear power plant, and mission-critical programs such as programs running an aircraft or spaceship—this is an essential criterion. *Readability* means that a language should support constructs and capabilities that make the programs readable for the ease of comprehension. For example, longer variable names should be allowed, so that programmers can use self-explanatory variable names. Ease of comprehension and overall simplicity are directly related to software development, evolution, and maintenance, and are dependent upon the amount of abstraction present in the language.

Industry executives ask an interesting question: What is a good language for automating my industry? This question emanates from the basic need of standardization, so that the companies have to spend fewer resources and finances to maintain software over a period of time. Unfortunately, there is no magic bullet; there is not a single language that can handle all the problem domains. Each domain has different requirements.

For example, a system programming language needs to be close to the assembly languages and needs to have instructions to efficiently access memory modules and I/O devices; inclusion of too many higher-level abstractions will not contribute to efficiency.

A scientific computing language needs to represent big matrices without duplicating them, because duplication will waste precious memory space and will cause computational overhead of copying large matrices from one memory space to another. A real-time language needs to be efficient to avoid losing real-time events. A user-friendly language will avoid any type declaration, but then it suffers from many problems such as lack of memory optimization, lack of efficient execution, and once in a while programs crash at run time.

Many of these requirements for a good language are contradictory. If we want to develop a good language that has many high-level abstractions, then the translation process will generate many redundant low-level instructions due to generalized mechanisms used in the translators. Similarly, if we make the language closer to assembly languages, then portability and consequently software maintenance suffers, as assembly languages vary for different architectures. We have moved from 16-bit instruction sets, to 32-bit instruction sets, and more recently to 64-bit instruction sets. If a program is strongly coupled to assembly-level instructions, then it will be difficult to port architecture changes every time. Technology keeps changing at a faster rate, necessitating adaptation of software solutions with minimal delay and investment. So the new languages need to move away from low language instructions for the sake of portability, modifiability, and maintenance of the software. Otherwise, we will be spending millions and billions of dollars just to port the software to new architectures.

1.8 HISTORY OF PROGRAMMING PARADIGMS AND LANGUAGES

Programming paradigm means style of programming. As programming matured, users and computer scientists found drawbacks and advantages in previous programming styles and evolved new programming styles. Modern-day languages are a combination of two or more programming paradigms, as shown in Appendix I.

We can classify programs into a combination of one or more programming paradigms: (1) imperative programming, (2) declarative programming, (3) object-oriented programming, (4) concurrent and distributed programming, (5) visual programming, (6) web-based programming, (7) event-based programming, (8) multimedia programming, (9) agent-based programming, and (10) synchronous programming. The list does not end here. However, the scope of this textbook is limited to the study of class of programming languages supporting these major programming paradigms.

1.8.1 Imperative Programming Paradigm

The basis of imperative programming is *assertion* or assignment statement that changes the state of the low-level von Neuman machine. A programmer manually translates the logic explicitly to tell the computer what to do. A variable in imperative programs is mapped onto a memory location in a computer, and the memory location can be modified repeatedly by using assignment statements. The effect of the assignment statement is that a new value is written into the memory location; the old value is lost. The advantage of assignment statement is *memory reuse,* since the same memory location has been used for storing multiple values. However, there are many disadvantages of writing into memory location multiple times as follows:

1. If a memory location is rewritten into, the old value is lost, making it impossible to use the old value in the future. In artificial intelligent programming, where a solution is searched in huge search space, backtracking—going back and undoing part of the search and trying out alternate search paths—becomes impossible.
2. If a memory space belonging to another procedure or function is written by the currently called procedure, then the old value is lost. If the calling procedure or the procedures is unaware of the modification or needs to use the old value, then the update has corrupted the memory location. The outcome would be to generate incorrect solutions. This problem is called *side-effect*, and has been discussed in detail in Section 4.8.
3. The interleaving of the assignment statement with logic of the program causes problems in program comprehension, as the control is very subjective to an individual programmer's thinking style and causes program maintenance problems in the long run.

Imperative programming paradigm was the first one to be developed in the late 1950s and 1960s in different versions of *FORTRAN* and in the early 1960s in the development block-structured languages such as *ALGOL*. Among imperative programming paradigms, four early languages were quite popular: FORTRAN, ALGOL and COBOL, and C, a descendant of Algol.

Early versions of FORTRAN were extensively used for scientific computing, made liberal use of jump statements, and did not support pointers. ALGOL developed a block-structured style of programming that supported control abstractions such as *while-do loop* and *do-while loop* that improved program comprehension. ALGOL also used pointers and "struct" that have been used to implement popular recursive data structures such as "linked-lists" and "trees". ALGOL is also known as the mother of modern-day imperative programming languages: most of the data abstractions and control abstractions described in ALGOL are still being used in modern-day languages.

C was originally a system programming language developed in Bell labs for writing the "UNIX" operating system. It was a subset of ALGOL 68 and gained popularity due to the popularity of Unix and its variations. There were other contemporary languages such as SDL (Burroughs corporation that supported the development of ALGOL 68) and BLISS 32 (developed by Digital Equipment Corporation) that lost out due to the popularity of public versions of Unix used by the industry and academia. Other noteworthy languages in this category are ADA, Pascal, and the Modula family of languages, including Modula-3 and Oberon, that supported high-level block-structured programming. The Modula family of languages supported the notion of modules and the concept of import and export of functions. All these languages were derived from ALGOL 68. However, except C and its variants, these languages remained in the academic world and never gained commercial popularity.

COBOL is a business programming language. It emphasizes on user friendliness, report writing, and handling financial data. There was also an attempt by IBM to develop a language, PL-I, that tried to integrate the constructs developed in FORTRAN, ALGOL, and COBOL. PL-I was projected as a language that could be used for multiple programming domains. However, due to its bulky nature, it lost the race.

Many imperative programming languages such as ADA, FORTRAN, and COBOL kept on evolving by incorporating proven constructs. Newer versions of FORTRAN and COBOL have many features such as block-structured programming, recursive programming, stack-based implementation, string processing, pointers, structures, and object-oriented programming. With the passage of time, language designers identified the advantages and disadvantages of contemporary languages and incorporated the useful constructs. When a language evolves, care must be taken to keep it compatible with older versions, so that evolved programs can use older libraries, and programs written using older versions of languages can be compiled using compilers for newer versions.

1.8.2 Declarative Programming Paradigm

In declarative programming, control has been taken out of the program. There is an abstract machine that takes care of the control part implicitly. A declarative program consists of *logic + abstraction* at the programmer level. The notion of variable in declarative programs is quite different from the notion of variables in imperative languages. A variable in declarative program is a value holder; once a value has been assigned to a variable, it cannot be altered by the programmer. The advantage of *assign-once* property is (1) there is less possibility of side effects, as a called procedure (or function) cannot rewrite into memory space of the calling procedure (or function) and (2) the old values of variables can be retained and used if needed. There are also many disadvantages of the write-once property as follows:

1. The memory location cannot be reused even if it is not needed such as for the use of index variable in the for-loop. Handling iterative computation and input–output processing such as printing a large array of data elements causes memory explosion.
2. Recursive programming is used more often, since traditional iteration using *mutable index variable* is not permitted due to the restriction imposed on the variable mutation. Recursive programming has both memory and execution overheads and is more difficult to program compared to iterative programming.
3. Global mutable variables that support storing of partial computations that can be used later by other parts of the program to improve the execution efficiency are not allowed. However, many declarative programming languages have tried to overcome this limitation: Lisp allows partial rewriting into memory location by allowing limited use of global variables; and Prolog, depending upon the implementations, allows constructs—such as "assert", "blackboard" or global mutable variables—to store the result of partial computations.

There are two major types of declarative programming languages: functional programming languages and logic programming languages. Functional programming languages are based on the use of mathematical functions, and logic programming languages are based on the use of predicate logic—Boolean logic combined with the notion of quantification, as explained in Chapter 2.

Lisp, a popular language that combined functional programming with some imperative constructs, was developed in the early 1960s for the implementation of artificial intelligence programming—a branch of computer science that tries to simulate human intelligence by computational means. Prolog, a popular logic programming language, was implemented in the 1970s for automated theorem proving and artificial intelligence programming. Both these languages matured in the early 1980s, when extensive compilation techniques were developed. Their compilers went through many innovations to overcome the difficulty of efficient execution and memory reuse over a period of time. In later years, other functional languages that were influenced by Lisp were developed and have become quite popular. Some of the popular examples are Scheme, ML, Miranda, and Haskell. Nowadays many multiparadigm languages such as Ruby and Scala support the functional programming paradigm. A complete list is given in Appendix I.

Declarative programming languages have been traditionally used in artificial intelligence. The key features of declarative languages needed in artificial intelligence are as follows:

1. The artificial intelligence systems need to dynamically develop and update knowledge. AI languages support abstractions for incorporating knowledge.
2. Most declarative languages treat the program as a first class object, which means that programs can be built at run time, as data then transformed into a program that can be executed. Artificial intelligence needs this property to compile new knowledge.
3. Declarative languages provide the capability of meta-programming—a program that reasons about another program in an abstract domain. This property of meta-programming is useful in developing reasoning and explanation capabilities in artificial intelligent systems.

1.8.3 Object-Oriented Programming Paradigm

As computers' memory size increased, executing large programs became possible. SIMULA was the first object-oriented language in the late 1960s. However, the notion of object-oriented programming caught on with the software development community in the early 1980s. As the program size grew, people started realizing the value of *modularity* and *software reuse* to develop even more complex software.

Many important interrelated concepts such as *modularity, software reuse, off-the-shelf library*, and *information-hiding* were conceived to provide ease of software development, maintenance, and evolution. *Modularity* means dividing large software programs into a set of interconnected modules, such that each one of them has clear functionality that does not overlap with the functionality of other modules. The major advantage of modular development is that the modification in one module does not adversely affect the functionality of other modules. *Software reuse* means that previously developed software can be reused by simply storing it in a file and by importing the needed modules or subprograms from the archive when needed. *Information hiding* means to make the part of the program module implementation inaccessible to other modules if it is not needed for interaction with other modules.

In order to provide modularity and information hiding, the notion of objects and classes were discovered. An object has both data abstractions and methods—related functions or procedures needed to manipulate the data abstractions in the object. The information inside objects can be *private, public*, or *protected. Public methods* are available to the outside world; *private methods* are specific to the object providing information hiding; and protected methods are visible to the objects in the subclasses of the current class. Information hiding is also related to the notion of modularity, as the hidden information cannot be used by other objects or classes. The notion of software reuse has been facilitated by the use of *inheritance*, where a subclass can inherit the methods from the parent class, and by the development of library of classes that can be developed and included in the software.

The first language to incorporate object-oriented programming was SIMULA. Smalltalk, developed by Xerox Parc and Eiffel, demonstrated the concept of objects. Since then many languages have been augmented with the object-oriented paradigm. For example, C++ is an integration of the language C, and the object-oriented paradigm, CLOS, is an integration of Lisp and the object-oriented programming paradigm. Many variations of Prolog have a library for object-oriented programming. Modern-day scripting languages, such as Python, Ruby, and PHP, have integrated object-oriented programming from their imperative ancestors. FORTRAN 2008 and COBOL 2002 are the latest evolution of FORTRAN and COBOL that have integrated the object-oriented programming paradigm with the earlier imperative versions. Java is a successor of C++ that integrates the Internet programming paradigm and the object-oriented programming paradigm. In recent years, X10 is a concurrent object-oriented programming language that has been developed by IBM researchers for high-level software development on massive parallel computers.

1.8.4 Concurrent Programming Paradigm

During the decade of 1980, as the software started getting bigger, the hardware technology was also changing very fast. Multiple fast processors were becoming available in a single computer, and the development in computer networking was allowing information exchange between multiple computers.

On the basis of this new advancement, independent subtasks could be mapped on separate processors to improve the execution efficiency of the programs. The development took two directions: (1) parallelizing compilers that could take a sequential program as input, and automatically transform it to a parallel version to execute concurrently and (2) incorporating high-level program constructs such as *threads, forking, busy-wait-loop*, and *remote procedure calls* in the existing languages. In all these constructs, a process—the active part of a program executing on a CPU—could start one or more concurrent subtasks to improve the execution efficiency.

The concurrent counterparts of many languages with the concurrent programming paradigm were given a generic name as "Concurrent X" or "Parallel X," where X was the name of an existing language. For example, integration of concurrent programming constructs in C was named "Concurrent C"; integration of concurrent programming construct in Pascal was named as "Concurrent Pascal"; integration of concurrent programming constructs in Prolog was called "Concurrent Prolog"; and integration of concurrent programming constructs in FORTRAN was called "Parallel FORTRAN."

The concurrent programming constructs are becoming common with the development of multicore-processor-based, modern-day personal computers. Modern-day programs and languages include many of these concurrency constructs to improve the execution efficiency of the programs. For example, the use of threads is quite common in modern-day languages, either as an interface to existing thread libraries or as a built-in library. For example, an extensive library has been provided for C, C++, and Java.

1.8.5 Visual Programming Paradigm

The textual programming paradigm uses one dimension. However, we humans are very good in perceiving the notion of vicinity instead of just sequential one-dimensional programming. Most of the programming languages developed are textual and suffer from this limitation of sequentiality caused by the single dimension present in the textual representation of programs.

In the late 1980s, starting with earlier work in Smalltalk to provide user-friendly interfaces, the visual programming paradigm took multiple directions. Some suggestions were to free the programming languages using symbolic representation for different data abstractions and control abstractions at low-level programming. However, the effort to incorporate visual programming was limited to drag-and-drop programming to provide user-friendly interfaces and animations. More recently, visual programming has been used in languages such as C# for event-based programming: the symbols corresponding to events and object attributes are dragged and dropped to make a complex scenario of interacting objects. At the low level, these events and interacting objects are automatically translated to low-level textual version of the language. Similarly, visual educational languages such as Alice use visual programming to develop code for animation, and languages such as SMIL, VRML, and Java3D use visual programming for multimedia web-based on-demand presentation and animation.

The use of the visual programming paradigm for large-scale general purpose low-level programming has remained dormant due to the following: (1) the parsing difficulty in two-dimensional planes, (2) difficulty in human comprehension of large-scale programs in two-dimensional planes, and (3) difficulty of presenting large visual programs to human programmers. There is also no standardization of symbols used in visual programming.

1.8.6 Multimedia Programming Paradigm

The multimedia programming paradigm means integration of multiple modes of visualization: text, images, audio, video, and gestures. Humans interact with each other using all these cues. Without these cues, our communications with each others and our perception of objects and real-world phenomena would be incomplete. In the early 1990s, with the advancement of web-based programming and the development of comprehensive formats for audio, images, and video, it became possible to embed multimedia objects and video clips into programs for human visualization and perception. Video can be represented by a sequence of frames where each frame is a set of possibly interacting objects. There are languages, such as Alice and Virtual Reality Markup Language (VRML), that create 3D animated objects and motion to model virtual reality. In recent years, many such 3D modeling languages such as X3D and Java3D have been developing that integrate computation and 3D modeling for real-time animation over the Internet.

1.8.7 Web-Based Programming Paradigm

The advent of Internet in early 1990 has provided us with a tremendous capability for sharing data, images, audiovisuals, database, and mobile–code located remote web sites. It has also provided us with the capability of code and data mobility. If a remote resource does not want to share the code, it can compute the data at the source, and transmit the resulting data. On the other hand, if the server does not want to get overloaded, it sends the code to the client to perform the computation at the client's end. Web-based programming has become a great engine for multimedia visualization and has made great impact on financial computing such as stock markets and banking. There are many Internet-based languages such as Java and SMIL, and web development languages such as PHP, Javascript, and XML. XML has become a popular intermediate language for representing databases, computations, and animations over the Internet. Java has a popular intermediate-level abstract machine called *JVM*.

Java programs are interpreted using a JVM, which is similar to an assembly language for zero-address machines. A zero-address machine does not use a register or a memory address in the instruction. Rather it is a stack-based machine, where the operands are placed on an evaluation stack, popped, evaluated, and pushed back on the stack. The reason for having a zero-address machine was that Java was designed to be implemented on any embedded computer in modern everyday devices such as microwaves, smart homes, and refrigerators. A zero-address machine is the most common abstract machine that could be run on all these embedded computers.

The problem with zero-address machines is that a high-level instruction is translated to many more instructions, wasting clock cycles in CPU. In order to speed up the execution, a new approach of *just-in-time compilation* has been developed, where known class libraries of the high-level language such as Java or C# are compiled to the native binary code. When a high-level language program-fragment is translated to the low-level equivalent, then there are two possibilities: (1) translate to the compiled native binary code for faster execution or (2) use an abstract machine interpreter such as JVM or .NET (for a Microsoft Windows operating system). Just-in-time compiled programs execute faster than interpreted codes on JVM.

1.8.8 Event-Based Programming Paradigm

Events are happenings that set up conditions that trigger some actions or computations. For example, clicking on a mouse is an event; moving the computer mouse over an image is an event; and an instrument reaching a threshold value is an event. The difference between traditional input-driven programming and event-based programming is that input-driven programming asks for some input, waits for the input, and then takes an action based upon the input value. In contrast, event-based programming never waits for any input from the external world. However, it responds to one or more events as soon as the event takes place. Event-based modeling can be used to model real-world phenomena, as events can become cause for further cascaded events resulting in additional actions. The concept of event-based programming has its roots in SIMULA developed during the early 1970s. However, it became popular in recent years due to graphical and web-based interactions. More than one modern-day language such as C#, uses event-based programming to model animation and for graphical user interfaces.

1.8.9 Integration of Programming Paradigms

Modern-day languages do not subscribe to one programming paradigm; multiple programming paradigms are embedded in a language. For example, C++ supports the imperative programming paradigm and the object-oriented programming paradigm. Visual C++ also supports the visual programming paradigm. C# supports the imperative programming paradigm, the object-oriented programming paradigm, the event-based programming paradigm, and supports extensive visualization capabilities—a multimedia programming paradigm feature. Java has many characteristics of the imperative programming paradigm, the object-based programming paradigm, the web-based programming paradigm, the concurrent programming paradigm, and the event-based programming paradigm. Scala and Ruby integrate functional programming and object-oriented programming paradigms. The programming language X10 developed by IBM integrates the imperative programming paradigm, the object-oriented programming paradigm, and the concurrent programming paradigm. Modern scripting languages PHP and Python integrate imperative programming paradigms, shell-based programming, and object-oriented programming paradigms into one. A detailed list of programming paradigms supported in some popular programming languages is given in Appendix I.

Each programming paradigm has both advantages and disadvantages. For example, the notion of objects in object-oriented programming, along with information hiding and the notion of modules, is suitable for large-scale software development and is being adopted by many older languages such as FORTRAN (in the latest version, Fortran 2008), that have evolved with time to incorporate many well-proven features such as stack-based implementation, recursion, and objects over a period of time. Similarly, COBOL (the latest version, COBOL 2002) has evolved to include objects. If a language stops evolving with time, programmers stop using the language for the lack of features, and old languages are replaced by new languages supporting the popular programming paradigms.

1.9 CLASSIFICATION OF LANGUAGES

Languages can be classified using multiple criteria such as programming paradigms, implementation models, the notions of types used in the programming languages, and their intended problem domains. However, a clear separation is missing in many modern-day languages, as they incorporate a combination of features and paradigms to utilize the underlying advantages.

1.9.1 Programming Paradigms–Based Classification

We have already discussed various programming paradigms. For example, early versions of FORTRAN, C, ALGOL 60, Pascal, ADA, and PL/I were imperative languages; early versions of Lisp, ML, and Scheme are popular functional programming languages; Prolog is a popular logic programming language; and C++, Java, Scala, Ruby, Python, and X10 are multiple programming paradigm languages. For example, C++ is an integration of the imperative programming paradigm and the object-Oriented programming paradigm; Common Lisp Object-Oriented System (CLOS) is an integration of the functional programming paradigm and the object-oriented programming paradigm; Concurrent C, Concurrent Pascal, and Parallel Fortran are integration of the concurrent programming paradigm and the imperative programming paradigm; and so on. Java's thread allows concurrent programming. So Java is an integration of the imperative programming paradigm, the object-oriented programming paradigm, the concurrent programming paradigm, the event-based programming paradigm, and the web-based programming. Microsoft C# is an integration of the imperative programming paradigm, the object-oriented programming paradigm, and the event-based programming paradigm. Combined with .NET, it is also suitable for web-based programming. Recent languages such as Scala and Ruby support multiple programming paradigms. In future, more such languages that would integrate multiple programming paradigms will evolve. Older languages have incorporated concurrent programming and object-oriented programming as they evolved.

1.9.2 Implementation-Based Classification

Languages can also be classified on the basis of implementation models. There are four broad implementation models: (1) static implementation, (2) stack-based implementation, (3) heap-based implementation, and (4) an integrated model of implementation that integrates static, stack-based, and heap-based memory allocation for execution efficiency. As discussed in Chapter 5, static implementation allocates memory for every declared data structure and variables at the time of compilation, and there is no chance of memory growth at run time. *Stack-based implementation* uses a stack, also called *control stack*, to allocate the memory for local variables and other run-time requirements such as various pointers needed to run a called procedure. The advantage of the stack-based implementation is that the stack can grow and shrink at run time, based on the memory needed by the chain of called subprograms. This runtime expansion allows dynamic memory growth, and supports the implementation of recursive procedures and memory reuse. The heap-based implementation model uses a central common memory area accessible to all procedures; the memory needed by the data structures can be borrowed from the heap as and when needed and released dynamically at run time, either manually by programmer's explicit action or automatically after the data structure is no more in use. Heap-based implementation allows the allocation of dynamic data structures such as objects in object-oriented programming, and recursive data structures such as linked lists and trees.

Static implementation does not support recursive procedures, recursive data structures such as linked lists and trees, and dynamic objects. However, static implementation uses absolute memory address, utilizing single-memory access to access a data element. Hence static implementation can access data elements faster than stack-based implementation and heap-based implementation that use pointer-based accesses, which are equivalent to more than one memory access.

Stack size in stack-based implementation is limited only by the memory allowed by the operating system. Memory locations needed by a procedure can be allocated at run time when the corresponding procedure is invoked. The memory locations are allocated on top of the stack based on the calling pattern of the procedure and are recovered after the called procedure is over. This allows *memory reuse*, as the same memory space can be used when another procedure is called. To summarize, there are two major advantages of stack-based implementation: execution of recursive procedure and memory reuse. However, the addressing mechanism of stack-based implementation makes use of pointers and offsets, as discussed in Chapter 5. Hence, in comparison to static implementation, it has additional overhead to access variables and data structures.

One major disadvantage of stack-based implementation is that recursive data structures such as linked lists and trees, and dynamic data structures such as dynamically created objects can still not be implemented. There are two major issues in recursive data structures and dynamic objects that prevent their implementation on a pure stack-based implementation. The first issue is uncertainty in the creation of the data structure at the compile time, and the second issue is the size of the data structure. In the case of recursive data structures, it is not clear at compile time how much memory will be needed at run time. Depending upon the invocation and input data, a recursive data structure may need to be invoked multiple times for varying numbers of cells. Similarly, it is not clear at compile time which object will be created at run time. In addition, recursive data structures and dynamic objects may have a lifetime beyond the life of the procedure in which they were created. For this reason, a common memory area called *heap* is used. Heap is a memory bank from which the processes can borrow memory as and when needed and return the memory when their need is over. One logical data structure may be distributed across the heap in different memory blocks connected through pointers. These memory blocks are allocated at different times in response to the request made during program execution. Deleting a recursive data structure or dynamic object is equivalent to releasing the memory in the heap. After the memory is released, a garbage collector recycles the memory, so that it can be used by other processes. Note that *this heap is different from a tree-heap you have learned in data structure.*

Heap-based implementation is slower than the stack-based implementation since there are (1) overheads of run-time allocation of objects, (2) overheads of excessive use of pointers to traverse the data structures in the heap, and (3) overheads of recycling the memory for reuse.

Modern-day languages use all three memory allocation techniques, since they want to give all the facilities along with maximum possible efficiency of execution. For example, a modern-day language such as C++ uses static variables that use static allocation, stack-based implementation for handling recursive procedures and memory reuse, and use of heap for dynamic data objects and recursive data structures.

1.9.3 Other Classifications

Programming languages can also be classified according to the type of system they support. There are two major classifications of type: monomorphic type and polymorphic type. As the name suggests, a function in a monomorphic type language can handle only one type of data structure. For example, if we write a function to count the numbers of elements in a linked list, then multiple similar functions will need to be written in a monomorphic language, depending on whether we are counting a list of integers or list of floating-point numbers of a list of another data type. However, just one generic function needs to be written in a polymorphic language that will adapt itself to different types of linked lists. Languages such as FORTRAN, ALGOL, Pascal, and C are monomorphic languages: they support a limited amount of polymorphism in the form of coercion and overloading, as discussed in Chapter 7. Many modern-day languages such as C++, Lisp, ML, Haskell, Java, and Prolog are polymorphic languages.

Another interesting way to classify programming languages is to identify them by the class of problems they are better at, which depends upon the language feature that a problem domain needs. A language that meets all the requirements of a problem domain can be classified as a language suitable for that domain. For example, C++ is a good programming language for large software development. However, it is not good for real-time programming. Similarly, FORTRAN is a good language for scientific programming.

1.10 SUMMARY

A program has three major components: *logic, abstraction*, and *control*. Different languages are good for different problem domains. Criteria for good programming languages are often contradictory. Different programming paradigms—imperative programming paradigm, functional programming paradigm, logic programming paradigm, object-oriented programming paradigm, visual programming paradigm, concurrent programming paradigm, event-based programming paradigm, multimedia programming paradigm, and web-based programming—can be characterized by different features. As different programming paradigms have evolved, language scientists have debated and identified the good features of the various programming paradigms. Modern-day languages combine these features and incorporate multiple programming paradigms. For example, Java is a language that incorporates the imperative programming, object-oriented programming, event-driven programming paradigm, the concurrent programming paradigm, and the web-based programming.

High-level languages have to support data and control abstractions for better comprehension, management, and evolution of software. The languages should also support modularity so that the effect of changing one module is localized to that module. We have discussed three different translation mechanisms: interpretation provided by the interpreters, compilation provided by the compilers, and just-in-time compilation provided by the just-in-time compilers. The compiled codes execute most efficiently because the translation process is not part of the execution. Interpreters execute the program slowly because translation and execution are interleaved during run time. Web-based languages use a virtual machine and just-in-time compilation to translate to low-level code for execution. The execution efficiency of just-in-time compilation is between a compiled code and an interpreted code.

To implement a programming language, various low-level abstract models are used. Static implementation uses direct access to memory. However, due to lack of support for memory growth, static implementation is unable to handle recursive procedures, memory reuse, recursive data structures such as linked lists and trees, and dynamic data objects. Stack-based implementation allows execution of recursive procedures and memory reuse. However, stack-based implementation cannot handle recursive data structures and dynamic data objects. In order to handle recursive data structures and run-time creation of dynamic data objects that can be allocated piecemeal, based on program demand, we need a *heap—a common memory area shared by all the subprograms.*

Languages can be classified using many different criteria such as (1) programming paradigm–based classification, (2) implementation-based classification, and (3) type-based classification. However, it is very difficult to classify modern-day languages. Modern programming languages benefit from a combination of features, and may not be classifiable to one single category.

1.11 ASSESSMENT

1.11.1 Concepts and Definitions

Abstraction; algorithm; alpha testing; assembly code; beta testing; binary code; code optimization; compiler; concurrency; concurrent programming; conditional jump; control; control abstraction; control flow; control flow diagram; data abstraction; data-dependent iteration; declarative programming paradigm; definite iteration; destructive update; event-based programming paradigm; functional equivalence; functional programming paradigm; goto statement; heap-based implementation; imperative language; indefinite iteration; information hiding; intermediate code; Internet programming paradigm; interpreter; just-in-time compilation; lexical analysis; linker; loader; logic programming paradigm; modularity; monomorphic type; multimedia programming paradigm; object-oriented programming paradigm; orthogonality; parsing; polymorphic type; portability; problem domain; program comprehension; readability; software development cycle; software engineering; software evolution; software maintenance; software reuse; stack-based implementation; static implementation; symbol table; unconditional jump; visual programming paradigm; writability.

1.11.2 Problem Solving

1. Write a program using for-loop and while-loop to construct the for-each loop as given below. Assume that "a" is an aggregate data structure and the function "size(a)" returns the number of elements in "a," and next(a) returns the next element of the data structure "a."

```
foreach x in a {
    y = f(x); print (x, y);
    }
```

2. Write a control flow diagram for the following code.

```
i = 0; j = 10;
for (k = 0; k = < 9; k++) a[k] = 0;
while (i < 5) {
    j = j - 1;
    for (k = 0; k < = j; k++) {
        a[k] = a[k] + 7;
        if (a[k] mod 2 == 0} print ('even number');}
        i = i + 1
            }
```

1.11.3 Extended Response

3. Explain the three major components—logic, control, and abstraction—in a program.
4. Explain the advantages of control and data abstractions in programming languages.
5. Compare and contrast various criteria of a good programming language.
6. Explain using examples from various problem domains and the use of contrasting criteria for good programming language, why there cannot be a universal good programming language.
7. How is the "goto" controversy related to program comprehension? Explain using a figure.
8. Explain the software development cycle and its effect on the development of programming languages.
9. Explain the characteristics of the various programming paradigms.
10. What are the various categories of memory allocation in programming language implementations? Explain the advantages and disadvantages of each category.
11. What do you understand by just-in-time compilation? How is it different from traditional interpretation and compilation? Explain using a figure.
12. Why do web-based languages use just-in-time compilation? Explain.
13. What are the schemes to categorize programming languages? Describe briefly each classification.
14. What do you understand by control flow diagrams? Explain the basic components of control flow diagrams.

FURTHER READING

Amber, Allen A., Burnett, Margret M., and Zimmerman, Betsy A. "Operational vs. definitional - A perspective on programming paradigms." *IEEE Computer* 25(9), September 1992, 25–42.

Brainerd, Walter S. *Guide to FORTRAN 2003 Programming*. Springer-Verlag London Ltd. 2009.

Dijkstra, Edsgar W. "GO TO statement considered harmful." *Communications of the ACM*, 11(3), 1968, 147–148.

Knuth, Donald E. "Structured programming with goto statements." *Computing Surveys* 6(4), December 1974, 261–302.

Kowalski, Robert. "Algorithm = Logic + Control." *Communications of the ACM*, 22(7), 1979, 424–436.

Kupferschmid, Michael. *Classical Fortran: Programming for Engineering and Scientific Applications*, 2nd edition. Chapman and Hall/CRC Press. 2009.

Ledgard, Henry F. and Marcotty, Michael. "A genealogy of control structures." *Communications of the ACM*, 18(11), November 1975, 629–639.

Peterson, Wesley W., Kasami, T., and Tokura, N. "On the capabilities of while, repeat, and exit statements." *Communications of the ACM*, 16(8), 1973, 503–512.

Stalling, William. *Operating Systems: Internals and Design Principles*. Pearson Publishers. 2011.

CHAPTER 2

Background and Fundamental Concepts

BACKGROUND CONCEPTS

Previous programming background; A course in data structures; A course in discrete structures

This chapter describes computer architecture concepts, mathematical concepts, data structure concepts, and abstract computational concepts needed for understanding the abstractions and implementation of programming languages. These concepts need not be studied in sequence. You may like to study the concepts as needed as background material in different chapters.

2.1 VON NEUMANN MACHINE

Present-day computers are based on a low-level abstract machine called the von Neumann machine after the mathematician John von Neumann proposed a stored-program computer model. Understanding the von Neumann machine is important for us, because low-level machine instructions are designed to execute on the von Neumann machine. Program translators translate high-level constructs to equivalent low-level instructions. High-level constructs—such as if-then-else statements, while-do loops, and for loops—are translated to a sequence of low-level instructions in the von Neumann machine. The application of the von Neumann machine and low-level abstract instruction sets will become clear when we discuss the meaning of high-level constructs in terms of low-level abstract instructions in Chapter 3 and translate control abstractions to sequences of low-level abstract instructions in Chapter 5.

The von Neumann machine has two major components: (1) memory that stores program and data and (2) a central processing unit (CPU) that fetches the instructions and processes the data based upon the instructions. The instructions are fetched one at a time by the CPU using a bus—a high-speed connection between the memory and CPU. A CPU interprets the instruction and loads the data from the memory using the same bus. Data

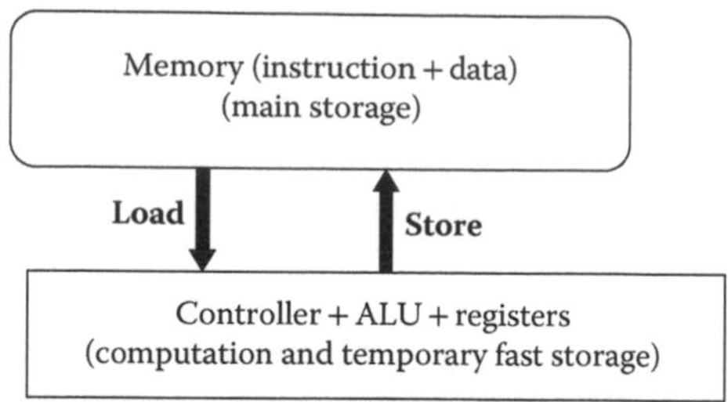

FIGURE 2.1 A von Neumann machine.

is processed in the arithmetic and logical unit (ALU) within the CPU, and the resulting data are stored back to the memory, as shown in Figure 2.1. Frequently used data items or their references are stored in hardware registers for faster access time. The von Neumann machine has a program counter, which contains the memory address of the next instruction to be executed. The program counter is incremented by one after fetching the current instruction. However, its value can be altered using conditional and unconditional jump statements.

The instructions could be categorized as loading from the memory (load); storing into the memory (store); performing arithmetic computations of addition, subtraction, multiplication, and division; logical computations such as logical-AND, logical-OR, exclusive-OR, and negation; comparing two values; conditionally or unconditionally jumping to another nonadjacent instruction by changing the value of the program counter and by checking the status of various system-level flags. The various flags are stored in a register called the *program status word* (PSW). In addition to these categories of instructions, instruction sets also support various addressing mechanisms to address the operands, depending upon the computer architecture.

2.1.1 Address Mechanisms

In the von Neumann machine, data or addresses of memory locations can be temporarily stored in the registers or in the memory (RAM). Similarly, a memory location can also hold data, as well as address of another memory location. If a memory holds the data, then it is accessed by an instruction using a single memory access called *direct access*. However, if a memory location holds the address of another memory location, then two memory accesses are needed to load the data into the CPU, and this method is called *indirect access*. In addition to direct and indirect accesses, an offset can be added or subtracted to an address stored in a register or another address to calculate a new address. This offset-based method is used to access subfields of complex data structures, where one needs to store the base address of the first memory location of the data structure and compute the address of the subfields by adding the offset of the subfield to the base address.

A computer may support 0-address, 1-address, 2-address, or 3-addresses, depending upon the computer architecture. As summarized in Table 2.1, the number of addresses is given by the maximum number of arguments the set of assembly-level instructions in

TABLE 2.1 Different Types of Addressing Mechanisms

Instruction Type	Type of Operations
Three-address machine instructions: *<instruction-name>* *<src$_1$>*, *<src$_2$>*, *<dst>*	An instruction in a 3-address machine. Instruction could be any arithmetic or logical dyadic operation
Two-address machine instructions: *<instruction-name>* *<src$_1$>*, *<src$_2$>*	An instruction in a 2-address machine. Instruction could be any arithmetic or logical dyadic operation. The destination is same as the second argument
One-address machine instructions: *<instruction-name>* *<src>*	Instruction could be any arithmetic or logical dyadic operation. One of the registers by default is *accumulator* that acts as destination
Zero-address machine instructions: *<instruction-name>*	Load, store, add, subtract, multiply, divide, and so on. Uses a stack-based evaluation. The arguments for operations are picked up from top of the stack

a low-level machine have when addressing memory locations or registers. For expressions involving dyadic operations, there are two input arguments and one output argument. For example, *add_integer* will require three addresses: two to hold the input arguments and one to hold the output argument. The arguments that hold the input values are called *source*, and the argument that holds the output value is called *destination*.

Most modern-day processors support 3-address instruction sets. However, during the early days, multiple addresses could not be accommodated with 16-bit word size. *Two address machines* have a special purpose register called *accumulator* that is always used as destination and is not mentioned as one of the instruction arguments to save the number of bits needed for the addressing mechanism. It can also be explicitly used as one of the source registers. *One address machine* implicitly has the first source argument as the accumulator, and the destination is invariably the accumulator. Accumulators are not mentioned in the instruction to save the number of bits; implicit use of accumulator for both source and destination further cuts down the instruction length.

Zero-address machines do not have any explicit argument and use a built-in stack to hold the argument. The instructions in zero-address machines are load, store, *push, pop, add, subtract*, and so on. For example, when an add instruction is given, the top two data elements from the evaluation stack are *popped* into ALU, evaluated, and the result is *pushed* back on top of the evaluation stack. The instructions *load* and *store* can pull the data from a local variable to the top of the evaluation stack or store the data from the top of the evaluation stack to a local variable. Zero-address machines can be implemented by a higher-address machine. An instruction in higher address machines can be translated to two or more lower-address instructions. Zero-address machines have been used to execute java programs in a Java Virtual Machine (JVM).

Example 2.1

This example illustrates the translation of an arithmetic expression $X = A + B$, where X, A, and B are integer variables. It is assumed that these variables are mapped into memory locations. The percent sign "%" starts a comment. An instruction in three-address machine is as follows:

```
integer_add A, B, X % Add data loaded from memory locations A
                    % and B, and store back in X.
```

Since 2-address machines do not have capability to have three addresses, 3-address instructions are split into a sequence of 2-address instructions. The second argument also acts as the destination. The corresponding low-level instructions in a 2-address machine are as follows:

```
load A, R0 % load the content of memory location A into the
           % register R0
integer_add B, R0 % add the contents of the memory location B
                  % and the register R0
store R0, X % store the content of R0 into the memory location X
```

One address machine implicitly uses accumulator as one of the input arguments and the destination. The corresponding set of low-level 1-address instructions are as follows:

```
load A % load the content of memory location A into the
       % accumulator
integer_add B % add the contents of the memory location B with
              % the content of the accumulator
store X % store the content of accumulator in the memory
        % location X
```

A zero-address machine (as given in JVM instruction set) loads the values of memory locations *A* and *B* into a built-in stack, adds the top two values of the stack, and leaves the sum on top of the stack, and then stores the top of the stack into the memory location *X*. The load operation is a sequence of two operations: first it loads the address of the memory location on top of the stack by using load_literal instructions, and then it loads the value of the memory location by using a load instruction.

```
load_literal 3 % Push the address of X on top of the
               % evaluation stack.
load_literal 1 % Push the address of A on top of the
               % evaluation stack.
load % Pop the address of A and push the content of A on top
     % of the evaluation stack.
load_literal 2 % Push the address of B on top of the evaluation
               % stack.
load % Pop the address of B and push the content of B on top
     % of the evaluation stack.
```

```
integer_add % Pop top two elements from the stack, add them,
            % and push the result on the stack.
store % Pop the result, pop the address X, and store result
      % into the memory location X.
```

It is evident from Example 2.1 that more instructions are needed in lower-address instruction sets to do the same task. Since each instruction means a memory fetch, lower address machines need more memory fetches when executed on higher-address machines. Thus the zero-address virtual machines run slower than the native compiled code on higher-address machines.

2.2 DISCRETE STRUCTURES CONCEPTS

In this section, we refresh selected concepts of discrete structures that are relevant to the abstract representation of data. Their relevance will become clear as we proceed with the discussions of programming language concepts in the following chapters.

2.2.1 Set Operations

The notion of sets and set-based operations is used in programming languages at multiple levels. Sets are used in (1) definition of the types in the type theory, as explained in Chapter 7; (2) set-based programming such as in languages like Pascal, SETL, Modula, Claire, Ruby, and Python; and (3) libraries for set-based programming in many popular languages such as Java, C++, and Prolog.

A *set* is a collection of unique data items. Shuffling the elements in a set does not alter the set. A *bag* or *multiset* is a collection of entities that can have multiple data items with the same value. For example, {*1, 2, 3, 2, 4, 5, 7*} is a multiset (or bag) where element *2* has been repeated three times. Still, shuffling the elements in the bag does not change the bag.

An interesting subclass of bags is *an ordered bag*. Each element of an *ordered bag* is associated with the corresponding position. In an ordered bag {*Meera, John, Li*}, *Meera* is associated with the position 1, *John* is associated with the position 2, and *Li* is associated with the position 3. Shuffling the elements in an ordered bag alters the ordered bag because the associated positions get disturbed. For example, an ordered bag {*Meera, John, Li*} is not the same as an ordered bag {*John, Meera, Li*}, because *Meera* is associated with the position 1 in the first ordered bag and is associated with the position 2 in the second ordered bag. Ordered bags play an important role in many programming language structures. Structures such as *sequence, string, files*, and so on are modeled as ordered bags as explained in Chapter 7.

Familiarity with subset, union, and intersection of the sets is assumed. A set of all the possible subsets of a given set is called the *power set* of the given set. For example, power set of a set {*X, Y, Z*} is given by {{} (empty set), {*X*}, {*Y*}, {*Z*}, {*X, Y*}, {*Y, Z*}, {*X, Z*}, or {*X, Y, Z*}}. To derive a subset, either an element can be selected or ignored giving two possibilities. Thus the size of a power set of a given set is given by 2^N, where *N* is the number of elements in the original set.

2.2.1.1 Cartesian Product

Cartesian product involves two or more sets and derives a new set of N-tuples, where N is the number of sets in the Cartesian product. Given a Cartesian product $A_1 \times A_{2\times}, \ldots, \times A_N$ ($N \geq 2$), the first field of an N-tuple is an element of the set A_1, and I_{th} element ($I \leq N$) of the N-tuple is an element of the I_{th} set. The number of elements in the resulting set is given by $|A_1| \times |A_2| \times |A_3| \times |A_4| \ldots |A_I| \times \ldots \times |A_N|$ where $|A_i|$ gives the number of elements in the I_{th} set.

Example 2.2

Let us consider three sets: A = {*a*, *b*, *c*}, B = {*x*, *y*, *z*}, and C = {*1*, *2*, *3*}. Cartesian product A × B × C has 3 × 3 × 3 = 27 possibilities as follows: {(*a*, *x*, *1*), (*a*, *x*, *2*), …, (*c*, *z*, *1*), (*c*, *z*, *2*), (*c*, *z*, *3*)}.

2.2.1.2 Mapping

Mapping involves associating the elements of one set to elements in another set, either in a one-to-one manner or many-to-one manner. Figure 2.2 shows mapping between the domain D and the codomain R. There are five elements in the set D, and there are six elements in the set R. If we define the mapping as the "square of the elements in D," then every element of the domain D is mapped to an element in R. More than one element in the domain D may map to the same element in the codomain R. However, one element of the domain D cannot map to more than one element in the codomain R.

2.2.1.3 Isomorphism

Two sets S_1 and S_2 are *isomorphic*, if there exist a pair of bijective functions f and g such that there is a mapping between S_1 and S_2: the function f maps all the elements of S_1 onto S_2 in a one-to-one manner, the function g maps the elements of S_2 to S_1 in a one-to-one manner, and the function g is an inverse function of the function "f."

Example 2.3

For example, let us take two sets S_1 = {*1*, *2*, *3*} and S_2 = {*4*, *5*, *6*}. There is a function *add_3* that maps all the elements of the set S_1 to every element of the set S_2 in a one-to-one onto manner by adding 3 to an element in the domain S_1, and there

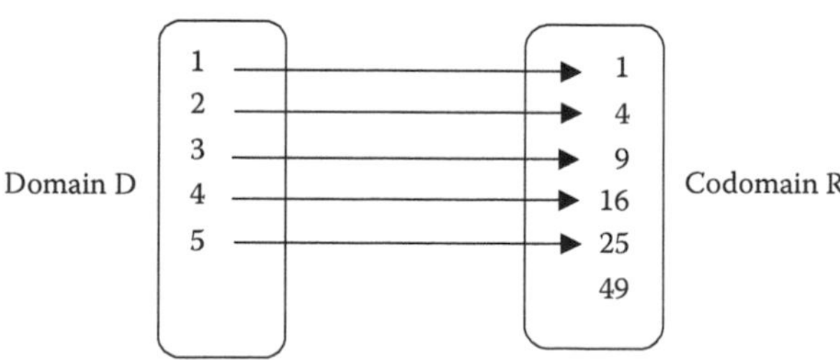

FIGURE 2.2 Mapping using the square function.

is a corresponding inverse function *subtract_3* that maps every element of S_2 to an element of set S_1 in a one-to-one onto manner by subtracting 3. Both add_3 and subtract_3 are inverse functions of each other.

2.2.1.4 Functions

A *function* f is a single-valued mapping from every element $x \in$ domain D to an image $f(x) \in$ codomain C. The set of images of $f(x \in D) \subseteq C$, and is called the *range* of function. An *identity function* maps the element to itself, and *a constant function* maps every element of a set to a fixed single element. A function is called *one-to-one* or *injective* if for all elements in the domain there exists a distinct image in the codomain that is if $y_1 = f(x_1 \in D)$ and $y_2 = f(x_2 \in D)$ and $x1 \neq x2$ and $y_1, y_2 \in C$ implies $y_1 \neq y_2$. A function is *onto* or *surjective* if for every element $y \in C$, there is an element $x \in D$ such that $f(x) = y$. A function f is *bijective* if it is one-to-one and onto. That means that for an element in the codomain, there is a unique image of the element in the domain. Given a bijective function f, an inverse function f^{-1} can be defined that maps the unique image in the codomain to the corresponding element in the domain that is $f^{-1}(f(x \in D)) = x$.

In programming, surjective functions become quite important, because absence of an image of a defined function is equivalent to an error condition. A nonsurjective function can be made surjective by introducing a *null symbol*, denoted by ⊥, in the range. The null symbol becomes an image of all the symbols in domains that do not map to any non-null symbol in the codomain. This idea of null symbol has been used to explain the denotational semantics of programming languages in Chapter 3 and functional programming in Chapter 9.

Programmatic representation of a function has three parts: *variables, expression body,* and *input parameters.* When an input value is given, the input value is bound to a variable and substituted everywhere in the expression where the variable occurs. After the simplification of the expression, an output value is returned. A function may be given a name to call the function from various locations of multiple program modules.

Example 2.4

Let us consider the well-known function *factorial(n)* that you have studied in data structure and programming courses. The function has a variable *n* that is bound to the given parameter. The *expression body* of the function has been given on the right-hand side of the definition.

```
factorial(n) =
        if (n == 0) return(1) % base case
        else return(n * factorial(n - 1))% recursive
                                         % definition
```

The body of the function *factorial(n)* consists of two expressions: "*if (n == 0) then return (1); if (n > 0) return (n * factorial(n – 1)).*" Both the expressions are connected through if-then-else control abstraction, and the test ($n > 0$) is implicit in the if-then-else abstraction. When we call *factorial*(3), the value 3 is bound to the variable *n*, and then it is substituted everywhere in the expression body of the function, and the function reduces to "*if (3 == 0) then return 1 else return (3 * factorial(3 – 1)).*" The condition is evaluated, and the function is further simplified to *return(3 * factorial(3 – 1))*. The expression *(3 – 1)* is simplified, and the function is further simplified to *return(3 * factorial(2))*. Now the evaluation of *function(3)* is suspended until *factorial(2)* has been computed.

2.2.2 Boolean Logic and Predicate Calculus

Boolean logic is based upon associating truth values to some statements. An axiom is a statement or proposition that can either be "true" or "false." These truth values can be combined using various logical operators such as logical-AND (denoted by the symbol "∧"), logical-OR (denoted by the symbol "∨"), implication (denoted by the symbol "→"), and negation (denoted by the symbol "¬"). Logical-AND and logical-OR combine two logical axioms A and B. As shown in Table 2.2, both A and B should be true for A ∧ B to be true, and both A and B have to be false for A ∨ B to be false. Logical operation implication states that A → B means that if A is true, then B is true. That B's truth value cannot be ascertained if A is false. Hence, if A is false, then B can be either true or false. Negation means if axiom A is true, then ¬ A is false, and if axiom A is false, then ¬ A is true.

Example 2.5

Let us take an axiom A that "people are innovative" and axiom B that "people are proud." Assume that Axiom A and axiom B are true; then A ∧ B is also true—that is, "People are innovative and proud." If we consider logical-OR, and if either axiom A is true or axiom B is true, or if either one of them is true, then the axiom "People are innovative or proud" would be true.

Logical-AND and logical-OR are commutative, associative, and distributive, as given in Table 2.3. This concept is used extensively in developing programs and concurrent execution of programs. There are many combinations of Boolean operations that are equivalent. For example, De Morgan's theorem relates logical-AND with logical-OR by making use of the negation. The De Morgan's law states (1) the negation of the conjunction of predicates is equivalent to the disjunction of negation of the predicates and (2) the negation of the disjunction of predicates is equivalent to the conjunction of the negation of the predicates.

TABLE 2.2 Truth Table for Logical Operators Used in Programming Languages

A	B	¬(A)	A ∧ B	A ∨ B	A → B
True	False	False	False	True	False
False	False	True	False	False	True
True	True	False	True	True	True
False	True	True	False	True	True

TABLE 2.3 Equivalence in Boolean Operations

Type of Operations	Equivalences
Negation	$\neg(\neg P_1) \equiv P_1$
Associativity	$P_1 \wedge (P_2 \wedge P_3) \equiv (P_1 \wedge P_2) \wedge P_3$
	$P_1 \vee (P_2 \vee P_3) \equiv (P_1 \vee P_2) \vee P_3$
Commutativity	$P_1 \wedge P_2 \equiv P_2 \wedge P_1$
	$P_1 \vee P_2 \equiv P_2 \vee P_1$
Distributivity	$P_1 \wedge (P_2 \vee P_3) \equiv (P_1 \wedge P_2) \vee (P_1 \wedge P_3)$
	$P_1 \vee (P_2 \wedge P_3) \equiv (P_1 \vee P_2) \wedge (P_1 \vee P_3)$
De Morgan's rule	$\neg(P_1 \wedge P_2) \equiv (\neg P_1) \vee (\neg P_2)$
	$\neg(P_1 \vee P_2) \equiv (\neg P_1) \wedge (\neg P_2)$

2.2.2.1 First-Order Predicate Calculus

Propositional logic handles only truth values of simple axioms and derivation of new axioms using logical operators. If we enrich propositional logic by two types of quantification—universal (denoted by the symbol ∀) and existential (denoted by the symbol ∃)—then the resulting logic is called "first order predicate calculus." Universal quantification associates a property for every element of a set. After associating a property with every element of a set, if an object is the member of the set with which the property has been associated, then the object will also be associated with the property. For example, if say all men like to live longer, and John is a man. A natural inference for us would be that "John likes to live longer." Let us map these statements using first order predicate calculus (FOPC). The FOPC rules are ∀x (*man*(x) → *likes_to_live_longer*(x)). If we read these rules, it says that for all x, if x is a man, that implies x likes to live longer. John is a man. Using the implication rule, John likes to live longer. An existential quantifier asks for an element that satisfies a specific property. For example, a rule ∀x ∀y (*sibling*(*x*, *y*) → ∃z (*parent*(*x*, *z*), *parent*(*y*, *z*), *not* (*x* == *y*) states that for all *x* and for all *y*, *x* is a sibling of *y* if there exists a *z* such that *z* is parent of *x*, *z* is parent of *y*, and *x* is not equal to *y*. The variable *z* is an existential quantifier, as it looks for an element in the set that is parent to both *x* and *y*.

There are many equivalent properties relating the two quantifiers as shown in Table 2.4. Both universal quantifiers and existential quantifiers are commutative. However, when a universal quantifier is mixed with existential quantifiers, then the situation is not as straightforward. For example, if a property is true for all the elements of a set (∀*x P*(*x*)), then it can be said that it is not the case that there exists an element for which this property is not

TABLE 2.4 Equivalence in Quantification

Operations	Equivalences
Commutativity	$\forall x \, \forall y \, P(x, y) \equiv \forall y \, \forall x \, P(x, y)$
	$\exists x \, \exists y \, P(x, y) \equiv \exists y \, \exists x \, P(x, y)$
Duality	$\forall x \, P(x) \equiv \neg \exists x \, (\neg P(x))$
	$\exists x \, P(x) \equiv \neg \forall x \, (\neg P(x))$

true (¬∃x ¬P(x)). Similarly, (∃x P(x)) is equivalent to ¬ ∀x (¬ P (x)). If we mix a universal quantifier with an existential quantifier, then law of commutation does not hold: ∀x ∃y is not equivalent to ∃y ∀x: a statement such as, "There exists a person y for every person x whom x likes" (∀x ∃y likes (x, y)) is not the same as saying that "There exists a person y whom every person x likes" (∃y ∀x likes(x, y)).

First-order predicate calculus forms the basis of the logic programming paradigm, as discussed in Chapter 10. One of the limitations of first-order predicate calculus is that it cannot express relations about relations. Higher-order predicate calculus can express relations about relations. However, discussion of higher-order predicate calculus is beyond the scope of this book.

2.2.2.2 Relations

Similar to functions, relations have been used extensively in programming and representation of types needed for programming. In this subsection, we review the basic properties of relations.

A binary relation R is defined as a proper subset of the Cartesian product of two sets: domain S_1 and codomain S_2. Mathematically speaking a relation $R \subseteq S_1 \times S_2$, and each element $x \in S_1$ is related to an element $y \in S_2$ through the relation R for an ordered pair of the form $(x, y) \in R$. A relationship between two entities can be represented by either as xRy or R(x, y).

A relation can be *reflexive, symmetric, antisymmetric,* or *transitive.* A relationship R is reflexive if xRx is true for every element of a domain. A relationship R is *symmetric* if for every ordered pair $(x, y) \in R$, there exists another ordered pair $(y, x) \in R$. In other words, xRy is equivalent to yRx. A relationship is *antisymmetric* if xRy means y is never related to x through the relationship yRx. A relationship is *transitive* if xRy and yRz imply that there exists an ordered pair $(x, z) \in R$. The notion of relations has been used extensively in the development of the fundamental properties of programming. A relation R is an *equivalence relationship* if R is reflexive, symmetric, and transitive.

Example 2.6

For example, the relation "greater than" (or less than) is *antisymmetric* and *transitive.* Similarly, the relationship "equality" is reflexive, symmetric, and transitive. The value of a variable x is equal to itself (xRx); if the value of a variable x is equal to the value of a variable y, then the value of a variable y is equal to the value of a variable x (xRy implies yRx); if the value of a variable x is equal to the value of a variable y, and the value of a variable y is equal to the value of a variable z, then the value of a variable x is equal to the value of the variable z (x == y and y == z implies x == z).

The properties of reflexivity, symmetry, and transitivity have been used implicitly in many comparison operators in regular programming, such as sorting a sequence of numbers. The property of transitivity has also been used in the data-dependency graph analysis in Chapter 8 and in the analysis of equivalence of aliased variables.

2.2.3 Recursion

Recursion means that the definition uses itself in the definition, although with a smaller data set or argument value. Recursion can be used to define a recursive function, such as *factorial* or *Fibonacci*, or a recursive data structure, such as linked list or trees.

A recursive definition has at least one base case and at least one recursive definition. The recursive definition progressively unfolds and approaches the base case(s). In a finite number of unfoldings, the recursion is terminated by the base case, and the result of the computation is passed back in the reverse order to the calling invocation of the recursive function. The number of invocations of recursive functions is decided by the input value. The previous invocation is suspended until the invocation caused by the next unfolding is evaluated. This need to store the suspended procedures needs an inherent use of a stack to store the memory locations needed to perform computations in the chain of invoked functions.

Example 2.7

```
factorial(0)  = 1. % base case
factorial(n)  = n * factorial(n - 1) % recursive definition
```

The function *factorial(n)* is defined recursively by *n* * *factorial(n* – *1)*, and the base case is *factorial(0)* = *1*. The function call *factorial*(4) is unfolded as *4* * *factorial*(3); *factorial*(3) is unfolded as *3* * *factorial*(2); *factorial*(2) is unfolded as *2* * *factorial*(1); and *factorial*(1) is unfolded as *1* * *factorial*(0); *factorial*(0) is the base case. After evaluating *factorial*(0), the value *1* is passed to *factorial*(1); *factorial*(1) passes *1* * *1* = *1* to *factorial*(2); *factorial*(2) passes *2* * *1* = *2* to *factorial*(3); and *factorial*(3) passes *3* * *2* to *factorial*(4). The final value, 4 * 6 = 24, is returned at the end.

Example 2.8

Another example of recursive function is the definition of "Fibonacci numbers." The definition of Fibonacci numbers has two base cases, as given below

```
Fibonacci(0)  = 1 % base case
Fibonacci(1)  = 1 % base case
Fibonacci(n)  = Fibonacci(n - 1) + Fibonacci(n - 2) % recursive
                                                     % definition
```

2.2.3.1 Tail Recursion and Iteration

Tail recursion is an important subclass of recursive definitions that needs special attention in the study of programming languages. In tail recursion, the recursive part comes at the end of the definition. For example, the function *gcd* can be expressed using tail recursion as follows:

```
GCD(x, y)  = gcd((bigger(x, y) modulo smaller(x, y)),
             smaller(x, y)). % tail recursive definition
GCD(0, x)  = x % base case
```

In the above definition, the base case returns the value of the second argument if the first argument is 0. It would occur when the left-hand side of the tail recursive definition has the bigger argument as a multiple of the smaller argument. The tail recursive definition calls the function *GCD* tail recursively with the first argument as the remainder of the division of the bigger and smaller arguments and the second argument as the original smaller argument that would become the bigger argument in the new invocation of the function *GCD*.

One of the properties of tail recursive definitions is that they can be modeled using indefinite iterations and avoid the memory overhead associated with invoking new procedure calls in recursive definitions. Note that neither the definition of *factorial* nor the definition of *Fibonacci* is tail recursive: factorial has "multiplication" as the last operation, and *Fibonacci* has two recursive calls in the recursive definition. *Tail-recursive-optimization* is a code optimization technique to allow memory reuse and reduce execution time overhead by transforming tail recursive programs to indefinite iteration. Iterative programs reduce the memory and execution time overhead present in stacks, which are much slower to implement and are wasteful of memory space.

2.2.3.2 Linear Recursion and Iteration

A large class of *linear recursive functions*—recursive functions with only one call to itself in the recursive definition—can be transformed to iterative programs using *indefinite iteration* and *accumulators*. An *accumulator* is an abstraction that keeps the last computed partial result. In the iterative version, the same set of variables are *reused*, and the invocations of recursive procedures are replaced by iterative code that starts from the base case and keeps accumulating the partial-output value after every cycle. Finally, the result is collected from the accumulator.

Example 2.9

An iterative version of the function *factorial* is given in Figure 2.3. The accumulator has been initialized to *factorial(0)* = 1—the base value of the function *factorial*; and the recursive calls have been substituted by iteration. After every step, the accumulator value is updated. This is equivalent to starting from the base case of recursion and building up; there is no unfolding of recursive procedures, making it more efficient.

```
Algorithm iterative_factorial
Input: value of n;
Output: accumulator value;
{ accumulator = 1;
  for (i = 1; i =< n; i++) accumulator = i * accumulator;
  return(accumulator);
}
```

FIGURE 2.3 An iterative version of the function *factorial*.

2.2.4 Finite State Machines

A *finite state machine* is an abstract machine to model a real-world phenomenon by modeling different situations (states) as nodes of a graph and the transitions between the states as directed edges between the nodes. A machine transits from one state to another based upon the input values. All the states may not be reachable from all the states. There are one or more initial states, and there are one or more final states. The machine starts from one of the initial states and ends up in a final state.

Example 2.10

A heating and cooling system can be modeled as finite state machine, as shown in Figure 2.4. There are three states: "room hot," "room cold," and "optimum temperature." All three states can be the initial state. However, only the state "optimal temperature" is the final state.

In "room hot" state, the thermostat starts cooling. The result is the transition to the state "optimal temperature." In "room cold" state, the thermostat starts warming. The result is the transition to the state "optimum temperature." No signal is sent if the difference between the preset temperature and the room temperature is within a threshold.

Example 2.11

The finite state machine in Figure 2.5 recognizes variable names in a program. A variable is defined as an English letter followed by any number of occurrences of English letters or digits. The machine contains the initial state S_0, the final state S_1, and a state S_2 to handle all error conditions. The machine starts in the state S_0 where

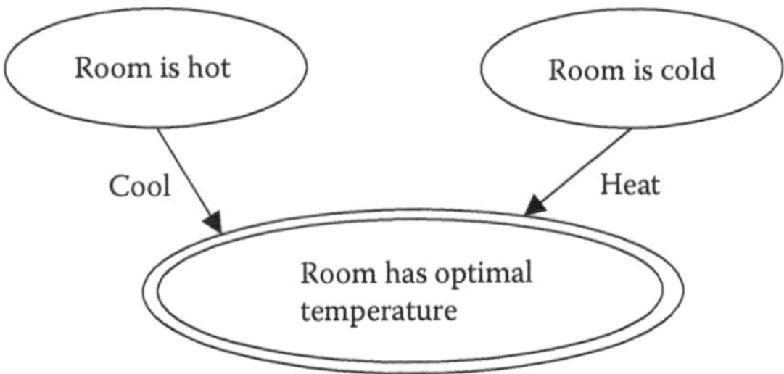

FIGURE 2.4 A finite-state machine modeling thermostat.

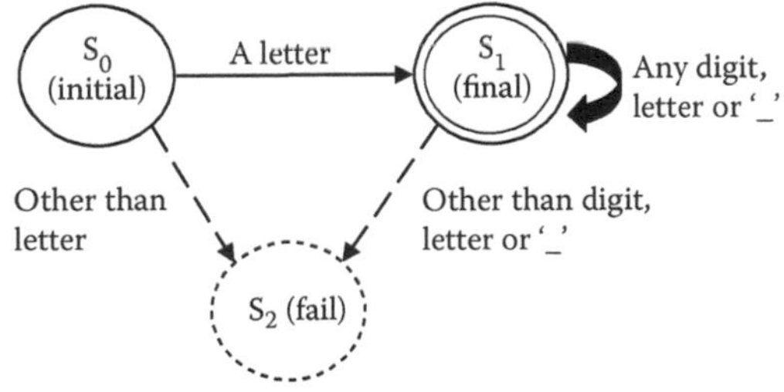

FIGURE 2.5 A finite-state machine to accept a variable name.

it can accept only an English letter to transit to state S_1; any other character will take the machine from state S_0 to S_2, where it will signal an error condition and gracefully terminate its execution. After accepting the first letter, the machine transits to the state S_1, where it can either accept a digit or another letter. The machine transits from S_1 to S_1 whenever it gets a digit, letter, or some special symbols such as "_"; otherwise, it transits to S_2—the fail state.

2.3 DATA STRUCTURE CONCEPTS

There are many concepts in data structures that are needed to understand the implementation of programming languages. Some of the important concepts are trees used in parsing, stacks used implicitly in recursion and implementation of programming language engines, queues and graphs used in some efficient garbage collection schemes, and hashing used to access the procedures and objects efficiently in the implementation of various programming paradigms.

2.3.1 Sequences

A *sequence* is an ordered bag of elements such that the immediately preceding element in the sequence is associated with a position that is less by exactly one from the position of the reference element. For example, a *string* is a sequence of characters, and a *file* is a sequence of data entities. Even abstract data types *stacks* and *queues* can be modeled using sequence as shown in the next subsection.

Let us denote a sequence with constituting elements within the angular brackets '<' and '>'. For example, *<x, y, z>* is a sequence where the data-element *x* is associated with position 1, the data-element *y* is associated with position 2, and the data-element *z* is associated with position 3. If we denote two functions—*predecessor* and *successor*—with a data element in a sequence, then *predecessor(y)* = *x*, and *predecessor(z)* = *y*, because the position of the data element *y* is greater than the position of the data element *x*, and the position of the data element *z* is greater than the position of the data element *y*. Similarly, *successor(x)* = *y* and *successor(y)* = *z*. The operations on sequences can be categorized as extracting an element, inserting an element, substituting an element by another element, joining two sequences, checking whether a sequence is subsequence of another sequence, extracting a subsequence, and substituting a subsequence by another subsequence. Major operations on sequences are listed in Table 2.5.

An element can be accessed either from the beginning, from the end, or by giving the index of an element. The operator *first* returns the first element of a sequence, the operator *second* returns the second element of a sequence, and the operator *last* returns the last element of a sequence. A complementary operation is to identify the remaining subsequence after an element has been taken out. The operator *rest* gives the remaining subsequence of the original sequence minus the first element, and the operator *butlast* gives the subsequence of the original sequence minus the last element. The operator *select* takes two inputs—the index of the element and the sequence—and returns the corresponding element of the sequence. The operator *cons* (abbreviation for construct) constructs a new sequence by inserting the given element in the beginning of the original sequence. The

TABLE 2.5 Major Operations Involving Sequence

Operation	Output	Explanation
$\texttt{first}(<x_1 \ldots x_n>)$	x_1	First element of a sequence
$\texttt{last}(<x_1 \ldots x_n>)$	x_n	Last element of a sequence
$\texttt{rest}(<x_1 \ldots x_n>)$	$<x_2, \ldots, x_n>$	Rest of the sequence
$\texttt{butlast}(<x_1 \ldots x_n>)$	$<x_1, \ldots, x_{n-1}>$	Subsequence except the last element
$\texttt{select}(i,\ (<x_1 \ldots x_n>)$	x_i	*i*th element of a sequence
$\texttt{cons}(a,\ <x_1 \ldots x_n>)$	$<a,\ x_1, \ldots, x_n>$	Constructs a new sequence by adding a at the beginning of the old sequence
$\texttt{insert}(i,\ a,\ <x_1 \ldots x_n>)$	$<x_1, \ldots x_{i-1},, a,, x_{i+1}, \ldots x_n>$	*a* is inserted as the *i*th element of a sequence
$\texttt{append}(<x_1 \ldots x_n>,\ <y_1 \ldots y_m>)$	$(<x_1 \ldots x_n,\ y_1 \ldots y_m>$	Concatenate sequences in order
$\texttt{subsequence}(<x_1 \ldots x_n>, i, m)$	$<x_i, \ldots, x_{i+m-1}>$	A subsequence starting from location *i* of length *m*
$\texttt{is\_subseq}(<x_1 \ldots x_n>,\ <y_1 \ldots y_m>)$	Boolean	Returns true if $<x_1 \ldots x_n>$ is included in $<y_1, \ldots y_m>$ otherwise returns false

operation *insert* inserts the given element at the specified position and generates a new sequence. The operator *cons* is a specific case of insertion. The operator *append* takes two sequences of the form $<x_1 \ldots x_n>$ and $<y_1 \ldots y_m>$ and generates a new sequence of the form $<x_1 \ldots x_n, y_1 \ldots y_m>$ by placing the elements of the second sequence after the elements of the first sequence. The operator *subseq* takes three input arguments: (1) the original sequence, (2) the starting position in the original sequence, and (3) the length of the subsequence, and extracts the corresponding subsequence. For example, *subseq(<4, 5, 6, 7>, 3, 2)* derives the subsequence <6, 7>. The predicate *is_subseq* takes as input two sequences and checks if the first sequence is a subsequence of the second sequence. For example, *is_subseq(<6, 7>, <4, 5, 6, 7>)* returns *true.*

A sequence can be implemented using a linked list, an array, or a vector. All three data structures have different characteristics. In a linked list, the elements are accessed sequentially generally from the beginning to the end; in an array, elements can be accessed randomly; and a vector can be accessed randomly and extended at run time.

2.3.2 Stacks and Queues

Both stacks and queues are important in the implementation of programming languages. Both stacks and queues are used in searching the complex search space modeled as trees or graphs. Stacks are used in (1) implementing programming languages, (2) handling recursion, and (3) managing and recycling dynamic data structures from the heap, as discussed in Chapters 5 and 6. Queues are used in efficient recycling of dynamic data structures allocated in the heap, as described in Chapter 6.

2.3.2.1 Stack

A *stack* is an abstract data type where data is entered only from one end. The inserted data is always put on the current top of the stack, and data element is removed from the current top of the stack; the other end is blocked. In addition, we can read the top element of the

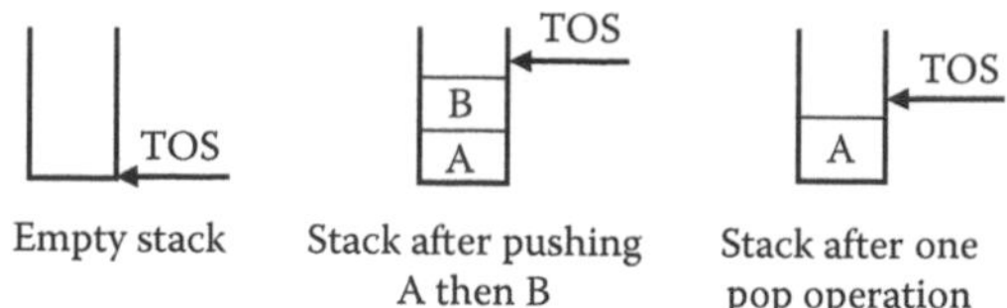

FIGURE 2.6 Stack operations.

stack without deleting the element from the stack, and we can check for an empty stack. The major advantage of a stack is its ability to work with the most recent data item. Stacks are also called last in first out (LIFO).

There are four abstract operations in a stack: (1) *push (stack, data element)*—pushing a data element on the top of a stack; (2) *pop(stack)*—reading and removing the top element from a stack; (3) *top(stack)*—reading the top element from the stack without changing the stack; and (4) *is_empty(stack)*—checking if the stack is empty. The instructions *push(stack, element)* and *pop(stack)* change the stack. In the case of the push operation, the element being inserted becomes the top element of the stack. In the case of *pop(stack)*, the top element is removed from the stack. The instruction *top(stack)* reads the top element of the stack without modifying the content of the stack. Stack-based operations are illustrated in Figure 2.6.

A stack can be modeled as a sequence of elements where data is inserted and taken out from one end. Pushing an element x in a stack $S = \langle a_1, a_2, \ldots, a_N \rangle$ gives a new stack S' as $\langle x, a_1, a_2, \ldots, a_N \rangle$, and the size of the stack is incremented by 1. Popping an element out of the stack $S = \langle a_1, a_2, \ldots, a_N \rangle$ gives a new stack $S' = \langle a_2, \ldots, a_N \rangle$, and the size of the stack is decremented by 1. An empty stack is modeled as an empty sequence $\langle\ \rangle$.

Implementation of stacks in computer memory requires two pointers: a start pointer and an end pointer. The top-of-the-stack pointer of an empty stack is equal to the start pointer, and the top-of-the-stack pointer of a full stack is equal to the end pointer. If the top-of-the-stack pointer tries to go beyond the end pointer, then a "memory overflow" error is indicated.

2.3.2.2 Queue

A *queue* is an abstract data type where data is retrieved from one end, and the data is inserted from the other end. Generally, a data-element is retrieved in the order it enters the queue. A queue needs two pointers: *front* and *rear*. The *front pointer* is used to retrieve the first data element, and the *rear pointer* is used to insert new data elements. A queue is empty when the front pointer catches up with the rear pointer. In order to insert an element in the queue, the data is inserted in the location pointed by the rear pointer, and the rear pointer is incremented by one. In order to retrieve the data element from the other end, the data element pointed by the front pointer is retrieved, and the front pointer is incremented by one. A queue is linear if the "rear" and "front" pointers move in only one direction, as shown in Figure 2.7. A queue is made circular by using modulo arithmetic, where the "front" or "rear" pointers traverse the first element of the queue after traversing

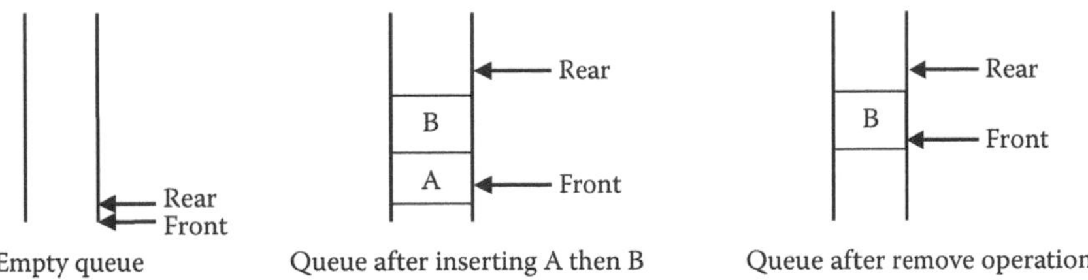

FIGURE 2.7 Queue operations.

the last element of the queue. Circular queues optimally reutilize the memory space freed after removal of the data elements.

Abstractly, queues can be modeled as a sequence of elements, where new data elements are inserted at the rear end, and the first data element is removed from the front end. Inserting an element x in a queue Q = $<a_1, a_2, \ldots, a_N>$ gives a new queue Q' as $<a_1, a_2, \ldots, a_N, x>$, and the size of the queue is incremented by 1. Removing an element out of the queue Q = $<a_1, a_2, \ldots, a_N>$ gives a new queue Q' = $<a_2, \ldots, a_N>$, and the size of the queue is decremented by 1.

2.3.3 Reference Mechanisms

Memory locations and registers can store two types of information: (1) data values on which computations are done and (2) memory locations pointing to other memory locations. *Pointers* are memory addresses of other memory locations that are stored in a register or a memory location. There are many advantages of the use of pointers:

1. Referencing complex data structures that are stored in some other part of the memory.
2. Avoiding the overhead of moving physically large data structures.
3. Delaying memory allocation for a variable until the run time. This property can be used for optimal memory allocation for extensible data structures such as linked lists, trees, vectors, and so on.
4. Allocating memory blocks for a complex logically contiguous data structure for which multiple physical memory blocks are allocated at different times and chained through pointers.
5. Using as a forwarding reference, providing independence of the program from data movement in the physical memory.
6. Sharing part of a complex data structure with other data structures.

Pointers have been used in the implementation of recursive data structures such as linked lists and trees, since the exact memory size needed for recursive data structures is unknown at the time of compilation. Moreover, depending upon the input data, the size of the recursive data structure may change and cannot be estimated at compile time.

System programming has a requirement to access and step through the memory locations. This requires that pointers should be provided with addition and subtraction capability, so that system programs can step through the memory space that is modeled as a single-dimensional array. This necessitates that the user-defined pointers should not cross the boundary of a *segment*—the area in the RAM allocated for the execution of a specific program.

Despite the advantages of the pointers, there are three major drawbacks with the use of pointers in high-level languages:

1. Arithmetic operations on pointers allow memory hopping. An arithmetic operation trying to access a memory location outside the segment causes a "segment-violation" error.

2. After the compilation of a program, different variables having different data types are mapped into a single linear memory space; all information about boundaries between these data types is lost. If we allow arithmetic on pointers, then pointers can step through different data types of objects at run-time reading or modifying the values incorrectly.

3. Memory locations used by multiple data structures cannot be reused until all the pointers pointing to the memory location are released.

Owing to these problems with pointers, programs can be error-prone and may crash at run time. It is difficult to figure out such pointer-related bugs.

2.3.4 Recursive Data Structures

There are classes of data structures such as linked lists, trees, and vectors that can be defined recursively. A linked list is defined recursively as a node of information followed by the rest of the list, and the base case is a null list.

```
<linked-list> ::= <information-node> <linked-list>
<linked-list> ::= null
```

A binary tree is defined recursively as an information node followed by a left subtree and a right subtree. The base case is an empty tree.

```
<binary-tree> ::= <binary-tree><information-node><binary-tree>
<binary-tree> ::= null
```

Pointer-based data structures have efficient insert and delete operations. However, recursive data structures cannot be allocated at compile time due to the lack of knowledge of run-time size of the recursive data structures. Recursive data structures grow at run-time, depending upon the program execution and input data. The memory for a recursive data structures is allocated in a piecemeal manner, based upon the programmer's demand during run time. We need a special common memory area called *heap* to implement recursive data structures. This heap is different from the heap studied in data structures. Recursive data structures are physically scattered in the heap, based upon the availability of the memory slots and the order in which program demands for memory allocation at run time. Different physical fragments of the same recursive data structure are connected through a chain of pointers.

2.3.5 Trees

A *tree* is a recursive data structure that consists of multiple information nodes organized in the form of an information node that contains one or more data elements and has possible subtrees rooted at the children of the information node. The base case of a tree is a "null" tree. A tree can have any number of branches. A *binary tree* has one cell for each information node, and has at most two subtrees: left subtree and right subtree. A tree is called *n-ary tree* if the maximum number of branches emanating from a node are "*n*." Trees can be implemented using two techniques: (1) indexible data structures such as arrays or vectors and (2) pointers connecting parent nodes to children nodes. Indexible data structures are suitable for representing complete binary trees or almost complete binary trees. An *almost complete binary tree* has all nonleaf nodes with two children except the rightmost nonleaf node that may have one left child, as shown in Figure 2.8.

Pointer-based implementation is suitable for any tree. However, they do not have the capability to compute the index of the parent node to traverse back. The trees implemented using pointers are directional in nature: trees can be traversed only from the parent node to the child node. To circumvent this directional property of pointers, we need either a stack to store the address of the parents before moving to a child or to use a reverse pointer stored in the child node in addition to pointers from parent node to child node. Both the schemes have memory overhead as well execution time overhead. However, in the case of a stack, memory waste is limited only by the depth of the tree.

Trees are used in programming languages and compilers during the parsing phase. A special kind of tree called "AND-OR tree" is the basis of implementing the logic programming paradigm and has been discussed in detail in Chapter 10. Trees can also be used to show the nesting structure of procedures in a program or blocks in a procedure and thus can help in understanding the extent of the scope of a variable declared in a procedure or within a block. N-ary tree are also used to implement data structures in a logic programming paradigm, as described in Chapter 10.

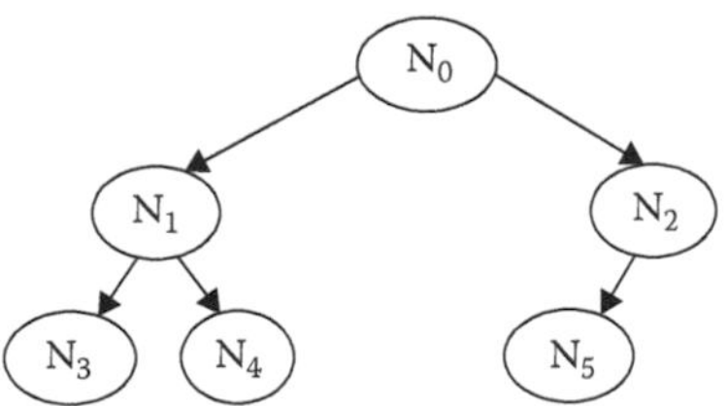

FIGURE 2.8 An almost-complete binary tree.

2.3.6 Graphs

A graph is formally defined as a pair (V, E) where V is the set of vertices, and E is the multiset of edges. In an undirected graph, edges are modeled as a pair (v_i, v_j), where $v_i, v_j \in V$ are the vertices. The pair $(v_i, v_j) \in E$ shows connectivity between the vertices v_i and v_j. There may be more than one edge between two nodes. A *path* is a sequence of edges that connect a *source node* to a *destination node*. There may be more than one path between two nodes.

In an *undirected graph*, the presence of an edge (v_i, v_j) implies the presence of a symmetric edge (v_j, v_i). In a *directed graph*—a graph where edges have directions, edges are modeled as a ordered pair (v_i, v_j), and the presence of (v_i, v_j) does not imply the presence of an edge (v_j, v_i). A weighted graph has weights associated with edges. The edges in a weighted graph are modeled as a triple of the form $(v_i, v_j, weight_{ij})$, where $weight_{ij}$ gives the weight of the edge. In a *weighted undirected graph*, the presence of a weighted edge $(v_i, v_j, weight_{ij})$ implies the presence of $(v_j, v_i, weight_{ij})$; the weights of symmetric edges remain the same. In a *weighted directed graph*, the presence of a weighted edge of the form $(v_i, v_j, weight_{ij})$ does not imply the presence of a symmetric edge in the other direction with the same weight. Graph connectivity is a transitive relationship. If a node v_i is connected to v_j and the node v_j is connected to v_k, then the node v_i is connected to v_k, although an explicit edge (v_i, v_k) may not exist. This property is important from the view point of reachability of data elements in recursive data structures.

Example 2.12

Figure 2.9 illustrates a model of road connectivity and distances between a set of cities in the United States using an undirected weighted graph. The cities could also be connected through rail links or air links to give multiple edges between two nodes. However, we consider only the road links in this graph. The graph shows that there are six cities represented as nodes (or vertices) of the graph, and the road connectivity between them has been shown by using weighted edges between the vertices, where the weight shows the road distance in miles between the cities.

A graph is *acyclic*—it has no cycle; if starting from a node, one cannot reach the same node using two different sequence of edges. A tree is an acyclic graph. A graph is *cyclic*, if starting from at least one node, we can reach the same node again using a path where edges in the path do not repeat. For example, there are many cycles in the connectivity graph given in Figure 2.9 such as <(*Seattle, Kent*), (*Kent, Washington DC*),

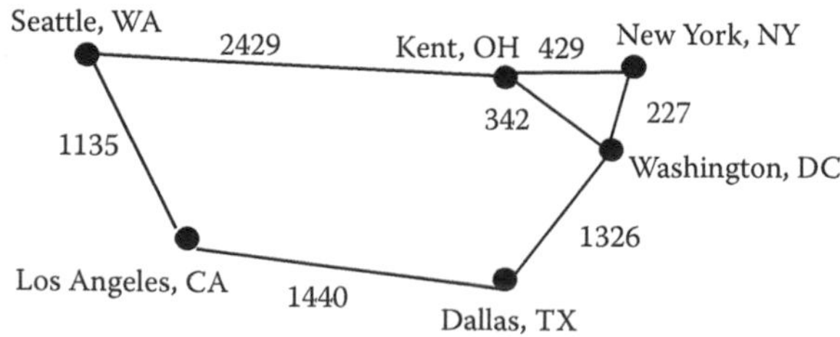

FIGURE 2.9 An example of a weighted undirected graph.

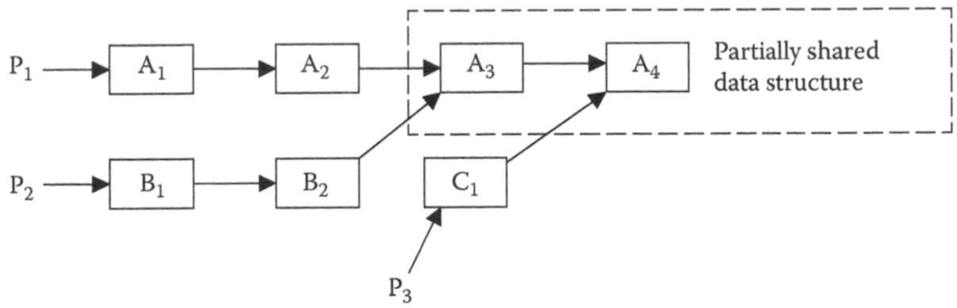

FIGURE 2.10 Modeling a directed acyclic graph with shared information nodes.

(*Washington DC, Dallas*), (*Dallas, Los Angles*), (*Los Angles, Seattle*)>. Another such cycle is < (*Kent, New York*), (*New York, Washington DC*), (*Washington DC, Kent*) >.

Identifying cycles in graphs is a major problem. In order to identify cycles, one has to store the previously visited nodes, so that a membership test can be used to check if the previously visited node, is being revisited. The membership test can be done efficiently by using hash functions as described in Section 2.3.9. In programming languages, graphs have been used to represent shared data structures, as explained in Example 2.13. Graphs are also used to (1) model information flow between program statements; (2) to find out which data cells are connected in heap during memory recycling; (3) modeling the solution space of a problem; and (4) to implement functional programming languages.

Example 2.13

Figure 2.10 shows three linked lists sharing parts of their data structures: the first linked list consists of four information nodes A_1, A_2, A_3, A_4; the second linked list consists of information nodes B_1, B_2, A_3, A_4; the third linked list consists of information nodes C_1, A_4. The first and second data structures share two information nodes: A_3 and A_4; the first, the second, and the third data structures share the common information node A_4.

2.3.7 Exhaustive Search

Computation can be modeled as a state-space search problem. A state-space problem has multiple computational states, and every computational statement takes the machine from one computational state to another computational state. The computation can be modeled as a state-space graph. Different search techniques can be classified into two major categories: (1) exhaustive search and (2) heuristics-based search. *Exhaustive search* schemes potentially search all the nodes in the state-space graph until a goal node is found. A goal node is a state that has the desired final condition. Exhaustive searches guarantee a solution if the search does not get caught in a cycle. There are two major approaches to exhaustively search a tree or a graph: *depth-first search* and *breadth-first search.*

Heuristic searches use mathematical functions to estimate the distance or closeness of the current state to the final state and move toward the final state in a focused manner. Heuristic searches are more efficient than exhaustive searches but do not guarantee a solution. Most of the programming language problems use exhaustive searches, while artificial intelligence problems use heuristic searches.

2.3.7.1 Depth-First Search

Depth-first search traverses to descendant nodes in a predetermined order, generally in left-to-right order. It uses a stack to store information to facilitate the traversal to the right subtrees. The search starts from the root node and selects the leftmost unexplored child for traversal. When the subtree rooted at the unexplored child is being explored, there is no way for the pointer to come back to the right children unless the pointer visits the current node again. To facilitate the traversal to the right children, the address of the current node is stored in a stack. In the absence of any unexplored child, there is no need to store the address of the current node in the stack. Since a stack stores the information in LIFO mode, the parent of the current node being traversed is popped first. After traversing a leaf node (node having no child), the next node is retrieved from the top of the stack. This process of retrieving from the stack and exploring the next right subtree for the retrieved node is repeated until the stack is empty and there is no more subtree to traverse. Since there is no more subtree to traverse, the search terminates.

Example 2.14

Consider the example in Figure 2.11. The pointer first traverses the node P_0. At this time, the stack is empty. However, the node P_0 has both left and right children. Next the node P_1—the left child of P_0—is traversed, and the address of the node P_0 is stored in the stack to facilitate traversal of the node P_2 later. Next the node P_3—the left child of the node P_1—is traversed, and the address of the node P_1 is stored on the stack, and the stack looks like <address(P_1), address(P_0)>. Since the node P_3 is a leaf node, the node P_1 is popped from the stack after traversing the node P_3, and the right child—the node P_4—is traversed. After popping the address of the node P_1, the stack looks like <address(P_0)>.

Since the node P_1 has no more children, no information is pushed on the stack while traversing the node P_4. After traversing the node P_4, the address of the node P_0 is popped from the stack, and the stack becomes empty. However, the right child of the node P_0 remains to be traversed. The stack remains empty, since the node P_2 has no right sibling. After traversing the node P_2, the address of the node P_2 is stored in the stack, and the node P_5 is traversed. The stack at this point looks like <address(P_2)>. After traversing the node P_5, the stack is popped. The node P_6, the right child of the node P_2, is traversed, and the stack becomes empty. After traversing the node P_6, the stack is empty. Since no additional node needs to be traversed, the search terminates.

The major advantage of the depth-first search is that the stack size is limited by the depth of the tree, and only a focused part of a tree is searched at a time. However, a naive depth-first search can get stuck indefinitely in the presence of cycles. Detecting a cycle requires considerable execution time and memory overhead to store all the visited nodes and to test the membership of a node among the visited nodes. A hash table can be used to identify the cycles efficiently. However, for large data structures, the memory overhead to store the traversed nodes is large.

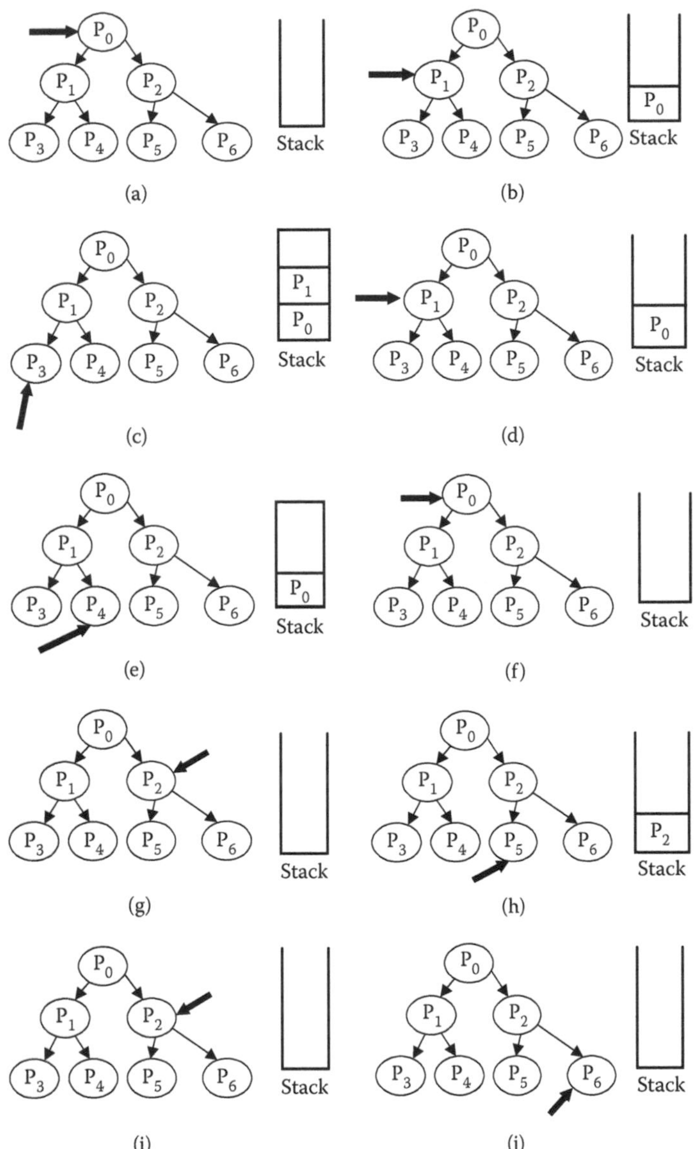

FIGURE 2.11 An example of depth-first search.

2.3.7.2 Breadth-First Search

Breadth-first search (see Figure 2.12) traverses a tree level by level from left-to-right order and uses a queue to store the children nodes to be visited. The root node (source node) is put in the queue initially. After that, every time one element is removed from the queue, all the children of the removed node are inserted at the rear end of the queue. The process is repeated until the queue is empty.

Example 2.15

Figure 2.12 illustrates breadth-first search. The pointer first visits the root node P_0 and inserts the addresses of its children P_1 and P_2 in the queue. Then it removes the

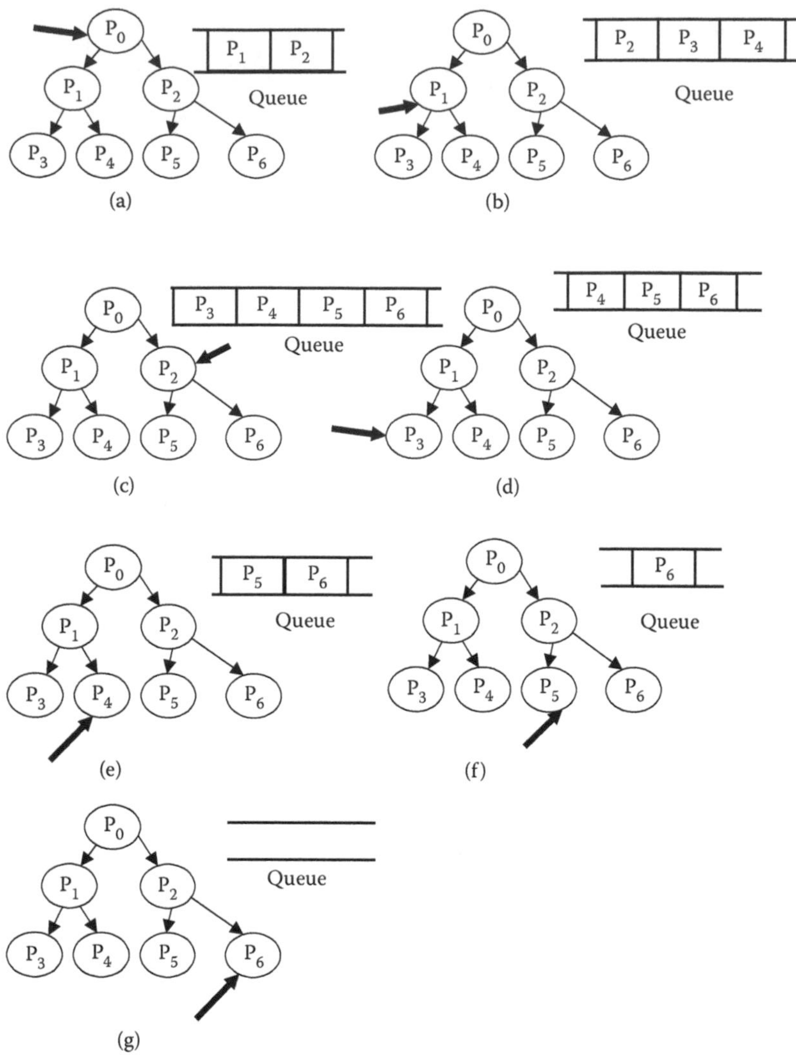

FIGURE 2.12 An example of breadth-first search.

address of the node P_1 from the queue inserts the addresses of its children P_3 and P_4 in the queue, and visits the node P_1. Then it removes the node P_2 from the queue and inserts the addresses of its children P_5 and P_6 in the queue. Since the nodes P_3, P_4, P_5, and P_6 are leaf nodes, nothing is inserted in the queue during their traversal. After the last leaf node P_6 is removed, all the nodes have been traversed, and the queue becomes empty.

The advantage of a breadth-first search is that it traverses a tree one level at a time.

There are many disadvantages to breadth-first search. The memory overhead of a breadth-first search is quite large. While traversing a tree, it holds as many nodes as there are nonterminal nodes in the previous level. This number can be as large as half the number of the leaf nodes in the case of a balanced tree. However, if the memory overhead of

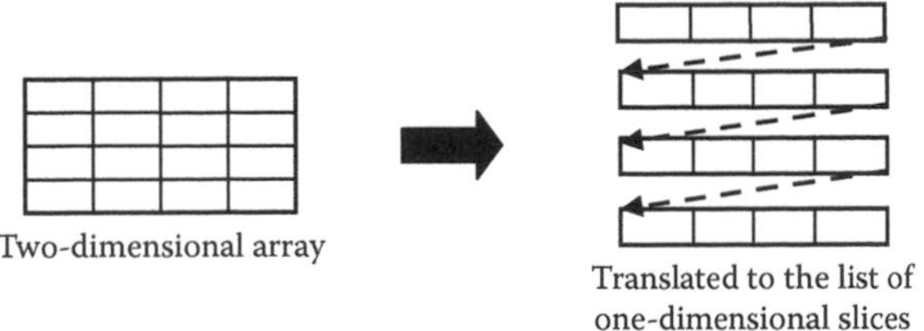

FIGURE 2.13 Translating a two-dimensional matrix to one-dimensional memory.

the queue can be managed efficiently, breadth-first search can be used quite effectively to traverse a graph and has been used in some efficient memory recycling schemes in a heap.

2.3.8 Mapping Data Structures in Linear Memory

A program can use *struct* (tuples with multiple fields) to model complex data entities or *multidimensional arrays* to model collection of data entities. Arrays could be either *static* or *dynamic. Static arrays* are allocated memory—once at the time of procedure invocation—and then no more additional memory allocation is required. The array size of the *dynamic arrays* changes at run time based upon the program execution. In addition to the arrays and structs, there are dynamic objects in object-based languages such as C++, Java, or C# and other modern programming languages. All these data structures have to be mapped on the RAM. A RAM can be abstracted as a single-dimensional array indexed by memory addresses.

In order to map multidimensional arrays on the RAM, multidimensional arrays have to be translated to single-dimensional arrays, and the addresses in the RAM have to be computed using an equation that translates the index of a data-element into an offset that can be added to the base address (start address) of the data structure. In order to understand the translation process of two-dimensional arrays, we take each row (or column) one by one individually, and put them one after another in one strip in a single dimension, as shown in Figure 2.13.

The equation to find out the address of an element a(*i*, *j*) for a two-dimensional array of size $M \times N$ is given by Equation 2.1, with the convention that the index of the array starts at 0. The first term on the right-hand side is the base address—address of a[0, 0]—and the second term produces the offset. The first element of the second term states that *i*-rows occur before the current row, and the second element of the second term shows the offset of the element in the current row.

$$\text{Address} = \text{Address}(\text{a}[0,\ 0]) + (\text{i} \times \text{N} + \text{j}) \times \text{Bytes in one element} \tag{2.1}$$

Example 2.16

Let us consider a two-dimensional matrix *m* of integers having the size 5 × 6 (5 rows and 6 columns). We want to find the address of an element located at the location m[2, 4]—row number 2 and column number 4. Let us also assume that the index starts from 0, and the base address of m[0, 0] is 1000, and a 32-bit integer needs four bytes. Using Equation 2.1, the location of m[2, 4] is 1000 + (2 × 6 + 4) * 4 = 1064. The integer stored in the element m[2, 4] will be in four bytes starting at location 1064.

The concept can be generalized to map any arbitrary dimensional matrix by progressively expressing a higher-dimensional matrix as a sequence of smaller-dimensional matrix, until it is reduced to a single-dimensional matrix. Given an N-dimensional matrix m having dimension values of the form $D_1, D_2, \ldots, D_N$, the address of an element located at $m[i_1, \ldots, i_N]$ is given by

$$\begin{aligned}\text{Address} &= \text{Address}\left(m\left[0,\ldots,0\right]\right) + \left(\left(i_1 \times D_2 \times \ldots \times D_N\right)\right. \\ &\quad \left. + \ldots + \left(i_{N-1} \times D_N\right) + i_N\right) \times \text{Bytes in one element}\end{aligned} \tag{2.2}$$

For a composite structure such as "struct" with multiple fields of the form (field_1, $\text{field}_2, \ldots, \text{field}_N$), the offsets of each field are computed during compilation. This offset is added to the base address to access the ith field.

2.3.9 Hash Tables

It is important to retrieve the data efficiently during the compilation and execution of programs. While static arrays have constant time operation to retrieve data elements, it is not suitable when the input given to search a data element is not an index but the key to a record. Hash tables support the key-based search in near-constant time using a hash function *f* that maps the key *<key>* to an index value: *f*(*<key>*) is equal to the index of the storage where the data element is stored. Hash tables are used in the implementation model of programming languages due to search efficiency.

The major limitation of hash-based search is *collision of the keys*—mapping of two keys on the same index value. Some of the popular techniques to handle index collision are (1) choosing the table size as a prime number, and using this prime number in the hash function to derive the index. The division by a prime number in a hash function evenly distributes the keys in the hash table, reducing the probability of collisions; (2) using a linked list attached to a particular index to handle multiple colliding keys; and (3) expanding the hash table, and rehashing all the elements, in case the collision or the density of elements in the hash table increases beyond a threshold value.

Example 2.17

Figure 2.14 shows an example of a hash table of size 11 being used to store a set of 6 variables and their attributes. There are three procedures: *p0*, *p1*, and *p2*. Procedure *p0* has variable *x*; procedure *p1* has variables *s*, *i*, and *j*; procedure *p2* has variables *a* and *i*. To avoid the naming conflict, procedure name is attached as a prefix to the variables. For example, the key for variable *x* in *p0* is *p0x*; the keys for variables *s*, *i*, and *j* in *p1* are *p1s*, *p1i*, and *p1j*, respectively; and the keys for the variables *a* and *i* in *p2* are *p2a* and *p2i*, respectively. All the keys are unique now, even for variables with conflicting names. Each data element is a triple of the form (*identifier*, *type*, *memory location*). The six triples are {(*p0x*, *int*, *L1*), (*p1s*, *Bool*, *L2*), (*p1i*, *int*, *L3*), (*p1j*, *int*, *L4*), (*p2a*, *int*, *L5*), and (*p2i*, *int*, *L6*)}.

The function to map a key is to (1) sum the ordinal value of characters '*a*' to '*z*' from left to right in the key; (2) add the digit value in the key; and (3) divide the sum by the table size 11, which is a prime number. Linked lists have been used to resolve key

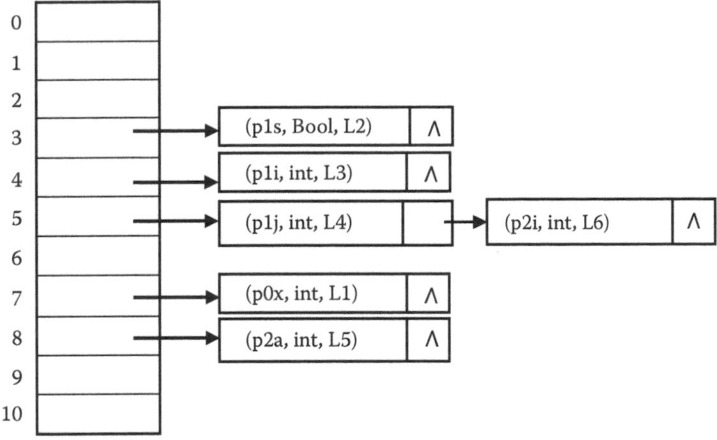

FIGURE 2.14 An example of a hash table with linked list to resolve conflicts.

collision. For example, the name *p0x* has three characters, starting with the character 'p' having an ordinal value 16, digit '0' having a value 0, and the character 'x' having an ordinal value 24. Summing up these values gives the value 40, and the division by the table size 11 gives the index value 7. The key *p1s* maps to the index value 3. The key *p1i* maps to the index value 4. The key *p1j* maps to the index value 5. The key *p2a* maps to the index value 8; and the key *p2i* maps to the index value 5. The collision of the keys *p1j* and *p2i* has been resolved using linked lists.

2.4 ABSTRACT CONCEPTS IN COMPUTATION

This section describes some background concepts used in the development of programming languages. Much of these concepts have been covered informally in various contexts in the previous programming and foundational courses such as data structures and discrete structures. These fundamental concepts are repeatedly used throughout this course and need more formal introduction from the program language design and implementation perspective.

A program is a *sequence of meaningful instructions.* Each such instruction is also called a *sentence* or a *statement.* Each *statement* is terminated by a delimiter in many languages or by line feed in some languages. A *reserved word* is part of the programming language similar to words in natural languages. Two reserved words or two variables or any two entities that have a separate meaning are separated by blank, comma, semicolon, colon, line feed, or reserved words. These separators are called *delimiters.*

The programs can have *literals, r-values, l-values, identifiers, definitions, variables, assignment statements, commands, expressions, strings, type declarations, operators, procedure invocations, parameters, labels*, and *sequencers.* A *literal* is an elementary expression that cannot be further split into smaller parts and that cannot be redefined. Literals are also known alternately as *atoms, constants*, and *denotations.* For example, a number is a literal, a character is a literal, a name within single quotes 'Arvind' is a literal. However, a string is not a literal: a string is a sequence of characters; special values *true* and *false* are not literals, since they can be redefined. An *r-value* is the right-hand side of an expression or the value stored in a memory location. An l-value is the memory location associated with a variable or an array element or a field within a structure. An *identifier* is a symbolic name used in the program that can be associated with another entity. For example, program names, function names, variable names, constant names, and user-defined types are identifiers. A *definition* denotes an associated value within a block or a procedure. The definition is substituted by the associated value during compilation.

A *variable* is associated with an *information unit.* The *information unit* can be a *simple value* in a *concrete domain, a complex object, or a value* in an *abstract domain* such as the *type domain.* A *concrete value* is an actual value the program works with. A *type variable* holds the type information. For example, a type variable can be associated with an abstract domain *integer.* The notion of type variables is used extensively in polymorphic languages and has been explained in Chapter 7. From now onward, we refer to a variable as a value holder of concrete values and type variables as a value holder of type information. Internally a variable is mapped to a memory location at compile time. The corresponding memory location can either hold a concrete value or a pointer to an object in the heap.

The *assignment* statement such as $x = y + 4$ is a programmer's assertion that reads the stored value from the memory location of the variables in the expression on the right-hand side of the assertion, evaluates the algebraic expression, and writes the resulting value in the memory location of the variable on the left-hand side of the assertion. In the above statement, the value of the variable y is read from the store, the expression *value-of*(y) + 4 is evaluated, and the result of the expression evaluation is stored in the memory location corresponding to the identifier x.

A *command* is a control abstraction that has at least one assignment statement embedded in it. The use of an assignment statement means that a new value is written in at least one memory location in a command. In contrast to a command, an *expression* does not include an assignment; an expression just reads the values from memory locations and evaluates them.

A *string* is a sequence of characters. A null string has no characters in it. *Type declaration* is used to create a user-defined data abstraction such that data can be stored in a user-specified way to model the entities in real-world problems. However, type declaration itself does not create a memory location. Memory locations are created only when a variable is declared.

An *operator* could be an arithmetic operator or a logical operator. An operator could be *monadic* or *dyadic. Monadic operators* have a single operand, and *dyadic operators* have two operands. Arithmetic operators could be '<' (less than), '>' (greater than), '= <' (less than or equal to), '> =' (greater than or equal to), '==' (equal to), '< >' (not equal to), '=/=' (not equal to as in Prolog). In addition, there are monadic operators such as '+' and '−' to provide the sign to a number. A dyadic logical operator such as logical-AND (∧), logical-OR (∨), exclusive-OR (⊕), and implication connect two Boolean expressions using truth values as described earlier, and a monadic operator 'not' takes one Boolean expression as an argument and returns *true* or *false.*

Both arithmetic and Boolean operators have *precedence*: given an expression, certain operators are reduced first. Monadic operators have higher precedence than dyadic operators; within dyadic operators, multiplication and division have higher precedence than addition and subtraction. Similarly, the Boolean monadic operator 'not' (¬) has the highest precedence, followed by logical-AND (∧), followed by logical-OR (∨), followed by implication (→).

2.4.1 Mutable versus Immutable Variables

In the imperative programming paradigm, the variables are *mutable*: a user can destructively update the content of the memory location corresponding to a variable at will, using *assignment operations*, such as $x = y + 4$. In contrast to imperative programming paradigm, declarative programming paradigm does not allow user-level mutations of memory locations; a variable is just an *immutable* value holder that can be associated with a value only once.

There are advantages and disadvantages of both the approaches of handling the variables. The major advantage of destructive update of a memory location is *memory reuse* such as in repeated use of statements involving index variables in for-loop. The major disadvantages of the destructive update of variable are (1) loss of the past values and (2) *side effects* resulting into undesired program behaviors caused by breakdown of fundamental mathematical properties of programs, as explained in Chapter 4. If past values cannot be recovered, then any problem-solving method will not be able to backtrack (comeback) to try out alternate possibilities to derive a solution, even though a solution may exist.

The major advantages of the assign-once property of an *immutable variable* are (1) reduced possibility of undesired program behaviors caused by side effects and (2) the use of the past value to try out alternate solutions such as in-logic programming, as discussed in Chapter 9. The major disadvantage is that memory cannot be reused, resulting in *memory explosion* when processing data elements involving large data structures, especially in an iterative manner. Many modern languages supporting multiple paradigms support both mutable and immutable definitions of variables to maximize the advantages in both the approaches.

2.4.2 Bindings and Scope Rules

A binding associates two entities or an entity with the corresponding attributes. For example, a variable name can be bound to a memory location, a memory location can be bound to a value, an identifier can be bound to a value, and a variable can be bound to a type. Similarly, a group of statements can be bound to an identifier and called a *function* or a *procedure*. A *scope rule* defines the use of a variable and the extent of its visibility. A variable is *visible* and can be used within its scope; it does not play any role beyond its scope. A binding may be temporary or persistent, depending upon the scope.

There are two types of scope rules: *static* and *dynamic*. As the name suggests, *static scope* is based upon a program structure such as block boundary, subprogram boundary or function (or method, in object-oriented programming languages) boundary, and class boundary; it does not change with the calling pattern of subprograms. A *dynamic scope* changes the scope of a variable, based upon the calling patterns of subprograms. An example of static scope is given in Example 2.18. An example of the dynamic scope is given in Example 2.19.

```
integer x, y, z;
main ( )
{x = 4; y = 10; z = 12;
  {integer temp, z;
       temp = x; x = y; y = temp; z = 5;}
print(x, y, z);
}
```

FIGURE 2.15 An example of static scope rule of a variable.

Example 2.18

The program in Figure 2.15 has three global variables *x*, *y*, and *z* in the outermost block, and two local variables *temp* and *z* in the inner block. The global variables have the scope of the whole program, while the scope of local variables is limited to the block in which they are declared.

The naming conflict of the variable *z* is resolved by giving preference to the variable that is closest to the use. The variable *z* in the print statement refers to the global variable *z*, while the variable *z* in the inner block refers to the local variable *z*. Both these variables are separate data entities, map to different memory location, and have different lifetimes. When the variable *z* is printed outside the scope of the inner block, the value of the global variable $z = 12$ is printed. This information about the scope of these variables can be identified by looking at the program structure, and has nothing to do with the calling pattern of the program.

Example 2.19

The program in Figure 2.16 uses the dynamic scope rule. The program has a function *sum* that has two variables: variable *x* is declared within the function *sum*, and the variable *y* is a free occurrence; that is, the variable y has not been declared in the function *sum*. The variable *x* picks up the value from the parameter passing mechanism. However, the variable *y* searches for the same name in the memory areas allocated to the chain of the calling procedures to find out the corresponding binding value.

The main program has two blocks. The first block has two local variables *y* and *z*, and calls *sum* with an argument using the variable *y*. The second block has three local variables *w*, *y*, and *z*, and calls the function *sum* with the actual parameter *z*.

```
integer sum(integer x);
return (x + y);
main ( )
{ {integer y, z; y = 4; z = 5; sum(y);}
  {integer w, y, z; w = 4, y = 5; z = 6; sum(z);}
}
```

FIGURE 2.16 An example of dynamic scope rule.

The first call to the function *sum(y)* will return the value 8, since the formal parameter *x* gets the value of the actual parameter *y* and the free variable *y* in the function *sum* is dynamically bound to the variable *y* declared in the first block of the program main. The second call to function *sum* returns the value 11, as the formal parameter *x* in the function *sum* gets the value of actual parameter *z* in the program main, and the free variable *y* in the function *sum* gets the value 5 associated with the variable *y* declared in the second block of the program *main*. It is the binding of the free variable and its binding to the latest declaration in the chain of the calling pattern of the subprograms that makes the scope rule dynamic.

2.4.3 Types of Variables

Variables have been categorized according to their visibility, based upon scope rules and lifetime. A *variable* that has the lifetime of a program is called a *global variable;* a variable that has the lifetime and scope of a called procedure is called a *local variable;* and the variable that may be declared in an outer nesting level of a nested procedural structure, but used in one of the nested procedure is called a *nonlocal variable.* In block-structured languages—languages supporting the notion of program organized as blocks of statement—variables can also be local to a block. The scope of a variables is limited to within the block where they have been declared.

There is a possibility of name conflict between nonlocal variables and local variables or global variables and local variables. The conflict is resolved by *shadowing* (making invisible) the nonlocal variables or global variables if a local variable with the same name is declared. The concept of shadowing also applies to handling the visibility of nonlocal variables if they have been declared at different nesting levels. In such cases, the variable that is declared at the nesting level nearest to the current procedure using the same variable name is seen, and other variables with the same name are shadowed.

A variable can be *static* or *dynamic. Static variables* are allocated in a fixed memory at compile time, and that memory location does not change during the execution of the program. The advantage of static memory allocation, as discussed in Chapter 5, is a faster direct memory access without the use of any pointer. A *static variable* can be a *static global variable* when the scope of the static variable is the whole program, or it could be a *static local variable* when the scope of the program is limited to within the procedure where it has been declared. However, due to local scope of the variable, the value is inaccessible outside the scope of the procedure and can be accessed only during subsequent calls to the procedure in which it is declared. *Dynamic variables* are allocated memory at run time and are placed on the control stack. Most of the local variable declarations are dynamic local variables and are allocated in a stack in a stack-based implementation of languages, as explained in Chapter 5.

Global variables are also allocated like static variables for efficient memory access. All other variables have limited visibility that depends either on the nesting pattern of the program units or on the blocks. Local variables are visible only within the scope of the block where they are declared. Local variables at the procedure level have visibility within the procedure.

In object-oriented languages, a variable can be declared within a class, and it is shared between all the object instances of that class. Such variables are called *class variables*. In contrast to the class variables, we can have *instance variables*, whose scope is limited to the created object.

Another classification of variables is based upon *mutability*, as discussed earlier. A *mutable variable* can be destructively updated any number of times, while an *immutable variable* has an *assign-once* property and cannot be altered or unbound programmatically once bound to a value. However, immutable variables can be unbound by the implementation mechanism such as backtracking as in the implementation of logic programming paradigm.

2.4.4 Environment and Store

The execution of a program utilizes two abstract components: *environment* and *store*. *Environment* is the surrounding where the computation takes place. Environment is defined as the set of bindings of pairs of the form (variable names ↦ memory locations or variable attributes) or (identifier name ↦ value), as in the case of constant declaration. New declarations change the environment by (1) creating new bindings of the form identifier ↦ memory locations or identifier ↦ value and (2) shadowing the bindings of the variables of nonlocal or global variables having conflicting names with the local variables.

After a call to a subprogram, the environment changes. As illustrated in Figure 2.17, the current environment of a subprogram consists of (1) local variable bindings to memory locations, (2) nonlocal variables binding to memory locations, (3) global variables bindings to memory locations, and (4) memory locations in the sequence of calling programs accessed using reference parameters. After the called subprogram is over, the environment reverts back to the environment available to the calling subprogram, augmented with the environment created by dynamic objects and recursive data structures that have life beyond the life of the called subprogram creating them.

Store is the set of bindings of the form (memory location ↦ value) and changes every time an assignment statement writes a value into a memory location, the value is initialized,

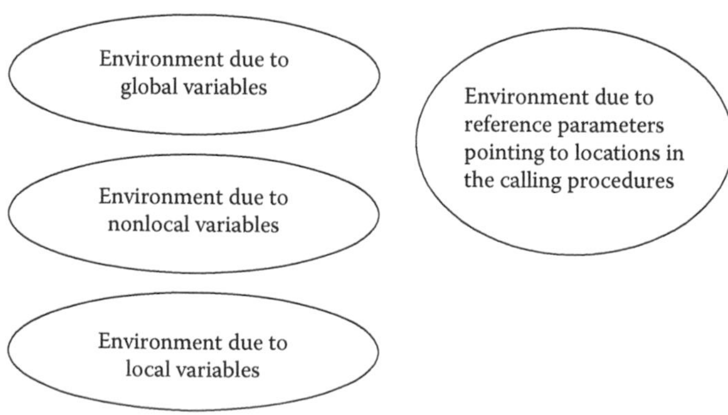

FIGURE 2.17 Environment for execution in a subprogram.

or the value is altered due to parameter passing. Assignment operations can update the value stored in the following:

1. The memory locations associated with local variables, nonlocal variables, or global variables.
2. The memory locations of the variables in the calling subprograms, when a parameter is passed as a reference parameter.
3. The memory locations of the formal parameters in the called program by initializing.
4. The memory location of the actual parameters by passing the results computed in the called subprograms.
5. The memory locations of a recursive data structure or dynamic data objects.

After a procedure is called, both the environment and the store get altered inside the called procedure. The store changes because of (1) passing of the parameter values or (2) initialization of the local variables. During the execution of commands in a subprogram, the store keeps getting altered, as described above. After the called procedure is over, the environment created by the local variables (in the called subprogram) is lost; the archived environment of the calling subprogram is recovered and augmented with the environment of the global variables and the new environment created for recursive data structures and dynamic objects that have a lifetime beyond the called subprogram.

The store may not be the same after the return from the called procedure due to the destructive updates of the global variables, the nonlocal variables, and the variables in the calling procedures, when passed as reference parameters. The assignment of new values to these variables will update the store of the calling procedure, and this change will be retained, even after the called procedure is over. Later we will study that this kind of change in the store can cause an unintended breakdown of the program behavior.

2.4.5 Functions and Procedures

We next discuss the difference between expression and commands, and function and procedures. An expression reads one or more values from memory and evaluates. However, it does not write the result of the evaluation back into the memory. This means that the *store* remains unaltered during the evaluation of an expression. In contrast, a command writes into a memory location, changing the store. For example, $x + y + 4$ is an expression, while the assignment statement $z = x + y + 4$ is a command. In the expression $x + y + 4$, the CPU reads the value of the variables x and y from the memory and then evaluates the expression. In contrast, the assignment statement $z = x + y + 4$ writes the result of evaluating the expression $x + y + 4$ to the memory location corresponding to the variable z.

A function has four components: (1) name of the function, (2) input parameters, (3) variable declarations, and (4) a set of connected expressions to map the input values to an output value. A function does not have an "assignment" statement to write back in the store of the program. In contrast, a procedure has one or more assignment statements.

In an imperative programming paradigm, it is inconceivable to do programming without the use of assignment statements. A function with assignment statements is actually a procedure that simulates the effect of a function.

2.4.6 Abstracting the Program Execution

Program execution can be abstracted as jumping from one computational state to another computational state until it reaches the final state that satisfies the final condition. There are two approaches to model a computational state. The first approach models a computational state σ as a triple of the form (σ^E, σ^S, σ^D), where the symbol σ^E is the *environment*, the symbol σ^S is the *store*, and the symbol σ^D is a stack of pairs of the from (part of the *environment*, in calling procedure not needed [or shadowed] in the called procedure, part of the store in the calling subprogram not shared by the called procedure) of the chain of calling subprogram invocations in the LIFO order. The second approach models the computational state as a Boolean expression connecting the logical axioms using logical operators such as logical-OR, logical-AND, negation, and implication. The Boolean expression changes each time an assignment statement is executed. The second approach is independent of the von Neumann machine and has been used to reason about the correctness of the program, as explained in Chapter 3, and stepwise program reconstruction starting from the final state, as described in Chapter 4.

2.4.6.1 Computational State as a Triple (Environment, Store, Dump)

In the first approach, the computational state σ changes after the execution of a statement due to (1) the change of environment σ^E, caused by declarations; (2) the change in store, denoted by σ^S, caused by an assignment statement or initialization; or (3) the change in dump, denoted by σ^D, due to function or procedure call. After a declaration statement, the current environment σ^E becomes σ'^E, and the computational state is altered. After a new assignment statement, the store σ^S changes to new store σ'^S, and the computational state is altered.

After a procedure call, a pair of the form (part of the environment of calling procedure, and part of the store in the calling subprogram not shared by the called procedure) is pushed on dump σ^D to give a new computational state (σ'^E, σ'^S, σ'^D). After the return from a function or procedure, the environment by the declaration of local variables in the called procedure is discarded, the saved part of the environment of the calling subprogram is popped, and union of the remaining part of the environment and the popped part of the environment gives the new environment σ^{Enew}. The new environment σ^{Enew} of the calling procedure would be the same as the old environment σ^E if no data structure with a lifetime greater than the lifetime of the called subprogram is created. Similarly, in the case of store, the store related to local variables in the called procedure is discarded, and the union of the remaining part of the store in the called procedure and the part of the calling procedure's store popped from the dump becomes the new store σ^{Snew} of the calling procedure. Note that σ^{Snew} may not be the same as the store σ^S. After the successful execution of the program, the final state is reached.

Example 2.20

Let us consider the example given in Figure 2.18. The program has three variables: *x*, *y*, *z*. The variables *x* and *y* have global scope, and the variable *z* is local to the main body. The main body assigns values to all three variables and calls the procedure swap by passing the references of the variables *x* and *y*. The denotation '*' in the body of swap dereferences the address and reads (or writes) into the memory locations corresponding to the global variables *x* and *y*. The procedure *swap* swaps the value of the global variables *x* and *y*.

The initial computational state is $\{< >, < >, < >\}$: the environment, the store, and the dump are empty. After the declaration of the global variables *x* and *y*, identifier *x* maps to an address $l\text{-}value_1$ and the identifier *y* maps to an address $l\text{-}value_2$, and the environment becomes $<x \mapsto l\text{-}value_1, y \mapsto l\text{-}value_2>$. The new computational state is $\{<x \mapsto l\text{-}value_1, y \mapsto l\text{-}value_2>, < >, < >\}$. Inside the main body, the declaration of the variable integer *z* changes the environment to $<(x \mapsto l\text{-}value_1, y \mapsto l\text{-}value_2, z \mapsto l\text{-}value_3>$. The new computational state is $\{<(x \mapsto l\text{-}value_1, y \mapsto l\text{-}value_2, z \mapsto l\text{-}value_3>, < >, < >\}$.

After the execution of each of the three assignment statements, the store changes altering the computational state. After the execution of the assignment statement $x = 4$, the new store becomes $<l\text{-}value_1 \mapsto 4>$, and the new computational state becomes $\{<(x \mapsto l\text{-}value_1, y \mapsto l\text{-}value_2, z \mapsto l\text{-}value_3>, <l\text{-}value_1 \mapsto 4>, < >\}$. After the execution of the statements $y = 10$ and $z = 12$, the new computational state would be $\{<(x \mapsto l\text{-}value_1, y \mapsto l\text{-}value_2, z \mapsto l\text{-}value_3>, <l\text{-}value_1 \mapsto 4, l\text{-}value_2 \mapsto 10, l\text{-}value_3 \mapsto 12>, < >\}$.

The call to procedure *swap* pushes the local part of the main program in the dump, and the identifiers *m* and *n* map to locations $l\text{-}value_4$ and $l\text{-}value_5$ respectively. The location $l\text{-}value_4$ maps to location $l\text{-}value_1$ due to the reference link; and the location $l\text{-}value_5$ maps to location $l\text{-}value_2$ due to the reference link. The environment and store due to the global variables *x* and *y* are retained, while the pair (*environment due to local variables, corresponding store*) of the main body is pushed into the *dump*. The new computational state before the start of the declarations in the procedure *swap* becomes $\{<x \mapsto l\text{-}value_1,$

```
integer x, y;
procedure swap( integer x, y);
main ( )
{ integer z;
x = 4; y = 10; z = 12;
swap(ref x, ref y)
print( x, y, z);
}
procedure swap(integer *m, *n)
{ integer temp;
temp = *m; *m = *n; *n = temp;
}
```

FIGURE 2.18 A program to illustrate computational states.

$y \mapsto l\text{-}value_2, m \mapsto l\text{-}value_4, n \mapsto l\text{-}value_5$>, <$l\text{-}value_1 \mapsto 4, l\text{-}value_2 \mapsto 10, l\text{-}value_4 \mapsto l\text{-}value_1, l\text{-}value_5 \mapsto l\text{-}value_2$>, < (<$z \mapsto l\text{-}value_3$>, <$l\text{-}value_3 \mapsto 12$>)>}. The declaration "**integer** temp" changes the environment, and the new environment becomes <$x \mapsto l\text{-}value_1, y \mapsto l\text{-}value_2, m \mapsto l\text{-}value_4, n \mapsto l\text{-}value_5, temp \mapsto l\text{-}value_6$>. The computational state changes accordingly. The assignment statement $temp = *m$ involves dereferencing the address: the memory location $l\text{-}value_6$ is mapped to the value stored in the memory location $l\text{-}value_1$, and the new store becomes <$l\text{-}value_1 \mapsto 4, l\text{-}value_2 \mapsto 10, l\text{-}value_4 \mapsto l\text{-}value_1, l\text{-}value_5 \mapsto l\text{-}value_2, l\text{-}value_6 \mapsto 4$>. *Dereferencing* uses transitivity of the map function between l-value$_4 \mapsto$ l-value$_1$ and l-value$_1 \mapsto 4$ to access the value 4. The assignment statement $*m = *n$ writes the dereferenced value of n into the location l-value$_1$, and the new store becomes <$l\text{-}value_1 \mapsto 10, l\text{-}value_2 \mapsto 10, l\text{-}value_4 \mapsto l\text{-}value_1, l\text{-}value_5 \mapsto l\text{-}value_2, l\text{-}value_6 \mapsto 4$>. The assignment statement $*n = temp$ writes the value of temp into the dereferenced location $l\text{-}value_2$. The new store becomes <$l\text{-}value_1 \mapsto 10, l\text{-}value_2 \mapsto 4, l\text{-}value_4 \mapsto l\text{-}value_1, l\text{-}value_5 \mapsto l\text{-}value_2, l\text{-}value_6 \mapsto 4$>. The computational state before the return from the subprogram swap is {<$x \mapsto l\text{-}value_1, y \mapsto l\text{-}value_2, m \mapsto l\text{-}value_4, n \mapsto l\text{-}value_5, temp \mapsto l\text{-}value_6$>, <$l\text{-}value_1 \mapsto 10, l\text{-}value_2 \mapsto 4, l\text{-}value_4 \mapsto l\text{-}value_1, l\text{-}value_5 \mapsto l\text{-}value_2, l\text{-}value_6 \mapsto 4$>, < (<$z \mapsto l\text{-}value_3$>, <$l\text{-}value_3 \mapsto 12$>)>}. After coming back from the procedure *swap*, local environment and the corresponding store for the procedure *swap* are removed from the environment and the store, the local environment and the store of the main body is popped from the dump, and the dump becomes empty again. The new computational state after the return from the procedure swap becomes {<$x \mapsto l\text{-}value_1, y \mapsto l\text{-}value_2, z \mapsto l\text{-}value_3$>, <$l\text{-}value_1 \mapsto 10, l\text{-}value_2 \mapsto 4, l\text{-}value_3 \mapsto 12$>, < >}.

2.4.6.2 Computational State as Boolean Predicates

In the second approach, a Boolean expression defining the computational state changes after the execution of an assignment statement. The assignment statement $x = 5$ makes the predicate $x == 5$ true. After the execution of the statement $y = 6$, the conjunctive predicate $x == 5 \wedge y == 6$ as true. This abstract approach to look at the program execution frees them from any specific architecture, and it has been used to analyze programs for their correctness.

2.4.7 Processes and Threads

Modern programming languages that support concurrent programming paradigm and scripting languages utilize concepts of *processes* and *threads*. The active part of a program or a subprogram is called a *process*. A process has its *own id, heap, stack*, and *memory area* to communicate with other processes and Boolean flags to store the execution state of the process. A process is stored in memory blocks. Memory blocks corresponding to a process are loaded into the RAM from the hard disk for the process execution. As illustrated in Figure 2.19, a process can be in five states: *created, ready, active, suspended,* and *terminated*. An *active process* is suspended to (1) give another process a chance based upon a scheduling strategy; (2) wait for an I/O to be completed; or (3) allow a process to sleep for a specified time, based upon a runtime command. A *suspended process* is reactivated after the data is available from I/O activity, or the scheduler gives another turn to the process.

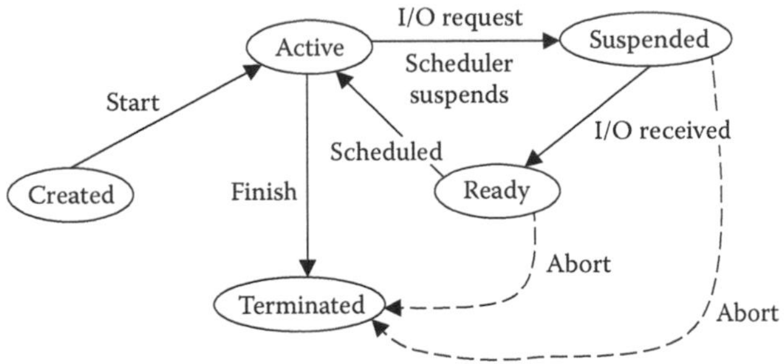

FIGURE 2.19 A state diagram of a process with labeled transitions.

A *thread* is a sequence of actions. The difference between a thread and a process is that a thread has smaller overhead of execution and reactivation, since it shares a major part of data area of the existing process that spawns the thread. A process can spawn multiple threads that join back to the spawning process after the subtasks are over. During the execution of a thread, the process may be running or suspended while waiting for the signal that the thread has terminated. A thread will terminate either after finishing a subtask or be aborted by an operating system action.

2.4.8 Buffers

When two processes or threads communicate with each other or transfer the data to each other, they may do so asynchronously at varying speed. A reusable memory space is needed to facilitate the asynchronous data transfer. This reusable memory space is called a *buffer*. There are two major operations in a buffer: (1) depositing the data and (2) retrieving the data.

Buffers are implemented using *circular queues*. As illustrated in Figure 2.20, circular queues continuously keep reusing the memory vacated by the removed data elements. Circular queues use modulo arithmetic on a linear array to go back to the beginning of the array after encountering the last memory element in an array.

There are four operations on a buffer: (1) to check if the buffer is empty, (2) to check if a buffer is full, (3) to remove an element from a nonempty buffer, and (4) to insert an element

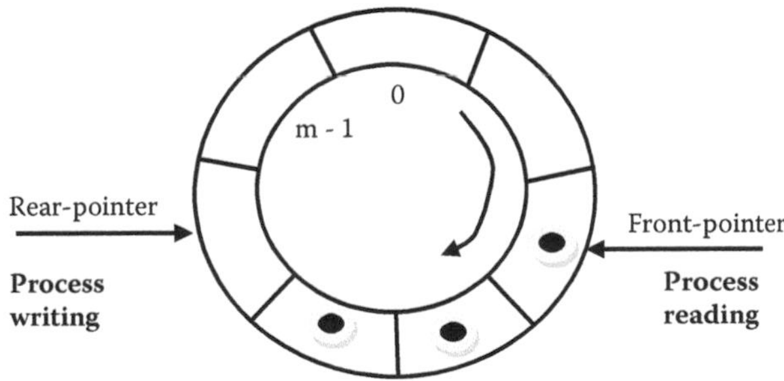

FIGURE 2.20 Schematics of buffer as a circular queue.

in a non-full buffer. Let us assume that the buffer is implemented as an array of size m; then the operations on a buffer are as follows:

```
is_empty_buffer(buffer) ::      if (front-pointer == rear-pointer)
                                return(true)
                                else return(false);
is_full_buffer(buffer) ::       if (((rear-pointer + 1) modulo m)
                                == front-pointer) return(true)
                                else return(false);
insert(buffer, element) ::      if not (is_full_buffer(buffer)) {
                                   buffer[rear-pointer] = element;
                                   rear-pointer = (rear-pointer + 1)
                                   modulo m;}
                                else raise_exception('buffer_full');
  remove(buffer)::              if not (is_empty_buffer(buffer)) {
                                   element = buffer[front-pointer];
                                   front-pointer = (front-pointer + 1)
                                   modulo m; return(element);}
                                else raise_exception('buffer_empty')
```

The operation *is_empty(buffer)* returns *true* if the value of the rear pointer is equal to the value of the front pointer. The operation *is_full(buffer)* checks if incrementing the rear pointer by 1 using modulo arithmetic will give the value of the front pointer. This means that the rear pointer has caught up with the front pointer, and no vacant cells are available. The operation *insert(buffer, element)* first checks if the buffer is not full. The data element is inserted only if the buffer is not full. The data element is inserted in the cell indexed by the rear pointer. After inserting the element in the cell, the rear pointer is incremented by 1 using modulo arithmetic. The operation *remove(buffer)* removes the data element from the location indexed by the front pointer if the buffer is not empty, and the front pointer is incremented by one using modulo arithmetic.

2.5 SUMMARY

Various fundamental concepts used in the design and implementation of programming languages have been described. These concepts are essential in the design and implementation of programming languages and will become relevant in the following chapters.

Discrete structure concepts needed to study programming languages include sets and multisets, Boolean logic, functions and relations, finite state machine, and set operations. Type theory makes extensive use of set theory, and many programming languages support the declaration of sets and set-based operations. Many relevant set operations, such as power sets, Cartesian product, finite mapping, and disjoint-union, were discussed.

Boolean logic is one of the pillars of programming. There is a class of functions (or sentences in logic programs) called *predicates* that return *true* or *false*, depending upon the values of the variables. Checking for complex Boolean conditions is used in many control abstractions such as selection (if-then-else statements; case statements, etc.) and indefinite iterations (while-do loop; do-while loop), exception handling, and conditional jumps such as getting out of nested blocks or graceful termination of execution if an event has occurred. Boolean conditions form the core of logic programming that is based upon connecting logical predicates using logical-AND, logical-OR, implication, and negation.

A *function* is a mapping between two sets: *domain* and *codomain*, such that every element of domain maps to only one element of a codomain. The mapping could be one-to-one or many-to-one. However, it can never be many-to-many or one-to-many. Functional programming paradigm exclusively uses functions, while other programming paradigms such as the imperative programming paradigm, use functions along with assertion. A functional program uses only expression evaluations. In the imperative programming paradigm, functions also use assertions in addition to expression evaluations.

A *relation* defines a property between two entities (or an entity and an attribute) x and y. A relation could be a combination of *reflexive, symmetric* (or *antisymmetric*), or *transitive*. The properties of relations are used in programming such as logic programming, aliasing of program variables, and data dependency analysis for translating a program to run concurrently.

Recursion is characterized by the presence of two types of definitions: *base case* and *recursive definitions*. In programming languages, recursion occurs both in the procedure definitions as well as data structure definitions. Recursive functions play a major role in programming. *Tail recursive functions* can be simulated using iterative programs. A class of *linear recursive* programs can be implemented using the notion of *accumulators* and iteration. Handling recursion needs a means for memory growth at run time, since the memory requirement of recursive procedures and recursive data structures cannot be determined at the compile time. Some of the overheads of recursive programming can be alleviated by transforming tail-recursive and linear-recursive programs to the corresponding iterative programs.

Finite state machines model transitions between different situations expressed as states, and have been used in programming languages during lexical analysis—a phase during compilation, which is discussed in Chapter 3.

Stacks and queues are abstract data types with different properties: stack supports last-in-first-out property, and queue supports first-in-first-out property. The elements are pushed and popped from only one end in a stack, while data elements enter from one end in the queue and are taken out from another end. Stacks play a major role in depth-first tree traversal, the execution of recursive procedures, and the implementation of programming languages. Queue is used in breadth-first tree traversal and where the fairness property of first-in-first-out is needed. Depth-first traversal traverses deeper into the tree, since the search reaches the leaf nodes in the left subtree before traversing to the right subtree. Breadth-first search traverses a tree level by level. Breadth-first search has been used in some efficient heap management techniques, as discussed in Chapter 6.

Hash tables are used for efficient insertion and retrieval of dynamic data by mapping the primary key to an index, using a hashing function. The use of a hashing function ensures that an index is identified in near constant time. Hash tables play a major role in the implementation of programming languages and process management, as the information can be retrieved in near-constant time.

A *variable* is the primary information holder in a program. A variable contains six major attributes: *name, type, scope, lifetime, memory location*, and the *assigned value*. A *scope* is the part of the program where variable can be read or written into, and the *lifetime* is the period when the variable can be used. *Global variables* have the same scope as the lifetime as of the program, while *local variables* are limited by the scope of the procedure in which they are declared. *Static variables* are allocated at fixed addresses at compile time and are accessed using direct memory access. *Mutative variables* can be destructively updated and are used in imperative programming paradigm for assignment statements.

During the program execution, there are three important concepts: *environment* that is altered by declarations; *store* that is altered by assignment statements; and *dump*—a stack of the environment and the store of the sequence of calling procedures. The execution of a program can be modeled as a sequence of transitions between computational states described abstractly either as the triple of the form (*environment, store, and dump*) or as *Boolean expressions* involving conjunction, disjunction, and negation of Boolean predicates. The second approach frees the definition of computational state from the von Neumann machine and has been used to reason about program correctness.

Process is an active part of a program or subprogram that runs on computer and occupies memory in RAM. A *thread* is a sequence of activities that acts like a process, except that it has less overhead of memory allocation than a process and uses the memory space of the parent process. Buffers are memory spaces used to transfer data between two processes, threads, between two I/O devices, or between CPU and I/O devices. A buffer is modeled as a circular buffer that uses modulo arithmetic to reuse the memory space vacated by removing the data elements from the buffer.

2.6 ASSESSMENT

2.6.1 Concepts and Definitions

Accumulator; acyclic graph; antisymmetric; binding; Boolean logic; breadth-first search; buffer; Cartesian product circular queue; command; computational state; cycle; cyclic graph; depth-first search; directed acyclic graph; directed graph; dump; dynamic scope rule; environment; existential quantification; expression; finite mapping; finite state machine; first-order predicate calculus; function; graph; hash function; hash table; immutable; implication; indexing; linear recursion; logical-AND; logical-OR; mapping; multisets; mutable; negation; one-address machine; ordered set; pointer; power set; predicate calculus; process; propositional calculus; quantification; queue, recursion; recursive data structure; recursive function; reference; reflexivity; relation; scope rule; sequence; stack; static scope; store; string; symmetric; tail recursion; thread; three-address machine; transitivity, tree; tuple; two-address machine; type variable; universal quantification; variable; visibility; von Neumann machine; weighted graph; zero-address machine.

2.6.2 Problem Solving

1. Write the Cartesian product of two sets: {"morning," "evening,""night," "afternoon"} and {"sun," "light," "rain"}.

2. Write the power set of a set given as {sun, light, rain}. Explain the size of the power set.

3. Given a function subtract-5 that subtracts five from a number in the natural numbers domain and maps to an element in the lifted natural numbers domain, show the mapping of the elements 10, 0, 3, 7, and 8. Note that the lifted domain also contains the bottom symbol $\perp$. Anything that does not map to regular elements in the range maps to the bottom symbol.

4. Write a predicate calculus representation for the following statement: "for every number N such that $N > 1$, there exists a number that is 1 less than the number."

5. Write a predicate calculus representation for the following statement: "for every person in this world, there exists a relation that connects the person to at least one person in this world."

6. Given a three-dimensional array of the dimensional size [0..4, 0..7, 0..9] and the base address of 10,000, give the starting address of the element located at the index (3, 4, 2), given that each element takes 2 bytes of memory. Explain your computation.

7. Given a quadruple (4-tuple) of the form (integer, character, floating point, integer), give the offset of the third field, given that the start address is 5000. Assume that an integer occupies four bytes, a floating point number occupies eight bytes, and a character occupies one byte.

8. Show the snapshots of a stack after every operation, starting with an empty stack, for the following sequence of operations: push(stack, 4), push(stack, 5), pop(stack); push(stack, 6). Represent a stack as a sequence of the form <$elem_N$, ..., $elem_1$>, where $elem_N$, ..., $elem_1$ are the elements that have been pushed in the stack in last-in-first-out order, and the elements are added and removed from the front-end.

9. Show the snapshots of a queue after every operation, starting with an empty queue, after the following operations: insert(queue, 4), insert(queue, 5), remove(queue); insert(queue, 6). Represent a queue as a sequence of the form <$elem_1$, ..., $elem_N$>, where $elem_1$, ..., $elem_N$ are the elements in the queue. The elements are inserted at the rear end of the sequence and removed from the front end of the sequence.

10. Write a program for breadth-first search using a queue and a depth-first search using a stack, and compare the average number of data elements stored in the queue versus stack. Implement the queue using a vector. Create at least 10 trees at run time using random number generators for the statistical analysis.

11. Write a simple program using at least five variables and a simple code with five simple statements, and show how environment and store changes after each declaration and the execution of each statement.

12. Take a program with at least five different procedures, with at least two procedures being called multiple times from different procedures. Represent each procedure by a labeled node in the graph. Draw a directed edge from a calling procedure to the called procedure. The weight of the edge is given by the number of procedure calls. If a procedure calls itself, then draw an edge from the node to itself showing a cycle. Describe the resulting graph.

2.6.3 Extended Response

13. What do you understand by predicate calculus? Explain. What are the various components of predicate calculus? Explain each one of them using a simple example.

14. What is the difference between first-order and higher-order predicate calculus? Explain.

15. Explain the differences between stack and queue abstract data types.

16. What do you understand by DAGs (directed acyclic graphs)? How are they different from trees? Give a real-world example of a problem that can be modeled by DAGs but not trees.

17. What are the advantages and disadvantages of the use of pointers in programming languages? Explain. Discuss three design solutions that would reduce the disadvantages of using pointers in programming languages.

18. Explain depth-first search, breadth-first search, and their differences. Explain using a clear example of a balanced n-ary tree ($n \geq 2$).

19. Explain static scope and dynamic scope using a simple example not given in the book.

20. Explain the difference between environment and store. Show how environment and store are modified under declaration, assignment statement, procedure call, and expression evaluation.

21. Explain how a hashing mechanism works for efficient search and retrieval of the archived data using a simple but complete example not given in the book.

FURTHER READING

Malik, Davender S. and Sen, Mridul K. *Discrete Mathematical Structures: Theory and Applications.* Thomson Course Technology. 2004.

Malik, Davender S. *Data Structures using C++*, 2nd edition. Course Technology Cengage Learning. 2009.

Patterson, David A. and Hennessy, John L. *Computer Organization and Design, the Hardware/Software Interface*, 5th edition. Morgan Kaufmann. 2012.

Silberschatz, Abraham, Galvin, Peter B., and Gagne, Greg. *Operating Systems Concepts*, 9th edition. John Wiley and Sons. 2010.

CHAPTER 3

Syntax and Semantics

BACKGROUND CONCEPTS

Abstract concepts in computation (Section 2.4); Boolean logic (Section 2.2.2); Control flow diagrams (Chapter 1); Discrete structures (Section 2.2); Environment and store (Section 2.4.4); Finite state machines (Section 2.2.5); Recursion (Section 2.2.4); Trees (Section 2.3.5); Graphs (Section 2.3.6); von Neumann machine (Section 2.1).

There are two major components to comprehend the constructs of programming languages: *syntax* and *semantics. Syntax* is about the validating structures of the statements in programming languages, and semantics is about understanding the unambiguous meaning of sentences in programming languages. Both the concepts are essential for program comprehension, compiler development, and software maintenance. Syntax is important, because unless a sentence structure is properly validated, it is not a part of a programming language and cannot be associated with the correct meaning. *Semantics* is important, because, unless we associate and communicate an unambiguous meaning for every word used in programs and constructs of programming languages, a sentence in a program cannot be comprehended and translated correctly to intermediate-level code for execution.

3.1 INTRODUCTION TO SYNTAX AND SEMANTICS

In most of the natural languages—with the exception of pictorial languages such as Chinese—there are four major components of syntax: *finite set of characters, words, sentences*, and *grammar rules* (also known as production rules) to construct and validate the sentences. The characters are categorized into *vowels* and *consonants*. Vowels are used to glue consonants and to help pronounce the words. The words are the basic unit for associating meaning; no meaning is associated at the character level. These words are further connected using grammar rules (production rules) to form sentences. If the meanings of individual words are unique, then a unique meaning for a given sentence can be derived.

In order to derive the meaning of a given sentence, first we have to apply grammar rules repeatedly to validate the structure of a sentence. The process of validating the grammatical correctness of a sentence using the production rules in the grammar is called *parsing*. During the validation, a sentence is progressively reduced to multiple *intermediate*

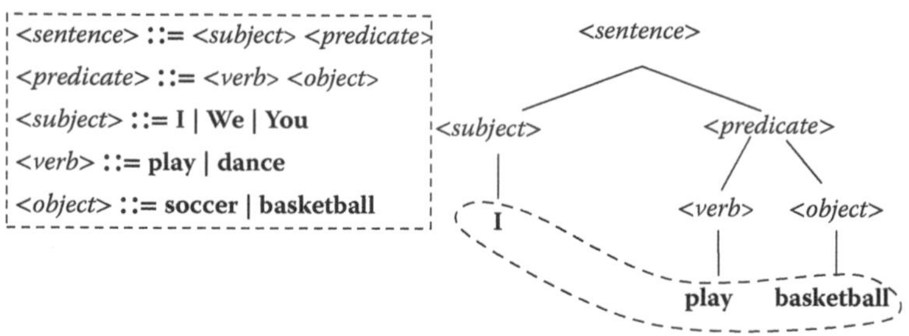

FIGURE 3.1 A simple English grammar and the corresponding parse tree.

forms using the production rules. The process terminates after the intermediate form is just a start symbol. The graphical representation of this progressive application of the production rules to reach to the start symbol beginning from the sentence is a tree and is called a *parse tree*. The details of parsing and parse tree are explained in Section 3.2.

Example 3.1

Consider an example of an English sentence: "I play basketball." Each word has a meaning. Before checking whether the sentence is grammatically correct, spelling of the word has to be corrected. After the spelling has been corrected, a simple subset of grammar rules of English (see Figure 3.1) is applied to validate the sentence structure. The sentence "I play basketball" is transformed to an intermediate form *<subject>* *<verb>* *<object>*, which in turn is transformed to another intermediate form *<subject>* *<predicate>*. By applying another production rule, the intermediate form *<subject>* *<predicate>* is reduced to the start symbol *<sentence>*.

Semantics derives the meaning of a syntactically correct sentence in a specific *domain*. For example, a string "214" means: (1) decimal number 214_{10} ($2 * 10^2 + 1 * 10^1 + 4 * 10^0$) in base-10 domain, (2) 214_8 in octal ($2 * 8^2 + 2 * 8^1 + 4 * 8^0$) = 148_{10} (*148* in *base-10* domain), and (3) "floor number two, isle number 1, and fourth room in the isle" in the building domain. The meaning depends upon the semantic domain in which the sentence has been used. In order to have a unique meaning, the semantic domain needs to be clearly fixed. For example, the meaning of a binary number "1011" in the base-10 domain is $1 \times 2^3 + 0 \times 2^2 + 1 \times 2^1 + 1 \times 2^0 = 11_{10}$, *13* in octal (base-8 semantic domain) and "D" in hex (base-16 semantic domain).

The meaning of a decimal number *<integer-part>*'.'*<floating-part>* in the number domain is *meaning-of* (*<integer-part>*) + *meaning-of* (*<floating-part>*. If the meanings of the nonterminal symbols *<integer-part>* and the *<floating-part>* are known individually, and the meaning of the dyadic operator '+' is understood, then the meaning of the decimal number can be derived. The operations that are used to derive the meaning of a sentence in the *semantic domain* are called *semantic algebra*. We need a semantic domain and the corresponding semantic algebra to understand the meaning of sentences in programming languages.

3.2 GRAMMARS

Programming languages are not as complex as the English language to avoid ambiguities in translating high-level instructions to equivalent low-level instructions. Many efforts have been taken to keep the grammar of programming languages simple as follows:

1. The basic units of programming languages are more like words in English language; there is no concept of 26 letters that make words. These basic semantic units are called *reserved words* and are associated with a unique meaning. The set of reserved words in programming languages is much smaller than the English language.
2. Programming languages avoid multiple meanings for the same word, which needs disambiguation based upon the context. The use of context-sensitive multimeaning words will make parsing very time consuming and will be prone to error while deriving the exact meaning for a sentence in the language. Since meaning of a sentence is related to low-level code generation, error in meaning will result in erroneous low-level code. However, a limited amount of vocabulary has multiple meanings, and their exact meanings are derived on the basis of the context. This concept is called *overloading* and is discussed in Chapter 7.

Grammars are used to derive a finite-length *sentence* in a language. Grammar has four components: *start symbol, set of terminal symbols (reserved words), set of nonterminal symbols,* and *set of production rules. Nonterminal symbols* are expanded to a combination of terminal and nonterminal symbols using the set of production rules. Nonterminal symbols are only part of the grammar and not a part of the sentences in programming languages. The set of all *terminal symbols* (reserved words) in a programming language is called the *alphabet* of the corresponding grammar. The sentences in a programming language are derived from the alphabet using production rules starting from the *start symbol.* An *alphabet* is traditionally denoted as the Greek symbol Σ. Formally, a grammar is defined as a 4-tuple of the form (*start-symbol S, set of production rules P, set of nonterminal symbols N, alphabet* Σ).

Given a sentence in a programming language, the production rules in grammar are also used to validate the structure of the sentence. This process of repeated application of the production rules in the grammar of a programming language to reduce the given sentence to the *start symbol* is called *parsing.* The right-hand side of a production rule is matched with the part of the sentence or the intermediate form and is substituted by the left-hand side of the production rule. This process of reduction using the production rules generates a tree structure called a *parse tree.* The sentence is located at the leaf nodes of the parse tree, the nonterminal symbols are located at the non-leaf nodes of the parse tree, and the start symbol is located at the root node of the parse tree.

Example 3.2

Figure 3.1 illustrates a simple example using a simple subset of English grammar rules that can drive the sentence "I play basketball" in English. The grammar has five production rules. The start symbol is *<sentence>*. Each production rule has a left-hand side and a right-hand side separated by the symbol '::='. The left-hand side of the rule has only one symbol enclosed within angular brackets '<' and '>', and the right-hand side of the rule is a combination of two types of symbols: (1) symbols enclosed within angular brackets and (2) symbols without angular brackets. The symbols within the angular brackets are called *nonterminal symbol*, and the symbols without angular brackets are *terminal symbols (reserved words)*. The set of nonterminal symbols is {*<sentence>*, *<predicate>*, *<subject>*, *<verb>*, *<object>*}, and the alphabet (set of terminal symbols) is {**I, We, You, play, dance, soccer, basketball**}.

Given the sentence "I play basketball," rules 3, 4, and 5 are respectively applied to the different parts of the sentence by matching the right-hand side of the production rules with the parts in the sentence and substituting the parts of the sentence by the corresponding left-hand sides of the production rules: rule 3 reduces the word "I" to the nonterminal symbol *<subject>*, rule 4 reduces the word "play" to the nonterminal symbol *<verb>*, and rule 5 reduces the word "basketball" to the nonterminal symbol *<object>*. The new intermediate form is *<subject><verb><object>*. In the second step, the nonterminal pair *<verb><object>* matches with the right-hand side of the rule 2 and derives the nonterminal symbol *<predicate>* (left-hand side of the rule 2). The new reduced intermediate form becomes *<subject><predicate>*, which matches with the right-hand side of the rule 1 and derives the start symbol *<sentence>*. Thus the applications of the production rules reduce the sentence "I play basketball" to the start symbol *<sentence>* validating the structure. A successful validation of grammatical structure does not always mean that the sentence will be meaningful. For example, the sentence "*I dance soccer*" is structurally valid using the given set of production rules. However, the sentence is not meaningful.

3.2.1 Types of Grammars

There are three types of grammars that can be used in the definition and implementation of programming languages: *regular grammar, context-free grammar*, and *context-sensitive grammar*.

The *regular grammars*, also called *type-3 grammars*, are the simplest type and represent a *nondeterministic finite-state machine*. A *nondeterministic finite state machine* is a finite state machine that has multiple possible transitions from one state to another state for the same input symbol. In Chapter 2, Section 2.2.4, we discussed a finite state machine that accepts a variable. The finite state machine can be expressed using a regular grammar:

```
<S1> → <letter> <S2>
<S2> → <letter> <S2> | <digit> <S2> | '_' <S2> | ε
```

Formally, *a regular grammar* is represented as 4-tuple (*N*, *Σ*, *P*, *S*), where *N* is the set of nonterminal symbols, Σ is the set of terminal symbols, *P* is the set of production rules, and *S* is the start symbol. In the above example, *N* is {*<S1>*, *<S2>*}, Σ is {*any letter, any digit,* '_'}, *P* is the set of production rules, and *S* is the start symbol *<S1>*. *Regular grammars* are used to accept a sentence and emit a signal in the form of a token upon acceptance. Given a sequence of characters, the above regular grammar will accept the sequence if it is a proper variable.

Regular grammars are used extensively in lexical analysis phase during the compilation process to generate streams of tokens—internalized representations for entities such as variables, literals, identifiers, and reserved words. Tokens are needed for the ease of parsing.

Context-free grammars, also called *type-2 grammar*, are represented as a 4-tuple of the form (N, Σ, P, S), where N is the set of nonterminal symbols, Σ is the set of terminal symbols, P is the set of grammar rules, and S is the start symbol. The major characterization of context-free grammar is that the left-hand side is a single nonterminal symbol. The grammar in Figure 3.1 is a context-free grammar. This restriction has advantages in time and space complexity when parsing a sentence. Since the left-hand side of a production rule is a single nonterminal symbol, it can be expanded without the presence of any terminal symbol. Context-free grammars are more expressive than regular grammars. For example, there is no regular grammar to accept a string a^nb^n. However, we can write a context-free grammar to accept a string a^nb^n as given below

```
<S> ::= a <S> b | ε
```

The above grammar is context free by definition, as it has a start symbol <*S*>, a set of nonterminal symbols {<*S*>}, a set of terminal symbols {a, b}, and one production rule. The rule has only one nonterminal symbol on the left-hand side. The Greek letter ε denotes a null symbol.

Context-sensitive grammars, also known as *type-1 grammar*, can have additional terminal symbols on the left-hand side in addition to nonterminal symbols. The only restriction is that the length of the string on the left-hand side of a production rule should be less than or equal to the number of symbols on the right-hand side. Consider the following grammar rules:

```
<S> ::= a<S>c | ε
<S> c ::= b<S>cc
```

The above example illustrates a context-sensitive grammar because the second production rule has a terminal symbol 'c' in addition to the nonterminal symbol <*S*> on the left hand side of the production rule. The grammar will accept a string of the form $a^nb^mc^{m+n}$ ($m \geq 0$, $n > 1$). However, the second rule is expanded only in the presence of the terminal symbol 'c'.

Context-sensitive grammars are more powerful than context-free grammars. For example, a context-sensitive grammar can be written that would accept a string $a^nb^nc^n$ ($n > 0$). While no context-free grammar can generate a string of the form $a^nb^nc^n$. However, there is a serious drawback of using context-sensitive grammar: context sensitivity significantly increases the number of production rules, which slows down the process of parsing a sentence. Owing to the efficiency reasons during parsing, programming language definitions use context-free grammars.

3.2.2 Representing Grammar Using Backus–Naur Form

In early days of development of compilers, two computer scientists, John Backus and Peter Naur, proposed a format to represent context-free grammars for programming languages. The proposed format is now popularly known as the *Backus–Naur form* or *BNF*.

In BNF, the left-hand and right-hand sides of a production rule are separated by a symbol ‘::=,’ the nonterminal symbols are enclosed within angular brackets, and the multiple definitions of the same nonterminal symbols are separated by a vertical bar ‘|.’ In addition, a grammar uses the Greek symbol ‘ε’ to represent the null value and uses tail recursion (see Chapter 2, subsection 2.2.3) to define zero or more occurrences of an entity. In tail recursion, the recursive part comes at the end of the definition.

Example 3.3

The nonterminal symbol *<sequence-of-statements>* is defined tail recursively. The base case is the null statement denoted by the symbol ‘ε,’ and the tail-recursive part is defined as a nonterminal symbol *<statement>* followed by a semicolon, followed by *<sequence-of-statements>* as follows:

```
<sequence-of-statements> ::= <statement> ';'
                             <sequence-of-statements> | ε
```

Example 3.4

Figure 3.2 illustrates a BNF grammar for a simplistic definition of an arithmetic expression. The context-free grammar has the start symbol *<expression>*, 13 production rules, the set of nonterminal symbols as {*<expression>*, *<A-expr>*, *<L-expr>*, *<comparison>*, *<identifier>*, *<characters>*, *<digit-or-letter>*, *<number>*, *<digit>*, *<A-op>*, *<L-op>*, *<comp-op>*}, and the alphabet as {**0 ... 9, ‘a’ … ‘z’, ‘A’ … ‘Z’, ‘+’, ‘−’, ‘*’, ‘/’, ‘&&’, ‘||’, ‘>’, ‘<’, ‘>=’, ‘=<’, ‘==’**}. The notation ‘0’ | ‘1’ |... | ‘8’ | ‘9’ gives the subrange including all the digits between 0 and 9.

An expression (*<expression>*) is a multidefinition. It could either be an arithmetic expression (*<A-expr>*) or a logical expression (*<L-expr>*) or an identifier (*<identifier>*). An arithmetic expression (*<A-expr>*) is defined as a number (*<number>*) or an arithmetic expression (*<A-expr>*) followed by an arithmetic operator (*<A-op>*) followed by another arithmetic expression (*<A-expr>*). A logical expression (*<L-expr>*) is *true;*

<expression> ::= *<A-expr>* | *<L-expr>*
<L-expr> ::= **true** | **false** | *<identifier>* | **not** *<L-expr>* | *<comparison>* | *<L-expr>* *<L-op>* *<L-expr>*
<comparison> ::= *<A-expr>* *<comp-op>* *<A-expr>*
<A-expr> ::= *<number>* | *<identifier>* | *<A-expr>* *<A-op>* *<A-expr>*
<identifier> ::= *<letter><characters>* | *<letter>*
<characters> ::= ε | *<digit-or-letter><characters>*
<digit-or-letter> ::= *<digit>* | *<letter>*
<number> ::= *<digit>* | *<number><digit>*
<digit> ::= **‘0’** | **‘1’** | **...** |**‘8’** | **‘9’**
<letter> ::= **‘a’**| **‘b’** | **...** | **‘z’**| **‘A’** | **‘B’**| **...**| **‘Z’**
<A-op> ::= ‘ + ’ | ‘−‘ | ‘/’ | ‘*’
<L-op> ::= ‘&&’ | ‘||’
<comp-op> ::= ‘>’ | ‘<’ | ‘>=’ | ‘=<’ | ‘ ==’

FIGURE 3.2 A BNF grammar for defining expressions.

false, a comparison of two arithmetic expressions, two logical expressions (*<L-expr>*) connected by a logical operator (*<L-op>*), or the negation of a logical expression. A comparison (*<comparison>*) compares two arithmetic expressions using comparison operators (*<C-op>*). An identifier (*<identifier>*) is a letter followed by the nonterminal symbol *<characters>*) that is zero or more occurrences of *<digit>* or *<letter>*. A number (*<number>*) is defined recursively as a number (*<number>*) followed by a digit (*<digit>*). The left-recursive definition of *<number>* will be used in a later section to explain denotational semantics in Section 3.5. A digit (*<digit>*) could be any element between 0 and 9. An arithmetic operator (*<A-op>*) is ‘+’, ‘−’, ‘*’, or ‘/’ and a logical operator could be logical-AND (‘&&’) or logical-OR (‘||’). The grammar is ambiguous due to the lack of operator precedence, and can yield two or more parse trees for the same sentence, as illustrated in Figures 3.12 and 3.13.

3.2.3 Extended Backus–Naur Form (EBNF)

Despite being expressive, BNF cannot capture some of the situations elegantly for human comprehension as follows:

1. BNF uses multiple definitions, even when small parts of the production rule change. This results in an unnecessary explosion of rules. For example, we can define an arithmetic expression as *<A-exp>* (‘+’|‘−’|‘*’|‘/’) *<A-exp>*, where the construct (‘+’|‘−’|‘*’|‘/’) shows one of the possibilities and merges two rules into one rule, with better comprehension.
2. BNF uses tail-recursion to handle zero or more occurrences of a symbol. Tail recursion is easily expressed using iteration. In Example 3.1, instead of defining *<sequence-of-statements>* tail recursively, it can be expressed as {*<statement>* ‘;’}* where the notation ‘*’ denotes zero or more occurrences of the nonterminal symbol *<statement>* followed by the terminal symbol ‘;’.
3. BNF does not capture the optional occurrence of a subpart of a production rule and describes them as definition or null symbol ε.

EBNFs have been used to remove this limitation present in the BNF. Table 3.1 shows the substitutions used in EBNF. Grouping for multiple definitions uses parentheses and a vertical bar of the form (alternative$_1$ | alternative$_2$). Optional features in the expansion on the right-hand side of a production rule are expressed using square brackets of the form [Optional-feature]. Repetition uses curly brackets around the repeated entity, augmented with different

TABLE 3.1 Transforming a BNF Grammar into an EBNF Grammar

BNF Representation	**EBNF Representation**
<NT> ::= Alternative$_1$ | Alternative$_2$	*<NT>* ::= (Alternative$_1$ | Alternative$_2$)
<NT> ::= ε | Optional-feature	[Optional-feature]
<NT> ::= Symbol *<NT>* | Symbol	*<NT>* ::= {Symbol}$^+$
<NT> ::= Symbol *<NT>* | ε	*<NT>* ::= {Symbol}*
<NT> ::= ‘0’|‘1’|‘2’|‘3’	*<NT>* ::= ‘0’–‘3’

symbols such as '*', '+' or a digit to describe the extent of repetition. For example, a number is defined as {*<digit>*}$^{+}$ meaning one or more occurrences of *<digit>*; and *<identifier>* is defined as *<letter>* {(*<letter>* | *<digit>*)}^{254} meaning a letter followed by a maximum of 254 characters (a letter or a digit). Owing to the use of parentheses, square brackets, angular brackets, vertical bars, '+' and '*' with special meanings, these symbols are put within quotes when used as terminal symbols in a programming language. The range is described in a shorthand using a hyphen between the lower bound and upper bound of the range.

Example 3.5

Figure 3.3 shows the EBNF version of the grammar described in Figure 3.2. The definitions of the nonterminal symbols *<A-expr>*, *<L-expr>*, and *<comparison>* has been modified by incorporating the "grouping" feature of EBNF, as described in Table 3.1. The definition of the nonterminal symbol *<L-expr>* uses the "optional" feature of EBNF, as the monadic operator '**not**' may occur optionally before *<L-expr>*. The corresponding multiple definitions in *<L-expr>* have been merged. The nonterminal symbol *<number>* uses a combination of "grouping" and "repetition," and is defined as one or more occurrences of grouping of digits 1 to 9. Some of the production rules that have been used to modify other top-level production rules are not required any more and have been merged with the top-level rules.

The EBNF grammar can be further expanded by defining *<A-expr>* iteratively as *<number>*(or *<identifier>*), followed by zero or more occurrences of the grouped operators ('+' | '−' | '*' | '/') and the group (*<number>* | *<identifier>*). The nonterminal symbol *<L-expr>* has been defined iteratively as the grouping (**true** | **false** | *<identifier>* | *<comparison>*), followed by repetition of grouping of comparison operators and grouping (**true** | **false** | *<identifier>* | *<comparison>*). The expression [**not**] *<L-expr>* on the right-hand side has been transformed to grouping [**not**] (**true**| **false**| *<identifier>* | *<comparison>*) by pushing down "[**not**]." The modified EBNF grammar is given in Figure 3.4.

Example 3.6

Figure 3.5 illustrates another realistic example of defining a grammar for *iterative statements* to study the syntax representation using EBNF representation. The names are self-explanatory. An *<iteration>* has multiple definitions: *<for-loop>*, *<while-loop>*, *<do_while-loop>*, and *<iterator>*. The nonterminal symbol (*<for-loop>*) has variable

<expression> ::= *<A-expr>* | *<L-expr>*
<L-expr> ::= **true** | **false** | *<identifier>* | *<comparison>* |
        [**not**] *<L-expr>* ('&&' | '||') [**not**] *<L-expr>*
<comparison> ::= *<A-expr>* ('>' | '<' | '>=' | '=<' | '==') *<A-expr>*
<A-expr> ::= *<number>* | *<A-expr>* (' + ' | '−' | '*' | '/') *<A-expr>*
<identifier> ::= *<letter>* {(*<letter>* | *<digit>*)}*
<number> ::= {*<digit>* }$^{+}$

FIGURE 3.3 An extended BNF grammar to define an expression.

<expression> ::= <A-expr> | <L-expr>
<L-expr> ::= [**not**] (**true** | **false** | <identifier> | <comparison>)
{ ('&&' | '||') [**not**] (**true** | **false** | <identifier> | <comparison>)}*
<comparison> ::= <A-expr> ('>' | '<' | '>=' | '=<' | '==') <A-expr>
<A-expr> ::= (<number>|<identifier>) {(' + ' | '–' | '*' | '/') (<number> | <identifier>)}*
<number> ::= {<digit>}+
<identifier> ::= <letter> {(<letter> | <digit>)}*
<letter> ::= ('a' – 'z' | 'A' – 'Z')
<digit> ::= '0' – '9'

FIGURE 3.4 An EBNF for defining an expression.

<iteration> ::= <for-loop> | <while-loop> | <do-while-loop> | <iterators>
<for-loop> ::= **for** '('<l-value> '=' <expressions> ';' <expressions>; <expressions> ')' <block>
<while-loop> ::= **while** '(' <L-expr> ')' <block>
<do-while-loop> ::= **do** <block> **while** '(' <L-expr> ')'
<iterator> ::= **foreach** '(' <identifier> **in** (<identifier> '|' <enumeration>) ')' <block>
<block> ::= '{' {<statement> ';'}* '}' | <statement> ';' | '{' '}'
<statement> ::= <assignment> | <if-then-else> | <iteration>
<assignment> ::= <identifier> '=' <expression>
<enumeration> ::= '{' <entity> { ',' <entity> }* '}' | <identifier>
<entity> ::= <integer> | <float> | <string>
<identifier> ::= <alphabet> {(<alphabet> | <digit>)}*
<string> ::= ""{(<alphabet> | <digit>)}* ""
<integer> ::= [(' + ' | '–')] {<digit>}*
<alphabet> ::= 'A' – 'Z' | 'a' – 'z'
<digit> ::= '0' – '9'

FIGURE 3.5 A grammar for iterative statements written in EBNF.

(<identifier>) initialized to the evaluated value of an expression (<expression>), followed by a terminal symbol ';' followed by the nonterminal symbol <identifier>, followed by the nonterminal symbol <op>, followed by the nonterminal symbol <expression> showing the final condition, followed by another delimiter ';' followed by an expression <expression> to show the step size in the index variable, followed by the right parenthesis, followed by a nonterminal symbol <block>, denoting a block of statements. The remaining production rules can be read similarly using the knowledge of EBNF gained in previous examples.

3.2.4 Attribute Grammar

The production rules in a grammar can be associated with different attributes for effective error handling, evaluation, constraint specification, and low-level code generation for efficient execution on a specific type of architecture. Different architectures impose different restrictions on the production rules such as word size, size of the strings allowed, maximum number of characters allowed in an identifier, and so on. In addition, each production rule has some meaning associated with it that is used to generate the low-level code from the parse tree. An attribute of a production rule captures the architecture restriction, or language-designer-imposed constraints, and the meaning of the production rule needed to generate low-level code from the parse tree generated during parsing.

For example, a language may be compiled for execution on a 16-bit machine, a 32-bit machine, or a 64-bit machine. For pragmatic and efficiency reasons, designers may like to limit the number of allowable characters in a variable. Similarly, based upon the word size, the value of an integer or floating point needs to be restricted to a maximum value that is architecture dependent. In syntax definitions, these restrictions (or attributes) are also part of the grammar and are defined along with the production rules.

During parsing, attributes associated with a production rule have to move up and down the parse tree to make sure that restrictions are properly enforced. The propagation of attributes is also useful in intermediate code generation and code for exception handlers in case the attributes are violated.

Example 3.7

Table 3.2 shows two production rules and the corresponding attributed production rules. The denotation <*int*>$_1$ denotes the left-hand side integer, and the denotation <*int*>$_2$ denotes the right-hand side integer. The function *value* gives the base-10 value of the integer, the function *length* denotes the number of digits in an integer, and the function *size* gives the number of characters in a string.

The attributes for rule 1 say that the value of an integer <*int*> is limited between -2^{31} to $+2^{31}-1$; the number of digits in the integer on the left-hand side, denoted by *length* (<*int*>$_1$), is one more than the number of digits in the integer on the right-hand side; and the value of the integer <*int*>$_1$ on the left-hand side is *10 * value of the integer* <*int*>$_2$ *on the right-hand side + value of the digit* <*digit*> *on the right-hand side.* Alternately if <*int*> is defined as <*digit*> then the *value of the* <*int*> is same as the *value of the* <*digit*>.

Rule 2 attributes impose a restriction on the size of an <*identifier*> for the efficiency reasons. The restriction is that the size of the identifier has to be less than or equal to 255, and the size of the left-hand side of the identifier is one more than the size of the right-hand side of the string.

TABLE 3.2 Examples of Production Rules in an Attribute Grammar

Production Rule	Production Rule with Attributes
1 <*int*> ::= <*int*><*digit*> | <*digit*>	<*int*> ::= <*int*><*digit*> **Attributes**: value(<*int*>) > -2**^{31}; value(<*int*>) < 2**^{31} - 1 value(<*int*>$_1$) = 10 * value(<*int*>$_2$) + value(<*digit*>) length(<*int*>$_1$) = length(<*int*>$_2$) + 1 <*int*> ::= <*digit*> **Attribute:** value(<*int*>) = value(<*digit*>)
2 <*identifier*> ::= <*letter*>{(<*digit*>|<*letter*>)}*	<*identifier*> ::= <*letter*> {<*digit*> |<*letter*>)}* **Attribute:** size-of(<*identifier*>) =< 255 size-of(<*identifier*>) >= 1

3.2.5 Hyper-Rules and Meta-Definitions

The syntax rules can be made more expressive by capturing the general pattern in the production rules. Two additional types of rules are used: (1) *hyper-rules* and (2) *meta-definitions. Hyper-rules* abstract multiple production rules by the general pattern, and *meta-definitions* specify multiple definitions to be substituted in hyper-rules. By substituting meta-definitions in hyper-rules, multiple production rules having similar patterns are derived. For example, a *<sequence>* is a general pattern across many production rules in grammars of programming languages and can be expressed as a hyper-rule as follows:

```
<sequence> : <definition> ';' <sequence> | ε -hyper rule
```

A meta-definition to define the nonterminal symbol *<definition>* is shown below. The names on the right-hand side are self-explanatory.

```
<definition> :: <formal-parameter> | <actual-parameter>|
                <declaration> | <statement> -meta-definition
```

By applying one meta-definition at a time in the hyper-rule, four production rules are generated. Only one meta-definition is applied at a time across all the definitions in a hyper-rule to generate a production rule; two or more meta-definitions cannot be applied on the same hyper-rule at the same time. The four production rules are as follows:

```
<sequence-of-formal-parameter> ::= <formal-parameter> ';'
                         <sequence-of-formal-parameter> | ε
<sequence-of-actual-parameter> ::= <actual-parameter> ';'
                         <sequence-of-actual-parameter> | ε
<sequence-of-declaration> ::= <declaration> ';'
                              <sequence-of-declaration> | ε
<sequence-of-statement> ::= <statement> ';'
                            <sequence-of-statement> | ε
```

In a grammar, there is a need to distinguish production rules from hyper-rules and meta-rules. Hyper-rules have ':' separating the left-hand side and the right-hand side, meta-definitions have '::' separating the left-hand side from the right-hand side, and production rules in BNF and its variations have '::=' separating the left-hand side from the right-hand side.

3.2.6 Abstract Syntax

In order to understand the properties of data abstractions and control abstractions for a class of programming languages, syntax rules are abstracted using data and control abstractions. For example, we can abstract a programming language construct by *program, blocks, iteration, selection statements, assignment, command, expression, declaration, formal parameters, actual parameters, identifiers, definitions, literals,* and *sequencers—goto statements, type expressions*, and so on. These abstractions are defined using the abstract syntax rules along with the important reserved words in the programming languages. A*bstract syntax rules* are different from the set of production rules for the grammar of a programming language. Some of the lower-level definitions such as the definitions of identifiers, numbers, integers, digits, expressions, strings, and operator precedence—that are universal across programming languages are ignored, as they do not add anything more to the understanding of the abstract constructs in the programming languages. Similarly, lower-level literals, delimiters, and white spaces are also ignored for the same reason. Abstract syntax is concise and explains the constructs using programmers' existing knowledge of abstractions. There are inherent ambiguities in the abstract syntax rules due to the omission of low-level details. However, it is concise, associated with control and data abstractions in the programming language in the form of nonterminal symbols, and carries the alphabet of the programming language.

Example 3.8

Abstract syntax rules for control abstractions of l-value, declarations, expressions, and commands can be defined for a major class of imperative programming languages, as shown in Figure 3.6.

The abstract syntax rules illustrate that an l-value abstraction could be an identifier, a specific field of a structure, or a subscripted variable (shown by *<l-value>* '[*<expression>*']'). An expression abstraction could be a literal, an identifier, an l-value, an expression within parentheses, two expressions connected by a dyadic operator, or a monadic operator followed by an expression.

The command abstraction in the given programming language can be a block within curly brackets, an assignment statement, a sequence of commands, an if-then-else

```
<l-value> ::= <identifier> | <identifier>.<l-value> | <l-value>' ['<expression>']'

<declarations> ::= variable <identifier> <type-expression> |
                   <type-expression>[<numeral>] |
                   structure {<type-expression>} <identifier> |
                   void <identifier> ( <formal-parameters> ) |
                   <identifier> function <identifier> ( <formal-parameters> )

<expressions> ::= <literal> | <identifier> | <l-value> | (<expressions> | <op> <expressions> |
                  <expressions> <op> <expressions>
<actual-parameters> ::= <identifier> ',' <actual-parameters>

<formal-parameters> ::= <identifier> ',' <identifier-sequence> ';' <formal-parameters> | ∈

<commands> ::= { <commands> } | <l-value> '=' <expressions> | <command> ';' <commands> |
               if <expressions> then <commands> else <commands> |
               if <expressions> then <commands> |
               while '(' <expressions> ')' <commands> |
               do <commands> while '(' <expressions> ')' |
               for '('<l-value>'=' <expressions> ';' <expressions>';' <expressions> ')' <commands> |
               <identifier> '('<formal-parameters> ')'

<sequencer> ::= goto <numeral>
<program> ::= main <identifier>';' <declarations> ';' <commands>
```

FIGURE 3.6 An example of abstract syntax rules.

statement, a while-loop, a do-while loop, a for-loop, or a procedure call. In addition, it tells the reserved words {**if, then, else, do, while, '{,' and '},' for**} and so on are part of the language constructs. However, other low-level details have been removed. For example, the definition of operator in the definition of abstract syntax rules for <*expression*> does not discriminate between various operator precedence and different types of expressions: logical versus arithmetic.

3.3 SYNTAX DIAGRAMS

Textual representations are good for processing the grammar by computers. However, humans are better in visualizing and comprehending simple pictorial diagrams. To visualize, comprehend, and communicate the syntax of programming languages to programmers, language designers use the pictorial version of the production rules. These pictorial versions are called *syntax diagrams.*

Understanding the correspondence between textual representation of the syntax grammar and the syntax diagram is quite important, since the language designers have to (1) write the syntax diagrams for the ease of comprehension by programmers and language designers and (2) write the textual version for developing parsers and code generators. The programmers also have to translate the knowledge of the syntax, gained by studying syntax diagrams to textual form during program development.

Formally a *syntax diagram* is a *directed cyclic graph* with terminal and nonterminal symbols as nodes and the concatenation between the terminal and nonterminal symbols on the right-hand side of the production rules as edges. The cycles model the tail-recursive definitions (or repetitions in EBNF). The left end of a syntax diagram describes the left-hand side nonterminal symbols in a production rule, and the remaining graph models the right-hand side of the production rule.

Syntax diagrams have three major components: *nonterminal symbols, terminal symbols,* and *directed edges.* For our convenience, the nonterminal symbols have been boxed in an oval shape to separate them clearly from terminal symbols. The arrow shows the direction of the flow. The leftmost symbol shows the nonterminal symbol on the left-hand side of the production rule that is being defined. Different conversions from syntax rules to syntax diagrams are illustrated in Figures 3.7 and 3.8.

Multiple definitions are represented as multiple forward paths in a syntax diagram. A null symbol is represented as a simple directed edge between the source and the destination that has no embedded symbol. Concatenation of multiple symbols on the right-hand side of a production rule is represented as a path connecting multiple nodes. Tail-recursive definitions are represented as a feedback loop (a cycle) to show multiple occurrences of a definition. The feedback loop with a definition in the forward edge describes one or more occurrences in EBNF. At the destination node, there are two options: use the backward edge to go through the cycle again or exit. An interesting path is to model zero or more occurrences of symbols. This is modeled as a feedback loop with a variation that the symbols occurs in the backward edge of the feedback loop, and the forward edge has no symbol.

For better visualization and comprehension, many production rules are merged into one syntax diagram. The merging is done to reflect abstract entities in programming languages such as *identifiers; variables; integers; decimal numbers; numbers; arithmetic expressions; logical expressions; actual parameters; formal parameters;* different types of statements, such as *if-then-else statements, while-loop, do-while-loop, iterators, and case statements; block of statements; type declarations; program;* and so on.

Example 3.9

Figure 3.7 illustrates syntax diagrams for multiple definitions (see Figure 3.7a), syntax diagrams for concatenation of symbols (see Figure 3.7b), syntax diagram for tail-recursive definitions showing one or more occurrences of a symbol(s) (see Figure 3.7c), and syntax diagram for zero or more occurrences of a symbol(s) (see Figure 3.7d).

As illustrated in Figure 3.7a, multiple forward paths between two nodes starting from one side and merging back on the other side show multiple definitions. The syntax diagram for the nonterminal symbol <*statement*> has multiple paths: (1) the first path containing the nonterminal symbol <*assignment*>; (2) the second path containing the nonterminal symbol <*if-then-else*>; and (3) the third path, containing the nonterminal symbol <*iteration*>. It is equivalent to a syntax rule of the form

```
<statement> ::= <assignment> | <if-then-else-statement> |
                 <iteration>
```

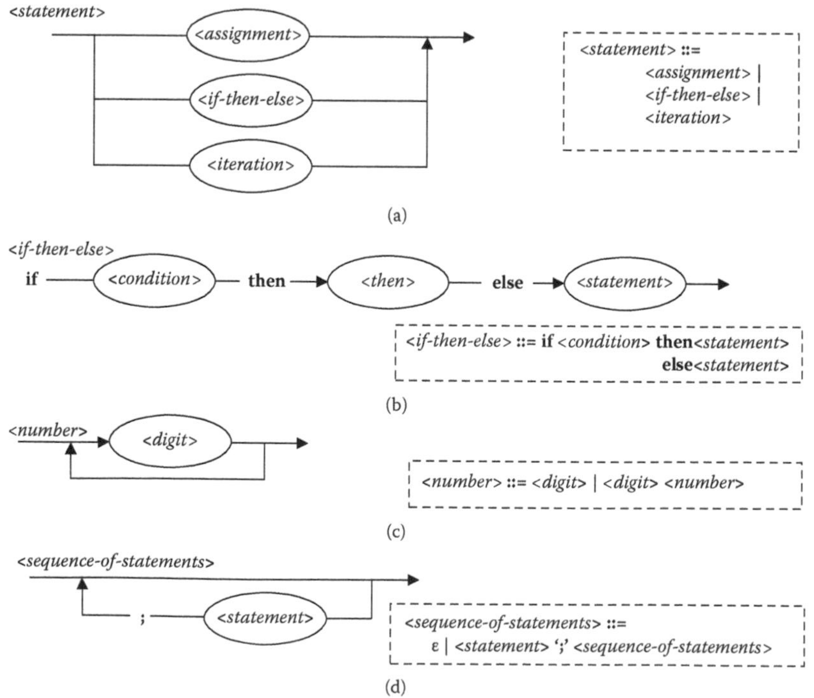

FIGURE 3.7 Syntax diagrams.

Figure 3.7b describes the syntax rule *<if-then-else>* ::= **if** *<cond>* **then** *<statement>* **else** *<statement>*. The right-hand side of the rule is a concatenation of the terminal symbols {**if, then, else**} with the nonterminal symbols {*<cond>*, *<statement>*} in a specific order. The syntax diagram preserves the order of catenation in the syntax rule.

Figure 3.7c illustrates the syntax diagram for the production rule *<number>* ::= *<digit>* | *<digit><number>*. The definition has a base case *<number>* ::= *<digit>* and the tail-recursive definition *<number>* ::= *<digit>* *<number>*. It represents one or more occurrences of *<digit>*, because the tail-recursive definition can be used to add additional *<digit>*, and the definition of the nonterminal symbol *<number>* can always be terminated by applying the base case that adds one additional *<digit>* before terminating the expansion. The definition represents one or more occurrences of *<digit>*. The EBNF version of the syntax rule is *<number>* ::= {*<digit>*}$^{+}$. The corresponding syntax diagram is modeled as a feedback loop, where the forward path contains a nonterminal symbol *<digit>*, and there is a backward edge from the destination node to the source node. The forward path can be traversed multiple times. Each time a forward path is traversed, an extra *<digit>* is added to generate a sequence of *<digit>*.

Figure 3.7d models the syntax diagram for a syntax rule for zero or more occurrences of statements. The corresponding production rule in BNF has a base part and a tail-recursive part. The base part is an empty symbol ε, and the tail-recursive part involves the nonterminal symbol <statement> followed by the terminal symbol ';' as shown below

```
<sequence-of-statements> ::= <statement> ';'
                              <sequence-of-statements> | ε
```

In EBNF, zero or more occurrences of symbols is written by enclosing the symbols within curly brackets and placing an asterisk "*" right after the right curly bracket, as shown below:

```
<sequence-of-statements> ::= {<statement>';'}*
```

Zero or more occurrences of symbols is represented in the syntax diagrams as a feedback loop, such that the forward edge is a straight arrow without a symbol, and the backward edge has the symbols. In Figure 3.7d, the backward edge has the nonterminal symbol *<sentence>* followed by the terminal symbol ';.'

3.3.1 Translating Syntax Rules to Syntax Diagrams

Syntax rules are translated to the corresponding syntax diagrams using the following rules:

1. The terminal and nonterminal symbols are distinct in syntax diagrams.
2. *A production rule with multiple definitions* in BNF or *grouping* in EBNF is modeled as parallel paths between two nodes and looks like a multilane road.
3. *Concatenation* of multiple symbols on the right-hand side is depicted as multiple symbols on the same path.

4. Tail-recursive definition is represented as a feedback loop, where the end of the right-hand side connects back to the start of the definition.
5. Empty symbol, denoted as ε, is written as a straight edge between two nodes.
6. An optional definition is modeled as a pair of paths, such that the first path contains a straight edge between two nodes, and the second path contains the definition.

Figure 3.8 describes the syntax diagrams for the corresponding components of syntax rules in BNF and EBNF. The ellipse denotes a symbol (terminal or nonterminal) in the textual representation.

The first row illustrates a corresponding syntax diagram for a production rule that has three symbols on the right-hand side. The three edges connecting the ellipse have been merged into one for convenience. The second row shows a production rule that has four different definitions in the same production rule as follows:

1. The first definition on the top edge has two symbols on the right-hand side.
2. The second and the third definitions have one symbol on the right-hand side.
3. The fourth definition, shown by the bottom edge, has two symbols on the right-hand side.

The third row shows a tail-recursive definition of one or more occurrences of a symbol on the right-hand side of a production rule. One or more occurrences are equivalent

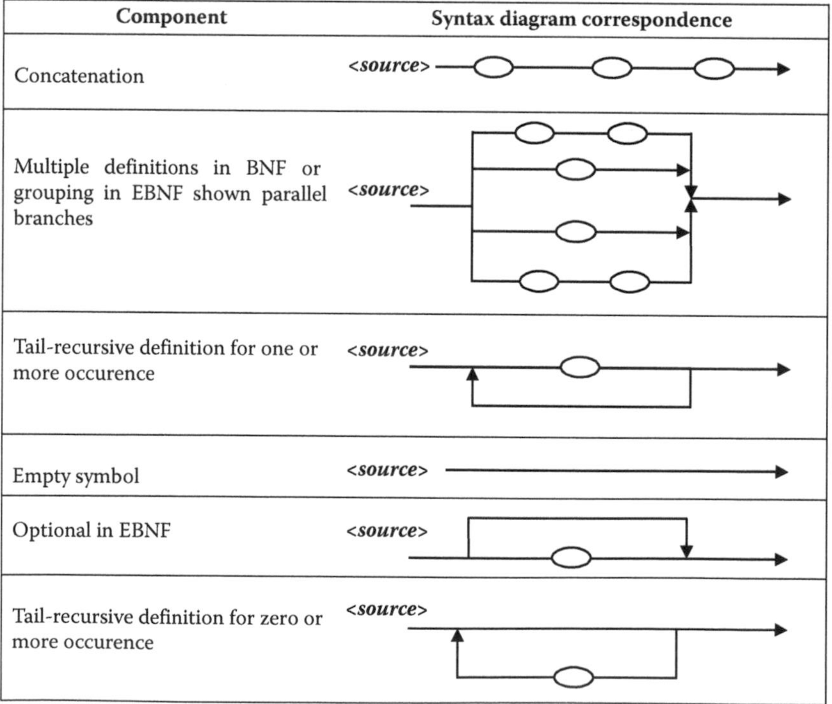

Component
Syntax diagram correspondence
Concatenation
<source>
Multiple definitions in BNF or grouping in EBNF shown parallel branches
<source>
Tail-recursive definition for one or more occurence
<source>
Empty symbol
<source>
Optional in EBNF
<source>
Tail-recursive definition for zero or more occurence
<source>

FIGURE 3.8 Correspondence of syntax diagrams and syntax rules.

to {<*symbol*>}$^+$ in EBNF and the tail-recursive definition <*definition*> ::= <*symbol*> <*definition*> | <*symbol*> in BNF, where <*symbol*> could be a combination of terminal or a nonterminal symbol. The fourth row shows the syntax diagram for an empty symbol 'ε.' Since there is no symbol, the corresponding syntax diagram is a straight arrow.

The fifth row illustrates the syntax diagram of an optional symbol. An optional symbol of the form [<*symbol*>] in EBNF is equivalent to a grouping (<*symbol*> | ε) showing two forward paths: one of the paths has <*symbol*> as an embedded node, and the other path has no embedded node. The sixth row shows zero or more occurrences of a symbol. The syntax diagram of zero or more occurrences also has a return edge, like one or more occurrences. However, the syntax diagram for zero or more occurrences has the symbol on the return path instead of the forward path. The definition of zero or more occurrences of <*symbol*> is written as {<*symbol*>}* in EBNF and as the tail-recursive definition <*definition*> ::= <*symbol*><*definition*> | ε in BNF, where <*symbol*> could be a terminal or a nonterminal symbol.

In order to translate a textual grammar into a syntax diagram, a grammar is grouped into the language-relevant functional units that have some meaning at the data or control abstraction level. The grammar is divided into a set of subsets of production rules, where each subset corresponds to a functional unit in the program development, such as *variables, formal parameters, block*, and so on. Then iteratively top-level rules, are translated into syntax diagrams, and the nonterminal symbols within the translated syntax diagrams, are expanded further to another syntax diagram, until the addition of more low-level syntax diagrams will not facilitate further understanding of the specific functional unit.

Example 3.10

The definition of <*identifier*> has three production rules using a set of three nonterminal symbols: {<*identifier*>, <*letter*>, <*digit*>}. However, <*letter*> and <*digit*> are used to define the nonterminal symbol <*identifier*>. Thus three productions rules are merged into a single syntax diagram.

```
<identifier>, <letter> and <digit> as shown below:
<identifier> ::= <letter> {(<letter> | <digit>)}*
<letter> ::= 'A' - 'Z' | 'a' - 'z'
<digit> ::= '0' - '9'
```

First the syntax diagram for the production rule defining nonterminal symbol <*identifier*> is constructed. The production rule defining the <*identifier*> has been chosen first, because the definition of the <*identifier*> includes the definition of other nonterminal symbols in its right-hand side, and these nonterminal symbols can be expanded later. In the second stage, the nonterminal symbols <*letter*> and <*digit*> are expanded to refine the syntax diagram. The production rules are converted to syntax diagrams using the correspondence between production rules classes. The resulting syntax diagrams is illustrated in Figure 3.9.

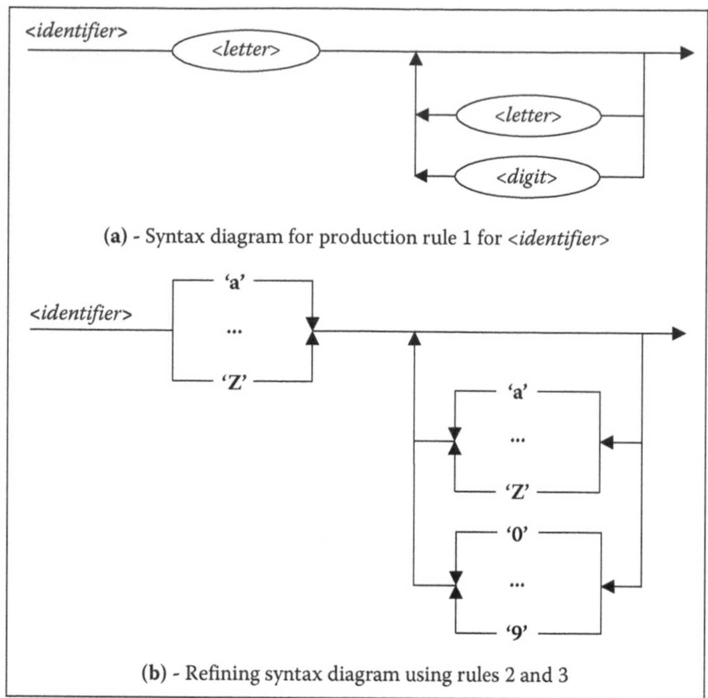

FIGURE 3.9 Building a syntax diagram using multiple production rules.

3.3.2 Translating Syntax Diagrams to Syntax Rules

Syntax diagrams have to be translated to textual grammar for use with parser generators. Thus a language designer or a programmer should be able to translate a syntax diagram to the corresponding set of syntax rules.

The translation of syntax diagrams to production rules is the reverse process of generating syntax diagrams from production rules. Those parts of the syntax diagrams that are embedded in the syntax diagrams, and that are defined only in terms of terminal symbols and already defined nonterminal symbols are converted to production rules using the correspondence between production rules and syntax diagrams, as given in Figure 3.8.

New meaningful nonterminal symbols are generated for the embedded syntax diagrams. After developing the production rule, the corresponding part of the syntax diagram is replaced by the newly defined nonterminal symbol, and the process is repeated. The process stops when the whole syntax diagram is reduced to a single nonterminal symbol.

Example 3.11

To generate a set of production rules for the syntax diagram in Figure 3.7, the most embedded parts—multiple parallel paths defining the grouping (**'0'** | **'1'** | ... | **'9'**) and multiple parallel paths defining the grouping *(***'a'** | ... | **'z'***)*—are translated as *<letter>* ::= **'a'** | ... | **'z'** and *<digit>* ::= **'0'** | ... | **'9'**. The nonterminal symbols *<letter>* and *<digit>* are now substituted in place of the multiple parallel paths, and the syntax diagram is reduced to a syntax diagram as given in Figure 3.7a. In Figure 3.7a, there is concatenation of the nonterminal symbol *<letter>* and the tail-recursive definition showing zero or more occurrences of the alternative symbols (*<letter>* | *<digit>*). Hence, a new production rule is formed as *<identifier>* ::= *<letter>* {(*<letter>* | *<digit>*)}*.

3.4 VALIDATING SENTENCE STRUCTURE

The program compilation process starts by validating the structure of the sentences. There are two steps in the validation of the grammatical structure: (1) converting the symbols in sentences into an internal format using "tokens" and (2) validating this internal format of the sentences using the grammar rules of the language. The first step is called *lexical analysis*, and the second step is called *parsing*. The output of the second step is a *parse tree*. A *parse tree* is a tree generated by repeatedly matching the right-hand side of the grammar rules and substituting the matched part of the intermediate form (or sentence) by the corresponding left-hand side nonterminal symbol of the matching production rule. This substitution generates a new intermediate form, where the part matching the right-hand side of the production rule has been substituted by the nonterminal symbol on the left-hand side of the production rule. The process is repeated until the start symbol is reached. The sequence of derived intermediate forms generates the parse tree. This parse tree becomes the input for the next level of code generation.

Both lexical analysis and parsing have been automated, and software tools have long been developed to generate an automated parser using *parser generators*. The input to a parser generator is the grammar of a language. Automated parsers use various top–down or bottom–up techniques along with symbol look-ahead to uniquely select the production rule for the reduction of the intermediate forms. Automated parsers have been described more in Section 3.4.5.

This section explains the basics of lexical analysis and parsing. We use our intelligence to identify the appropriate rules to reduce progressively intermediate forms to the corresponding nonterminal symbols, so that we can generate the optimum parse tree. The study of parser generators and different techniques for automated parsing are within the scope of a course in compilers.

3.4.1 Lexical Analysis

Lexical analysis is the first phase of program translation, where the reserved words, identifiers, and literals are recognized, archived in a symbol table, and translated to tokens. Literals are constant values in a programming language that cannot be altered. For example, *10* and *4.7* are literals. The lowest syntactic unit that forms a reserved word, identifier, or a literal is called a *lexeme*. A *lexeme* is a string of characters that forms the lowest-level syntax units in a programming language. During the lexical analysis, white spaces are removed; tokens are created; and the stream of token is passed to the parser for the sentence validation. For example, the assignment symbol '=' is assigned a token *assigned_to*; the symbol '<' is assigned a token *less-than*.

A finite state automaton (FSA) is used to recognize words, identifiers, and literals and to generate the tokens. The FSA is defined by the corresponding regular grammar that is further divided into small interconnected FSAs that generate the tokens, as shown in

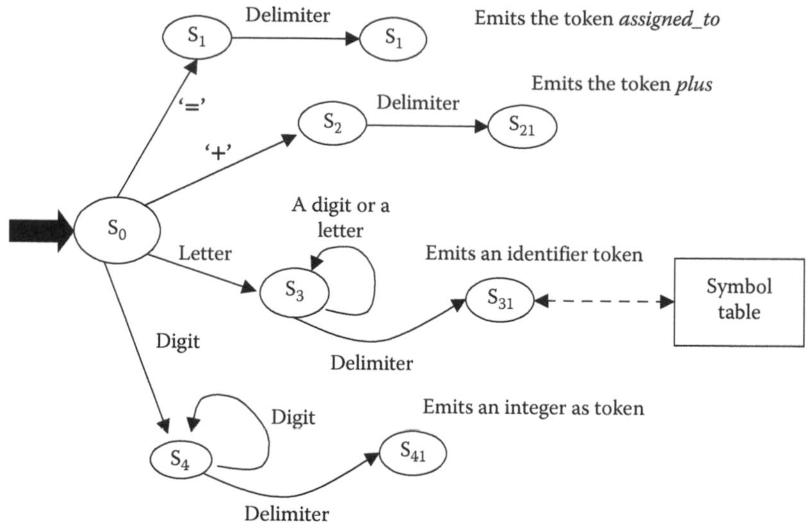

FIGURE 3.10 A simplified scheme of finite-state automaton for lexical analysis.

Figure 3.10. The input to an FSA is a sequence of characters in a program, and the output is a tokenized version of the program.

The lexeme recognition process starts from the initial state S_0, and FSA transits to different states during a lexeme recognition. After identifying a delimiter or a look-ahead character that resolves the ambiguity in the transition to the next state, the corresponding token is emitted. After the token is emitted, the FSA transits back to the start state S_0. Possible delimiters are blanks spaces, end of the sentences such as a semicolon, end of block of sentences such as curly brackets, end-of-line in some languages, or end-of-file. The definition of delimiters is language dependent. After the emission of the token, the pair (*lexeme, emitted token*) is archived in the symbol table. Next time, when the same lexeme is identified in the program, the symbol table is looked up to generate the same token.

Example 3.12

The FSA in Figure 3.10 can accept identifiers, unsigned integers, and reserved words '=' and '+'. It has four subparts that respectively generate the corresponding tokens for the reserved words '=' and '+', an identifier, and the internal number representation for a number. The FSA has an initial state S_0. The reserved word '=' takes the FSA to state S_1, the reserved word '+', takes the FSA to state S_2, a letter takes the FSA to the state S_3, and any digit takes the FSA to the state S_4. The FSA keeps cycling in S_3 if the next symbol is a letter or a digit, and upon finding a delimiter converts the identifier to a token, and it transitions back to the state S_0. The FSA keeps cycling in S_4 when it gets a digit as the next symbol, and upon finding a delimiter, outputs a token that is an internalized representation of an unsigned integer representing the sequence of digits, and transitions back to state S_0.

Let us consider a sentence *sum* = *sum* + 10. The symbol *sum* is an identifier, the symbols '=' and '+' are reserved words, the symbols 10 is a number, and the symbol ';' is a delimiter ending the sentence. The blank character acts as a delimiter to separate the identifiers and reserved words. There are ten characters in the input string "*sum* = *sum* + 10" in addition to the blanks.

The lexical analyzer processes one input character at a time. Initially, the FSA is in the state S_0. The first letter is 's'. The state machine transits to state S_3—a state that recognizes identifiers. The next two symbols are letters 'u' and 'm'. The machine keeps cycling onto the state S_3. The next symbol is a blank that acts as a delimiter for the identifiers. The FSA transits to a new state S_{31}, and the identifier *sum* is internalized, put in a lookup table and outputs a token value '_1'. Upon the subsequent occurrences of the identifier, FSA will search in the lookup table and output the internalized token '_1'. After emitting the token '_1', the FSA transitions back to the start-state S_0. The next symbol is a reserved word '='. It takes the FSA to the state S_1. The next look-ahead symbol is a blank that acts as a delimiter. The FSA transits to another state, S_{11}, and emits the corresponding token *assigned-to.* After emitting the token *assigned-to*, FSA transitions back to the initial state S_0. The next letter 's' takes the FSA to the state S_3, and the process is repeated, with a difference: the lookup table has an entry for the identifier *sum,* and the token '_1'. is emitted again. After emitting the token '_1', the FSA transitions back to the initial state S_0. Upon seeing the next look-ahead character '+' the FSA transitions to the state S_2. Upon seeing the delimiter blank, the FSA transitions to state S_{21}, emits the token *plus*, and transitions back to the initial state S_0. The next character is digit 1. The automaton transitions to state S_4. The next character is the digit 0. The FSA cycles back to state S_4. After seeing the delimiter ';', the FSA outputs the token—an internalized version of unsigned integer 10, and transitions back to the initial state S_0. The output of the lexical analyzer will be a stream '_1' *assigned_to* '_1' *plus* 10. This tokenized sequence is the input to the parser.

Many times it is not possible to identify the lexeme just by looking at the next look-ahead character, due to the ambiguities in transitions to more than one state. In such cases, more than one look-ahead character is used to resolve the ambiguities. For example, '=' and '==' and '=<' are three different symbols with three different tokens: '=' will get a token *assigned_to*, '==' will get a token *is_equal_to*, and the sequence of characters '=<' will get a token *less_than_or_equal_to.* If we just look at the first input character '=' then there is an ambiguity. After we look at the next character, the ambiguity is resolved: a blank as delimiter will emit a token *assigned_to*, the second look-ahead character as '=' will result in a token *is_equal_to*, and the second look-ahead character as '<' will result in a token *less_than_or_equal_to.*

FSA for a lexical analyzer is modeled as a two-dimensional transition table of size *M* rows *X N* columns, where *M* is the number of states in FSA, and *N* is the number of possible input characters. Each cell in the table contains the state (or set of states) to which the FSA will transit to upon seeing the next input character. In the case of an ambiguity, there will be more than one state in the cell. The resolution is done using a stack, storing the pairs of the form (*look-ahead-character position, set of remaining states to be explored*) in the stack, and moving ahead with the next look-ahead character in the input stream. If the next look-ahead character gives a unique lexeme, then the stack is reinitialized, and the token is generated. Otherwise, the stack is popped, and remaining possible states are explored.

3.4.2 Parsing

A parser takes the sequence of tokens in a sentence and the grammar of the language as input and generates an unambiguous parse tree. The process of parsing repeatedly applies one of the grammar rules on a subsequence of the *reduced-intermediate-form* of the sentence being parsed. A *reduced-intermediate-form* contains a combination of terminal and nonterminal symbols. The next rule is picked by matching a part of the *reduced-intermediate-form* with the right-hand side of a production rule and replacing the matched part with the left-hand side nonterminal symbol, until the start symbol is reached. If the repeated application of the production rule does not lead to the start symbol, or there is no subsequence of the reduced-intermediate-form on which a production rule can be applied, then parsing fails, resulting in an error. A simplified scheme to understand parsing is given by an algorithm in Figure 3.11.

The input to the algorithm is a set of production rules $\{p_1, \ldots p_n\}$, the sentence S that is a sequence of symbols $s_0, \ldots, s_m$, and the start symbol *root*. The output of the algorithm is a parse tree T. Initially the reduced-form is the original sequence S. The while-loop keeps iterating, while the reduced-form is not equal to the start symbol root and there is no parsing error. Inside the iterative loop, it checks for a production rule p_i the form *<non-terminal-symbol>* ::= $s_i \ldots s_j$ such that the subsequence $s_i \ldots s_j$ is included in the current *reduced-form*. After the match, the subsequence *si* ... *sj* is replaced by the left-hand-side-of (*<production-rule* p_k*>*), and the variable *reduced-form* is updated. The tree T is also updated accordingly by adding a new edge of the form ($s_i \ldots s_j \rightarrow$ *left-hand-side*(p_k)), and the process is repeated until the start-symbol root is found or parse-error becomes true. The Boolean variable *parse-error* becomes true if there is no matching rule.

```
Algorithm bottom-up-parse-sentence
Input:   1. A set of production rules R = {p_i, ... p_n} of the grammar;
         2. A sentence as a sequence of symbols S = s_0, ..., s_m;
         3. The start symbol root;
Output: A parse tree T;
{   reduced-form = S;
    parsing-error = false;
    T = null-tree;
    while ((reduced-form ≠ root ) && not(parsing-error) )
              { If there exists a subsequence s_i .. s_j in reduced form, such that
                s_i... s_j == right-hand-side(p_k ∈ R) where 1 =< k =< n {
                nonterminal = left-hand-side(p_k);
                reduced-form = substitute(reduced-form, s_i... s_j, nonterminal);
                T = T + edge(s_i ..s_j → left-hand-side(p_k));
                }
                else parsing-error = true;
              }
    If not(parsing-error) return(T); else print('parsing-error');
}
```

FIGURE 3.11 A simplified scheme for bottom–up parsing a sentence.

The process of finding out an appropriate subsequence that matches the right production rule is a tricky problem and has been solved using different automated parsing techniques, such as LL(K) and LR parsers. Automated parsers have been described briefly in subsection 3.4.5.

Example 3.13

Let us parse the sentence "x + 3 * 4," using the grammar given in Figure 3.2. The sentence contains five symbols: 'x', '+', '3', '*', and '4'. These symbols will be converted by the lexical analyzer to the corresponding tokens. However, for the sake of convenience, we use the symbols in their original forms instead of tokens. Any symbol or a subsequence of five symbols is a candidate for matching with the right-hand side of the production rules. Using the grammars given in Figure 3.3, we get two alternate parse trees, as shown in Figures 3.12 and 3.13. *A grammar that gives more than one parse tree for the same sentence is ambiguous and should be avoided.*

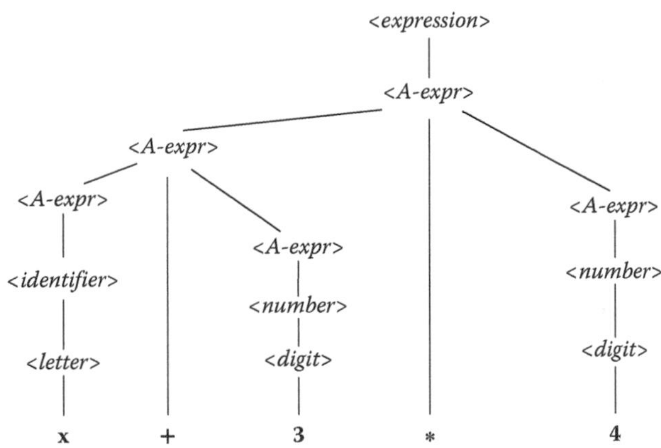

FIGURE 3.12 An incorrect parse tree due to ambiguity in the grammar.

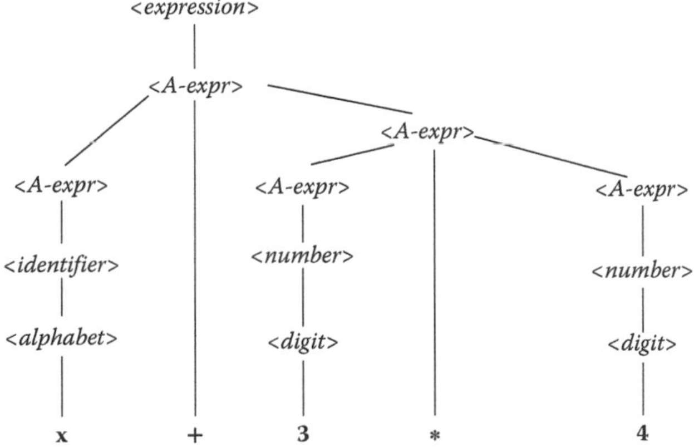

FIGURE 3.13 The correct parse tree for the expression x + 3 * 4.

3.4.3 Handling Grammar Ambiguities

Ambiguous grammars violate one of the fundamental principles of programming language design that each sentence should have a unique meaning. Two different parse trees for the same sentence imply that the sentence can be translated to two different sets of low-level instructions, leading to two different computations. There are two major classes in computation where grammar ambiguities are caused: (1) when a production rule does not take care of the precedence of operators and (2) matching the constructs in nested structures if the nesting levels are not explicitly separated by reserved words.

Example 3.14

The grammar to define *<expression>* in Figures 3.2 through 3.5 are ambiguous due to the grouping of operators in the same production rule using multiple definitions for arithmetic expressions and logical expressions. In arithmetic expressions, dyadic operators '*' and '/' have higher precedence than the dyadic operators '+' and '–'. Similarly, in logical expressions, the monodic operator 'not' has higher precedence than '&&' (logical-AND) and '&&' (logical-AND) has higher precedence than '||' (logical-OR).

One way to handle the ambiguities in expressions is to explicitly separate the subexpressions using parentheses. However, the order of precedence can be coded in the grammar rules by splitting the production rule into multiple production rules. A multidefinition production rule is split into multiple production rules by introducing new nonterminal symbols that are progressively defined in terms of other nonterminals with increasing precedence of operators, and the last production rule uses the highest precedence operator. The operators with higher precedence are parsed first, because rules involving higher-precedence operators are applied first when parsing bottom–up. The corresponding unambiguous grammar for expressions has been shown in Figure 3.14.

The production rule for the arithmetic expression *<A-expr>* has been split into three production rules, using two additional nonterminal symbols *<expr-mult>* and *<A-term>*. Similarly, the production rule for the logical expression *<L-expr>* has

<*expression*> ::= <*A-expr*> | <*L-expr*> (1)
<*A-expr*> ::= <*A-expr*> (' + ' | '-')<*expr-mult*> | <*expr-mult*> (2)
<*expr-mult*> ::= <*expr-mult*> ('*' | '/') <*A-term*> | <*A-term*> (3)
<*A-term*> ::= '(' <*A-expr*> ')' | <*identifier*> | <*number*> (4)
<*L-expr*> ::= <*L-expr*> '||' <*expr-and*> (5)
<*expr-and*> ::= <*expr-and*> '**&&**' <*L-term*> (6)
<*L-term*> ::= [**not**] ('(' <*compare*> ')' | '(' <*L-expr*> ')' | <*identifier*> | **true** | **false** (7)
<*compare*> ::= <*A-expr*> ('>' | '<' | '>=' | '=<' | '==') <*A-expr*> (8)
<*identifier*> ::= <*letter*>{(<*letter*>|<*digit*>)}* (9)
<*number*> ::= [(' + '|'–')] {<*digit*>} + (10)
<*letter*> ::= '**a**'| '**b**' | ... |'**z**'| '**A**' | '**B**' | ... | '**Z**' (11)
<*digit*> ::= '**0**' | '**1**'| ... |'**9**' (12)

FIGURE 3.14 An unambiguous grammar for expressions.

been split into three production rules using two additional nonterminal symbols: *<expr-and>* and *<L-term>*. The nonterminal symbol *<A-expr>* has been defined using *<expr-mult>*. The nonterminal symbol *<expr-mult>* has been defined in terms of *<A-term>*. The nonterminal symbol *<A-term>* has been defined in terms of (*<A-expr>*) or *<identifier>* or *<number>*. The nonterminal symbol *<L-expr>* is defined using nonterminal symbol *<expr-and>*. The nonterminal symbol *<expr-and>* is defined using nonterminal symbol *<L-term>*. The nonterminal symbol *<L-term>* has been defined in terms of the optional terminal symbol '**not**' followed by a grouping of logical expressions within parentheses, or comparison within parentheses, or the nonterminal symbol *<identifier>*, or the terminal symbol **true** or the terminal symbol **false.**

Example 3.15

Let us parse the sentence "x + 3 * 4" using the unambiguous grammar given in Figure 3.14. The literals '3' and '4' are reduced to nonterminal symbol *<A-term>* using a sequence of rules: rule 12, rule 10, rule 9, and rule 4. The reduced form "*<A-term>* – *<A-term>*" is reduced to the nonterminal symbol *<expr-mult>* using rule 3, and the new reduced form is "x + *<expr-mult>*." The symbol x is reduced to the nonterminal symbol *<identifier>* using a sequence of rules: rule 11 and rule 9, and the new intermediate form is "*<identifier>* + *<expr-mult>*." The nonterminal symbol *<identifier>* is further reduced to the nonterminal symbol *<A-expr>* using a sequence of rules: rule 4, rule 3, and rule 2. The new reduced form is "*<A-expr>* + *<expr-mult>*" that is reduced to the nonterminal symbol *<A-expr>* using rule 2. The nonterminal symbol *<A-expr>* is reduced to the start-symbol *<expression>* using rule 1. The resulting parse tree is unique and correct, as shown in Figure 3.15.

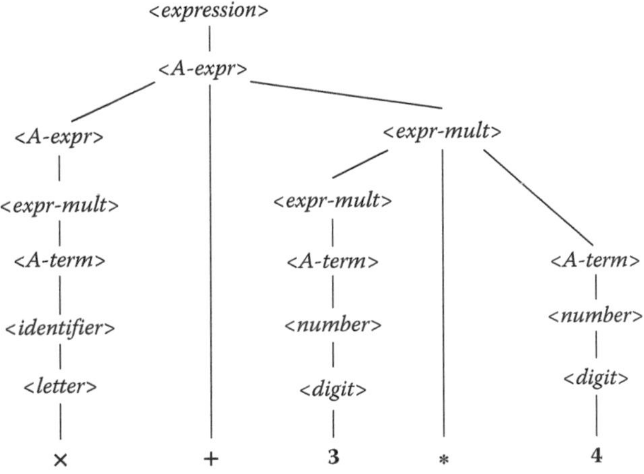

FIGURE 3.15 Parse the expression "x + 3 * 4" using an unambiguous grammar.

TABLE 3.3 Possible Interpretations of Unmatched Nested if-then-else Statement

Incorrect Interpretation	Correct Interpretation

`if (x > 4) then`
`if (y > 0) then return(1);`
`else return(0)`
	`if (x > 4) then`
	`if (y > 0) then return(1);`
	`else return(0);`

(a) else part is matched with outer if	(b) else part matched with the nearest if

3.4.3.1 *Ambiguities in Nested Structures*

If-then-else statements can have two options: (1) **if** *<statement>* **then** *<statement>* or (2) **if** *<statement>* **then** *<statement>* **else** *<statement>*. The first option is unmatched, as the else part is missing, and the second option is matched, as the else part is present. A combination of matched and unmatched options to form a nested if-then-else statement causes ambiguity, as it would be unclear whether the else part matches with the outer if or with the inner if, as shown below in Table 3.3.

Both the statements have two occurrences of "if" and one occurrence of "else." The rule for programming is that else goes with the nearest if occurrence. In order to interpret it correctly, there are two options: (1) put the inner matched part using delimiters to separate them clearly from the outer blocks or (2) write an unambiguous grammar for handling a nested if-then-else statement. An unambiguous grammar is given below

```
<if-then-else-statement> ::= <matched-if-then-else> |
                              <unmatched-if-then-else>
<matched-if-then-else> ::= if <condition> then
                              <matched-if-then-else>
                           else <matched-if-then-else>|
                              <other-statements>
<unmatched-if-then-else> ::= if <condition> then
                                <if-then-else-statement> |
                              if <condition> then
                                <matched-if-then-else>
                              else <unmatched-if-then-else>
```

3.4.4 Abstract Syntax Tree

A parse tree made out of abstract syntax rules gives an *abstract syntax tree*. An *abstract syntax tree* is free of all the nonterminal symbols in the concrete syntax rules that are not in the abstract syntax rules. The expressions are rooted at the operators. For example, the abstract syntax tree for the expression "x + 3 * 4" is given by Figure 3.16.

Concrete syntax trees (trees developed using concrete syntax rules) are reduced to abstract syntax trees by removing low-level and redundant nonterminal symbols not contributing directly to the meaning of the sentence. For example, the concrete syntax tree for the expression "x + 3 * 4" as shown in Figure 3.15 contains many low-level nonterminal symbols, such as {*<A-term>*, *<expr-mult>*, *<digit>*}, that do not directly contribute to the semantics and have been removed from the abstract syntax tree in Figure 3.16.

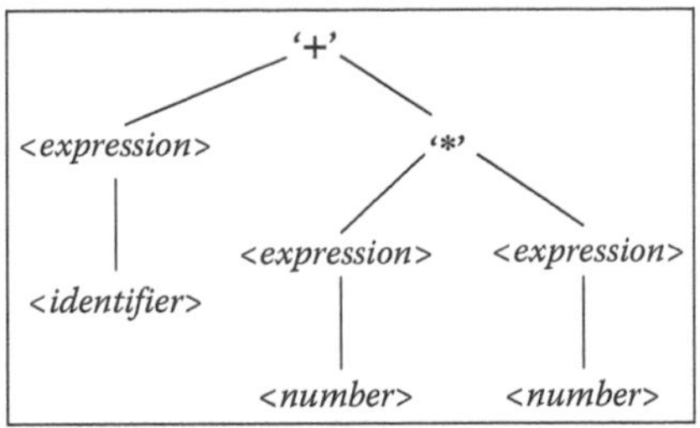

FIGURE 3.16 Abstract syntax tree for the expression "x + 3 * 4."

Abstract syntax trees are useful in understanding the language constructs in programming languages, as they hide unnecessary details of the concrete syntax trees. Abstract syntax trees are also used in syntax-directed code generation of programs: after parsing, a concrete syntax tree is reduced to the corresponding abstract syntax tree, and then the semantic analysis is done on the abstract syntax tree to generate an intermediate-level code. Abstract syntax trees have also been used in program analysis and understanding abstract properties of programs.

3.4.5 Automated Parsing

In the previous section, while generating parse trees, we assumed that proper subsequence that would match the right-hand side of a production rule would be automatically identified, and it would be reduced to the left-hand side nonterminal symbol of the matching production rule. While humans use their intelligence and other cues to identify such subsequences, computers need automated parsing software to identify unambiguously such subsequences that would match the right-hand side of the production rules.

The automated parsing techniques are classified under two major categories: *top-down* parsing and *bottom-up* parsing. In *top-down parsing*, also known as *recursive descent parsing*, the leftmost nonterminal symbol is expanded to the right-hand side of the production rule and matched with the corresponding part of the given sentence. The parsing engine *backtracks*—goes back to find out alternative solutions—and tries alternative rules in case the matching fails.

Backtracking is a computationally inefficient technique. In order to improve the efficiency and remove the need for backtracking, *terminal symbol lookahead* is used. *Predictive parsers* use a $M \times N$ size parsing table, where M is the number of nonterminal symbols and N is the number of terminal symbols (used for look-ahead) in the grammar. Each cell of the parsing table contains the reference to one or more production rules that show the definition. If the cell refers to a unique production rule, then the production rule is picked, and the look-ahead process is stopped. Otherwise, more symbols are looked ahead, and the corresponding cells are looked up in the table until only one production rule is possible.

Bottom-up parsing, also known as *shift-reduce parsing*, starts from the given sentence and works upward. The parser identifies the matching subsequence in the intermediate reduced form and reduces it to the left-hand side nonterminal symbol of the production rule. A general class of shift-reduce parsing *is LR(k) parsing*: *k* stands for the maximum number of look-ahead symbols needed to make an unambiguous parsing decision. LR(k) parsing constructs the parse tree in a reverse order, starting from the rightmost part of the sentence. There are many advantages of LR parsing: (1) it is nonrecursive, (2) it is nonbacktracking, (3) it can parse all the programming constructs, and (4) it can identify the ambiguities in the grammar. One specific subclass of LR(K) parser, called LALR (Lookahead LR parser), is popular for parsing programs in programming languages and can be generated using automated parser generators. A further detailed study of automated parsers is part of a course in compilers.

3.5 SEMANTICS

In programming languages, there are five major ways to define the semantics: *operational semantics, axiomatic semantics, denotational* semantics, *action semantics* and *behavioral semantics.* The following subsections explain each of the semantics.

3.5.1 Operational Semantics

Operational semantics, first described as part of ALGOL 68 and later formalized by Plotkin, is concerned about giving the meaning of a sentence by describing its effect on an abstract computation state in an abstract machine using the corresponding abstract instruction set. The meaning of a high-level control abstraction is modeled as the sequence of *small step abstract instructions* that potentially modifies the computational state to a new computational state. The computational states depend upon the underlying abstract machine that in turn is based upon programming paradigms and languages. The transitions in the computational states for different abstract instructions describe the operational semantics. The operational semantics that involve *small steps abstract instructions* is called *small-step operational semantics.* The operational semantics that reasons about transition of computation states involving high-level control and data abstractions, such as *for-loop, if-then-else* statement, or *while-loop,* is called *big-step operational semantics.*

In our case, to understand machine translation of high-level programs to low-level abstract instructions, we consider an abstract machine that is rooted in the von Neumann machine described in Section 2.1. The abstract machine can model the imperative programming paradigm. The computational state is a triple of the form (*environment, store, dump*), and a program is a sequence of *declarations, expressions*, and *commands.* A *declaration* alters the computational state by changing the environment, a *command* changes the computational state by changing the store, and an expression reads the store without changing the computational state. A call to a subprogram potentially changes both the environment and store. A statement can be *composite* and may change both environment and store. For example, a statement *integer* $x = 10$ uses both a declaration and an assignment statement: declaration *integer* x changes the environment, and assignment $x = 10$ changes the store.

As described in Section 2.4, a computational state is denoted by the Greek symbol σ, and a statement is denoted by the symbol S to model transition between computational states. The operational semantics of a statement is defined as $(S, \sigma_0) \rightarrow \sigma_{1.}$ A *composite statement* is of the form <*S*; *Ss*>, where the symbol *S* denotes the first statement, and

$$
\begin{array}{l}
(\text{no-op}, \sigma) \rightarrow \sigma \\
(\text{literal}, \sigma) \rightarrow \text{literal} \\
(\text{identifier}, \sigma) \rightarrow \text{r-value(identifier)} (\sigma) \text{ if } ((\text{identifier} \mapsto \text{l-value}) \in \sigma^E \text{ and } (\text{l-value} \mapsto \text{r-value}) \in \sigma^S \\
(\textbf{\textit{new}} \text{ identifier}, <\sigma^E, \sigma^S, \sigma^D>) \rightarrow <\sigma^E \oplus (\text{identifier} \mapsto \text{l-value}), \sigma^S \oplus (\text{l-value} \mapsto \text{undefined}), \sigma^D> \\
(\text{exp}_1 \text{ op exp}_2, \sigma) \rightarrow \text{value}_1 \text{ op value}_2 \text{ and } \sigma \text{ does not alter} \\
\qquad\qquad\qquad \text{where } (\text{exp}_1, \sigma) \rightarrow \text{value}_1 \text{ and } (\text{exp}_2, \sigma) \rightarrow \text{value}_2, \text{ and} \\
\qquad\qquad\qquad\qquad \text{op} \in \{\text{add, subtract, multiply, divide}\} \\
(\text{identifier} = \text{exp}, <\sigma^E, \sigma^S, \sigma^D>) \rightarrow <\sigma^E, \sigma^S \oplus (\text{l-value(identifier)} \mapsto \text{value}, \sigma^D> \\
\qquad\qquad\qquad \text{where } (\text{exp}, <\sigma^E, \sigma^S, \sigma^D>) \rightarrow \text{value}
\end{array}
$$

FIGURE 3.17 Operational semantics of simple abstract instructions.

the symbol *Ss* denotes the rest of the statements. In order to understand the meaning of composite statements, we have to understand the operational semantics of the sequence of statements. The operational semantics of the sequence of statements is given by (<*S*; *Ss*>, σ_0) → (*Ss*, σ_1) →* σ_{final}, where (*S*, σ_0) → σ_1. The symbol "→*" denotes the transitions equal to the number of statements in the remaining sequence of statements, and σ_{final} denotes the final computational state.

Figure 3.17 gives the operational semantics of some common small-step abstract instructions such as *no-op, evaluating a literal, looking up an identifier, updating the value of an identifier, evaluating a simple expression, evaluating a composite expression, declaring a new identifier*, and *assignment statement.* To explain the effect of a statement, we denote the computational state σ as a triple of the form $<\sigma^E, \sigma^S, \sigma^D>$, where σ^E denotes the environment—mappings of the set of identifiers to the set of l-values, σ^S denotes the store—the mappings of the set of l-values to the set of r-values, and σ^D denotes the dump—a sequence of the partial environment and store in the chain of the calling procedures that are shadowed (not visible) or archived during the execution of the called procedure.

The computational state of the abstract machine remains the same after the no-op instruction, evaluation of a literal, identifier lookup, and evaluation of an expression. When an identifier is encountered, its l-value is looked up in the environment σ^E, and the corresponding l-value is looked up in the store σ^S to retrieve the corresponding r-value.

Evaluation of a composite expression does not change the computational state. Two expressions are evaluated with respect to the current computational state σ, and the dyadic operator is applied on the two values to derive the result. The declaration of a new identifier <*ident*> changes the environment: a new binding (<*ident*> ↦ l-value) is added to the environment σ^E. In addition, a new binding (l-value ↦ undefined) is added to the corresponding store σ^S.

The assignment statement is a composite statement. The meaning of an assignment statement is described as follows: (1) evaluate the right-hand side expression using the current computational state σ_0 and (2) update the current store σ^S by changing the binding of the l-value of the identifier with the evaluated value of the right-hand side expression.

Since the meanings of small-step abstract instructions in an abstract machine are well understood, the composite meaning of the high-level abstractions becomes clear with respect to our understanding of the sequence of small-step abstract instructions that are equivalent to high-level abstract instructions. The high-level control abstractions are translated to a sequence of these small-step abstract instructions by representing high-level control abstractions as a control flow diagram and then translating the control flow diagram to a sequence of equivalent small-step abstract instructions that would give the same effect on the computational state derived by high-level control abstractions. The details of the translation of high-level control abstractions, such as if-then-else statements, while-loops, for-loops, procedure calls, and their combinations to low-level abstract instructions, is given in Chapter 5.

3.5.2 Axiomatic Semantics

Programs can be developed using mathematical logic such as predicate calculus independent of any underlying architecture or abstract machine. The meaning of a program is understood at a logical plane instead of a sequence of computational state transformations in an abstract machine based upon the von Neumann model. *Axiomatic semantics*, based upon predicate calculus first explained by C.A.R. Hoare, uses Boolean expressions to define a computational state at an instant of computation. The execution of a statement alters the computation state by changing the Boolean expression. The meaning of a statement is derived by the difference between *postcondition* and *precondition.* A *precondition* is the Boolean expression immediately before executing a statement, and a *postcondition* is the Boolean expression that will become true immediately after executing the statement. In the case of conflict in axioms derived by the statement due to the assertion of new condition as in assignment statement, the axiom in the precondition is dropped in favor of the newly derived axiom. Let us assume that the initial condition before the execution of a command is {**P**}, and the postcondition after executing the statement is {**Q**}. Then the notation to express the axiomatic semantics is {**P**} *S* {**Q**}, where *S* denotes the statement.

Example 3.16

For example, after the execution of an assignment statement $x = 4$, the Boolean condition $x == 4$ becomes *true.* If the previous condition $x == 10 \wedge Y == 9$ was *true* before executing the statement, then the execution of the assignment statement $x = 4$ substitutes the Boolean subexpression $x == 10$ with the Boolean subexpression $x == 4$, and the Boolean expression $x == 4 \wedge y == 9$ becomes *true.* The execution of a statement $y = 5$ generates the postcondition $x == 4 \wedge y == 5$ is *true.* The meaning of the statement $y = 5$ is equal to the difference of the postcondition $x == 4 \wedge y == 5$ is *true* and precondition $x == 4 \wedge y == 9$ is *true.*

A typical programming language has control abstractions such as *command, sequence of commands, conditional statements* such as *if-then-else statements, iterative constructs* such as *while-loop*, and *procedure calls.* In order to understand the meaning of a program and to derive final conditions using axiomatic semantics, we should be able to derive postconditions after executing these control abstractions. The effect of an assignment statement has already been illustrated in Example 3.15. In the case of a sequence of statements, the postcondition after executing the previous statement becomes the precondition for the current statement, and the postcondition after executing the current statement becomes the precondition for the next statement (declaration, command, or expression). In the following rule, the symbol **{P}** denotes the precondition, the symbol **{Q}** denotes the postcondition, the symbol *S* denotes the current statement, and the symbol *Ss* denotes the sequence of statements after executing the current statement. The rule says that after the execution of the statement *S*, the new precondition for the rest of the statements *Ss* is **{Q}**, where **{Q}** is the postcondition after executing the statement *S*, given that the precondition before executing the statement *S* was **{P}**.

```
{P} (S; Ss) → {Q} (Ss) where {P} S {Q}
```

The axiomatic semantics of control abstraction *if B then* S_1 *else* S_2 is given by the composite postcondition **{**Q_1**}** V **{**Q_2**}**. The postcondition **{**Q_1**}** is derived after executing the statement S_1 if the Boolean expression *B* is true in precondition **{P}**, and the postcondition **{**Q_2**}** is derived if the Boolean expression *B* is *false* in the precondition **{P}**. We can write the axiomatic semantics rule for if-then-else constructs as follows:

```
{P} if B then S₁ else S₂ {Q₁ ∨ Q₂} where {P} S₁ {Q₁} if B is true,
                                    or {P} S₂ {Q₂} if B is false
```

The disjunction (logical-OR) joins the then-part and the else-part. Either the postcondition realized by the then-part is true or the postcondition realized by the else-part is true due to the inherent mutual exclusion in the execution of the statements in the then-part and the else-part.

Example 3.17

Let us take a program code as follows:

```
x = 10; z = 4; if (x > 4) then value = x; else value = z;
```

Let us assume that the initial condition is **{P}**. After executing the statement $x = 10$, the postcondition becomes $\{P \wedge x == 10\}$ is *true*. This postcondition becomes the precondition for the next assignment statement, $z = 4$. After executing the statement $z = 4$, the postcondition becomes $\{P \wedge x == 10 \wedge z == 4\}$ is *true*, and this becomes the precondition for the if-then-else statement. The postcondition of the if-then-else statement is $\{P \wedge x == 10 \wedge z == 4 \wedge (((x > 4) \wedge (value == x)) \vee (x \leq 4) \wedge (value == z))))\}$.

The axiomatic semantics of the control abstraction "**while** *B S*" is tricky. The rule says that the precondition $\{I \wedge B\}$ before executing a while-loop is split into two parts: an invariant part **I** and the remaining Boolean expression **B**, and the postcondition $\{I \wedge \neg B\}$ contains the same invariant part **I** and the negation ¬ **B**. An *invariant condition* remains the same throughout the execution of the iteration. Identifying the invariant condition **I** is not straightforward. The axiomatic semantics of while-loop is given by the following:

```
{I ∧ B} while B S; {I ∧ ¬ B} where {I} S {I}% I is an invariant
 condition
```

The major advantages of axiomatic semantics are as follows:

1. It can be used to derive the final condition after the execution of a program without actually executing it. This property can facilitate, in a limited way, checking program correctness without executing the program. The idea is to derive the final condition F^C derived using axiomatic semantics, and compare it with the intended condition F^I. If the final derived condition $F^C \subseteq F^I$, then the program is *correct* (see Appendix VII), and if the final derived condition $F^C \subseteq F^I$ and $F^I \subseteq F^C$, then the program is both complete and correct. While the scheme works for smaller programs, it becomes computationally prohibitive for the bigger programs due to (1) the computational cost of matching the equivalent Boolean expressions, (2) the overhead of identifying that a condition expressed by a Boolean expression is subsumed by another Boolean expression, and (3) the overhead of identifying invariant conditions.

2. By knowing the intended final condition F^I, a program can be constructed stepwise by reasoning in backward manner: postconditions and the effect of program constructs are used to derive progressively the preconditions that are true. These preconditions become postconditions for the previous statements. By repeating this process of backward heuristic reasoning, a program can be constructed stepwise, as described in Chapter 4.

3.5.3 Denotational Semantics

Denotational semantics derives the meaning of a control or a data abstraction by mapping the syntax rule into a *semantic domain* using *mathematical functions* and *semantic algebra* in the semantic domain.

In order to derive the meaning of a sentence, the same parse tree constructed during syntax analysis is used with a difference: the semantic rules corresponding to syntactic rules are applied on the edges of the parse tree, and the meanings of the parts of the sentence are derived at the internal nodes. The complete meaning of the sentence is derived at the root of the parse tree. The meaning of a composite sentence is derived using functional composition of meanings of individual statements.

The difference between operational semantics and denotational semantics is that operational semantics describes the changes in a computational state as an effect of abstract instructions in an abstract machine, while denotational semantics uses abstract syntax trees and composition of semantics rules expressed as mathematical functions to derive the meaning of a sentence. There is no concept of computational states and abstract machines in denotational semantics.

There are some similarities in denotational semantics and operational semantics:

1. Denotational semantics also uses the environment and store like operational semantics does.

2. Both operational semantics and denotational semantics use abstractions in their definitions.

Example 3.18

Let us understand the notion of denotational semantics using a simple grammar that parses only unsigned integers in a base-10 domain. The grammar to accept base-10 unsigned integers is given below

```
Syntax rule # 1: <integer> ::= <integer><digit> | <digit>
Syntax rule # 2: <digit> ::= '0' | '1' | '2' | … | '9'
```

The semantic domain to derive the meaning for the whole number is integer base-10 denoted as $\mathbb{Z}_{10}$, where the subscript denotes base-10 domain. In order to define the meaning, we need multiplication and addition operations. Thus the semantic algebra for the semantic domain is given by *plus, times*: $\mathbb{Z}_{10} \times \mathbb{Z}_{10} \rightarrow \mathbb{Z}_{10}$ that says that given a input pair of integers, the result is also an integer in base-10 domain under the addition and multiplication operation. The symbol "×" denotes the Cartesian product. The semantic rules use a function 'ĭ' to map the syntax rule to the corresponding meaning in the semantic domain $\mathbb{Z}_{10}$.

```
Semantic Domain: ℤ10
Semantic Algebra: plus, times: ℤ10 × ℤ10 → ℤ10
Semantic rule # 1: ĭ (<integer>LHS) ::= ĭ (<integer>RHS) multiply
                   ten plus ĭ(<digit>) | ĭ (<digit>)
Semantic rule # 2: ĭ (<digit>) ::= zero | one | two | … |
                                   nine
```

There is an implicit understanding that we know the meanings of ten, hundred, and so on. Under this assumption, we can derive the meaning of any sequence of digits in the base-10 domain. Let us parse a sequence of three digits, 237, using the grammar, and then use the same parse tree to apply the corresponding semantic rules to give a meaning *two hundred and thirty seven* in the semantic domain $\mathbb{Z}_{10}$. The parse tree and the derivation of the meaning using the corresponding semantic rules are illustrated in Figure 3.18.

As shown in the parse tree in Figure 3.18a, each of the symbols is parsed to *<digit>* using syntax rule 2. By using syntax rule 1b, we can make the leftmost *<digit>* as *<integer>*, and then use syntax rule 1a twice to join the other two digits in the definition of *<integer>*. At the root of the parse tree is the start symbol *<integer>*.

The Figure 3.18b shows the application of the corresponding semantic rules on the parse tree to give the meaning of a subtree at the internal nodes and the meaning of the sentence at the root node. The semantic rule 2 gives the meaning *two, three*, and *seven* to the symbols '2,' '3,' and '7', respectively. Applying the semantic rule 1b transforms the digit to integer value *two*. Application of the semantic rule 1a derives the meaning *two times ten plus three = twenty three.* Application of the semantic rule 1a again derives the meaning *twenty three times ten plus seven = two hundred and thirty seven.*

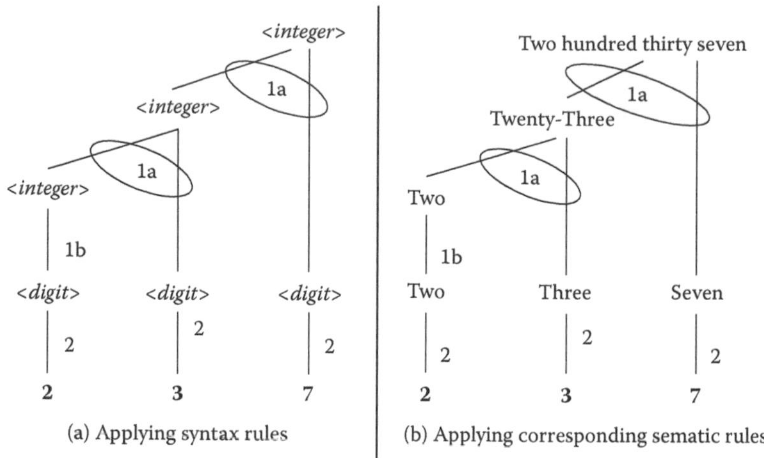

FIGURE 3.18 Deriving meaning by applying semantic rules on a parse tree.

Example 3.19

Let us understand denotational semantics using an example of defining numbers in programming languages. The number domain is a composite domain consisting of a disjoint union of an integer domain and real number domain. However, integer domain can be coerced to real domain when an arithmetic operation involves both integer and real number. The syntax rules for numbers are given in Figure 3.19. The syntax rules are easy to read, except for one definition: the definitions of both the nonterminal symbols <*whole-number*> and <*float*> generate a sequence of <*digits*>. However, they have been written differently for better semantics interpretation: while <*whole-number*> generates the sequence of digits using a left-recursive definition, the definition of <*float*> uses a tail-recursive definition. Although the end product is the same, the way meaning associated with the two syntax rules is different. The meaning of the nonterminal symbol <*whole-number*> when shifted left is *multiply by ten*, and the meaning of the nonterminal symbol <*float*> when shifted right is *divided by floating point value* 10.

The meaning of <*whole-number*> is different in the integer domain and real domain: in the integer domain, it is treated as an integer; while in the real domain, the value is *coerced* to

Rule 1: <*number*> ::= <*integer*> | <*real*>
Rule 2: <*integer*> ::= <*sign*> <*whole-number*> | <*whole-number*>
Rule 3: <*real*> ::= <*sign*> <*unsigned-real*> | <*unsigned-real*>
Rule 4: <*unsigned-real*> ::= <*whole-number*> '.' <*float*> |
<*whole-number*> '.' <*float*> 'E' <*integer*>
Rule 5: <*whole-number*> ::= <*whole-number*><*digit*>|<*digit*>
Rule 6: <*float*> ::= <*digit*><*float*> | <*digit*>
Rule 7: <*digit*> ::= '0' | '1' | '2' | ... | '9'
Rule 8: <*sign*> ::= ' + ' | '–'

FIGURE 3.19 Syntax rules for validating numbers.

the corresponding number in the real domain. The semantic domain for the interpretation rules is the *number domain in base* 10, denoted by $\mathbb{N}$. Note that number domain $\mathbb{N}$ is a disjoint union of integer domain $\mathbb{Z}$ and real number domain $\mathbb{R}$. However, every element $z \in \mathbb{Z}$ has a unique image $r \in \mathbb{R}$ that has the same value. This mapping from the integer domain to the real domain is an example of *coercion*—type transformation without any loss of information, which is discussed in Chapter 7. The semantic algebra involves *integer-add* ('$+^{I}$'), *integer-multiply* ('$\times^{I}$'), *real-add* ('$+^{R}$'), *real-multiply* ('$\times^{R}$'), *real-divide* ('$/^{R}$'), and *exponentiation* ('^'). The operations *addition* and *multiplication* are overloaded: an addition becomes an integer addition in the integer domain and a floating point addition in the real number domain. Similarly, the operation multiplication becomes integer multiplication in the integer domain and floating-point multiplication in the real number domain.

The semantic rules are given in Figure 3.20 in the same order as the syntactic rules. The function ň denotes the interpretation of the syntactic rule in the number domain $\mathbb{N}$, the function ř denotes the interpretation of the syntactic rule in the real number domain $\mathbb{R}$, and the function ǐ denotes the interpretation in the integer domain '.

Let us now understand each semantic rule. Rule 1 states that the meaning of a number denoted by ň(<*number*>) is equal to the meaning of the real number denoted by ř(<*real*>) or the meaning of the integer denoted by ǐ(<*integer*>). Rule 2 states that the meaning of the nonterminal symbol <*integer*> is the same as the meaning of the nonterminal symbol <*sign*> in the integer domain multiplied by the meaning of the nonterminal symbol <*unsigned-integer*> or the meaning of the nonterminal symbol <*unsigned-integer*>. Rule 3 states that the meaning of the nonterminal symbol <*real*> is derived by multiplying the meaning of the nonterminal symbol <*sign*> in the real number domain with the meaning of nonterminal symbol <*unsigned-real*> or is the meaning of the nonterminal symbol <*unsigned-real*>.

Semantic Domains: integer domain $\mathbb{Z}_{10}$
real number domain $\mathbb{R}_{10}$
number domain: $\mathbb{N}_{10} = \mathbb{Z}_{10} \uplus \mathbb{R}_{10}$ % $\uplus$ denotes disjoint-union
Semantic algebra in $\mathbb{R}_{10}$: real-add '$+^{R}$'; real-multiply '$\times^{R}$' : $\mathbb{R}_{10} \times \mathbb{R}_{10} \rightarrow \mathbb{R}_{10}$
Semantic algebra in $\mathbb{Z}_{10}$: int-add '$+^{I}$'; int-multiply '$\times^{I}$' : $\mathbb{Z}_{10} \times \mathbb{Z}_{10} \rightarrow \mathbb{Z}_{10}$
Semantic algebra in mixed domain: exponent ∧': $(\mathbb{R}_{10} \times \mathbb{Z}_{10}) \rightarrow \mathbb{R}_{10}$
Semantic functions: ň; ř; ǐ;

Rule 1: ň(<*number*>) ::= ǐ(<*integer*>) | ř(<*real*>)
Rule 2: ǐ(<*integer*>) ::= ǐ(<*sign*>) $\times^{I}$ ǐ (<*whole-number*>) | ǐ(<*whole-number*>)
Rule 3: ř (<*real*>) ::= ř(<*sign*>) $\times^{R}$ ř(<*unsigned-real*>) | ř(<*unsigned-real*>)
Rule 4: ř (<*unsigned-real*>) ::= ř(<*whole-part*>) $+^{R}$ ř(<*decimal-part*>) |
(ř(<*whole-part*>) $+^{R}$ ř(<*decimal-part*>)) ×(10.0 ∧ ǐ(<*integer*>))
Rule 5a: ř(<*whole-part*>1) ::= ř(<*whole-part*>2) $\times^{I}$ 10.0 $+^{R}$ ř(<*digit*>) | ř(<*digit*>)
Rule 5b: ǐ(<*whole-number*>1) ::= ǐ (<*whole-number*>2) $\times^{I}$ ten $+^{I}$ ǐ(<*digit*>) | ǐ (<*digit*>)
Rule 6: ř(<*decimal-part*>1) :: ř(<*digit*>) $+^{R}$ ř(<*decimal-part*>2)) / 10.0 | ř(<*digit*>) / 10.0
Rule 7a: ř(<*digit*>) ::= **float zero** | **float one** | ... | **float nine** % float 1 is 1.0
Rule 7b: ǐ(<*digit*>) ::= **zero** | **one** | **two** | ... | **nine** % interpretation of digits
Rule 8a: ř(<*sign*>) ::= **plus float-one** | **minus float-one** % + 1.0 or –1.0
Rule 8b: ǐ(<*sign*>) ::= **plus one** | **minus one**

FIGURE 3.20 Semantic rules corresponding to the syntax rules.

Rule 4 says that the meaning of the nonterminal symbol *<unsigned-real>* is the sum of the meanings of the nonterminal symbol *<whole-part>* and the nonterminal symbol *<decimal-part>*. Alternately, Rule 4 states that the meaning of the *<unsigned-real>* is the sum of the meanings of the nonterminal symbol *<whole-part>* and the *<decimal-part>* multiplied with the meaning of the exponent part. The meaning of the exponent part is the real-domain image of the 10 raised to the power of the meaning of the integer value in the exponent part.

Rule 5 has different interpretations in the real number domain and in the integer domain. Rule 5a states that the meaning of the nonterminal symbol *<whole-part>* on the left-hand side of the production rule is the sum of the (1) multiplication of the floating point 10 with the meaning of the nonterminal symbol *<whole-part>* on the right-hand side of the production rule and (2) the meaning of the nonterminal symbol *<digit>* in the real number domain derived by rule 7a. Rule 5b states that the meaning of the left-hand side of the nonterminal symbol *<whole-number>* is the sum of (1) the multiplication of integer 10 with the meaning of the right-hand side occurrence of the nonterminal symbol *<whole-number>* in integer domain and (2) meaning of the nonterminal symbol *<digit>* in the integer domain derived by the rule 7b.

Rule 6 says that the meaning of the nonterminal symbol *<decimal-part>* is the real-domain image of the sum of the meanings of the nonterminal symbol *<digit>* on the right-hand side and the nonterminal symbol *<float>* on the right-hand side divided by floating point 10. Rules 4 and 6 are different, despite generating a sequence of digits. The meanings associated with both the rules are different. In the definition of the nonterminal symbol *<whole-number>* in rule 5, moving the *<whole-number>* to the left is equivalent to the multiplication of the meaning of nonterminal symbol *<whole-number>* by integer 10, while in the definition of *<float>*in rule 6, moving the *<float>* to the right is equivalent to division of the meaning of *<float>* by float 10. Besides, two meanings are interpreted in different domains: the nonterminal symbol *<whole-number>* is interpreted in the semantic domain $\mathbb{Z}_{10}$, and the nonterminal symbol *<float>* is defined in the semantic domain $\mathbb{R}_{10}$.

Rule 7 has two interpretations: one in the real number domain and one in the integer domain. Rule 7a states that the meaning of the nonterminal symbol *<digit>* is floating point zero, floating point one, floating point two, floating point three, and so on. Rule 7b states that the meaning of the nonterminal symbol *<digit>* could be integer one, two, three, and so on.

Rule 8 says that the meaning of the sign '+' is *plus one*, and meaning of the sign '–' is *minus one* as we understand. The image of *minus one* in the semantic domain $\mathbb{R}_{10}$ is *floating minus one*, and the image of *plus one* in the semantic domain $\mathbb{R}_{10}$ is *floating plus one.*

In order to understand the denotational semantics of control and data abstractions, the semantic domains should be clearly defined. In addition, the notion of error is handled in functional definition of denotational semantics by including a bottom element, denoted by the symbol ⊥, in the semantic codomain. The inclusion of the bottom symbol transforms a semantic codomain to the corresponding *lifted semantic domain.* If the application of a semantic function does not find an image in the *lifted semantic domain*, then a function maps to the bottom symbol $\perp$. The inclusion of the bottom symbol $\perp$ allows handling of programmatic error conditions in function-based meaning in the denotational semantics.

Some of the semantic domains are as follows: (1) *set of basic values denoted by the symbol B*, (2) set of identifiers denoted by the symbol *Ide*, (3) set of l-values denoted by the symbol *L*, (4) set of r-values denoted by the symbol *R*, (5) set of storable values denoted by the disjoint union $B \uplus L$, (6) store denoted by the symbol *S* which is equal to $L \mapsto R \uplus$ *unused locations*, (7) set of procedures denoted by the symbol *P*, (8) set of denotable values (l-values and storable values) denoted by the symbol *D* which is equal *to* $L \uplus R \uplus$ *{undefined}*, (9) environment denoted by the mapping $Ide \mapsto D$, (10) set of expression results denoted by the lifted domain $R^{\perp}$, (11) set of command results denoted by the lifted domain $S^{\perp}$, and (12) the set of returned values from a program denoted by the lifted domain $B^{\perp}$.

Traditional denotational semantics suffers from many problems: (1) semantic domains change as new control and data abstractions are added in programming languages; (2) the semantics of iterative-loop is explained in terms of complex mathematical concepts such as *fix-point semantics* despite programmers having a good intuitive understanding of the control abstraction without any need for such mathematical concepts; (3) the addition of "goto" commands necessitates the notion of *continuation*, the sequence of commands actually executed at runtime following the current command; and (4) complex notations which are difficult to comprehend for programmers and language designers.

The use of functions to define the meaning in denotational semantics makes them suitable for functional programming paradigm, and not for other programming paradigms such as object-oriented programming paradigm, imperative programming paradigm, and event-based programming paradigms which are better represented by semantics based upon computational state transformations.

3.5.4 Action Semantics

Semantics models are used to explain the meaning of the abstractions to program developers, language designers, and compiler developers. Individually, the three major semantics models (operational, axiomatic, and denotational) are restricted, and make the semantic description of programming language abstractions difficult to comprehend. However, a pragmatic combination of the three semantics gives a better picture to the language designers and compiler developers.

Action semantics integrates the advantages of denotational semantics, axiomatic semantics, and operational semantics to explain the meaning of control and data abstractions using rules in natural English language. These rules are called *actions*. The major advantage of action semantics is that it defines the meanings of realistic programming language abstractions in a comprehensible way.

Action semantics borrows the idea of context-free grammars to derive abstract syntax trees, and then uses semantic equations from denotational semantics. There are three major components of action semantics: *action, data*, and *yielders*. The *actions* represent computational entities responsible for the stepwise specification of the dynamic program behavior. The information processed by actions is called *data*. The *yielders* retrieve the information but do not process them. There are two types of actions: *primitive* and *combinatory*. A *primitive action* is responsible for the single step of information processing, and *combinators* take a composite action. An action can be *completing, diverging*, or *fail*, resulting in exception handling.

Specification of action semantics consists of (1) specification of nodes to be constructed in the abstract syntax tree using the notation [[...]] and (2) grouping the components together such as sequence of statements. For each syntactic rule, the corresponding action semantics is given in simple English instead of complex functional notations. For example, the semantic rule corresponding to the syntax rule *<identifier>* '=' *<expression>* would be as follows:

```
execute [[<identifier> '=' <expression>]]
action: evaluate <expression> giving value then store the
  resulting value in the l-value(<identifier>)
Similarly, the action semantics of the if-then-else statement is
  given by
execute [['if' <expression> 'then' <statement>1 'else'
  <statement>2]]
action: evaluate <expression> giving truth value B then
          ((check truth-value of B and execute <statement>1)
          (check not (truth-value of B) and execute
          <statement>2))
```

The overall format of the action semantics contains the declaration of modules, declaration of semantic variables representing the semantic domains, and action rules for each abstract syntax rule. *Yielders* are used whenever data is retrieved after some action.

3.5.5 Other Models of Semantics

There are other behavioral models of semantics to provide meaning to *domain-specific modeling languages* and *object-oriented languages*. As software development becomes bigger and more complex, more than one domain is needed for the development of a software, and each domain may have a *domain-specific language* (DSL). The semantics of a DSL can be modeled as the state transition of a group of active objects in one state to another state using the objects' reaction to messages as means of transition from the current state to the future state. The state transition diagram is expressed as the behavioral semantics of the language. Each object is modeled by a tuple of the form (*set of attributes, methods, the messages that it can receive, the messages it can emit, the transitions that occur in response to a message*, and *the triggers that start the reaction in response to a message*). A *system state* is modeled as a collection of the state of the objects. A message from one object to another object may trigger a new transition in the receiving object if the trigger condition is satisfied, and changes the state of the receiving object, thus changing the overall computational state. The changed object emits new messages that can cause a change in state transitions in other objects, resulting in the state change of the overall system.

3.6 SUMMARY

This chapter describes two major components of programming languages: *syntax* and *semantics*. Both syntax and semantics are necessary to understand the constructs, their meanings, and their translation to low-level intermediate code. *Syntax* is concerned about the validity of the sentence structure, and *semantics* is concerned about deriving meaning of the statements in programming languages. The process of sentence analysis for structure validity consists of two phases: *lexical analysis* and *parsing*. Lexical analysis is concerned about internalization of the input stream of symbols by generating tokens. Parsing is concerned about validating the structure of tokenized sentences.

A grammar is a 4-tuple of the form (*start-symbol, set of production rules, set of nonterminal symbols, and set of terminal symbols*). *Set of terminal symbols* form the *alphabet* of a programming languages. The set of *nonterminal symbols* is used in the grammar to generate intermediate forms of sentences during parsing. Automated parsers can be *top–down* or *bottom–up*. Automated parsers use transition tables and symbol look-ahead to resolve the ambiguity in the application of appropriate production rules. Automated parsing repeatedly substitutes a part of the intermediate form of a sentence by the corresponding left-hand side nonterminal symbol of the production rule in the grammar with the matching right-hand side. The process of parsing keeps simplifying the intermediate form until only the start symbol remains.

Production rules of a grammar are represented textually either as BNF or EBNF, or visually using *syntax diagrams*. BNF uses tail-recursive rules to model repeated occurrence of a group of symbols. EBNF removes the complexity of tail-recursive definitions by using curly brackets and special symbols, such as '*' or '+'. EBNF also reduces redundant multiple definition by grouping the multiple definitions using parentheses, and uses square brackets to model optional definitions.

Syntax diagrams are pictorial and are comprehended better by humans. Syntax diagrams can be transformed to syntax rules by breaking up the parts of syntax diagrams using (1) simple correspondence of tail-recursive definitions by a feedback loop, (2) multiple definitions by parallel edges, and (3) the occurrences of concatenated multiple symbols by multiple adjacent nodes in the same path.

There are two types of rules: *concrete syntax rules* and *abstract syntax rules. Concrete syntax rules* contain the full set of nonterminal symbols. *Abstract syntax rules* use only control and data abstractions as nonterminal symbols, and avoid low-level nonterminal symbols. The parse tree generated by abstract syntax rules is called an *abstract syntax tree* and can be generated from a *concrete parse tree* by removing redundant nonterminal symbols and keeping terminal symbols and the control and data abstraction symbols. Production rules are also associated with various attributes to help the translation of parsed sentences to intermediate-level codes. The attributes in the attribute grammar state the constraints imposed by the architecture and language designers to handle efficiency and pragmatic issues. Attribute grammar also contains the semantic rules to generate low-level code from the parse tree.

Semantics can be divided into five major types: *operational, axiomatic, denotational, action*, and *behavioral. Operational semantics* models the control and data abstractions as transformation from one computational state to another computational state in an abstract computational machine. *Small-step operational semantics* describes how the state of the underlying abstract machine is altered when a low-level abstract instruction is executed. *Large-step operational semantics* describes how the state of the underlying abstract machine is altered by high-level data and control abstractions. Axiomatic semantics uses Boolean expressions as preconditions and postconditions of a sentence and gives the meaning of the sentence as the difference between postconditions and preconditions. Preconditions and postconditions are expressed as Boolean expressions. The denotational semantics provides the meaning of abstract syntax rules in a variety of semantic domains using mathematical functions. Both axiomatic and denotational semantics are free from any underlying abstract computational machine. However, both axiomatic and denotational semantics are quite complex for full-scale programming languages for various reasons. Axiomatic semantics finds it difficult to model invariant conditions in loops and to match Boolean expressions. The semantic domains in denotational semantics changes when new, high-level constructs are introduced; the use of "goto" statements forces the use of *continuation-based semantics*, and uses complex fixed-point semantics to explain the meaning of the while-loop that is intuitively clear to the programmers. Neither denotational semantics nor axiomatic semantics can explain event-based programming constructs and object-oriented programming.

Operational semantics, along with the control flow diagrams, can be used to translate high-level control abstractions to equivalent low-level abstract instructions, so that the resulting effect on the computational state is the same. Axiomatic semantics has been used in a limited way to reason about program correctness and the scientific development of programs using predicate transformation and progressive backward reasoning. Denotational semantics has been used for syntax-directed translation and execution of the programs.

Semantics of language constructs is necessary for the programmers, language designers, and compiler designers for the code generation part of the compilers. In order to facilitate comprehensible semantics, *action semantics* has been developed. *Action semantics* has three components: *actions, data*, and *yielders*. Actions work on data to transform the current state, and yielders yield the data from the current state. Actions are written in English for better comprehension.

To handle domain-specific and object-oriented languages, behavioral semantics based upon the state transition model is used. The computational state is an aggregate of the state of the individual objects, and the behavior semantics is modeled as a transition from one state to another state using reaction to the incoming messages. The model is independent of any underlying implementation.

3.7 ASSESSMENT

3.7.1 Concepts and Definitions

Abstract machine; abstract syntax; abstract syntax tree; action semantics; alphabet; ambiguous grammar; attribute grammar; automated parsing; axiomatic semantics; behavioral semantics; BNF; bottom symbol; bottom–up parsing; context-free grammar; context-sensitive grammar; denotational semantics; extended BNF; grammar; hyper rules; lexical analysis; lifted domain; meta-definitions; nonterminal symbols; operational semantics; parsing; parse tree; precondition; postcondition; production rules; program correctness; regular grammar; semantic algebra; semantic domain; semantic rule; sentence; start symbol; syntax; syntax diagram; syntax rule; top–down parsing.

3.7.2 Problem Solving

1. Give a finite state automata that accepts the reserved words 'int', 'float', 'char', 'Bool', and ';' in addition to numbers and identifiers assuming that the reserved words, numbers, and identifiers can be separated by a single space, and all other spaces after the first space need to be trimmed.

2. Write a simple BNF representation of a grammar that accepts an octal number. Convert the grammar to extended BNF.

3. Make a syntax diagram for the following syntax rules:

```
<block> ::= '{' <decl>; <statements> '}'
<statements> ::= <statement> ';'<statements> |
                <statement> '; '| ∈
<statement> ::= <cond-statement> | <while-loop>|
               <assignment> |
               <for-loop> | <procedure-call> | return
                 (<expression>)
<while-Loop> ::= while <Boolean> do <statements>
<cond-statement> ::= if <Boolean> then <statement> else
                    <statement>
<assignment> ::= <identifier> = <expression>
<Boolean> ::= <Boolean> <l-op> <Boolean> | <predicate> |
             true | false
<l-op> ::= '&&' | '||' | 'not'
<predicate> ::= <identifier> <c-op> <identifier>
<c-op> ::= '==' | '>' | '>' | '>=' | '=<'
<expression> ::= <expression> <a-op> <expression>
<a-op> ::= '+' | '-' | '*' | '/'
<decl> ::= <type> <identifier-seq>; <decl> | ∈
<type> ::= int | float | Bool | string
<identifier-seq> ::= <identifier> ','<identifier-seq> |
                    <identifier>
```

4. Write a simple BNF for while-loop including the Boolean expression part and the statement part, and draw the corresponding syntax diagram.

5. Study the parameter passing in C++, and give the corresponding syntax rules. Convert the syntax rules to syntax diagrams, and optimize the syntax diagram.

6. Write a simple unambiguous grammar in EBNF for an if-then-else construct, including a complete Boolean expression in C++ or Java, and draw the corresponding syntax diagram.

7. Make a parse tree for the floating point number 23. 416 using the syntax grammar given in Figure 3.14, and use the semantic rules given in Figure 3.15 to derive the meaning of the number.

8. Give the preconditions and the postconditions for the following code-fragment, starting from a state "undefined."

```
x = 4; y = 6; z = 7;
if (x > y) then max = x else max = y;
```

9. Assuming an environment $\sigma^E = [x \mapsto 1, y \mapsto 2, z \mapsto 3]$, and the store $\sigma^S = [1 \mapsto 43, 2 \mapsto 5, 3 \mapsto 10]$, and dump $\sigma^D = []$, and assuming that the computational state is modeled as the triple $(\sigma^E, \sigma^S, \sigma^D)$, give the computation state after the declaration "int w = 5;" followed by the command "x = x + 4; y = z;".

10. Write a grammar to accept a Hex number, and write the corresponding denotational semantics in integer domain. Define the domain, the semantics algebra, and the evaluation function clearly.

3.7.3 Extended Response

11. Compare three types of grammars: regular grammar, context-free grammar, and context-sensitive grammar.

12. Why is regular grammar preferred over other types of grammar for lexical analysis? Explain.

13. What is the role of the symbol table during lexical analysis? Explain using a simple example.

14. What is the difference between abstract syntax and concrete syntax? Explain by using a simple but illustrative example.

15. Discuss the potential source of ambiguity in the if-then-else statement, and how it can be solved.

16. What is the difference between operational semantics and denotational semantics? Explain.

17. Why is logical-OR introduced in axiomatic semantics? Explain by using a simple example.

18. Compare action semantics with other types of semantics, and state the advantages of action semantics.

19. What is the need of behavioral semantics in programming languages? Explain.

FURTHER READING

Aho, Alfred V., Lam, Monica S., Sethi, Ravi, and Ullman, Jeffrey D. *Compilers: Principles, Techniques, and Tools,* 2nd edition. Addison Wesley. 2007.

Gries, David. *The Science of Programming*. Springer Verlag. 1981.

Hoare, Charles A. R. and Shepherdson, John C. *Mathematical Logic and Programming Languages.* Prentice Hall. 1985.

Liang, Sheng and Hudak, Paul. "Modular denotational semantics for compiler construction." In *6th European Symposium on Programming Languages, LNCS 1058*. Springer Verlag. 1996. 219–234.

Meyer, Bertrand. *Introduction to the Theory of Programming Languages*. Prentice Hall. 1985. 447.

Mosses, Peter D. *Theory and Practice of Action Semantics.* BRICS Report Series RS-96-53. Department of Computer Science, University of Aarhus. 1996.

Plotkin, Gordon D. *A Structural Approach to Operational Semantics.* Technical report. Department of Computer Science. Aarhus University, Denmark. 1981.

Poetzsch-Heffterand, Arnd and Schafer Jan. "A representation-independent behavioral semantics for object-oriented components." In *Proceedings of the International Conference on Formal Methods in Object-Oriented Based Distributed Systems, LNCS 4468*, Springer Verlag, 2007, 157–173.

Rivera, Jose E. and Vallecio, Antonio. "Adding behavioral semantics to models." In *Proceedings of the 11th International Enterprise Distributed Object Computing Conference.* 2007. 169–180.

Schmidt, David A. *Denotational Semantics—A Methodology for Language Development.* Allyn and Bacon. 1986.

CHAPTER 4

Abstractions in Programs and Information Exchange

BACKGROUND CONCEPTS

Abstract syntax (Section 3.2.6); Abstract concepts in computations (Section 2.4); Grammars (Section 3.1); Control flow (Section 1.4.2); Data structure concepts (Section 2.3); Discrete structures (Section 2.2); Graphs (Section 2.3.6); Programming background; Trees (Section 2.3.5); Syntax diagrams (Section 3.3).

A program is about the manipulation of structured data to achieve an intended computational state that satisfies the final intended condition. Programming can be done at multiple levels: machine-level programming, assembly-level programming, low-level procedural programming, high-level procedural programming, declarative programming, and so on. The amount of abstraction and the level of explicit control separate high-level programming languages from low-level programming languages. One of the goals of abstractions is to make the software reusable with minimal change, so that software evolution has minimal overhead as the needs evolve or technology changes.

There are two types of abstractions as described earlier: *data abstractions* and *control abstractions. Data abstractions* are used to model real-world entities using a required subset of their attributes, and *control abstractions* are used to structure the sequence of instructions for better comprehension and maintenance. A data entity can be characterized by a set of attributes and can be abstracted by a relevant subset of those attributes needed to solve a problem at hand.

Data entities are generally expressed in the declaration part of a program, and the control part is expressed in the body of the program. As described earlier, the declaration part of the program modifies the program environment, and the control part of the program that involves assignment statement modifies the program store.

Depending upon a programming language, the distinction between data and control can be well defined or become fuzzy as follows:

1. The program unit itself can be treated as data, and a program could be created dynamically at run time as data and later converted into a program unit.
2. A program may treat another program as data, as in meta-programs.
3. A class may encapsulate both data and control abstractions together into one package and regulate the interaction with other computational objects by declaring the encapsulated entities public, protected, or private.

As we have discussed earlier, a program consists of multiple subprograms or modules that exchange information with each other. Programmer-regulated information hiding by encapsulation is also part of abstraction. Encapsulation provides a natural boundary for information hiding. Encapsulation can be provided by subprograms, objects and classes, modules, or a combination of these. Classes provide information hiding by encapsulating both data abstractions and control abstractions. The visibility of these control and data abstractions can be explicitly regulated using *import–export mechanism* or using *inheritance. Export–import mechanism* consists of two components: selectively making encapsulated data or program units visible to other modules or program units using *export mechanism*, and pulling the data or program unit within a module or program unit using an *import mechanism.* The export mechanism enhances the scope from private to public, and the import mechanism allows the visibility of an exported entity in the local environment. Import and export are explicitly declared by the programmer.

In imperative programming, the basic unit that contains data abstractions and control abstractions is a subprogram or a module, while in the object-oriented programming paradigm, the basic unit is a subclass or an object—an active instance of a class or subclass. In module-based languages, a subprogram or data entity may be encapsulated within a module, and the usage of data entity or subprograms may be regulated *using the export–import mechanism.*

In this chapter, we study the data and control abstractions, information exchange between program units, and different abstract programming constructs. However, some of the concrete examples related to programming paradigms other than the imperative programming paradigm have been deferred to the corresponding chapters described later after the paradigms have been discussed in detail.

4.1 DATA ABSTRACTIONS

One of the important concepts in data abstraction is to create an abstract data entity that can be used to hold information to be processed, can be used repeatedly at multiple locations in the program module, and can be easily modified. The definition could be *specific* or *generic* that could be instantiated at run time to specific data abstractions through a parameter passing mechanism or redefinition using an override mechanism.

The data entity could be a *single entity* such as a "literal" or an "atom" that cannot be divided further in general, or the entity may represent an *aggregation* of entities having some set of common properties on which common, well-defined operations are performed. The set of common properties should not be confused with having the same type of data objects. The only restriction is some common property that allows a repeated common operation on different elements of the aggregate entity. If they are of the same data types such as integers, then the common operation is related to the operation on integer types. However, many times the operations are *structure related* and need not involve specific data type. This distinction will become clear when we study theory of types in Chapter 7.

A data entity has a scope where it is visible, and computations are performed within the scope. To ascertain visibility, a programming language has to define program units with clear boundaries within which a data entity is visible. There is a need to set up the visibility boundary of data entities if we want to allow multiple data entities with the same name within a program. Otherwise, there would be naming conflict, as the same name may refer to different data entities. There are two approaches to access a data entity within a program-unit: (1) by giving a name to the data entity and (2) by defining the extent of the position where it would be visible from the declared position. The use of name facilitates accessing a data entity at multiple locations of a program unit, and the extent of position gives the boundary of visibility of the data entity.

A data entity may be referred by multiple names in different parts of a program. Multiple names referring to the same data entity are called *aliases* and will be described in detail later. When we declare a data entity, there is a need for generic structuring template, as many data entities may conform to a similar data template. These structures are called *data abstractions*.

For a subprogram to work generically on an indefinite number of data entities confirming to some well-defined properties, we need the concept of *constants* and *variables*. A constant is a value holder that can be mapped to a data entity at compile time and does not alter afterward. A variable maps to a memory location that can be rewritten one or more times dynamically at run time. A variable can be created anonymously by the compiler or created at run time by language's implementation engine or named by the programmer. The mapping of the variable to a data entity could be altered repeatedly at run time, as in a destructive assignment in the imperative programming paradigm, or it could be fixed after the value is assigned once, as in the declarative programming paradigm. The imperative programming paradigm allows a programmer to change the mapping at will by introducing the concept of memory location and store. Owing to the introduction of memory locations such that the identifier $\mapsto$ memory location and memory location $\mapsto$ data entity, the variables in imperative programming paradigm are *mutable* and are referred as *mutable objects*. In contrast, declarative languages allow a programmer to map an identifier to a data entity once: identifier $\mapsto$ data entity. After the value is assigned, a programmer needs to create a new variable to assign new values. Assign-once objects are called *immutable objects*.

In modern programming languages, certain classes of data abstractions are common. All modern programming languages support *single entities* and *aggregation*. Aggregation consists of *composite types*, *collections*, and *extensible data structures*. Multi-paradigm languages that support both declarative and imperative programming distinguish and support both *mutable* and *immutable* objects.

4.1.1 Single Data Entities

Abstraction to define data entity is related to the need of modeling entities present in the real world and their interconnection to other entities. A real-world object has multiple attributes, and these attributes themselves are divided further. These attributes can be represented using basic mathematical domains such as integers, floating point numbers, characters, and strings, or computational domains such as bits, bytes, and semaphores, or user-defined enumerable domains. Different problem domains have different types of entities. Most of the languages treat strings as an indexible sequence of characters, while some languages such as Java and C# also allow "string" as a basic declaration that can be treated as sequence of character if needed.

4.1.2 Composite Data Entitics

A composite object has more than one attribute. In a composite entity, each field itself could be a composite data entity. Tuples are used to model composite object. Tuples can be named or unnamed. Most of the programming languages use named tuples such as record or struct to model a composite data entity. The only difference between tuples and records or structs is that records are named templates of tuples. The advantage of named templates of tuples in type declaration is that they can be used to generate multiple tuples bound to different variables for ease of programming. The use of name is that the template can be reused for defining multiple complex user-defined types. Irrespective of syntactic sugar used in different programming languages, all definitions of composite objects represent a tuple. A recursive definition of tuple is used to model an extensible data entity. For example, a linked list is modeled as a pair (info-field, linked list); and a tree can be modeled as a triple (info-field, left-subtree, right-subtree).

Example 4.1

An example of a composite entity is the definition of "class" given in Figure 4.1. Class is a 6-tuple of the form (*course number, course name, instructor, students, location, time*). The attributes *course number, course name,* and *instructor* are single entities; *location* and *time* are pairs; and the attribute *students* is a collection of data entities.

As shown in the definition of "class," a composite data entity consists of name of the data entity, number of attributes, and the description of various attributes. Each attribute can be a tuple comprising data abstractions or a collection of data abstractions or a single entity. For example, *location* is a data abstraction that is modeled as a pair. Similarly, *students* is a collection of *students.*

Student is 4-tuple (quadruple) of the form (*student_id, student_name, department-name, years-in-college*). Each of the fields is a single entity. Location has been abstracted as a pair of the form (*building, room*). Time has been abstracted as a pair of the form (*start time, duration*).

```
data abstraction: class
    abstraction-type: tuple
    attribute-size: 6
    begin attribute-description
         attribute1: single-entity course-number
         attribute2: single-entity course-name
         attribute3: single-entity instructor
         attribute4: bag of student
         attribute5: tuple location
         attribute6: tuple time
    end attribute-description
end data abstraction

data-abstraction: student
    abstraction-type: tuple
    attribute-size: 4
    begin attribute-description
                 attribute1: single-entity student-id
                 attribute2: single-entity student-name
                 attribute3: single-entity department-name
                 attribute4: single-entity years-in-college
    end attribute-description
end data-abstraction

data-abstraction: location
    abstraction-type: tuple
    begin Attribute-description
                 attribute1: single-entity building
                 attribute2: single-entity room
    end attribute-description
end data-abstraction

data-abstraction: time
    abstraction-type: tuple
    begin Attribute-description
                attribute1: single-entity start-time
                attribute2: single-entity duration
    end attribute-description
end data-abstraction
```

FIGURE 4.1 Abstracting a data entity class.

Pairs are special types of tuples, consisting of two attributes that have been used in abstractions of mathematical entities such as complex number, rational numbers, and time interval. For example, a complex number is represented as a pair of the form (real part, imaginary part), and a rational number is represented as a pair of the form (numerator, denominator). Abstract concepts such as events can be modeled as a pair (action name, occurrence), where occurrence is a pair of the form (time, location). Many programming languages such as ADA allow complex and rational types as built-in types. Polynomials are modeled as a bag of triples, where each triple is of the form (coefficient, variable name, power).

4.1.3 Collection of Data Entities

A collection of entities is a bag of entities and is quite different from composite object. In a composite object, each attribute is a part of the abstraction of the same object and is not indexible. In a set or bag, each object is a different entity and is indexible. There are many ways a collection of data entities can be modeled: (1) sequence—an ordered bag or (2) a bag of pairs of the form (key, entity), where key is a unique identifier associated with the data entity.

In an ordered bag, each data entity is associated with a specific position, and can be modeled using an indexible sequence such as array representation or vector representation.

Sequences are implemented using arrays, linked lists, or vectors. Arrays and vectors are indexible. Linked lists are not indexible; to reach the i_{th} element, first $(i - 1)$ elements have to be traversed. Arrays can be *static arrays*, *semidynamic arrays*, or *dynamic arrays*. The size of the *semidynamic arrays* are passed as parameters from the calling subprogram and can change between different calls. Dynamic arrays are allocated in heap, are implemented as recursive data structures such as trees, and are extensible. For example, one way to implement dynamic arrays is to use a quad-tree. A quad-tree has four branches. If the size of the bag is between 1 and 4, then one needs a tree that is one-level deep. If the size of bag is between $4 + 1$ and 4^2, it needs two levels of trees, and if the size is between $4^2 + 1$ and 4^3, it needs three levels of trees. All the data entities are stored at the leaf node. The advantage of the representation is that at any time to extend the bag, looking at the index value, the exact position of the data entity in the tree can be computed, and it can be traversed in logarithmic time, since the tree grows in a balanced way.

The collection of (key, data entity) pairs is called *association lists* in declarative languages, and is called a *map* in Scala—a modern multiprogramming language that integrates functional programming and object-oriented programming. The use of the key makes the data entity free of specific implicit ordering at the expense of extra memory to keep the key; data elements can be placed randomly at any position in the bag. Retrieval of a data entity requires a search mechanism such as hashing or binary search to match the key and then retrieve the corresponding data entity.

Any dynamic collections of data entities need to have two major properties: (1) efficient retrieval of data entities based upon the index value or the key value and (2) optional run time extensibility of the collection size of the data abstraction to accommodate new data entities that are provided either from the input–output mechanism or through programmatic generation. Two major search schemes have been used for the efficient retrieval of data entities: (1) hash function maps the key in near-constant time to find an index value, and then the index value can be used to retrieve the data from the corresponding array and (2) the use of tree-based search in logarithmic time. Depending upon the type of tree-based representation such as binary tree or tertiary tree, the key can be searched using an appropriate search algorithm.

4.1.4 Extensible Data Entities

Recursive definitions and tuples can be used to model extensibility. Certain data aggregates are abstracted recursively using the notion of tuples that contains two types of fields: data entity and a reference field to indefinitely extend the data aggregate at run time.

A set can be modeled using a hash table or a tree. While there is no need to order the key or the values in a hash table, a tree-based implementation of a set has to be in sorted order for efficient logarithmic time search.

A sequence is defined recursively as *<sequence>* ::= *(<data-entity> <sequence>)* | *null.* In this representation, data entities are extended linearly, and, after any number of extensions, the collection can be terminated by the base case *null.* The size of the aggregate is indefinite and can be extended at run time. Linked lists, arrays, and vectors are used to implement a sequence.

Another example of recursive abstraction using tuples is a binary tree that can be modeled recursively as *<binary-tree>* ::= *(<data-entity>, <binary-tree>, <binary-tree>)* | *void.* The left-subtree and right-subtree are again tuples, and the nesting level of the tuple gives the level of the tree. The base case of the binary-tree is a void-tree. A node can be added at a leaf node of the tree to extend the original tree. For example, a one node binary tree can be modeled as *(info, null, null)*, where info denotes a data entity, and *null* represents a void-tree; a two element bag can be modeled as *(info, (info, null, null), null)*; and a three-element bag can be modeled as *(info, (info, null, null), (info, null, null)).*

Recursive data structures and extensible dynamic data structures such as vectors are implemented using *pointers. Pointers* are addresses that can be used to join physically separated chunks of memory. A pointer can point to the next data entity or could be null: not pointing to any object. This property of pointers is used to implement extensibility in the implementation of linked lists, vectors, and trees. A tree is extended by adding a pointer in the leaf node to add new data entity. A vector is extended by keeping a header that keeps the information of the last cell. New elements are added at the end of the vector. Pointers can be strongly associated explicitly with recursive data structures as in Pascal. In many languages such as ADA, ALGOL, and C, pointers are supported as an independent entity that can be associated with any data structure.

4.1.5 Ordering and Accessing Data Entities

In a composite object, attribute values are projected using the field name such as *<data-entity>.<attribute-name>*. If the field is composite, then its attributes are retrieved as *<data-entity>.<attribute-name>.<nested attribute-name>*. At the time of compilation, the various attributes of a single entity are allocated in contiguous memory locations, and the code generator calculates the offset of each attribute name with respect to the base address of the data entity. The calculated offset is added to the base address to access the corresponding attribute.

In a collection of data entities, the index value or the key value plays a major role to access a data entity. If the collection is modeled as an array or vector and size of each data entity is fixed, then the offset of the given data entity is calculated as *size-of(individual data-entity)* * *index-value* under the assumption that the index value starts with zero. If the collection is modeled as a tree, then the key is searched using some algorithm that is dependent upon the type of tree representation: binary, tertiary. If the bag is represented using a hash table, then the access is near-constant time: a hash function is applied on the key to generate the appropriate index, and the data entities associated with that index are searched sequentially for the matching key to identify the corresponding data entity.

4.1.6 Interconnected Data Entities

Multiple data entities can be embedded inside a single aggregate data entity. An aggregate data entity can be a tuple (composite data entity), a collection of data entities, or an extensible data entity. An interesting aggregation is the *semantic network* used in Lisp that is a graph of entities connected through relations between them. Each entity is modeled as a *frame* that is a pair of the form (object name, property list), where the property list is a bag of pairs of the form (attribute name, value). The value could be a single entity, an aggregation, a function call to compute a value, or a reference to another data entity. The semantic network is an interesting abstraction, as it can be used to model relationship between multiple real-world objects. By using the connecting relational edges between the objects, the graph can be traversed to find attributes of other related objects.

4.1.7 Persistence

Persistence is associated with two important properties: visibility across program units and lifetime that goes beyond the life of the program. The advantage of persistence is that different program units can share the information and the result of computation that may be generated in other program units at a different time. Persistence can be grouped into two types of data entities: (1) global transient data entities, whose lifetime is limited to the lifetime of the program and (2) persistent data entities, whose lifetime goes beyond the lifetime of the program such as databases, files, and archived objects.

A blackboard or tuple space is a global transient data entity that is modeled as a bag of (key, value) pairs, where key is a user-defined unique identifier. The operations on the blackboard are retrieving a value given the key, deleting a (key, value) pair by giving the key, updating the value bound to a key, and inserting a new (key, value) pair.

Files and databases are persistent data entities where the information stored can be reused even after the program has ended, or the information can be passed to other programs. Files are sequence of characters that are formed by linearizing the data entities. All the properties of sequences that we studied about the sequences apply to files. In addition, a file is a logical entity that can be opened to read or write. However, file is a very large sequence, and processing the data after reading all the elements of the files into working memory will cause tremendous memory wastage and delay in loading the data.

Most of the time, a program unit needs part of the data from an input file, processes it, and then repeats the process with the remaining data as shown below

```
my-sequence = file;
while (not is_empty (my_sequence)) {
    data-entity = first(my-sequence);
    process data-entity;
    my_sequence = rest(my-sequence);}
```

Since the rest of the file is not being processed when the current data entity is processed, there is no need to have all the data entities in the working area before the first data entity is processed; the second data entity can be retrieved from the file while the first data entity is being processed, and the third data entity can be retrieved while the second data entity is being processed. The resources such as memory and processor used by the previous data entity can be reused by the next data entity.

A stream is a data abstraction that allows this resource reuse while avoiding the wastage of memory. A stream is a sequence that is partially available at a time. If the stream is modeled as $<d_0, d_1, d_2, \ldots d_N>$, then only subsequence of the form $<d_I, \ldots d_J>$ is available in memory at a particular instance of time. If K data entities are processed in one instance, then the subsequence $<d_{I+K}, \ldots d_{J+K}>$ would be available at the next instance.

The stream abstraction is important in large data processing, specially in Internet-based programming, multimedia programming, and mobile computing, where video files and audio files are very large and cannot be made available for local processing due to delays in the communication channel and the large amount of computation involved to render the video or audio files. In such cases, a stream of data is being received continuously. A subsequence of data is processed while next subsequence of data is being transmitted through the communication channel.

4.1.8 Declarations and Environment Change

Declaration can have the following five components: binding an identifier to a constant; macro definition—binding of an identifier to a text block; declaration of a user-defined type—binding of an identifier to a data abstraction that contains type of data values; declaration of a variable—binding of an identifier to memory locations; and declaring a sequencer—a label that can be used for jump. Declaration is also used to bind an identifier to a procedure or a function. In imperative languages, declaration of labels used for jump and declaration of variables have no effect on the subsequent declarations. However, binding an identifier to a constant, declaration of user-defined types, and declaration of macro definitions can be used in the following definitions to form more complex declarations.

Declaration of a variable changes the environment by binding the identifier to the corresponding l-value. We model the state of computation as a triple of the form $(\sigma^E, \sigma^S, \sigma^D)$, where σ^E denotes the environment, σ^S denotes the store, and σ^D denotes the dump—stack of calling procedures. Under this model, the computational state changes every time (1) a declaration changes the environment, (2) execution of an assignment statement changes the store, and (3) dump is changed when a procedure is called. During parameter passing, both environment and store may change, because the formal parameters are mapped to memory locations, and there memory locations are initialized to a value. A command executes in the environment that is created by the declarations.

A sequence of declaration of the form D_1 *<delimiter>* D_2 can have two different meanings: *sequential* and *parallel.*

1. In *sequential* declarations, D_1 updates the environment σ^E and changes to σ'^E, and D_2 uses the new environment σ'^E in its definition to pick up the binding of the variables bound in declaration D_1. For example, in functional programming language *Lisp*, the declaration _(let* ((X 4) (Y (+ X 4))))_ is equivalent to binding the variable *X* to a value 4

and binding the variable *Y* to a value 8, because the first declaration becomes part of the environment, and when *Y* is declared, it picks up the binding of *X* from the environment and adds 4 to the value of *X* to derive the value 8. After *X* is bound to 4, the environment $\sigma'^{E} = \sigma^{E} \uplus \{X \mapsto 4\}$, and after the second declaration the environment becomes $\sigma^{E} \uplus \{X \mapsto 4\} \uplus \{Y \mapsto 8\}$.

2. In *parallel* declarations, both D_1 and D_2 pick up the bindings from the old environment σ^{E}, and D_2 is not affected by the change caused by D_1, as illustrated in Example 4.2.

Example 4.2

Let us take the following example from the programming language Lisp again.

```
(defun my_square(b)                    % define a function
       (let ((b (+ b 5)) (c (+ b 6)))  % parallel declaration
            (+ (* b b) (* c c)) % compute b² + c²
       ) % close the scope of let declaration
)     % close the scope of the function declaration
```

The let construct is used to create a local environment for the local variables *b* and *c*. A function call *my_square(4)* computes the value of *b* as the value(*b*) + 5 = 9 and computes the value of c as value of *b* + 6 = 4 + 6 = 10. Although, the value of *b* in the local environment has changed to 9, the second initialization still takes the old value b as 4 due to parallel assignment semantics of *let construct*. The final answer is $9^2 + 10^2 = 181$.

When a procedure or function is called, then the environment changes, because many bindings that are local to the calling procedures are shadowed—become invisible during the scope of the called subprogram—and only the bindings of the global variables, nonlocal variables, imported variables, and bindings of the formal parameters are retained in the new environment of the called subprogram.

4.2 CONTROL ABSTRACTIONS

Depending upon a programming paradigm, the instructions that manipulate the data are abstracted. Irrespective of the programming paradigm, we can categorize the instructions as follows: (1) *constructors*—that create a new data entity; (2) *mutators*—instructions that modify the values that an identifier is bound to; (3) *selectors*—instructions that access the data elements from a composite data object or an aggregate data object; (4) *conditionals*—instructions that pick up one of the alternatives, depending upon the evaluation of a condition; (5) *iterators*—instructions that repeatedly perform similar operation on a collection of data objects in an organized manner; (6) *evaluators*—instructions that evaluate an expression; (7) *sequencers*—instructions that pass the control through jump to a given label; and (8) *invocations*—instructions that activate a function or procedure to compute some value.

A *mutator* such as an assignment statement modifies the store by changing the value stored in a memory location. In some languages, the store is also modified during initialization with a value when a new data entity is constructed or declared. Given a sequence of mutators, each mutator progressively modifies the store to a new store. The store is also modified when the value of an actual parameter is bound to the memory location corresponding to the formal parameter, or when a variable is initialized to a specific value during declaration.

The sequence of commands can be altered either by conditional instructions, jump statements, or by iterators. The actual sequence of command that follows the current command during run time is called *continuation. Continuation* is essential to reason about the behavior of the program and is an essential part of defining the denotational semantics of the realistic programming languages. The continuation of a selection function such as an if-then-else statement of a case statement has multiple alternatives that only can be decided at run time, based upon the evaluation of the condition. Similarly, the continuation of the iterative statement involves unfolding the loop multiple times when the condition evaluates to be true. The continuation of an iterative loop and conditional statements are difficult to predict, since it is dependent upon the outcome of the evaluation of the condition.

Example 4.3

Let us discuss the continuation for the following program.

```
x = 4; z = 6;
goto L;
z = 8;
L: y = 5;
while (z > 4)
     {x = y + 5; z = z - 1;}
```

The continuation of the statement $x = 4$ is {$z = 6$; $y = 5$; **if** $z > 4$ **then exit;** $x = y + 5$; $z = z - 1$; **if** $z > 4$ **then exit;** $x = y + 5$; $z = z - 1$; **if** $z > 4$ **then exit**}. Note that the sequencer "**goto** L" alters the order of execution from the textual order. Similarly, the statements within the while-loop are executed twice before Z becomes equal to 4, and the control exits the while loop. Thus the continuation is computation dependent and is difficult to predict just by looking at the program.

4.2.1 Assignment and Command Sequence

Assignment statement binds the memory location corresponding to an identifier to the evaluation of the expression written on the right side of the assignment statement. The binding is done after the evaluation. Generally the assignment consists of a single binding that means that statement is of the form *<identifier>* = *<expression>*. However, some languages, such as ALGOL-68, C++, Python, and Ruby, support *chained assignment* (also called *multiple assignment*) that has multiple variables bound to the evaluation of the same expression. In the syntax of a modern language, Ruby, that supports functional and object-oriented programming, multiple assignments are written as follows:

```
x = y = 4 + 5 + 6 (supported in Python and Ruby)
x, y = 4 + 5 + 6 (supported in Scala)
```

The statement performs the evaluation of the expression "4 + 5 + 6," and binds the value 15 to the memory locations of the variables *x* and *y*.

Some languages such as Ruby, C++, Python, Perl, and Lua also allow *simultaneous assignment* or *parallel assignment*. In simultaneous assignment, multiple distinct nonaliased variables are assigned the values of the corresponding expressions simultaneously, and the expression evaluation takes the original store values. The basic requirement for simultaneous assignment is that all the variables should be writing into separate memory locations. A simultaneous assignment is of the following form:

```
var₁, var₂, var₃, …, varN = exp₁, exp₂, exp₃, …, expN
```

The semantics of the above statement is that each expression exp_i $(1 \leq i \leq n)$ is evaluated with respect to original store σ^S, and the evaluated values are bound to the memory locations of the corresponding variables var_i $(1 \leq i \leq n)$. For example, in the syntax of interactive Ruby, the following multiple assignment will bind x1 to 4, x2 to 5, and x3 to 6. However, if two variables are aliases, then the store contains the last binding.

```
x1, x2, x3 = 4, 5, 6 % x1 ↦ 4; x2 ↦ 5; x3 ↦ 6
y, y = 8, 9 % y ↦ 9
```

The sequence of assignment statements of the form C_1; C_2 can change the store sequentially or concurrently, depending upon the programming language paradigm. In *sequential update* of the store, the previous assignment statement updates the store first, and the following assignment statements work on the updated store. In concurrent programming paradigm, during the *concurrent update* of the store, two or more assignments work concurrently on the same original store. The update should occur on different memory locations; reading or writing the same subset of memory locations would cause *racing condition*—a condition where store is inconsistent and may have different values if the same set of statements are executed multiple times. Racing condition is discussed in detail in Chapter 7, on concurrency.

Example 4.4

For example, sequential assignment of the form "$x = 4; y = x;$" will first update the store σ^S destructively to create a new store $\sigma'^S = \sigma^S \bigoplus (x \mapsto 4)$. Note that here the symbol "$\mapsto$" denotes the destructive update for imperative programming paradigm. Now the assignment statement "$y = x$" will first read the value of the variable x from the store σ'^S and generate a new updated store $\sigma''^S = \sigma'^S \bigoplus (y \mapsto 4) = \sigma^S \bigoplus (x \mapsto 4) \bigoplus (y \mapsto 4)$.

In the programming languages that support the imperative programming paradigm, the assignment statement is a *mutator*, which means that the identifier is bound to a memory location that can be destructively updated. As we have discussed earlier, the advantages of destructive update is memory reuse at the cost of losing the results of the past computation and side effects—undesired programming behavior that may violate the basic mathematical principles essential for programming.

The declarative programming paradigm uses assign-once property and assigns the value of an expression to the variable only once, and the binding cannot be destroyed explicitly by the programmer. However, some languages such as Prolog allow backtracking—going back to the previous computation in search of an alternate solution. During backtracking, the underlying implementation engine of Prolog can unbind the variable from the previous value.

A variation of the assignment statement is the concept of *unification* used in logic programming. In unification, two logical terms are equated. A logical term is a composite structure containing variables, constants, literals, and other embedded logical terms. During the process of equation, no evaluation is done as terms are not expressions. Instead, logical terms on the left-hand side and the logical terms on the right-hand side are matched position by position. If one of them is a variable, then every occurrence of the same variable in both the logical terms is bound to the corresponding logical term using the assign-once property of the declarative programming paradigm. If both the corresponding terms are literals, then they are matched literally.

For example, the unification of the term $X + 4 + 3$ with the term $5 + Y + 3$ gives $5 + 4 + 3$, the variable X is bound to literal 5, the variable Y is bound to literal 4, and the literals 3 and 3 are matched successfully. In contrast, the unification of $X + 4 + 3$ and $5 + Y + 2$ fails, because the corresponding literals 3 and 2 do not match.

Unification is discussed in detail in Chapter 10 on logic programming. However, here it has been mentioned for the sake of completeness of the concept of assignment statement.

4.2.2 Conditional Statements

The conditional constructs are almost universal across programming languages. The basic construct of if-then-else statement is given by the following abstract syntax rule:

```
<if-then-else> ::= if '('<condition>')' then <statement>
                                        [else <statement>];
<condition> ::= <condition> && <condition> | <condition> ||
                <condition> |not <condition> |
                <arithmetic-exp> <c-op> <arithmetic-exp> |
                true | false
<c-op> ::= '>' | '<' | '>=' | '=<' | '=='
```

Different languages use different reserved words. For example, '**.and**' may be used in place of '**&&**' or '**.or**' may be used instead of '||'.

In an if-then-else statement, the then-statement and else-statement are equally probable, and changing the order of execution after checking the negation of the condition will be functionally equivalent to the original if-then-else statement; that is, "if (*<condition>*) then *<then-statement>* else *<else-statement>*" is functionally equivalent to if (not *<condition>*) then *<else-statement>* else *<then-statement>*, because *<condition>* and not *<condition>* are mutually exclusive.

Both the then-statement and else-statement can be any statement, including another if-then-else statement. As discussed in Chapter 3, in case of nested if-then-else statement, the else part always goes with the nearest then-statement.

The other selection construct is *case statement*. The structure of case statement is of the form as given below

```
case (<expression>) of :
    <value-set_1>: <command-sequence_1>;
    ...
    <value-set_N>: <command-sequence_N>;
    otherwise: <command-sequence_{N+1}>
end case
```

In the case statement, an expression is evaluated. The outcome is possibly multivalued, and a different sequence of activities can be taken for each different output value or a range of output values. All the cases that have a different sequence of activities are handled, and in the end there is a catch-all statement that is executed when all other possibilities fail. Note that the case statement is deterministic, since only one choice is selected on the basis of the evaluation of the expression.

Lisp uses a set of conditional statements of the following form:

```
(cond  ((<predicate_1>  <expression_1>)
         ...
         (<predicate_N>  <expression_N>)
         (t  <catch-all-expression>))
)
```

This function checks from top to bottom: if any of the *<predicate*$_i$*>* is true, then the corresponding function is evaluated; otherwise, *<catch-all-expression>* is evaluated.

Another conditional construct used in some functional languages such as Lisp is *when-construct*, which is written as *(***when** *<predicate> <expression>)*. The construct evaluates the expression *<expression>* if the predicate *<predicate>* is true. Another conditional construct is the use of *unless* construct used in the programming language Ruby. It has the same semantics as "if not." The construct is as follows:

```
unless '('<condition>')' <command-sequence_1> else
  <command-sequence_2> is equivalent to saying
if not '('<condition>')' <command-sequence_1> else <command-sequence_2>
```

4.2.3 Iterative Constructs and Iterators

Iterative constructs are classified in three major categories: (1) definite iteration, (2) indefinite iteration, and (3) iterators. Both definite and indefinite iteration assume that the collection of data entities has been modeled explicitly using indexible data structures, such as arrays and vectors. In contrast, the iterator treats the data entity as a list, such that individual elements and the mechanism to access the individual element may not be available to a programmer. An iterator retrieves the next element of the list and performs operations on every element of the list in an ordered manner. All three classes of iterative constructs have single-entry and single-exit points, unless the execution is aborted due to some error condition or exception.

Definite iterations do not allow programmer to alter the loop parameters such as initial-bound, final-bound, the value of the index variable, and the step size during the execution of the statement block. All of them are fixed in advance. However, the implementation engine automatically alters the value of the index variable by the step size in the next iteration-cycle. An abstract syntax for the definite iteration is as follows:

```
for (<index-variable> = <initial-expr>';' <final-expr>';' <step-expr>)
        <block>
```

In the above construct, the initial expression *<initial-expr>* gives the initial bound, final-expression *<final-expr>* gives final-bound, and the step-expression *<step-expr>* gives the steps. The expressions are evaluated before entering the loop, and loops are single-entry and normally single-exit, unless an exception condition forces an abort or an exit out of the loop.

A definite iteration can take an index value from a list of expressions. The definite iteration construct would be of the form:

```
for '('<index-variable> in <list-of-expressions>')'
                <block>
```

A definite iteration can also pick up an index from an unordered set. The advantage of using a set is that the index can occur in any order, and there need not be a pattern to them. The definite iteration construct would be of the following form:

```
for '('<index-variable> in <set>')'
         <block>
```

In an indefinite iteration, the embedded statement block keeps executing until the given condition is satisfied. Unlike definite iteration, which regularly changes the value of index in predetermined manner, the condition could be any predicate in indefinite iteration. Moreover, the value of the variables involved in the predicate can be freely altered in the embedded block of statements. Owing to this property, indefinite iterations can loop indefinitely.

As described in Chapter 1, two types of indefinite iterations have been shown to be able to be functionally equivalent to model any problem that requires repeated manipulation of data abstractions. The two indefinite iterations are while-loop and do-while loop, which is also called repeat-until-loop. The abstract syntax for while-loop is as follows:

```
<while-loop> ::= while '('<condition>')' '{'<statement-block>'}'
```

The abstract syntax for the repeat-until-loop is as follows:

```
<repeat-until-loop> ::= repeat <statement-block> until <condition> |
                        do <statement-block> while <condition>
```

As described in Chapter 1, while-loop executes the embedded block <*block*> zero or more times, and repeat-until-loop executes the embedded block <*block*> at least once. The repeat-until construct and do-while construct are equivalent and achieve the same effect, with a difference: repeat-until construct executes the block repeatedly until the condition becomes true, while do-while syntax executes the block repeatedly while the condition is true. A generalized multiple-exit iterative loop is used in ADA language. The abstract syntax for the multiple-exit iterative loop is as follows:

```
<multiple-exit-iteration> ::= loop {<conditional-exit>';'
                                    <command-sequence>}*
                              end-loop
<conditional-exit> ::= if <condition> then exit;
```

The abstract syntax means that there are multiple exit conditions interspersed within the block of the loop, and if a condition is satisfied, then the control jumps out of the loop. While it may be a convenience to some programmers, it had been shown in the 1970s that the functional power of a nested while-loop and do-while loop along with Boolean variables is the same as multiple-exit loops. Hence, they do not add any additional programming power.

In order to avoid the explicit use of index that supports only array or vector-based implementations, *iterators* support *list-based* and *set-based* abstractions. Declarative languages that support list-based programming or languages that use extensible data abstractions use iterators. An iterator steps through an extensible bag of data entities, picks up the next data entity, and performs some operation on the data entity. The process is repeated for every data element in the collection. The major advantage of the use of iterator is that explicit index is not needed, and the implementation level details of data abstractions are hidden from a programmer; the only restriction is to retrieve the next element in the next cycle. An abstract syntax for iterator is as follows:

```
<iterator> ::= foreach <variable> in <ordered-bag> <block>
```

In functional languages, <*block*> is replaced by a function definition that is applied on the next data element. The disadvantages of basic scheme of iterators is that an iterator steps through every element of the data abstraction, unlike iterative constructs, where index can be used to access the elements selectively. Languages such as C++ and Java treat iterator as a mutable data structure that can produce data element one at a time and has the capability to check if the next element exists. An iterator construct in Java will look like the following:

```
for (iterator i = data-object.iterator(); i.hasNext();)
                  … visit i.next() …
```

4.2.4 Block Structure

The blocks are embedded inside a programming unit. Unlike programming units, blocks are not bound to an identifier and cannot be invoked simply by calling. Block structures are generally *single-entry* and *single-exit* structures and have a specific functionality. Random grouping of statements is discouraged to provide a better program structure for better comprehension. Block structuring allows only a restricted number of "go to" statements to exit out of the nested blocks or for exception handling to handle the error conditions.

Blocks include a set (possibly null) of declarations followed by a set (possibly null) of commands. The scope of the declarations is within the block, which means all the variables and data abstractions are visible only within the block boundaries clearly marked by reserved words like **begin** ... **end** pair or a pair of curly brackets. These reserved words are specific to programming languages. Multiple blocks can be nested inside a block. Nested blocks fall within the scope of the outer block, which means all the declarations in the outer block that do not have conflict with declarations inside the inner blocks are visible inside the inner block. In the case of name conflicts, the declarations in inner blocks get the preference, and the corresponding declarations in the outer blocks are shadowed in the inner block. The sibling blocks (and their descendants) that are nested at the same level inside an outer block do not share the environment. The nesting level of the blocks can be modeled as a tree structure, with the program unit as the root node and the nested blocks as the descendants.

This property of not sharing the environment with the siblings has been exploited in memory reuse during the execution. The blocks (and their descendants) that are defined after their sibling blocks (and their descendants) can share the same memory space, since the scope of the sibling defined earlier will be over before the following sibling.

Example 4.5

Let us consider the following nested block structure:

```
{declaration A
     ...
     {declarations B
          ...
          {declarations C
               ...}
               {declarations D
               ...}
          ...
}
     {declarations E
... }
}
```

The corresponding pictorial representation of the scope of the declarations and the corresponding tree structure is given in Figure 4.2.

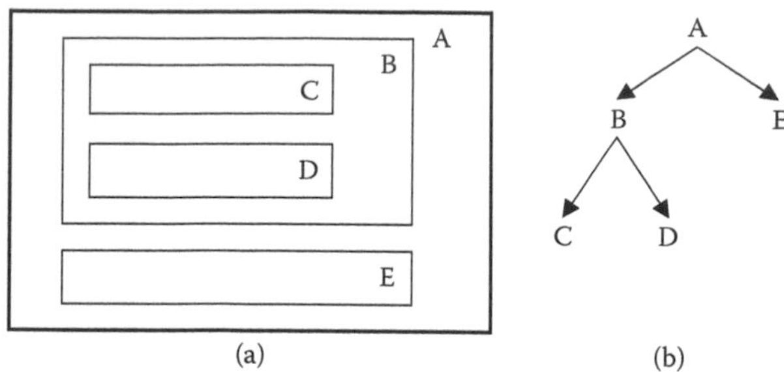

FIGURE 4.2 Nesting of (a) block structure and (b) tree representation.

Blocks B and E are siblings, and are nested inside the block A, and blocks C and D are nested inside the block B. Since B and E are siblings, and blocks C and D are nested inside B, the blocks B, C, and D do not share their local environments with the block E. Similarly, blocks C and D do not share their local environments. Owing to tree hierarchy, the environment of B includes the environment that is declared inside block A. Similarly, blocks C and D inherit the environment inherited by B and the local environment of the block B. The memory allocations of block B and the blocks nested inside the block B will share the memory allocations with the block E and the blocks nested inside the block E. The details are deferred to Chapter 5, on implementation models of programming languages.

4.2.5 Program Units and Invocations

Program units are a group of declaration and commands that have been bound to an identifier and can be called at run time by other program units using the identifier the program that the units are bound to. These two properties separate the program units from the blocks.

Many programming languages such as the Modula family allow nested declaration of programming units. The property of environment sharing by nesting is similar to the blocks: nested program units inherit the local and the inherited environments of their parents, and siblings do not share their local programming environments.

The abstract syntax (given in EBNF) of a programming unit in a block-structured language is given in Chapter 3. A limited version of the abstract syntax rules has been described below

```
<program-unit> ::= (program | function | procedure) <identifier>
  <block>
<block> ::= '{' [{<declaration>}*] [{<command>}*] '}'
<declaration> ::= <sequencer-decl>| <type-decl> | <variable-decl>
<command> ::= null | <assignment>| <conditional> |
            <definite-iteration> | <indefinite-iteration> |
              <iterator> | <subprogram-call>|
            '{'<block>'}' | <sequencers>
```

The above abstract syntax states that a program unit could either be the main program, a function, or a procedure that has been bound to an identifier. The body of the program unit is a structured block. The reserved words for the program unit could be different for different languages. For example, Lisp uses the reserved word "defun" to define a function, and Scala and Ruby use the reserved word "def" to define a function.

A program unit can invoke another program unit to solve a complex problem. By embedding an invocation of a program unit inside an iterative construct or a recursive invocation, a program unit can be invoked multiple times, as invocation of a program itself is a command abstraction and is part of the command sequence inside the program unit. The compilation process of a program unit is independent of the number of invocations; each program unit is translated only once during compilation and just-in-time compilation. During just-in-time compilation, after the program unit is compiled at run time for the first time, it is cached for future retrieval.

The invocation of program units can be modeled as a directed graph, with the direction showing the calling pattern of the program units. The execution of a program looks like a graph, since the same program unit can be called from multiple program units. The graph can be acyclic or cyclic. An acyclic graph means that program units are invoked nonrecursively, and a cyclic graph means that program units are invoked recursively.

Each invocation that a program unit creates has a separate unique environment that is different from the previous invocations of the same program unit, and the new invocation works in its own environment. As described in Chapter 2, the environment is a sum of the declaration of global variables, nonlocal variables—variables declared in the ancestor program units if the current program unit is nested, local declarations, and reference parameters.

4.2.5.1 Self-Recursive Procedures

A recursive program is *self-recursive* if it invokes itself directly. For example, a definition of *factorial* function is self-recursive. Let us assume if the initial invocation is P_0; then the subsequent invocation will be P_1, P_2, P_3, and so on. The invocation stops when the base case is evaluated, and the result is passed back in last-in-first-out order. In a graph model of procedure invocation, the length of the self-recursive invocation is zero, since the procedure invokes itself. Note that the environments of the various invocations of the self-recursive procedures are separate from each other.

Example 4.6

For example, a call to factorial(4) will start a chain of invocation: factorial(4) → factorial(3) → factorial(2) → factorial(1) → factorial(0). Each invocation will have its own environment, its own store, and its own dump that will include environment and store of the previous invocation added to the dump of the previous invocation. The dump is modeled as a stack, and previous invocation is added to the stack by abstractly pushing the environment and the store on the stack. Note the use of the word "abstractly pushing." In reality, when the dump is created, a part of the previous environment can be retrieved by the use of a pointer called a *dynamic link*. However, the implementation related discussion is deferred to Chapter 5.

After the invocation *factorial(0)* is over, the result is passed back in the last-in-first-out way updating the store of previous invocations. After passing the result, the corresponding environment and stack of the calling invocation are popped out of the dump, and the control is passed back to the next instruction of the previous invocation.

4.2.5.2 Mutually Recursive Procedures

A program unit may support *mutual-recursion*, which means that the program unit may start a chain of invocations of one or more program units other than itself, and eventually the last invocation in the sequence invokes the first program unit again. Mathematically, let the invocation chain be P_0, P_1, ... P_N ($N > 0$); then program unit $P_{i(0 < i < n)}$ is not the same as program unit P_0 or program unit P_N, and program unit P_0 is the same as program unit P_N. There are four types of structures: (1) a hierarchical structure–like tree; (2) a structure like directed acyclic graphs; (3) a zero-length cycle, denoted by a circle, showing self-recursion, and (4) a nonzero-length cycle showing mutual-recursion.

Example 4.7

There are three types of structures: (1) a hierarchical structure like tree, (2) a zero-length cycle, denoted by a circle, showing self-recursion, and (3) a nonzero-length cycle showing mutual-recursion. An example of a graph of procedure invocations has been illustrated in Figure 4.3. The top-level subprogram calls two subprograms B and C, in that order. The subprogram B calls two subprograms D and E. Both the subprograms D and E form terminal nodes, which means they do not call any other subprogram. The subprogram C calls the subprograms F and G. The subprogram F is self-recursive, as shown by the circle; the subprogram G is mutually recursive, as G calls H; and the subprogram H calls back the subprogram G. The length of the cycle in the mutually-recursive subprogram is 1. Note that the calling pattern of the program units is very different from the nested structure of programming units. Calling patterns of the program units are run-time properties, while nested structure are static program properties.

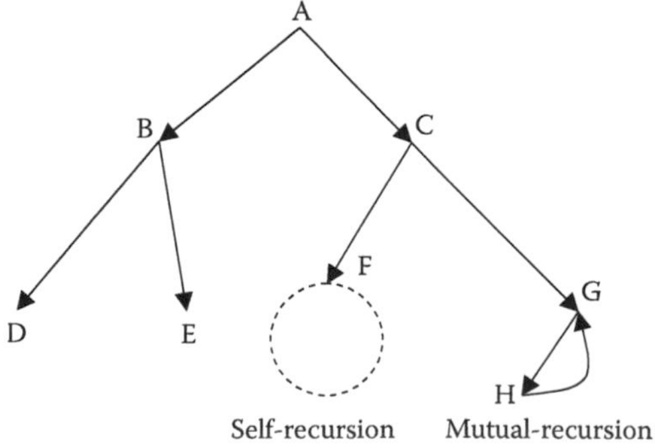

FIGURE 4.3 Calling pattern of program units.

4.2.6 Modules

A module is an abstraction that provides logical encapsulation around a declaration, modeling one or more data abstractions and the corresponding subprograms (or methods in object-oriented programming terminology) that perform some useful operation on the data abstractions. Declarations could be (1) *importing* some information or function from another module; (2) declaration of constants or user-defined types or variables; or (3) definitions of procedures, functions, objects, and interface to the rest of the system or other embedded modules. The definition of module provides an explicit logical boundary that keeps external declarations outside the local scope. A declaration inside a module has a natural scope inside the module, and is not visible outside the module unless *exported*. A module can be replaced by another equivalent module with the same interface without changing the semantics of the system.

Explicit modules generally include all the common operations to manipulate an abstract data type that can be compiled and stored in the form of a library to be used later. When the module is loaded or imported within another module or program unit, the exported abstract data types and all its library functions also become available. The basic purpose of module abstraction is to provide (1) multiple naming systems; (2) independence from the rest of the system; and (3) regulated interaction with the rest of the system through the module interface using import–export of the declared entities such as variables and subprograms.

Unlike subprograms in imperative languages or inheritance in many object-oriented languages, the interface is strictly regulated using *export–import* mechanism. An entity declared in the module is not available outside unless explicitly *exported*, and another module cannot use an entity declared and explicitly exported from other modules unless it explicitly *imports* it. There are two approaches to import the information from the modules. The first technique imports every entity that has been exported from the specified module. The second technique is to export a specified list of entities from the specified module. The export–import mechanism of modules provides a better regulation than nonlocal variables and inheritance as described later in subsection 4.3.5. Within a module, information between subprograms and objects is passed using local variables, global variables, and parameter passing.

A module can be developed, compiled, and archived independently, without the need to compile the rest of the system. An abstract syntax for the module is given below

```
<module> ::= module <identifier>
             [export {<program-unit>}*]
             [{import {<program-unit>}* from <identifier>}*]
             {<declaration>}*
             {<program-unit>}*
             {<module>}* % definition of nested modules
             end <identifier>
```

Modules become active when loaded into the memory by other programs. A module can temporarily pass the control to another module to perform a computation with some request or intended expectation of a result. A programming unit in a module executes on the basis of the available information in the environment within the module. If a complex function is too big, it can be broken across multiple modules, and the higher-level function can call embedded functions in other modules.

One of the limitations of the modules is the lack of information passing between two modules using parameter passing mechanisms. The export list can also become quite big. Although modules are compiled separately, type compatibility between the variables and operators must be guaranteed by the programmers. Types of the variables are stored in a *symbol file*—a meta file used to provide interface across modules. Each time a client module—module that uses the included module—is compiled, the type compatibility is checked in symbol file.

Many languages such as ADA, C++, CLU, Euclid, Fortran 90 onward, Java, the Modula family of languages, and Ruby support the concept of explicit module. ADA and Java call modules as *packages*. While the above languages support a nested structure of modules, other languages such as Mesa support a flat definition of modules. In Mesa, modules are simply functional units that break up a big program into multiple simpler units based upon functionality.

4.2.7 Objects and Classes

Another important category of encapsulation is the notion of objects. As we have seen that the basic purpose of the encapsulation is to regulate the visibility of the entities declared within a boundary so that one module can become independent of another in terms of declaration, compilation, and the low-level information it carries. Objects do exactly this. However, unlike explicit definition of modules that are passive units, an object is an active unit that can be created at run time, invoked, and have a run time computational state. Objects are created using a constructor that allocates memory space in the heap.

Each object encapsulates data entities, and methods—the subprograms and functions that operate on these data entities. Encapsulation provides the information hiding. Many programming languages support the notion of objects such as ADA, Modula-2, and other object-oriented languages. The data entities inside the object can be accessed using an abstract syntax of the form *<object-name>.<data-entity-name>*, and the methods can be accessed using *<object-name>.<method-name>*.

An interesting category of programming languages called object-oriented programming is based upon *object classes*. Generally a *class definition* follows a hierarchical structure that means that at the root of the tree is the top-level class definition, at level 1 are the class definitions that are subclasses of the root class, and this property of defining subclass can be repeated at level 2, level 3, and so on. An object is an instance of a class or a subclass that can be created and invoked at run time. In addition, class hierarchy allows for inheritance of the methods—subprograms and functions declared inside the parent class—and need not be redefined or explicitly imported. However, each subclass may add its own definition of new methods, or *override* the definition of the methods given in the parent class or ancestor class. Class definitions are different from explicit module definitions, because classes have objects as an instance of classes, and support inheritance; explicit definition of modules supports neither instances nor inheritance. However, modules can have methods, as supported in the multiparadigm language Ruby. An abstract syntax of class is given below

```
<class> ::= class <identifier>
         [subclass-of] <identifier>
         [private {<data-declaration>}*]
         [protected {<data-declaration>}*]
         [public {<data-declaration>}*]
         [private {<method-declaration>}*]
         [protected {<method-declaration>}*]
         public {<method-declaration>}+
```

A *public method* is visible to other objects, private method is invisible to objects from other classes and subclasses; and protected methods are visible only to the subclasses unless explicitly sealed inside the class. Depending upon the language, either public declaration or private declaration could be implicit.

An object of a subclass is created at run time, using a *constructor* that allocates the memory location for the object, binds the identifier to the memory location, and initializes the data entities in the object to initial values. Each object has its own state, and objects communicate to each other by sending a message to invoked public methods.

Object-oriented languages support data abstractions as classes. Some of the classes supported are hash tables—(key, value) pairs, where key can be searched using hash functions; arrays—indexible collection of data entities, matrices—collection of collections of data entities, structures—named tuples, vectors—extendible collection of data entities and maps—collection of (key, value) pairs.

Programming languages such as C++, Eiffel, CLOS (Common Lisp Object System), Modula-3, Ruby, Scala, and Java support object classes. The object-oriented programming is based upon this concept of hierarchical object classes and has been discussed in detail in Chapter 11.

4.3 INFORMATION EXCHANGE

Each program unit executes commands including invocation of other program unit. A program unit can call another program unit using an invocation or be called by another program unit. A program unit can be a program, a function, or a method inside an object. In many languages that support nesting of program units, program units can be arranged in a hierarchical manner, which means that a program unit may be embedded inside another program unit. A program unit may also be passed as an argument to another program unit.

Whenever a program unit invokes another program unit, part of the environment is saved, most of the time some information is passed to the called program unit by the calling program unit, and some result is passed back to the calling program unit from the called program unit. This information exchange between program units is essential for the successful solution of a problem.

4.3.1 Mechanisms of Information Exchange

There can be multiple ways to exchange the information. In order to exchange the information, a subset of the environment and store available to the calling program unit has to be made visible to the called program units. This visibility can be achieved by (1) making the subset visible to all the program units; (2) making the subset visible to a subset of program units based upon nesting level of program units and; (3) making the subset visible using point-to-point visibility between the program units by sharing one of the three major variable attributes: *name, memory location,* or *value.* Visibility can be achieved in multiple ways: (1) by specifically creating a copy of the value from the calling program unit; (2) by passing the reference to the memory locations; (3) by copying the reference to the first memory location of a data entity; or (4) by passing the name of the variables.

FORTRAN uses a declaration named "COMMON" between the subprograms to globally share the memory space. The advantage of the common shared memory space is that only one copy of the memory space is created, and both the calling subprogram and called subprogram can directly write into and access from the shared memory space. This saves both excessive copying cost and memory allocation cost. Common blocks are assigned a name, and the data entities in two subprograms having the same common blocks are matched location by location for the correspondence. There are two major disadvantages of common block declaration from the programming viewpoint:

1. Common block has to be declared in every subprogram that uses it.
2. The alignment of multiple variables declared in the same common block is error prone.

Block structured languages use global variables to make the information visible to every program unit and nonlocal variables to the nested program units. Nonlocal variables are available only to the program units that are embedded inside. The information to outer program units can be passed using a variable local to the outer block that acts as a nonlocal variable in the program units nested inside the outer program unit. In addition, block structured languages use variable-name or memory location reference or value of the variables to pass the information between the calling program unit and the called program unit. During parameter passing, the arguments in the calling program unit are called *actual parameters*, and the corresponding arguments in the called subprogram are referred to as *formal parameters.* During parameter passing, the parameter correspondence can be achieved in four ways:

1. The actual parameters and the formal parameters are aligned from left to right, and they have position-by-position correspondence. This is a common mode of parameter correspondence in most programming languages.
2. The actual parameter and formal parameter names are matched using name associations, as in the programming language ADA. If name associations are used, then there is no need for position-based matching. The correspondence between

actual and formal parameter is specified in the call to the subprogram. In case of aliasing—the same actual parameter being associated with two different formal parameters— position ordering is used for binding the actual parameter attributes to the formal parameters.

3. In case the number of arguments in the procedure call and the number of formal parameters do not match, then the remaining formal parameters after matching the formal parameters with actual parameters are initialized to the default values.
4. The formal parameters may be extensible types, such as list, and indefinite number of arguments can be passed to the called subprogram such as in C# using "param" declaration.

Object-oriented languages can have nested structure of classes. Thus a variable can be described in a class, and it is visible in the nested subclasses. A variable declared inside a class is called *class variable* and is visible among all the methods and data entities within that class. A variable can be a *static global variable* that means it is available across all the classes and thus across the objects. A variable can be specific to a method. The information can be exchanged across the objects using global variables, using class variables, using parameter passing between the methods within a class, and using message passing between the objects. If the language supports modules in addition to classes and objects such as Modula-3, then the methods and variables can be used across the modules using an import–export mechanism.

Functional programming languages have a powerful mechanism of information exchange. They can pass the whole function as a parameter that can be invoked in the called function. The dual nature of function as data and back to function is a special property in functional languages. Functional programming also uses textual substitution of formal parameters by the name of the actual parameter before evaluating an expression. If the evaluation of an expression is delayed until needed, then it cannot be evaluated before passing the information to the called function. Rather, it has to rely on textual substitution. This property of information passing by textual substitution of formal parameter by actual parameter name is called *call-by-name* and is discussed in the next section.

4.4 PARAMETER PASSING

A variable in imperative language is defined as *name* ↦ *memory location* ↦ *r-value*. There are three attributes that are associated with the variable: *name*, *memory location*, and *value*. By giving either of the three, the final r-value can be derived. Parameter passing just does that. If the name is passed as an argument in the actual parameter, then the parameter passing mechanism is called *call-by-name*. If memory location is passed as an argument, then the parameter passing mechanism is called *call-by-reference*, and if the r-value is passed, then the parameter passing mechanism is called *call-by-value* or *call-by-copying*.

Most of the time, the called subprogram passes some results derived from the local computation back to the calling program. Results can be passed using global variables,

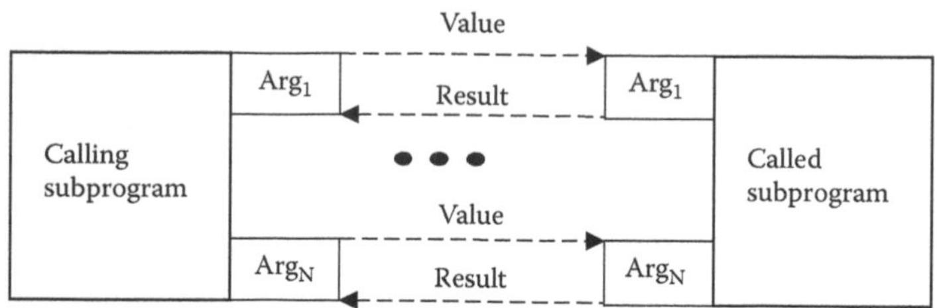

FIGURE 4.4 Parameter passing by copying the r-value of arguments.

nonlocal variables, sharing of common memory space, or using parameter passing. Call-by-name and call-by-reference implicitly handle the memory locations bound to the variables in the calling subprogram. Call-by-value has three options, as shown in Figure 4.4.

The calling program may pass the information from the actual parameter to the formal parameter. However, no result is passed back from the called subprogram to the calling subprogram. This type of parameter passing is called *call-by-value.*

The calling program may pass the information from the actual parameter to the formal parameter. After the called subprogram is done, the result is passed back to the calling program. This mechanism is called *call-by-value-result.* In *call-by-value-result*, the communication of information is two way. However, result is passed back only after the called program is over. The information exchange occurs in a specific order. Generally, it is a left-to-right order based upon the position correspondence.

The last parameter passing mechanism, where the r-value is explicitly exchanged, is called *call-by-result.* In call-by-result, no information is passed from the actual parameter to the formal parameter when the called subprogram is invoked; the formal parameters are initialized to default value. However, after the called subprogram is over, the result is passed back to the calling program using the formal parameter and actual parameter correspondence.

Generally, the number of arguments passed as parameters are well defined and fixed. However, some languages allow a varying number of actual parameters to be passed to the calling routine. A varying number of actual arguments can be modeled as an extensible data abstraction in a formal parameter that can be processed by an iterative loop that derives the size of the extensible data abstraction and processes every element of extensible data abstraction. Languages such as C# and Java support this feature. For example, C# will map three arguments such as "Mike," "Karen," and "Ambika" in a calling program using a formal parameter declaration "param string[] names." The reserved word "param" instructs the calling routine to treat the remaining arguments of the calling program as an extensible data abstraction. Here, "names" is a vector that has three data entities: "Mike," "Karen," and "Ambika." The advantage of an extensible parameter passing as a vector is that it can be represented and processed like a vector.

The parameter passing mechanism in the logic programming paradigm is *unification.* The difference between unification and other parameter passing mechanisms is that unification allows two-way information passing and assignments and does not evaluate any expression. However, unification is discussed in detail in Chapter 10.

4.4.1 Call-by-Value and Variations

In *call-by-value*, also known as *call-by-copying* or *call-by-in-mode*, the expression in the actual parameter is first evaluated, and then the resulting value is passed to the called subprogram and is bound to the corresponding formal parameter. The formal parameter has the same status as the local variable, and binding the formal parameters with the value of the expressions in the actual parameter is similar to assignment statement. When the called subprogram is invoked, new memory locations are created in the environment of the called subprogram, and the memory locations are assigned the value of the evaluated expression in the corresponding actual parameters. However, copying is strictly one way: from the memory location corresponding to actual parameter to the memory location of the formal parameter. After the values are copied, there is no more communication between the formal and actual parameters. The called subprogram performs computation and terminates without passing back any information.

Call-by-value is used for calling functions that do not return the results using parameter correspondence. However, functions return the value explicitly using "return (expression)" as illustrated in Example 4.8.

Example 4.8

Let us take the following code. The program reads the value of the variables, *x* and *y*, passes the value of the variables *x* and *y* by using call-by-value to a function "square_sum" that returns the computed value to the calling program main. The syntax is generic to explain the concept.

```
program main
{ integer x, y, z;
  read(x, y);
  z = square_sum(x, y)
  print("square sum of the numbers: ~d and ~d is ~d", x, y, z);
}


function integer square_sum(a, b)
{ return(a*a + b*b);}
```

The formal parameter "a" corresponds to actual parameter "x," and the formal parameter "b" corresponds to the actual parameter "y." Note that formal parameters are of the same type as actual parameter. Both "a" and "b" are copies of the variables "x" and "y" causing allocation of additional memory space in the environment of the function *square_sum*.

A major advantage of call-by-value is that since no result is passed back to the corresponding actual parameter, the called subprogram does not destructively update the corresponding memory locations, which will avoid side effects—undesired program behavior due to unexpected destructive updates in the environment of the calling procedure. Call-by-value is also used where only initial parameter values are needed by the called subprogram to perform computation, as illustrated in Example 4.9.

Example 4.9

The following program has two program units: *main* and *my_print*. The program *main* reads two arrays "a" and "b" each of size 100, and calls the subprogram *my_print* to add them element by element and print them out.

```
program main
{       integer x[100]; y[100];
        for (i = 0, i = < 99; i++) read(x[i], y[i]);
        call my_print(x, y);
}

subprogram my_print(integer a[100], b[100])
{         integer c[100];
          for (i = 0; i = < 99; i++) {
                      c[i] = a[i] + b[i];
                      for (i = 0; i = < 99, i++)
                      print("c[~d]= " ~d~n", i, c[i]);
                      }
}
```

There are two disadvantages of call-by-value: (1) the copying requires an additional amount of memory locations, the same as the memory needed to hold the actual parameters. If the actual parameter is large such as a 10000 × 10000 matrix as is common in large-scale scientific computing, then it would require an additional 100 million × size-of(single data entity) memory locations to run the called subprogram, and (2) the copying cost of the large data structures from the memory area of the calling subprogram to the memory area of the called subprogram would be quite high.

4.4.1.1 Call-by-Value for Sharing Complex Objects

Complex objects, extensible data structures and dynamic objects in object-based programming languages are stored in the heap—a shared common global space. These objects are accessed from the environment of a program unit using a reference link. The reference link points to the base address of the data object stored in the heap. When the calling subprogram needs to share an object with the called subprogram, then the reference of the object to be shared is copied to the environment of the called subprogram using call-by-value, as shown in Figure 4.5. The left side shows the scenario before the called subprogram is invoked. Since the copied information is an address, the formal parameter carries the address of the object. This mechanism is used in every language that supports dynamic objects stored in the heap.

4.4.2 Call-by-Reference and Variations

Call-by-reference, also called *call-by-access*, passes the l-value or the memory location of the actual parameter to the formal parameter. The formal parameter is a reference to the actual parameter. In the called program, the memory location of the actual parameter is accessed by dereferencing the formal parameter, and the memory location corresponding

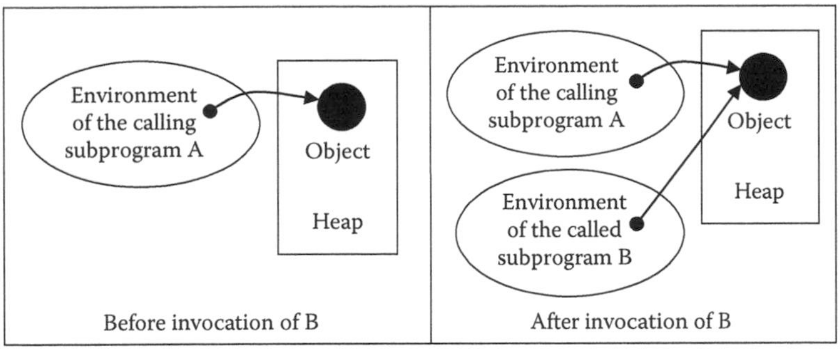

FIGURE 4.5 Using call-by-value to copy reference links to objects.

to the actual parameter is read or updated. In the case of a collection of data entities such as an array or vector, the memory location of the first data entity is passed, and all other data entities are accessed using the offset method that calculates the address of the *i*th data entity by adding the offset *i* * size(one data entity) and adding it to the memory location stored in the formal parameter.

It is clear from the discussion that changing the dereferenced value of the formal parameter is the same as accessing and updating the memory location of the actual parameter. There is no need for explicitly passing the result back to the calling subprogram from the called subprogram.

There are many advantages of use of the *call-by-reference* as follows:

1. A formal parameter is just a pointer to the first memory location of the complex data structure and is independent of the size of the data structure.

2. There is no need to explicitly copy the values from the actual parameter, saving the overhead of copying cost both in terms of memory locations and execution time.

3. There is no need to explicitly transfer the result back to the calling subprogram, as the memory locations of the actual parameters are continuously being updated whenever the corresponding formal parameters are modified.

Example 4.10

The following program has two program units: *main* and *count*. The program unit *main* uses a random number generator in the program library to create an integer between the value 1 to 200, and the called subprogram *count* counts the number of elements in a slice of array *d* that have value greater than 100. We also assume that an integer occupies 4 bytes, and a pointer occupies 4 bytes.

The program *main* has an array of integers *d*, and three additional integer variables: *i*, *j*, and *final_count*. The program *main* calls subprogram *count* using three actual parameters: array *d* using call-by-reference; variable *j* using call-by-value; and the variable *final_count* using call-by-reference. The argument variable *j* stores the value of the last index of the array *d*. The notation "&" denotes that address is being passed to the

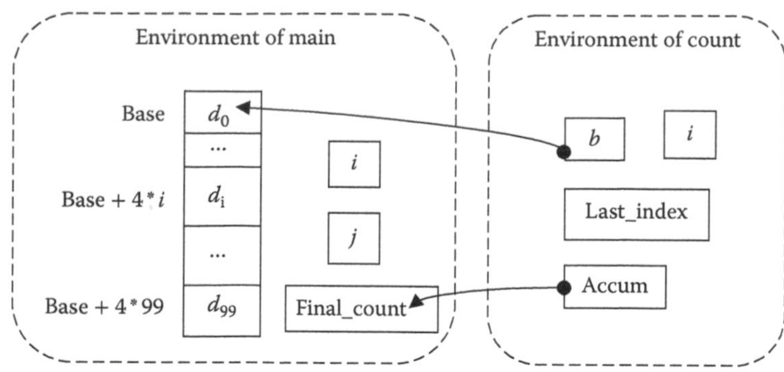

FIGURE 4.6 A schematic of call-by-reference.

called subprogram, and the notation "*" means that the following symbol is a reference to the actual parameter and there is a need to dereference to access the r-value.

The program unit *count* has three formal parameters: the variable *b* that is a reference to the data element d[0], the variable *last_index* that is a copy of the actual parameter *j* and the reference variable *accum* that is a reference to the actual parameter *final_count*. The reference variables *b* and *accum* occupy 4 bytes (32 bits), irrespective of the size of data elements they are pointing to. The address of the element *b[i]* in the program is computed using *address(d[0])* + 4 * *i*. Note that the actual parameter *j* is passed as call-by-value, and the formal parameter *last_index* gets a copy of the value of *j*. Figure 4.6 illustrates the reference links in call-by-reference.

```
program main
{     integer d[100], i, j, final_count;
      for (i = 0; i = < 99; i++) d[i] = random_number(1, 200);
      j = 50;
      call count (&d,j, &final_count);
}
subprogram count (integer *b, last_index, *accum)
{     integer index;
      *accum = 0;
      while (index =< last_index)
           {       if (*b[index] > 100) *accum = *accum + 1; %
                     end_if
                   index = index + 1;
           }% end_while
}
```

There are some disadvantages of call-by-reference as follows:

1. The called subprogram updates the store of the calling subprogram. After the called subprogram is done, the store of the calling subprogram is not the same as before. If the update was not needed by the calling subprogram, then calling subprogram may behave erroneously, because it will be reading the values from the corrupted store to perform the computation.

2. Every time an actual parameter is used, two memory accesses take place: one to access the memory location of the actual parameter, and then using the memory location of the actual parameter to access the r-value.

3. In distributed computing, if the calling subprogram and the called subprogram are residing on different processors (or computers), then they may have different address space. Accessing different address space requires the use of communication protocol and packing and unpacking of data that has significant overhead in addition to the unreliability of the communication link.

In order to alleviate the problem of inadvertently updating the store of the calling program due to misuse of the reference parameters, some languages such as C++ and Modula-3 support an option of *read-only call-by-reference*. In a *read-only call-by-reference*, the actual parameter value can be read only by using the pointer stored in the formal parameter memory location; the memory location corresponding to the actual parameter cannot be destructively updated.

Another variation combines first parameter passing as call-by-reference followed by subsequent parameter passing using call-by-value. The effect is that all the called subprogram sequences will have access to the original actual parameter: the first reference link when copied subsequently using call-by-value creates reference links in every subsequent called subprogram. For example, let us take a scenario when a program unit "A" uses call-by-reference to pass an actual parameter "x" to the called subprogram "B". The formal parameter "y" in the program unit "B" will be a reference link to "x" to the called subprogram "B". If program unit "B" calls another program unit "C" and passes "y" as call-by-value to a formal parameter "z" in program unit "C," then the formal parameter "z" will also be a reference link to "x." Both program units "B" and "C" can access the actual parameter "x" through their reference links. This scheme is allowed in many languages, especially object-based languages such as C#, Java, and C++, and is called *call-by-sharing* by CLU.

4.4.3 Call-by-Result

Call-by-result, also called parameter passing by *copying out mode*, is just the reverse of the call-by-value. In call-by-result, the actual parameter is not copied to the formal parameter location at the time of invocation. Instead, the formal parameter is initialized to a default value according to the type of the object. At the end of the called procedure, the value of the formal parameter is copied back to the actual parameter. Like call-by-value, call-by-result also treats the formal parameters as local variables. There is no communication between the formal parameter and the actual parameter during the execution of the called program.

4.4.4 Call-by-Value-Result

Call-by-value-result, also called *parameter passing by in–out mode*, passes the r-value in two ways: after evaluating the expression in the actual parameter to formal parameter, it passes back the result after the called subprogram is done. The passing back of the result is done to the same memory locations in a specific order. During the execution of the called subprogram, there is no communication between the calling subprogram and the called subprogram.

Like call-by-value, call-by-value-result also treats the formal parameters as local variables and creates memory locations in the local environment of the called subprogram. The memory allocation overheads in the call-by-value-result are the same as the call-by-value. However, the copying overhead of the call-by-value-result is twice as high as call-by-value, since call-by-value-result also passes back the result to the actual parameters.

Example 4.11

The following program creates local variables x[100], y[100] and z[100, 100], and copies the values of a[100] into the corresponding memory locations of x[100], b[100] into y[100] and c[100, 100] into z[100, 100]. The final result is passed back after the termination of the subprogram "multiply," such that from x[100] is copied to a[100]; y[100] is copied to b[100]; and z[100, 100] is copied to c[100, 100].

```
Program main
{    integer a[100], b[100], c[100, 100], i;
     for (i = 0; i =< 99; i++) read(a[i]);
     for (i = 0; i =< 99; i++) read(b[i]);
     call multiply (value-result a[100], b[100], c[100, 100]);

}
subprogram multiply(integer x[100], y[100], z[100, 100])
{    integer i, j;
     for (i = 0; j = < 99; i++)
          for (j = 0; j = < 99; j++)
                z[i, j] = x[i] * y[j];
}
```

Although call-by-value-result passes back the outcome of the computations to the calling program, there is a significant difference between call-by-reference and call-by-value-result as follows:

1. Call-by-reference keeps modifying the actual parameters as the computation proceeds in the called subprogram and retains the last value due to continuous destructive update. While the final value in the call-by-value-result is dependent upon the arguments' order in the procedure call and the actual parameters are updated only after the successful termination of the called subprogram. The final value of the actual parameter in the call-by-value-result need not be the same as call-by-reference, as explained in Example 4.12.

`main ( );`
`integer i;`
`{i = 1; sub(& i, & i);}`
 	
`void sub(integer *j, *k);`
`{*k = 4; *j = 2}`
	`main ( );`
	`integer i;`
	`{i = 1; sub(value-result i, value-result i);}`
	
	`void sub(integer j, k);`
	`{k = 4; j = 2}`

FIGURE 4.7 Comparison of *call-by-reference* and *call-by-value-result.*

2. *Call-by-reference* needs only one memory location for storing the reference link to access a large data structure, while call-by-value-result creates a copy of the actual parameter. Thus memory allocation overhead for large data structures is almost negligible in call-by-reference compared to call-by-value-result.
3. Call-by-reference has additional overhead of accessing the actual values, due to additional memory accesses. If the calling program is performing computation-intensive operations on large arrays, vectors, or any large collection of data entities, then the memory access overhead in call-by-reference may be much more than the copying cost in call-by-value-result.
4. In distributed computing, where the calling program and called program reside on two different processors or computers, the overhead of memory access in call-by-reference can be significant, and working on a local copy of the actual parameter is advantageous.

Example 4.12

Let us understand the difference between call-by-value-result and call-by-reference using the program given in Figure 4.7. The left-hand column shows the call-by-reference version of the program, and the right-hand column shows the call-by-value-result version of the same program. The sign "&" shows the address of the variable, and the sign "*" is used to dereference the pointer to access the value of the actual parameter. In the call-by-reference version of the program, the actual parameter's memory location is updated in the order of the execution of the statements. Hence the final value of the variable *i* would be 2. In the call-by-value-result version of the program, the result is passed back at the end of the called procedure in the left-to-right order. Hence the final value of the variable "*i*" would be the final value of the formal parameter "*k*"; that is, 4.

4.4.5 Call-by-Name

Call-by-name is the third major category of parameter passing, where the formal parameter is substituted literally by the whole expression-text of the actual parameter without any evaluation before substitution, and this substituted body of the called subprogram is executed in the environment of the calling procedure on a demand basis using a technique called *thunking.* Thunk is a parameter-less procedure with an unevaluated expression that is evaluated every time the actual parameter is accessed in the environment of the calling procedure. A thunk returns the address of the actual parameter every time the expression is evaluated in the environment of the calling procedure. If the body of the called subprogram after substitution contains a local variable that has a name conflict with a variable declared in the calling program, then the name of the local variable in body of the called subprogram is altered to avoid the naming conflict. The difference between call-by-name and call-by-value is that the expression in the actual parameter of call-by-name is not evaluated immediately before the substitution. Rather, it is delayed until after the substitution and is evaluated every time afresh when the actual parameter is accessed.

Example 4.13

The program in Figure 4.8 shows the *call-by-name* parameter passing. The left-hand side shows the actual program, and the right-hand side shows the run-time behavior after the called subprogram "sub" is invoked. The formal parameters *a*, *b*, and *w* are substituted by the expression-text $x + y$, $x + z$, and *w*, without any evaluation.

The variable *z* in the right-side expression of the assignment statement corresponds to the variable *z* declared in the environment of program *main* and is in name conflict with the local variable *z* in the subprogram *sub*. Thus, the local variable *z* is renamed to *z1* and the substituted body in the called subprogram is logically substituted in place of the call. The effect is shown on the right side of Figure 4.8. Essentially, after the substitution of the expression, the body of the subprogram *sub* acts as a block of the calling program. The execution of the program binds *z1* to the evaluation of (3 + 4) * (3 + 4) + (3 + 5) * (3 + 5), and the variable *w* gets the final value. Call-by-name is a powerful mechanism. However, it has two drawbacks:

1. Delayed evaluation using thunking has implementation and computational overhead.
2. Resolution of name conflict is an additional overhead.

The delayed expression evaluation in thunk means that the mapping of the identifier to the memory location can change at run time, and this can cause serious problems with

<table>
<tr><th>Program</th><th>Run time behavior after call to sub</th></tr>
<tr><td><pre>program main
{   integer x, y, z;
      real r;
      x = 3; y = 4; z = 5;
    call sub(x + y, x + z, w);
}

subprogram sub (name a, b, w)
{   integer z;
    z = a * a + b * b;
    w = square _ root(z);
}</pre></td><td><pre>program main
{   integer x, y, z;
      real w;
      x = 3; y = 4; z = 5;

      {   integer z1; % rename the variable
            z1 = (x + y) * (x + y) + (x + z)
              * (x + z);
            w = square _ root(z1);
      }
}</pre></td></tr>
</table>

FIGURE 4.8 An example of call-by-name mechanism.

<pre>program main
{   integer i, j, k; integer a[5];
    k = 3; j = 2;
    for (I = 0; I =< 4; i++) a[i] =0;
    swap(name j, k);
    swap(name k, a[k];
}
 	
subprogram swap(name m, n);
{   integer temp;
    temp = m; m = n; n = temp;
}</pre>
	<pre>program main
	{ integer i, j, k; integer a[5];
	    k = 3; j = 2;
	    for (I = 0; I =< 4; i++) a[i] =0;
	    { integer temp;
	        temp = j; j = k; k = temp;
	    }
	    { integer temp;
	       temp = k; k = a[k]; a[k] = temp;
	    }
	 }</pre>

FIGURE 4.9 Problem with memory locations mix-up *in call-by-name.*

run-time behavior of the program that may have different behavior than the program comprehension based upon program structure. It is very difficult to reason about the programs with call-by-name, specially when indexed elements such as a[i] are involved, because the value of "i" can be computed as an expression at run time, changing the memory location bound to the identifier "a[i]," while the programs may be written with a[i] being bound to a fixed memory location in mind, as illustrated in Example 4.14 and Figure 4.9.

Example 4.14

Figure 4.9 shows a program that swaps the values of variables using multiple calls to the subprogram *swap*. In the first instant, it swaps the value of *j* and *k*; and in the second instance, values of *k* and *a[k]* are swapped. The swap procedure is standard. The right-hand side of the figure shows the behavior of the program execution under call-by-name.

The first call to *swap(j, k)* works correctly, giving the value of $j = 3$ and $k = 2$. However, the second call gives the value of $k = a[2]$; that is, 0. Since *k* is equal to 0, instead of updating *a[2]*, the program updates *a[0]* to the old value of *k*; that is, 2.

Owing to this mix-up of the memory locations, *call-by-name* has found limited use in imperative languages. *Call-by-name* was originally proposed in ALGOL-60. It has been dropped later from other imperative programming languages. However, a variation of *call-by-name,* called *call-by-need,* is used in the functional programming languages that delay the evaluation of an expression until needed. Haskell is one such language that uses call-by-need.

4.4.6 Call-by-Need

Call-by-need is a variation of *call-by-name.* Unlike call-by-name, where the address of the actual parameter is computed every time the expression is evaluated, the call-by-need caches the value generated by the expression the first time it is evaluated and retrieves the cached value every time the expression is evaluated. The first evaluation is delayed. However, subsequent evaluations are not delayed.

In most of the cases, where the index is not being recomputed such as in the example of Figure 4.9, the address does not alter, and call-by-need acts in theory like an efficient *call-by-name*, where the value need not be computed every time. The effect of caching becomes like evaluate once, and then copy every time. *Call-by-need* can be seen as *call-by-name* followed by repeated *call-by-value*, where value is copied from the cache.

Example 4.15

Let us compare the efficiency issue in call-by-name and call-by-need using a simple example. In Figure 4.8, two expressions $x + y$ and $x + z$ are evaluated two times in the right side of the assignment statement "z1 = $(x + y) * (x + y) + (x + z) * (x + z)$;." Call-by-name will evaluate the expression both times, while *call-by-need* will evaluate the expression $x + y$ once and store the resulting value in cache. Next time it encounters the expression $x + y$, it retrieves the cached value. Similarly, call-by-name will evaluate the expression $x + z$ both times, while call-by-need will evaluate the first occurrence of $x + z$, cache the resulting value, and the next occurrence of the expression retrieves the value from cache.

Call-by-need is used in functional programming languages such as Haskell, which defers evaluation of expression until needed, and improves the execution speed by storing the value of a subexpression and utilizing the resulting value to avoid evaluating the same subexpression occurring at another location in the expression.

4.4.7 Passing Subprograms as Parameters

Languages supporting the functional programming paradigm such as Lisp, Scheme, Haskell, Ruby, and Scala, and many imperative programming languages, such as ALGOL and Pascal, support passing functions/subprograms as parameters. Passing functions as parameters involves passing the reference to a function's first instruction and the reference to the corresponding environment of the function to check for the number and types of the argument at run time. This is quite difficult for statically typed languages—languages where type of the data entities is declared at compile time—because different subprograms can be invoked by different calls at run time, and the compile-time analysis has to make sure that the environment, including argument types received by the called function, is same as the environment and the argument types of the function passed as the parameter. In dynamically typed functional programming languages, this is not a problem, as type checking is done at run time.

4.4.8 Parameter Passing for Distributed Computing

Distributed computing requires invocation of remote procedures on different processors. That means that different subprograms execute in different address spaces. Distributed computing uses three types of parameters passing: (1) *call-by-moving*, also called *call-by-copy*; (2) *call-by-reference*; and (3) *call-by-visit*, also called *call-by-copy-restore*. These parameter passing mechanisms are distributed address-space counterparts of call-by-value, call-by-reference, and call-by-value-result in single-processor implementation of programming languages. The invocation of the remote procedure involves evaluation of the expression, relocating a copy of the object to the remote site and binding it to the formal parameter, executing the remote procedure, and returning the result or object back to the calling procedure.

Call-by-visit makes a copy of the object temporarily on the remote processor and copies back the object after the called procedure successfully terminates. *Call-by-moving* makes a copy of the object for the execution of the called procedure on the remote processor. However, the object is not copied back. Call-by-visit is analogous to call-by-value-result, and call-by-move is analogous to call-by-value. Call-by-reference just copies the reference of complex objects.

Passing the address of the object on the processor of the calling procedures to called procedure on a remote processor has got significant overhead in distributed computing, due to communication overhead and going through system routines to distributed address space. Despite this overhead, there is an advantage of call-by-reference that the reference of an object can be easily passed between the distributed processors without excessive data transfer overhead.

Distributed programming language Emerald uses all three parameter passing schemes, namely, call-by-reference, call-by-visit, and call-by-move. The parameter passing for distributed computing is revisited in Chapter 8 on concurrent programming.

4.5 SIDE EFFECTS

A procedure has access to stores other than the ones created by local variables, due to the presence of global variables, nonlocal variables, reference to the environment of the calling procedures, and persistent data objects. *Side effect* is defined as an effect that outlives the called subprogram. Side effect is caused by (1) the updates that are made in the store that do not correspond to the environment created by local variables of the currently executing procedure; (2) an observable interaction with the outside world, such as writing into a file or a stream; or (3) raising an exception. While changes caused in the store corresponding to the local variable expire after the termination of the procedure, the changes caused in other stores or persistent objects or dynamic objects and recursive data structures that outlive the subprogram that created them remain even if the currently executing procedure terminates.

In imperative languages, *side-effect-based programming* is used to send the result back to the calling subprogram. However, if the store is altered in an indiscriminate way by some *scratch pad computation* in the called subprogram, then other program units, unaware of the memory locations being corrupted, may use corrupted values inadvertently. This causes an unexpected behavior, resulting in an incorrect computational outcome.

One of the serious problems of the side effect is the *loss of commutativity*. Commutativity is a fundamental property of many operators in expression evaluation, such as addition and multiplication. Given an expression $e_1 + e_2$, both subexpressions e_1 and e_2 are evaluated under the same original store by reading the value of the variables and ideally should not modify the store. However, if the evaluation of the expression e_1 or e_2 has a side effect, then the original store will change after evaluation of the subexpression e_1 or e_2, and the expression evaluated next in the sequence will read the values from the new store, giving a different final value of the expression, as shown in Example 4.16.

Example 4.16

Let us consider the following program that uses call-by-reference to pass the parameters to the called function *square_sum*. The function *square_sum* squares the two variables *x* and *y*, adds them, and returns the result back to the main program. However, it destructively updates the actual parameters *x* and *y* to 9 and 16. After the return, when the value of *x* is added to the result of the function *square_sum*, it gives the value of *B* as 34, which would be different if the subexpressions were commuted to *x* + *square_sum*(*x*, *y*), which would derive 28.

```
program main
{    integer A, B, x, y;
     x = 3; y = 4;
     A = square_sum(&x, &y) + x; % A becomes 34 instead of 28
     B = x + y;% B becomes 25 instead of 7
     print(A, B, x, y);

}

function integer square_sum(integer *x, *y);
{         *x = *x * *x; *y = *y * *y; % Assignment updates
             the variables x and y
          return (*x + *y);
}
```

4.5.1 Aliasing and Side Effects

Aliasing is defined as two identifiers mapping to the same memory location, or two pointers pointing to the same memory location. If one of the variables is assigned a value, the other variable gets automatically updated. This may have a drastic effect if the programmer is unaware of the aliases, or the called subprogram has been compiled separately from the calling subprogram. Let us see the effect using Example 4.17.

Example 4.17

The following program describes a combination of aliasing and call-by-reference to illustrate the unexpected behavior that can be caused by aliasing. Independently, both the program units—*main* and *swap*—behave correctly. These two units may be compiled separately and linked later, or swap may be loaded from another module, without any knowledge of its source code.

The main program calls the subprogram *swap* twice: (1) in the first call, it swaps the value stored in two different memory locations and (2) in the second call, it swaps the value of the same memory location. The swap routine is unconventionally written to avoid the use of any local variable to swap the value. Instead it uses arithmetic expressions. The new value of x is equal to $x_{old} + y_{old}$. The new value of y is $x_{new} - y_{old} \rightarrow x_{old} + y_{old} - y_{old} \rightarrow x_{old}$, and the new value of x is $x_{old} + y_{old} - y_{new} \rightarrow x_{old} + y_{old} - x_{old} \rightarrow y_{old}$.

Clearly, the subprogram swaps the value of y and z correctly. However, when *swap(&x, &x)* is called, both the formal parameters refer to the same memory location, and the first assignment statement $*x = *x + *y$ doubles the value of the actual parameter x. The second assignment statement subtracts the value with itself, storing 0 in the memory location of actual parameter x.

```
program main
{       integer x, y, z;
        x = 3; y = 4; z = 5;
        swap(&y, &z); % swaps the value of y and z
        swap(&x, &x); % stores 0 in the memory location
          x - an unpredictable behavior
        print(x, y, z) % x becomes 0; Y = 5 and z = 4
}
subprogram swap(integer *x, *y);
{       *x = *x + *y;
        *y = *x - *y;
        *x = *x - *y;
}
```

When multiple pointers are pointing to the same data structure in the heap, then releasing one pointer marks the memory location for recycling or garbage collection, as shown in Figure 4.10.

In Figure 4.10, two pointers P and Q point to different parts of a linked list: P points to the beginning, and Q to a few data cells. After the pointer P is released, then the whole linked list will become recyclable, the garbage collector will recycle it, and the memory locations will be allocated to other dynamic data structure. Thus the memory location will be corrupted, and the program will not behave correctly.

Another problem that causes undesired change in the store is the violation of a data abstraction boundary during run time, caused either by pointer arithmetic or the value of

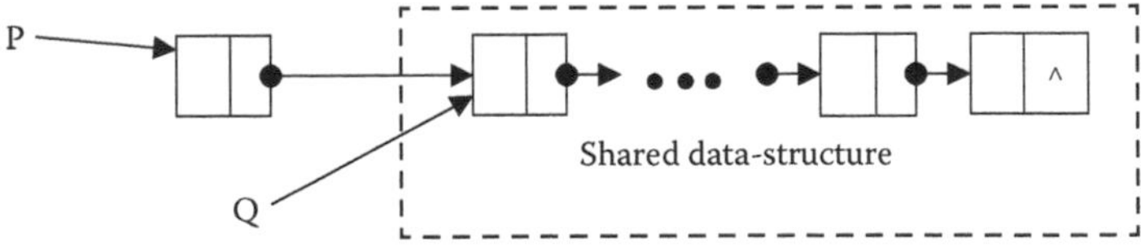

FIGURE 4.10 Pointers sharing a data structure.

an index accessing an array or vector that exceeds the upper bound of the data structure. For example, if we declare an array *a[100]*, and during execution the value of the index variable *i* is equal to 130, then *a[i]* will cross the boundary of the memory allocation for the data structure *a[100]*, and retrieve an incorrect value from some other memory location.

4.5.2 Regulating Side Effects

There are various causes of the overt side effects resulting in incorrect run-time behavior of the programs. The different causes are as follows: (1) scratch-pad computation in variables whose scope is not local, (2) pointer arithmetic, (3) independent allocation of pointers and allowing them to point independently to a data structure, and (4) destructive update of variables or data objects whose scope is not local. There have been many approaches to handle the problem of the side effects. The approaches have been to provide restriction at the language level as well as the disciplined approach to programming. The approaches are as follows:

1. *Programmer's discipline:* Use local variables for scratch-pad computations. Variables with scope outside the local environment should be modified only for information transfer. This is by far the most popular approach to handle undesirable side effects caused due to modification of variables whose scope is beyond the local environment.
2. *Disallow pointer arithmetic:* The pointers are not allowed any arithmetic. This takes care of type violation caused by a pointer crossing over the boundary of a data entity during run time.
3. *Disallow independent pointers:* Pointers can be declared only when describing a recursive data structure and internally to point to an object. This solves the problem of inconsistent operation on different data types.
4. *Disallow destructive updates:* Variables can be associated programmatically with value only once. Since the destructive update is a major cause of updating the store in the calling subprogram, this source of side effect will be removed. This solution has been tried in the earlier version of declarative languages. However, the assign-once property: (1) restricts the memory reuse (2) causes excessive generation of variables; and (3) forces recursive style of programming due to lack of support for iteration that needs destructive update of index variables in every cycle. The restriction on the use of mutable objects has resulted in some interesting recursive programming styles and the use of iterators in declarative programming languages. Iterators were later borrowed in most modern languages, as they provide a level of information hiding about the implementation level details.

4.5.3 A Comprehensive Example

This section describes a hypothetical programming example that exploits call-by-value, call-by-reference, call-by-value-result, and the mix-up caused by aliasing. Call-by-reference uses "&" in front of the actual parameter to denote that the address is being passed and uses "*" in front of the formal parameter to show dereferencing to access the value in the actual parameter. Presence of the symbol "#" in front of the actual parameter denotes that parameter passing uses call-by-value-result. Placing a symbol "$" denotes that the parameter is passed using call by-result. Figure 4.11 illustrates the effect of various parameter passing mechanisms using a program *main*, and a subprogram *messy*.

After the call to subprogram *messy*, the variable *a*[*1*] is passed using call-by-value, the variable *a*[2] is passed using call-by-reference, the variable *j* in the third and fourth argument is passed as call-by-reference, the fifth argument *a*[3] is passed as call-by-value-result, and the variable *k* is passed as call-by-result. The formal parameter *A* gets assigned the value 10—the value of the variable *a*[*1*], the formal parameter *B* is a pointer to the actual parameter *a*[2], and the formal parameters *C* and *D* are pointers to the memory location of the actual parameter *j*. Since formal parameters *C* and *D* point to the same memory location, changing the value of the memory location using the pointer stored in the memory location of the formal parameter *C* will also change the value **D* and vice versa. The formal parameter *E* gets a copy of the value of *a*[*1*] = 10. The formal parameter *F* gets initialized to 0, because the value is not copied in call-by-result.

The first statement retrieves the value of the actual parameter *a*[2] = 10 and the value of actual parameter *j* = 0 and adds them up to modify the value of *A* as 10. The second statement retrieves the value of the actual parameter *j* = 0 and adds the value of *E* that has been initialized to 10 (copy of the value of *a*[3]) and the value of *F* that has been initialized to 0 to derive the value 10 that is written back into the memory location of the actual parameter *a*[2]. The third statement reads the values of the variables *A* and *E*, adds them to derive the value 20, and stores in the actual variable *j*. The fourth statement reads the value of the actual parameter *j* and subtracts the value of the actual parameter *j* to derive 0 that is stored in the memory location of the actual parameter *j*. The fifth statement reads the value of the actual parameter *a*[2] and the value of actual parameter *j* and adds them to derive the value 10 that is assigned to the variable *E*. The last statement reads the value of the variable *E* is assigned the value 10,

```
program main( )
{       integer i, j, k, a[6];
        i = 0; j = 0; k = 2;
        for (i = 1; i =< 5; i++) a[i] = 10;
        messy(a[1], &a[2], &j, &j, #a[3], $k);
}

subprogram messy(integer  A, *B, *C, *D, E, F)
{       A = *B + *C;  % A = value(a[2]) + value(j) = 10 + 0 = 10
        *B = *D + E + F; % a[2] = value(j) + value(E) + value(F) = 0 + 10 + 0 = 10
        *C = A + E; % j = value(A) + value(E) = 10 + 10 = 20
        *D = *C - *D; % j = value(j) - value(j) =  20 - 20 = 0
        E = *B + *C;  % E = value(a[2]) + value(j) = 10 + 0 = 10
        F = E + A; % F = value(E) + value(A) = 10 + 10 = 20
}
```

FIGURE 4.11 Illustrating the combined effect of parameters and aliasing.

adds the value of the variable A that is 10, and assigns the resulting value 20 to the variable F.

After the procedure *messy* terminates, only the result for the variables E and F are copied back. The value of the variable E is copied back to the memory location of the actual parameter $a[3]$, and the value of the variable F is copied back to the memory location of actual parameter K. The final values are $i = 0$, $j = 0$, $k = 20$, $a[0] = 10$, $a[1] = 10$, $a[2] = 10$, $a[3] = 10$, $a[4] = 10$, and $a[5] = 10$.

4.6 EXCEPTION HANDLING

One of the major concerns in executing a program is the successful termination and not a sudden abortion due to unrecoverable errors such as "file not found," "invalid memory access," "divide by zero," "unable to open a file," or "array out of bounds." Many times, errors are data dependent and cannot be attributed to as the logical error. The errors may show once in a while depending upon the input data or system condition. In such cases, the program should be able to terminate, releasing all user-allocated resources such as files, buffers, and i/o devices, rather than abort prematurely, blocking the resources.

Exception could be at the system level, such as "file not found"; or the exception could be a user-defined error condition. Exceptions can be handled by the operating system using the concepts of the *traps*—software interrupts checked at the end of an instruction. However, traps are at the operating system level and will not allow graceful handling of the error condition. In order to gracefully handle the error condition, so that program does not abort abruptly, a provision has to be made for programmer-defined exception handlers in a programming language.

Exception handling is a programming language level abstraction to initiate an action in response to error conditions or undesired conditions that may potentially abort a program or cause a pathological situation, where the result of computation is erroneous. Exceptions can be built-in or user-defined. The languages that support exceptions also support user-defined exceptions. The built-in exceptions set a predefined Boolean condition that is tested in the exception handler. User-defined exception handlers need to be declared as an *exception type*. An *exception type* is a static Boolean variable that can be tested in a subprogram invocation in the chain of invoked subprograms. Different languages use different mechanisms to handle the scope of the exception conditions between the procedures that raised them and the exception handlers. One scheme is to use call-by-value to pass the exception condition to the exception handler routine.

If the exception handler can verify the exception flag, then the corresponding subroutine or the sequence of specified instruction is executed. In the absence of the appropriate exception handler, the control is passed back to the next higher level calling routine. Before the control passes to the next level, it executes the optional block, followed by the reserved word "*finally.*" Note that the declaration of final block is optional.

If the control finds the appropriate exception handler at this level, then the exception handler is executed. Otherwise, the process is repeated, and the control keeps going up to the next level of calling subroutines in the chain of calling subprograms, until the control reaches the main program. If no appropriate exception handler is found, the program terminates after reaching the main body of the program.

Different languages have different syntaxes for exception handlers. An abstract representation of an exception handler is as follows:

```
<extended-statement> ::= try  <statement>
       if <expression_1> raise <exception_1>;
       if <expression_2> raise <exception_2>;
       ...
       if <expression_M> raise <exception_M>;
       exception-handlers
       {      when <exception_1>: <block_1>
              when <exception_2>: <block_2>
              ...
              when <exception_N>: <block_N>;
       }
[finally <block_F>]
```

An exception handler can take many actions such as (1) passing the control to another routine; (2) releasing the resources; (3) correcting the source of error condition and passing the control back for retrying the actions; (4) returning the control to the next instruction after the exception handler, after notifying the programmer about the error condition; (5) raising the exception handler and returning to the calling routine to be processed at higher level; (6) skipping the data that caused the exception, and (7) returning to the calling subroutine.

Different parts of the program executing the same statement may have different exception handlers associated with the statement. Different statements may have the same exception handler.

Exception handling was first started in PL I. Most modern-day languages such as ADA, Java, C++, and Ruby have robust exception-handling capabilities.

Example 4.18

The program in Figure 4.12 illustrates exception handling using an intuitively clear syntax. The program opens a file "myfile," and the corresponding stream "mystream" is read for the debt amount for a credit card company. The program raises a built-in exception "file-not-found" if the file is missing. There is a user-defined exception "incorrect_debt" that is raised if the account value is greater than zero. The first exception handler writes a statement "Account-file missing," and returns to the calling subprogram. The second exception handler writes a statement, closes the stream, and returns.

```
subprogram illustrate_exceptions
integer  i;
real deposit, account;
file myfile;
stream mystream;
exception incorrect_debt; % user defined exception
open_file(myfile, mystream, read);
exception-handler {
        when file-not-found: write('Account file missing'); return }
read(mystream, account);
if (account > 0) raise incorrect_debt;    % raise the user defined exception
exception-handler {  % handle the user defined exception
        when incorrect_debt: write('Incorrect debt'); close(mystream); return;
        }
close(mystream);
return
```

FIGURE 4.12 An illustrative example of exception handling.

4.7 NONDETERMINISTIC COMPUTATION

Nondeterministic computation allows for alternate control flow of the program during program execution. The only condition for nondeterministic computation is that the final state should give the correct solution to the problem. If we represent the execution of a program as a graph traversal problem, where each node of the graph is a computational state, then a program is nondeterministic if we can find out more than one path from the initial state to the final state. Part of a computation is nondeterministic if the final state is independent of the order of execution of the statement. There are many fundamental properties of programming that support nondeterministic computation as follows:

1. Commutativity of operators in arithmetic expression and logical expression supports nondeterministic evaluation of the expression at the lower level. By definition, "commutativity" is independent of the order of evaluation. Some of the operators that support "commutativity" are addition, multiplication, logical-AND, and logical-OR. For example, if we evaluate an expression 4 + 5 + 3 + 9 using an intermediate-level instruction set, then it can be translated to multiple ways of adding two numbers at a time. The final result would be the same, because addition is associative and commutative in nature. Similarly, if we solve a logical expression ($exp_1 \wedge exp_2 \wedge exp_3$) involving logical-AND, it is immaterial in what order the expression is evaluated, the result will be the same, since "logical-AND" and "logical-OR" are individually associative and commutative operators.

2. If a store σ^S can be decomposed into disjoint union of sets of the form $\sigma_I^S \uplus (\sigma_1^S \uplus \sigma_2^S)$ and a sequence of commands C_1 ; C_2 works, such that C_1 works on σ_1^S and C_2 works on σ_2^S, and σ_I^S is invariant during the execution of C_1 and C_2. Furthermore, σ_1^S is independent of σ_2^S; the effects on them would not interfere with each other. After the command C_1, σ_1^S would be mapped to a new store $\sigma'{}_1^S$; and after the command C_2, σ_2^S would be mapped to a new store $\sigma'{}_2^S$. Thus the total effect of the command

sequence C_1; C_2 will give the new store $\sigma'^S = \sigma_I^S \uplus (\sigma'^S_1 \uplus \sigma'^S_2)$. If we commute the order of the commands, then they will produce the same final store $\sigma'^S = \sigma_I^S \uplus (\sigma'^S_2 \uplus \sigma'^S_1)$. For example, if we take two assignment commands, $x = 4$; $y = 5$, such that x and y are not aliases, then they will be altering two different memory locations and can be executed in any order.

3. If we take a selection command, such as if-then-else statement, and write it into a form such that all conditionals statements are freed from control flow dependency and are treated as equally likely to be executed, then changing the order of execution does not alter the final condition.

Let us take the following example from Dijkstra's famous article on guarded commands:

```
if (x >= y) smaller = y;
elseif (y >= x) smaller = x;
```

The above code finds the smaller of two values and can be transformed to two semantically equivalent codes as illustrated in Table 4.1.

The statement is nondeterministic, because, there are two possibilities of executing the statement and getting the same final condition. If $x > y$, then also the program can be executed in the order: the Boolean condition $y >= x$ will fail if tried first, and the first statement will be executed. In general, a nondeterministic conditional command can be written in the following abstract syntax form:

```
if { <Boolean-Expression_1> → <Command_1>  |
     <Boolean-Expression_2> → <Command_2>  |
     ...                                  |
     <Boolean-Expression_n> → <Command_n>
   }
```

In the above abstract construct, there is no control flow dependency, which means all conditional commands of the form (*<Boolean-Expression$_i$>* → *<Command$_i$>*) ($1 \le i \le N$) are equally likely. Each conditional statement is connected to the other using logical-OR denoted by the symbol '|'. The symbol '→' implies that the command part follows the successful execution of the Boolean expression. At least one of the Boolean expressions *<Boolean-Expression$_i$>* ($1 \le i \le N$) has to succeed to generate a solution. If all the Boolean expressions fail, then an error is generated. After the execution of the first successful Boolean expression, the corresponding command is executed to get a possible solution, and other Boolean expressions are not tried.

TABLE 4.1 Semantically Equivalent Programs

`if (X > = Y) then smaller = Y │`	`if (Y > = X) then smaller = X │`
`if (Y > = X) then smaller = X`	`if (X > = Y) then smaller = Y`

4.7.1 Guarded Commands

Dijkstra proposed a model of the nondeterministic programming where each Boolean expression was a *weakest necessary condition*, which needed to be true for the execution to commit to the conditional statement, and, once the execution was committed to one of the conditional statements, other conditional statements were not tried, although other Boolean expressions may also result in successful execution. Dijkstra called these Boolean expressions *guards*, and the conditional statements were called *guarded commands*. A guarded command consists of two components: *guards* and *commands*. Guards are Boolean expressions that do not modify the store. Commands contain the assignment statements and change the *precondition* to *postcondition*.

Guarded commands have the following properties:

1. Guards are the weakest preconditions for the commitment to a guarded command.
2. Guards are necessary but not sufficient conditions. Guarded commands can be tried in any order.
3. Once a guard succeeds, the corresponding command is committed, and other guards are aborted.
4. Owing to commitment to one of the guarded commands, and guard being only a necessary and not a sufficient condition, a guard may succeed without yielding a solution in the command part. A solution may exist in another guarded command. However, the other guarded commands are not tried after commitment to the first successful guard. Thus a program with guarded command is incomplete: it may not yield a solution, although a solution may exist.

Given the postcondition Q and the statement S, the notion $wp(S, Q) = P$ means that P is the *weakest precondition* for a nondeterministic statement, so that Q is the *postcondition*. Alternately, given a precondition P and a statement S, Q is the *strongest postcondition*. There are many properties of the weakest preconditions: for example, if $wp(S, Q) = P_1$ and $wp(S, R) = P_2$ then $wp(S, Q \wedge R) = P_1 \wedge P_2$. Similarly $wp(S, Q \vee R)$ is equal to $P_1 \vee P_2$; $wp(skip, Q) = Q$ for all the postconditions; and $wp(abort, Q) = fail$, which means that initial conditions must be *fail* to derive the final condition *abort*.

Given a sequence of statements $S_1; S_2$ and the final condition Q, the postcondition of S_1 is the weakest precondition of S_2. Thus the equation can be written as $wp(S_1; S_2, Q) = wp(S_1, wp(S_2, Q))$.

There are two types of constructs in guarded commands: guarded command selection and guarded command iteration. A guarded command selection is of the following form:

```
if
{     <guard_1>  →  <command_1>  |
      <guard_2>  →  <command_2> |
      ...
      <guard_N>  →  <command_N>
}
```

The symbol '→' separates a guard from the corresponding command in each guarded command, and the symbol '|' separates the guarded commands. The above construct tries out the guards in any order. If the guard succeeds, then the corresponding command is executed, and the remaining guarded commands are not tried. If a guard fails, then another guard is tried, until all the guards have been tried or one of the guards succeeds.

The guarded command iterative construct is described below

```
loop
{     <guard_1> → <command_1>  |
      <guard_2> → <command_2>  |
      ...
      <guard_N> → <command_N>
}
```

The iterative construct keeps looping until all the guards fail. If any of the guards succeeds, the corresponding command is executed, and the next iterative cycle starts again. When the control comes out of the iterative constructs, all the guards have failed, which means that the postcondition Q is equal to $P \wedge \neg\, guard_1 \wedge \neg\, guard_2 \wedge \ldots \wedge \neg\, guard_N$. Using De Morgan's theorem, the final condition can be rewritten as $P \wedge \neg\, (guard_1 \vee guard_2 \vee \ldots \vee guard_N)$. The iterative construct adds the Boolean condition $\neg\, (guard_1 \vee guard_2 \vee \ldots \vee guard_N)$ to the precondition.

Example 4.19

In the following example taken from Dijkstra's seminal work on nondeterministic programs, we see a toy example of nondeterministic iteration. Suppose we want to sort in a sequence of four numbers *a0, a1, a2*, and *a3*. We can use the nondeterministic iteration to solve this problem.

```
loop
{         a0 > a1 → swap(a0, a1)  |
          a1 > a2 → swap(a1, a2)  |
          a2 > a3 → swap(a2, a3)
}
```

For an initial condition *P*, after the execution of the iterative loop, the final condition is $P \wedge not\ (a0 > a1) \wedge not\ (a1 > a2) \wedge not\ (a2 > a3)$, which can be rewritten as $P \wedge (a0 \leq a1) \wedge (a1 \leq a2) \wedge (a2 \leq a3)$. By comparing the postcondition and precondition, the iterative construct makes $(a0 \leq a1) \wedge (a1 \leq a2) \wedge (a2 \leq a3)$, which is the same as the sorted ascending order $a0 \leq a1 \leq a2 \leq a3$.

4.7.2 Stepwise Program Construction

The intended final condition, and the axiomatic semantics of the statements, can be used to find out the weakest precondition that needs to be tested for the previous statement, and the process can be repeated backward to develop a nondeterministic program in a stepwise manner.

Given the final condition and the initial condition, the difference is taken. If the difference can be represented using conjunctive form $B_1 \wedge B_2 \wedge \ldots \wedge B_n$, then iterative construct could be used as follows: the guards are identified using the De Morgan's law by taking negation and using the rules about weakest preconditions described in the previous section. By using De Morgan's law, *not* $(B_1 \wedge B_2 \wedge \ldots \wedge B_n)$ gives guard$_1$ as *not* (B_1), guard$_2$ as *not* (B_2), and so on. Similarly, if we have the difference in the disjunctive form as $B_1 \vee B_2 \vee \ldots \vee B_n$, then the guarded command selection could be used such that guard$_1$ is $wp(command_1, B_1)$, guard$_2$ is $wp(command_2, B_2)$, and so on, such that each weakest precondition $wp(command_i, B_i)$ is included in the initial condition P.

Example 4.20

Let us take the case given in Example 4.2, and reason backward to develop the nondeterministic program. The final condition for the sorted sequence is $a0 \leq a1 \leq a2 \leq a3$, which can be written in conjunctive form as $(a0 \leq a1) \wedge (a1 \leq a2) \wedge (a2 \leq a3)$. Looking at the conjunctive form, we know that nondeterministic iterative construct can be used. Guard$_1$ will be *not* $(a0 \leq a1)$ that gives the guard$_1$ as $a0 > a1$. Similarly, other guards can be derived as $a1 > a2$ and $a2 > a3$.

4.8 PROGRAMS AS DATA

In programming languages, generally data is separated from the command, and data is something that is transformed, while command remains the same. However, many times, such as in artificial intelligence, there is a need to build a program as a data object and then convert into a program at run time that can act on data.

If a program is built as data at run time and then converted into a program, the program is being treated as a *first class object.* A *first class object* can be constructed at run time, bound to an identifier, passed as a parameter, or returned as a result of a computation.

Similarly, many times a program needs to analyze or process another program as data. For example, an editor needs to treat another program as data. A language can be written that can reason about its own execution in an abstract domain. All these are examples of when a program is treated as data. The process of representing program as data using some data structure is called *reification.*

4.8.1 Functions as First-Class Objects

Declarative languages such as ML, Haskell, Lisp, Scheme, and Prolog support programs as first-class objects. For example, most of the functional programming languages have a function-forming function that takes a function name and the corresponding arguments as data, and maps the data to an active function. The function (apply 'first '(Arvind Programs)) in the programming language Lisp will take two data arguments: function-name *first* and the list *'(Arvind Programs)*, and transform the data to a function *(first '(Arvind Programs))* that returns the data element *'Arvind.* Note that data elements have an additional quote marker in the Lisp family to separate them from the functions. The syntax used here has been used in Lisp and Scheme and is different for other functional programming languages.

4.8.2 Meta-Programming and Reflexivity

Meta-programs are the programs that manipulate or reason about other programs as their data. If a language can write meta-programs for its own programs, then it is called a *reflective language*. The declarative languages that support first-class objects support meta-programs. When a declaration is found, then the environment is updated by the meta-programs, and when a command string is found, then it is transformed to the program and executed. An example of meta-programs is an *abstract interpreter* that reasons about another program in an *abstract domain* to analyze the run-time behavior of the program. An example of an abstract domain is type domain, where the concrete values are transformed to their corresponding types. Another example of a meta-programs is writing an engine that executes the program of a language. Meta-programs can also be used to generate another program. Meta-programming can also be used to generate templates of programs that can be easily filled by the programmers, saving coding time and effort. Languages such as Ruby, Scala, Lisp, and Prolog provide natural support to meta-programming.

4.9 SOFTWARE REUSE

As software size is growing due to complex automation problems, it has become imperative that, everytime software is developed, the previously developed software be reused to avoid duplication, to improve the software development time, and to reduce errors in the software development process. Different subproblems can be coded efficiently using different paradigms. A complex software development requires integration of the programs developed using multiple programming paradigms.

Different languages support different programming paradigms, as shown in Appendix I. Different programming paradigms differ in the implementation of the supported data and control abstractions. For example, imperative languages traditionally use mutable indexible arrays or vectors to implement collections. Declarative languages traditionally use immutable sequences to implement collection of data entities. Paradigms even differ in the data and control abstractions they support. Some data abstractions and control abstractions are more suitable for different classes of problem domains needed in the development of complex software. An integration of multiple paradigms requires interoperability—interface with other languages, and capability to handle libraries developed in other languages.

Modern-day software uses both the approaches: use of the external off-the-shelf library, and interoperability with the software developed in other languages. The library is a set of executable functions that has been made public and can be imported in the software being developed. A library can be imported by a command "import *<library-name>*."

4.9.1 Interoperability Revisited

Language interoperability provides the ability to call a program written in another language, programming features present in another language, and to improve the execution efficiency.

The major problem in language interoperability is to interface the arguments in the calling program to the called program. There are two approaches to solve this problem: (1) develop a common middleware language, and all the languages are translated to this common language or (2) develop a data-format conversion interface across two interoperable languages. Both approaches have been used.

The *common middleware language* provides interoperability by specifying a *type system* and *metadata*. The type system of every interoperable language is translated to this *common type* to provide the interface. *Metadata* provides a uniform mechanism to translate the type of information from other languages to the common type in common interface language and to store and retrieve the information about common types. The *metadata* contains tables and data structures that hold information about the compiled program. The table contains information about the classes, fields and their types, exported types, and references to other metadata tables in a language-neutral interface, and removes the need for a specific interface definition language and customized interoperability code.

Despite the support provided by a common type system and metadata to translate one type to another type, different languages support different data abstractions, and it is difficult to map one implementation of data abstraction completely to another implementation of the same data abstraction exploiting full features.

The common language approach has been taken by .NET framework for many programming languages supported by Microsoft such as C#, C++, and Visual Basic. The .NET framework describes a common language specification (CLS) that describes the features and rules about the common language. XML has become a standard common language interface for web-based languages and databases.

4.10 CASE STUDY

In this section, we see multiple modern programming languages and discuss the abstractions supported by them. It should be noted that not all languages support all the data abstractions discussed in this chapter. We take simple examples from the languages ADA 2012, C++, C#, Fortran 2009, Java, Modula-3, Ruby, and Scala. ADA is a rich language that supports an imperative style of programming along with objects and modules. Fortran 2009 is a scientific block structured imperative programming language that supports the notion of objects. C++ is a combination of imperative and object-oriented programming. Java also supports the imperative and object-oriented programming paradigm. Lisp supports the functional programming paradigm, along with a limited imperative style of programming and is rich in control and data abstractions. Modula-3 supports explicit modules and object-oriented programming. Ruby integrates the notion of modules and class, and supports an integrated functional and object-oriented programming paradigm.

Scala is a modern multiparadigm language that integrates functional and object-oriented programming paradigms. The language Prolog representing logic programming language has been left out, as it requires explaining many concepts such as unification and AND-OR tree, which are different from abstractions in other programming languages and is discussed in detail in Chapter 10.

4.10.1 Data Abstraction in Programming Languages

In modern programming languages, many data abstractions are quite common. Almost all the modern languages support composite types, collections, and extensible data structures. Multiparadigm languages that support both declarative and imperative programming separate between mutable and immutable objects and support both types of objects. Mutable objects can be destructively updated, and immutable objects support assign-once property.

4.10.1.1 Single-Entity Abstraction

Single-entity objects such as integer, Boolean, real, and char are common and come from the mathematical world. In many languages such as ADA, Pascal, Modula, C, and C++, a user can define an enumerable set. However, enumerable sets were dropped from Oberon—a successor of Modula-2. The declaration of subrange and its use in arrays are standard in ADA, Fortran, ALGOL, Pascal, Modula-2, and ADA. However, range type has been dropped from C, C++, and Oberon due to limited use of this facility, and the lower bound of an array was fixed to "0."

Set type is supported in Pascal, Modula-2, and in a more recent language, Scala. However, set-based programming has not been adopted in many languages due to the limited use by programmers during commercial software development. An example of set type using Modula-2 syntax is given below

```
Type week = (Mon, Tue, Wed, Thu, Fri, Sat, Sun)
Var workdays: set of week;
```

First enumeration type *week* is declared. The variable *workdays* is declared as a set of type *week*. The variable workday can take any value that is a proper subset of week.

Scala uses the reserved word 'Set' to create a set. For example, a statement *var major = Set("CS," "Math," "Finance")* will create *major* as a set of three elements.

Enumeration type could also be a subrange, as shown below in an example from ADA:

```
Type        week is (Mon, Tue, Wed, Thu, Fri, Sat, Sun);
Type        year is 1..12;
Subtype     weekend is week range Sat..Sun;
Subtype     workingdays is week range Mon..Fri;
Subtype     fall_semester is year 8..12
Subtype     summer is year 6..8
```

ADA also supports a subtype that inherits all the properties of the original type and is discussed in detail in Chapter 7. Oberon extends Modula-2 type system by inclusion of multiple arithmetic types shortint ⊆ integer ⊆ longint ⊆ real ⊆ real. The value of an included type can be assigned to variables of including type. For example, if N is an integer and M is a longint, then M = N will be valid. This is inherently a case of coercion, as discussed in Chapter 7.

Scala or its extended variation Escala, a modern multiparadigm language that supports integration of functional programming, object-oriented programming, and event-based programming (supported in Escala), supports basic types such as integers and Booleans. In addition, it treats string declaration.

In addition to regular single-entity abstraction, many languages support types needed for concurrent execution such as monitors, coroutines, and semaphores, which are discussed in Chapter 8.

4.10.1.2 Composite-Entity Abstraction

Composite objects are tuples. Named tuples are called "struct" in ALGOL, C, and C++, and are called "record" in ADA, Pascal, and the Modula family of languages. The "struct" in C and C++ is written under a reserved word 'struct' followed by an identifier naming the structure followed by various fields. The group of fields is placed under the left and right curly brackets. For example, let us take a record *student* containing three fields: *name, department, degree.* Assuming that all three fields are strings, the composite abstraction is written in C++ using 'struct' as follows:

```
struct student {
          char* name;
          char* department;
          char* degree;}
```

where char* is used to represent an array of characters. The above declaration says that name is a pointer that points to an array of characters. Similarly, department is a pointer pointing to an array of characters. The pointer points to the first memory location of the array. The tuple can be initialized by associating a tuple of concrete values with a variable of type student as follows:

```
student plstudent = {"John", "Computer Science", "BS"};
```

ADA represents the named tuple using 'record' as follows:

```
Type PERSON is
record
   Name :         STRING(1..30);
   Department  : STRING(1..20);
   Degree  :  STRING (1..20)
end record;
```

The user-defined type can be used in another type definition. For example, the following definition of record *student* can be extended to a new record definition *alumni* by adding a field *graduationYear*. In Oberon, it can be written as follows:

```
student  = RECORD
              name: ARRAY 30 of CHAR;
              department: ARRAY 20 of CHAR;
              degree:     ARRAY 20 of CHAR;
           END
alumni  = RECORD
              person: student;
              graduationYear: INTEGER;
           END
```

Many languages, such as ADA, Pascal, and Modula-2, support *variant record. Variant record* is a composite structure that has two parts: *fixed part* and *variable part.* Variable part is a mutually exclusive groups of fields that can be allocated to the same memory space. Variant record interprets the shared memory space in the variant part of the record based upon the value of an enumerated variable. In case the value of the enumerated variable changes at run time, the same memory space may be interpreted as a different type of data. Incorrect interpretation can cause type-violation error, as discussed in Chapter 7.

Example 4.21

The following example representing composite entity in ADA syntax shows a variant record declaration. The type *Date* is a record having three fields of the type integer with different ranges. The type *assignment* is a variant record that contains the record name as one of the fields. The fields *problem, solved,* and *date_assigned* are fixed fields, and the field starting with the enumerated type *submission* is a variant field. There are three possibilities: *submitted, tobe,* and *missed.* If the value of *submission* is *submitted,* then the variant field is of the type integer with a range 1..100. If the value of submission is *tobe,* then the field is treated as the record type *date.* If the value of submission is *missed,* then the type of the field is integer with a score 0. Note that the variant record has different ranges and different types. However, the interpretation of the variant field is dependent upon the value of the field *submission.*

```
type Date is
        record
            month: INTEGER range 1..12;
            day:  INTEGER range 1..31;
            year: INTEGER range 2012..2015
        end record
type status is (submitted, tobe, missed);
```

```
type assignment (submission : status) is
  record
          Problems: Integer range 1..10;
          solved: Integer range 1..10;
          date_assigned: DATE;
          case submission is
                   when submitted => score:
                     Integer range 1..100;
                   when tobe => expected: Date;
                   when missed => score: Integer := 0
          end case
  end record
```

ADA, Pascal, and Modula-2 support variant record. However, variant record was dropped from Oberon, and is not explicitly present in many languages due to type violation error in the variable part as discussed in Chapter 7.

Prolog and Scala support dynamic tuples of the form (1, 2, 3). Scala can create tuples using the command *new tuple<no-of-entries>* '('*<tuple-values>*')'. For example, a statement could be **val** instructors = **new Tuple2**("Arvind," "Paul"), and individual elements can be accessed using instructors._1 and instructors._2.

4.10.1.3 Collection of Data Entities

Collections are modeled as sequences that are implemented using indexible arrays, linked lists, association lists, or vectors. The arrays can be declared by (1) using the size of an array, (2) giving the values stored in different dimensions, or (3) creating a jagged array that may carry indefinite extensible size data elements in each dimension.

Example 4.22

For example, let us take the various array declaration schemes in C#.

```
int[] x = new int[5];//single dimensional array of size 5
int[] y = new int[] {1, 30, 44, 33, 8};//size is number
                                           //of elements
int[] y = {1, 30, 44, 33, 8}; //size is number
                                  //of elements
int[,] z = new int[2, 3]; // Two dimensional array
int[,] w = {{1, 2, 3}, {4, 5, 6}}; //2 × 3 matrix
int[] [] jagged1 = new int[6] []; //jagged array with 6 rows
jagged1[0] = new int[3] {10, 20, 30};
jagged1[1] = {1, 30, 44, 33, 8};
```

Similarly, Ruby arrays can be declared for definite size or indefinite size that can be extended dynamically as follows:

```
x = Array.new # create a new array dynamically of indefinite size
x = Array.new(3) # create an array of size 3
x[0] = Array.new # first element is again an array
x[1] = [4, 5, 6, 7] # second array is of size four
```

Ruby also uses abstract operations such as *pop, push, shift*, and *unshift* as array is nothing but an indexible sequence. *Pop operation* removes the last element from the sequence. *Push operation* inserts an element at the end of the sequence. *Shift operation* removes the first element from the sequence and moves all the elements of the array one position toward the first element to fill in the empty space. Unshift operation moves all the elements toward the last position and inserts the given element at the beginning of the array.

Example 4.23

Let us implement the following statements in interactive Ruby:

```
p = [10, 20, 30, 40] # Create an array of four integers
p.pop # Remove the last element and p will have value [10, 20,
  30]
p.push(80) # Insert 8- at the end and p will have value [10,
  20, 30, 80]
p.shift # Remove the first element and p will have value [20,
  30, 80]
p.unshift(100) # Insert 100 as the first element and p will be
  [100, 20, 30, 80]
```

Languages supporting imperative programming paradigms such as ADA, Fortran, C, C++, C#, Java, Pascal, Modula-2, and Modula-3 support arrays. ADA also supports semidynamic arrays. Object-oriented languages such as C++, C#, Scala, and Ruby support dynamic arrays, as arrays can be represented as any other object that can be created dynamically. Dynamic languages such as Lisp and many implementations of Prolog also support dynamic arrays. Dynamic arrays are allocated in the heap of a program. Some of the syntax of declaring arrays in different programming languages is given in Table 4.2.

Languages such as Modula-3 also support ordinal type in array declaration. For example, an array in Modula-3 can also be represented *as ARRAY ['A'..'D'] OF INTEGER*, where *'A'* maps to the index 1, and *'D'* maps to the index 4. This array is equivalent to *ARRAY[1..4] OF INTEGER*.

TABLE 4.2 Array Declaration in Selected Languages

Language	Array Description
ADA	**array**(1..5) **of INTEGER**
C, C++:	**int** x[5]
C#	**int**[] x = **new int**[5];
Java	**int**[] x = **new int**[5];
Modula-3	**VAR** x : = **ARRAY** [1..5] **OF INTEGER**
Ruby	x = **Array.new**(5)
Scala	**var** x = **new Array** [**integer**](30)

4.10.1.4 Extensible Data Entities

Almost all the modern languages that support pointer, explicitly or implicitly at the implementation level, can create any recursive data structure such as linked lists, trees, and graphs through the use of pointers. Pointers can be strongly associated explicitly with recursive data structures, as in Pascal and Modula-3. In many languages such as ADA, C, and C++, pointers are supported as an independent entity that can be associated with any data structure. A data element in pointer-based data structure is accessed using pointer *dereferencing*—a mechanism to read the memory location pointed by the pointer. Declarative languages such as Lisp and Prolog embed the pointers in the language implementation and do not allow programmers to directly manipulate the pointers; data structures can be extended using higher-level kernel operators, such as *cons, insert*, and *append*. These operators are discussed in Chapters 9 and 10.

Linked List declaration in ADA:

```
type List_Pointer is access My_List;
 type My_List is
  record
    Info : INTEGER;
    Next : List_Pointer;
  end record;
 Start : List_Pointer;  – Always points to start of the list
 Last : List_Pointer;  – Points to the end of the list
```

Linked list declaration in C and C++:

```
struct My_List
{  int info;
   My_List* next;
}
```

Linked list declaration in Modula-3:

Modula-3 supports reference type, and extensible data entities are built using pointers.

```
TYPE
My_List = RECORD
```

```
   info: INTEGER;
   next: Ref My_List
END
```

Linked list declaration in Scala:

Scala is a high-level multiparadigm language and hides the use of pointers at the programmer's level. Its list is more like Lisp and Prolog. A list in Scala can be represented in many ways, as shown below. A signature may be declared for the list, or it can be inferred automatically by the value given.

```
val num : List[Int] = List(1, 2, 3) // num is a list of integers
val num = List(1, 2, 3) // The type of the list is inferred
val num = List()// Num is an empty list that can be extended later
val nested = List (List (1, 2, 3), List(4, 5, 6)) // Nested list
```

The concatenation of a data entity with a list is given by the operator '::'; thus, List(1, 2, 3) can also be represented as *Val num = 1 :: List(2, 3)*. Many languages such as Lisp, Ruby, and Scala support key-value pairs. These key-value pairs are called associative maps in Lisp, map in Scala, and hash in Ruby.

4.10.2 Control Abstractions in Programming Languages

4.10.2.1 Mutations

Languages that support imperative programming paradigms support mutable objects and destructive updates. For example, languages such as ADA, Java, C++, C#, Fortran, and Pascal support destructive update. Functional programming languages support immutable objects. In addition, many functional programming languages such as the Lisp family, Scala, and Ruby also support mutable objects and destructive updates in addition to immutable objects. Lisp supports destructive update of global variables and also allows index variables inside do-loops to be bound to different values inside the iteration. Multiparadigm languages such as Ruby and Scala support destructive updates and both immutable objects and mutable objects. Generally, linked lists and strings are treated as immutable objects, and indexible sequences such as arrays are treated as mutable objects.

4.10.2.2 Conditional Statements

Conditional statements in programming language are generally standard. Almost all languages support the nested if-then-else statement. Most of the languages also support the case statement. A few languages such as Haskell and some concurrent logic programming languages such as Parlog and GHC support guards. Functional programming languages have a functional version of the if-then-else statement. Lisp also has a general purpose conditional statement that groups multiple conditions within the same conditional construct.

4.10.2.3 Iteration and Iterators

Almost all programming languages that support imperative programming paradigm explicitly support index-variable based for-loop and indefinite iterations, such as while-loop and do-while loop. Some languages such as C simulate for-loop using while-loop. Many functional programming languages such as Lisp, Ruby, and Scala support indefinite iterations such as while-loop and for-loop for invoking functions repeatedly. Functional languages that support destructive update of variables also support iterative loops. Ruby and Scala are examples of such multiparadigm languages. Scala supports while-loop, do-while loop, for-loop, and iterators; and Ruby supports while-loop, for-loop, indefinite-loop, and iterators. An indefinite loop keeps iterating until exited using an exit condition. Since every data element in Ruby is an object, iterators and loops are treated as methods in Ruby. Following is an example of iteration constructs in Ruby.

```
[4, 5, 6].foreach {|i| puts i} will write 4, 5, and 6 in that other.
10.times {|i| puts i} will write 0, 1, 2, 3, 4, 5, 6, 7, 8, 9 in
  separate lines
```

4.10.3 Information Exchange in Programming Languages

Different languages support different parameter-passing mechanisms. For example, the language "C" uses call-by-value, and call-by-reference is simulated by copying the address of the data objects. The language "C++" uses call-by-value and call-by-reference and "const call-by-reference". The parameter-passing mechanism "const call-by-reference" allows read-only access and does not allow writing in the actual parameter. The access to a dynamic object stored in heap is always passed using call-by-value. This allows the objects to be shared. Java uses call-by-value. Since objects in the heap are accessed using pointers, call-by-value allows sharing of the objects by copying the pointers.

ADA uses call-by-value (in mode), call-by-result (out mode), call-by-value-result (in out mode), and call-by-reference (access). The entities are matched position wise, and access to the actual parameters in the calling program can be protected against modification by using the reserved word "protected" in front of the actual parameter. ADA also allows the use of named parameters of the form formal parameter → actual parameter to provide flexible binding of the actual and formal parameters; named parameters need not have position matching.

C uses call-by-value. However, C can pass the address of a data structure that is equivalent to call-by-reference. C++, C#, F#, Modula, Modula-2, Modula-3, Pascal, Fortran 90 onward, PHP, and Python use two parameter passing mechanisms: call-by-value and call-by-reference; call-by-reference is tagged with a reserved word to separate from call-by-value. All the heap objects' reference is passed using call-by-value, and any reference to objects stored in control stack is passed using call-by-reference. In addition, smaller data structures and single data entities stored in the control stack are copied using call-by-value.

Ruby passes the parameters by call-by-value. Every data entity in Ruby is an object stored in a heap; every variable is a reference to the corresponding object. Assigning a value to a variable is equivalent to creating a new object and pointing to that new object.

Assigning the value of a variable to another variable is equivalent to copying the address of the object. Call-by-value in Ruby copies the address of the object, so that the formal parameter starts accessing the object. The following program in interactive Ruby clears up the concept:

```
def entityId(Object);
     puts("passed parameter: #{Object.object_id}\n");
end
         y = 4.5;
puts("actual address: #{y.object_id}\n");
entityId(y)
```

The above code will give the same address in both the cases: the identifier *y*.**object_id** gives the address of *y* before passing the parameter, and the identifier *Object*.**object_id** gives the address of object pointed by the formal parameter *Object*.

Programming language Emerald supports parameter passing call-by-object-reference that is same as call-by-value for object sharing, since Emerald variables store pointers to objects stored in heap. Emerald supports distributed computing, and passing the reference of an object to the remote processor makes the memory access slow. Hence Emerald also uses call-by-visit and call-by-moving to create a local copy of the objects that can be processed by the called procedures on the remote processors.

Scala supports both call-by-name to pass complex expression for functional programming and call-by-value to pass the reference of the objects. Functions are first-class objects and can be passed as parameters. Programs can be developed using a functional style as well as an object-oriented style.

Haskell uses call-by-need instead of call-by-name to handle complex expressions during parameter passing. The advantage of call-by-need is that complex expressions are evaluated once and cached, and in the subsequent occurrence, the cached value is used.

ADA, C++, Clojure, Common Lisp, Fortran 90 onward, Python, Ruby, and F# allow for the default values in the parameter. This means that if the corresponding actual parameter is missing in a particular call to the subprogram, then the formal parameter will take the default value in the body of the subprogram.

4.11 SUMMARY

In this chapter, we discussed various program abstractions. Program abstractions include data abstractions, control abstractions, abstractions encapsulating data and control, programs as data, meta-programs, information exchange mechanisms, and interoperability between languages.

Data abstractions are single entities, composite entities with multiple attributes, collection of data entities, run-time extensible collection of data entities, network of data entities, and transient objects with global scope and persistent objects. Single data entities are modeled using a basic type declaration such as integer, float (or real), character, string, or an element in an enumerated set. A composite attribute contains multiple different fields that may be of different types and is abstracted as a *named tuple*. Programming languages use *named tuples* with *named fields* for better access to the tuple and the field. A collection of data entities can be abstracted as a set, an ordered set, a bag, an ordered bag, or a bag of (key, value) pairs. A key is unique for every data entity in the collection of (key, value) pairs. All these collections of entities are associated with names for ease of access and update during a program execution. Different data structures can be used to implement a collection of data entities. It can be implemented using linked lists, indexible arrays, vectors, and hash tables. Extensible collections can be implemented using linked lists, vectors—indexible as well as extensible—trees, and hash tables.

Control abstractions are assignment statements, including multiassignment, sequence of commands, block of statements, conditional statements, iterative constructs involving index variables, iterators, recursion, and subprograms like functions and procedures. Blocks are identifier-less structures of declaration and command sequence, such that the scope of the declaration is limited to the block. Blocks provide natural boundaries to regulate the visibility of the data and code. The regulation of visibility is important to allow flexibility to have same identifier name within blocks and modules, and to avoid naming conflicts.

Modules are another layer of abstractions that provide a natural boundary for embedded subprograms and data abstractions. The difference between block and module is that blocks are contained within one subprogram, and modules may contain more than one subprogram. Modules can be separately compiled and archived for future use. Subprograms and data abstractions from other modules can be used using export–import mechanism. An entity in a module becomes visible to other modules only after it is exported from the source module, and can only be used in a module after the entity is imported using an explicit import declaration. Class is a passive template containing both data abstractions and subprograms that operate on the data abstractions; objects—instance of classes are active components, and have a run-time state.

Information between subprograms can be exchanged using (1) global variables, (2) nonlocal variables, (3) class variables in object-oriented languages, (4) common memory space in low-level implementation and earlier FORTRAN implementation, and (5) parameter passing in the block-structured languages. Since variable is a value holder, all the attributes that can derive the r-value of the variables can be used in parameter passing. Broadly, there are three attributes: name, memory-location, and the r-value itself. On the basis of this, we discussed five major types of parameter passing mechanisms: call-by-name, call-by-reference, call-by-value, call-by-value-result, and call-by-result.

Call-by-value evaluates the expression in the actual parameter, and copies the value to the formal parameter and treats the formal parameter as a local variable. The communication is one-way only, and computation results in the called subprograms not being passed back. Call-by-reference passes the address of the memory location of the actual parameter to the formal parameter. In case of composite data entity or collection of data entities, the base address of the data structure is transferred to the formal parameter. The data structure remains in the environment of the calling procedure and is accessed using the base-address + offset of the individual element (or subfield in the case of composite entity).

Object-based languages use heap for storing the objects. Complex objects and dynamic objects stored in the heap are referenced using a reference link. Using *call-by-value*, the reference links to the objects can be copied. We also discussed briefly two variations of call-by-reference: read-only call-by-reference, and call-by-sharing. In read-only call-by-reference, actual parameters can be read using the reference stored in the formal parameter, but not updated. It is important to avoid any accidental destructive update of the actual parameters that can cause incorrect program behavior. Call-by-sharing is a combination of first call-by-reference and then on subsequent procedure invocation in the chain of invocation using call-by-value to share the reference link with other called subprograms.

We discussed the unpredictable run-time behavior of call-by-name, due to the run-time computation of indexed variable that may map an identifier to a different memory location. Call-by-need is a variation of call-by-name, where a subexpression is evaluated once, and the value is cached for subsequent occurrence of the same expression. Instead of evaluating the subsequent subexpressions, cached value is used to improve the execution efficiency.

We discussed side-effect as modifying the store that is accessible and modifiable by the called program but has lifetime beyond called programs. The outcome of side-effect is that the effect of modification remains even after the called subprogram is over. If the effect is unintended, then fundamental principles of programming such as commutativity of the expressions can be violated causing program to behave incorrectly at run time. Aliasing combined with reference links can also cause unpredictable behavior of the programs.

First-class objects can be built as data at run time, and then transformed into a program. Meta-programs treat another program as data in some abstract domain to derive the properties of the program or to interpret the program behavior.

Exception handling makes the programs to be more robust and allows for graceful exit from the exceptions. Interoperability can be handled by interfacing data types and data abstractions of the calling and called programs. It can be done using specific interface definition languages or using a common specification language and metadata that specifies the conversion details between data types of two different programming languages.

Finally, we looked at some examples of data and control abstractions in various languages such as ADA, C, C++, C#, Java, Modula-3, Ruby, and Scala. More examples from paradigm-specific languages have been deferred to the chapters on specific paradigms: the programming examples from Lisp, Scheme, and Haskell have been deferred to Chapter 9 on functional programming paradigm; programming examples and information exchange mechanism from Prolog have been deferred to Chapter 10, on logic programming paradigm; detailed examples showing object-oriented features from C#, C++, Java, Ruby, Scala, and Modula-3 have been deferred to Chapter 11 on object-oriented programming paradigm; programming examples from scripting languages such as Python, Perl, and Lua have been deferred to Chapter 14 on scripting languages.

4.12 ASSESSMENT

4.12.1 Concepts and Definitions

Actual parameter; aggregate; aliasing; assertion; blackboard; block; call-by-copy; call-by-move; call-by-name; call-by-need; call-by-reference; call-by-result; call-by-sharing; call-by-value; call-by-value-result; call-by-visit; chained assignment; class; collection of data entities; common middleware language; common type; composite data entity; conditional statement; control abstraction; data abstraction; declaration; destructive update; exception handling; extensible data entity; first-class object; formal parameter; global variable; guards; guarded commands; immutable object; import–export; information exchange; information hiding; inheritance; iteration; iterator; map; meta-data; meta-programming; module; multiple assignment; mutable object; mutator; mutual recursion; named tuple; nondeterministic programming; nonlocal variable; object; object class; override; parameter passing; persistence; postcondition; precondition, private method; protected method; public method; read-only call-by-reference; recursion; reflective language; reflexivity; reification; self-recursion; sequence; sequential assignment; shadowing; side-effect; single data entity; single-entry; single-exit; symbol file; thunking; tuple; type system; unless; visibility; weakest precondition; when.

4.12.2 Problem Solving

1. Write a quicksort program in programming languages Modula-3, C++, and Java, and compare the various abstractions used in the programming languages.

2. Write a program in Modula-3 and ADA 2012 to perform a merge-sort in a module called "my_sort," and then use it in the main program to read and write a sequence of data elements.

3. A hotel is abstracted as a two-dimensional array hotel[5, 120], where the first dimension shows the floors, and the second dimension shows the rooms within a floor. Each data entity occupies *b* bytes. Give an equation to compute the offset of hotel[i, j], and apply it to derive the offset for the location 4_{th} floor room number 18. Assume that the numbering starts from 1 and not 0.

4. A class is modeled as an array of 30 composite data entities called students. Each composite data entity is modeled as a tuple of the form (*name, age, department, year*). Assume that name is modeled as a fixed size string having 32 characters, age is modeled as an integer with 4 bytes, department is modeled as an integer with 4 bytes, and year is modeled as an integer with 4 bytes. Compute the offset of various fields in the following declarations such as class[i].name.

5. Assume that a language supports four types of parameter passing: call-by-value, call-by-reference, call-by-value-result, and call-by-result. For the following program, show the trace after the execution of every statement in the form of a two-dimensional matrix where the rows show a statement, and columns show various variables and formal parameters. Assume that & is the actual parameter that passes the reference, # is the actual parameter that passes the parameter using call-by-value-result, and $ in the actual parameter that is used to pass the parameter using call-by-result.

```
Program main ( )
integer i, j, k, a[6];
{       i = 0; j = 0; k = 2; a[0] = 1
for (i = 1; i =< 5; i++) a[i] = a[i-1] * 2;
messy(a[3], &a[4], &j, &j, #a[3], $a[4]);
}

void messy( integer a, *b, *c, *d, e, f)
{        f = 2;
         a = *b + *c + e;
        *b = *d + f;
        *c = a + *b;
        *d = *c - *d;
         e = *b + *c + e;
         f = e + a;
}
```

6. Write a program to perform bubble sort in ADA, Modula-3, C#, and Java, and compare the control abstractions and parameter passing used in four programs.

4.12.3 Extended Response

7. How are extensible data abstractions implemented? Discuss.
8. What is the advantage of export–import in the development of a reusable library? Explain.
9. What is the difference between the concept of class and modules? Explain clearly.
10. What are the problems of *call-by-name*? Explain using simple examples.
11. What is *call-by-need*? How does it improve the execution efficiency over *call-by-name*? Explain using a simple clear example.
12. What is *call-by-sharing*? How is it similar to *call-by-value*? Explain using a simple figure.
13. What are the differences between *call-by-value* and *call-by-reference*? Explain in terms of mechanism and access efficiency.
14. What are the advantages of modules and export–import mechanism over nested procedures that support nonlocal variables? Explain.
15. Under what circumstances is *call-by-value-result* preferred over *call-by-reference* and why?
16. Explain and compare various parameter passing mechanisms for distributed computing.
17. What are the different mechanisms to provide interoperability across different programming languages? Explain.

18. What is the mechanism to handle exception? Explain.
19. Compare the export–import mechanism with nonlocal variables and inheritance.
20. What is the difference between module, class, and objects? Explain.

FURTHER READING

Abelson, Harold, Sussman, Gerald J., and Sussman, Julie. *Structure and Interpretation of Computer Programs*, 2nd edition. MIT Press. 1996.

American National Standard Institute. *Programming Language ADA*. CSA ISO/IEC 8652:201z. 2012. Available at http://www.adaic.org/ada-resources/standards/ada05/

Birrell, Andrew D. and Nelson, Bruce J. "Implementing remote procedure calls." *ACM Transaction of Computer Systems*, 2(1). 1984. 39–59.

Black, Andrew, Hutchinson, Norman C., Jul, Eric, and Levy, Henry M. "The development of the Emerald programming language." In *Proceeding HOPL III Proceedings of the Third ACM SIGPLAN Conference on History of Programming Languages*. 2007. 11-1–11-51.

Collingbourne, Huw. *The Book of Ruby*. No Starch Press. 2011.

Dijkstra, Edsger W. *Discipline of Programming*. Prentice Hall. 1976.

Hoare, Charles A. R. "An axiomatic basis for computer programming." *Communications of the ACM*, 12(10). 1969. 576–583.

Hudak, Paul, Hughes, John, and Jones, Simon P. "A history of Haskell: Being lazy with class." In *Proceedings of the Third ACM SIGPLAN Conference on History of Programming Languages*. 2007. 12-1–21-55.

Kennedy, Ken, Koelbel, Charles, and Zima, Hans. "The rise and fall of high performance fortran: An historical object lesson." In *Proceedings of the Third ACM SIGPLAN Conference on History of Programming Languages*. 2007. 7-1–7-22.

Klein, Peter. "Designing software with Modula-3." *Technical Report 94-16*. Department of Computer Science III, Aachen University of Technology. 1994.

Liskov, Barbara and Guttag, John. *Abstractions and Specification in Program Development*. MIT Press. 1986.

Liskov, Barbara and Guttag, John. *Program Development in Java: Abstraction, Specification, and Object-Oriented Design*. Addison-Wesley. 2000.

Odersky, Martin, Spoon, Lex, and Venners, Bill. *Programming in Scala: A Comprehensive Step-by-step Guide*, 2nd edition. Artima Incorporation. 2011.

Stroustrup, Bjarne. "Evolving a language in and for the real world: C++ 1991-2006." In *Proceedings of the Third ACM SIGPLAN Conference on History of Programming Languages*. 2007. 4-1–4-59.

Watt, David A. *Programming Language Concepts and Paradigms*. Prentice Hall, 1990.

Wirth, Nikolas. "Modula-2 and Oberon." In *Proceedings of the Third ACM SIGPLAN Conference on History of Programming Languages*. 2007. 3-1–3-10.

CHAPTER 5

Implementation Models for Imperative Languages

BACKGROUND CONCEPTS

Abstraction and information exchange (Chapter 4); Abstract concepts in computation (Section 2.4); Data structure concepts (Section 2.3); Program and components (Section 1.4); Recursion (Section 2.2.4); von Neumann machine (Section 2.1).

In this chapter and the following chapters, we study the abstract implementation of the programs using a low-level abstract machine. An abstract machine abstractly explains the step-by-step execution of high-level programming constructs at an intermediate level. Using an abstract implementation model, we will be able to understand and analyze low-level execution behavior of data and control abstractions that is necessary for (1) writing better and efficient programs and (2) developing a code generator for a compiler.

The low-level translated code can be executed using four types of memory allocation schemes: (1) *static allocation*; (2) *stack-based allocation*; (3) *heap-based allocation*; and (4) *hybrid allocation*, which integrates the first three allocation schemes for optimum memory allocation. In this chapter, we discuss static allocation, stack-based allocation, and hybrid allocation. Our focus would be the hybrid allocation that (1) allocates local dynamic variables in a stack, (2) allocates recursive and dynamic data structures in a heap, and (3) uses *static allocation* for the efficient direct memory access of static variables.

Static allocation schemes allocate the needed memory up front, at the compile time, and do not support memory growth at run time. The advantage of this scheme is that every data object is mapped to a unique memory location and can be directly accessed using a single memory access. However, there are drawbacks too. Static allocation does not support (1) recursive procedures, since recursive procedures can call themselves an indeterminate number of times, which needs run-time memory growth; (2) extensible recursive data structures such as linked lists, trees, and vectors, since recursive data structures can extend indefinitely and need run-time memory growth; and (3) dynamic creation of objects at run time because it requires mapping dynamic objects to memory locations at run time.

All three limitations are caused because run-time growth of memory is not supported by static allocation. Static allocation is also wasteful of memory, because every object gets mapped to a unique memory location, and allocation does not allow reallocation of memory locations after the lifetime of the currently allocated object is over. So static allocation scheme is not good for (1) block-structured programs that support multiple calls to functions and procedures within a subprogram, as it does not support memory reuse; (2) an object-oriented programming paradigm that creates dynamic objects; (3) recursive data structures that can be extended at run time; and (4) language features such as recursive procedures that require run-time growth of the memory.

Stack-based allocation uses a stack, called *local stack* or *control stack*. There are many advantages of the use of a stack. The stack-based allocation supports recursive procedures, because the stack grows with the procedure invocation and is not fixed at compile time. The size of the stack is limited only by the operating system. The local store of the called procedure is pushed on the control stack upon a subprogram call during run time, because the stack can grow. After the called subprogram is over, the local store of the called subprogram is popped, vacating the memory locations that are *reused* when another subprogram is called. The major disadvantage of stack-based allocation is the lack of support for dynamically extendible data structures, such as linked lists, trees, vectors, or dynamically created objects, that outlive the lifetime of the subprogram that created them.

Heap-based allocation uses a common memory area called heap that is seen by all subprograms, has the same lifetime as the program, and can accommodate all types of data structures including extensible data structures and dynamically created objects. Heap-based allocation uses a pointer from the processor registers or from the control stack to point to the first data cell of a data structure stored in the heap, and then uses internal pointers between the data cells to traverse to other data cells in the same logical data structure. Since each pointer is a memory address, traversal of recursive data structures requires multiple memory accesses. Heap space can be extended at run time either automatically by the operating system or by the programmer directives embedded inside a program. After an allocated data structure is deleted, the memory is released and marked for recycling. After the allocable memory space has been consumed completely, there are two options: (1) extend the memory space using the operating system or (2) recycle the released memory for reallocation. The first approach is dependent upon the available memory space with the operating system. The second approach reutilizes memory and is known as *garbage collection*.

All three models have their advantages and drawbacks. The major advantage of the static implementation is that the memory location of a variable can be accessed directly using a single memory access. The major advantages of a stack-based allocation are (1) the ability to handle recursive procedures and (2) memory reuse due to memory reclaim after a called subprogram is over. The major advantages of a heap-based allocation are (1) allocation of extensible data entities and dynamically created objects and (2) memory reuse when the lifetime of dynamically allocated extensible data structures or dynamic objects is over.

Modern programming languages use a hybrid allocations model that exploits the advantages of all three approaches. The hybrid allocation model uses a static allocation for static variables and global variables, stack-based allocation for dynamic local variables in a program unit, and heap-based allocation for recursive data structures and dynamically created objects.

In this chapter, we discuss the translation of control abstractions to low-level code, static allocation scheme, and stack-based allocation scheme. We also discuss how control stacks handle various parameter passing mechanisms. In the following chapter, we will discuss dynamic memory management using heap.

5.1 ABSTRACT COMPUTING MACHINE

Abstract implementation of a program on von Neumann machine is composed of five major components: *data area*, *code area*, *instruction pointer to step through the instructions in the code area*, *registers to perform scratch pad computation*, and *program status word* (PSW)—a set of flags stored in a special register. The flags in the PSW are set accordingly after every instruction and form an important part of the computation state of an executing program. The memory area where the code is stored is called *code area*, and the memory area where the data is stored is called *data area*. For the languages that do not support first-class objects, the code area is *fixed* and *reentrant*, which means that every time a called subprogram starts from the beginning in the same initial state.

Data area is a triple of the linear structures (*heap*, *control stack*, *static-area*). Heap and control-stack grow in opposite directions starting from the opposite end. They grow toward each other for maximum utilization of the available memory space. A *control-stack* is a sequence of frames reflecting the calling pattern of the subprograms in last-in-first-out order. Each frame, also called an *activation record*, is an indexible sequence of memory locations that store (1) information needed for computation in the current subprogram, (2) information to return the control back to the calling subprogram, and (3) frozen state of the calling program when the called subprogram was invoked. The information in a frame includes (1) memory locations for the local dynamic variables—variables that are not static variables; (2) image of the registers that would be updated in the called subprogram; (3) various pointers to the data entities in the calling program, the code area, and top of the control-stack; (4) the archived part of environment and store of the calling program that are shadowed in the currently executing subprogram; (5) simple dynamic objects that are created once in a subprogram and do not outlive the subprogram; (6) the frozen state of the calling subprogram at the time of the call to the current subprogram; and (7) parameters passed to the called subprogram.

Each invocation of a subprogram pushes the corresponding frame on top of the control stack. The frame is discarded to release the memory locations after the called subprogram is over. A frame is indexible to support access to different data entities stored within the frame and through references to other frames or heap. After the called subprogram is over, the state of the calling subprogram is retrieved, and the execution of the code area in the calling program resumes from the next instruction following the call to the executed subprogram.

Code area is the sequence of low-level instruction blocks. Each instruction block is a sequence of low-level instructions representing a subprogram. The control flow uses a stepwise movement to the next instruction or jumps from one instruction to another instruction using a jump instruction. Control moves from one subprogram to another subprogram using jump statements. The control flow in the code area is controlled using an *instruction pointer* that is similar to the *program counter* in assembly-level instructions. There is one major difference between the instruction pointer described here and the program counter used in processors: the instruction pointer is incremented after the current instruction has been executed successfully, while the program counter in processors is incremented right after fetching the current instruction. This difference is only for our convenience, to explain the concepts, and reduces the confusion in handling offsets. A *sequencer*—jump instruction—takes the control to an instruction specified by the user-specified value rather than the adjacent instruction in the sequence. A jump instruction is essential for the low-level translation of control abstractions such as conditional constructs and iterative constructs. In the next section, we discuss how various control abstractions are translated to low-level abstract instructions using a control flow diagram.

Low-level instructions are categorized as subsets of instructions: (1) to load values stored in a memory locations, (2) to load constant into a register, (3) to store values from registers to a memory location, (4) to store constant value in a memory location, (5) arithmetic and logical operations to evaluate expressions, (6) operations involved in comparing arithmetic and logical expressions, (7) conditional and unconditional jumps to an instruction, (8) pushing a data on the control stack, and (9) popping a data from the control stack.

We assume three types of memory accesses: (1) *direct memory access*, (2) *indirect memory access*, and (3) *offset-based memory access*. *Direct memory access* performs a single memory access to read or write data from (to) the given memory location and is fastest. Direct memory access is suitable for static variables. *Indirect memory access* uses a pointer to access a memory location that stores another memory location or value. Indirect memory access may go through a chain of pointers before accessing the value and needs more than one memory access. Indirect memory access provides independence from fixed memory locations. However, it is slower due to multiple memory accesses. Many times, pointers for frequently used indirect accesses are stored in processor registers to improve the efficiency. Offset-based memory access is used to access a data entity in a collection of data entities or a subfield inside a composite data entity. Offset-based addressing requires two units of information: base address and an expression that shows the offset of the data entity or the subfield. Base address is the address of the first memory location in RAM where the object is located.

The *status flags* in PSW are needed to handle the outcome of expression evaluation. *Negation-bit N* is set to “1” when the comparison yields a negative value and “0” if the comparison yields a positive value. *Zero-bit* is set to “1” when the comparison yields a zero and “0” when the outcome is a nonzero value. Evaluation of a logical expression sets a *Boolean flag* to “1” when the outcome is *true* and 0 when the outcome is *false*. Conditional branch statements need these flags to make the conditional jumps to an instruction area. Some of the conditional jump statements are *brlt* (branch on less-than), *brgt* (branch on greater-than), *breq* (branch on equal-to), *brne* (branch on not-equal to), *brle* (branch on less-than-or-equal-to), and *brge* (branch on greater-than-or-equal-to).

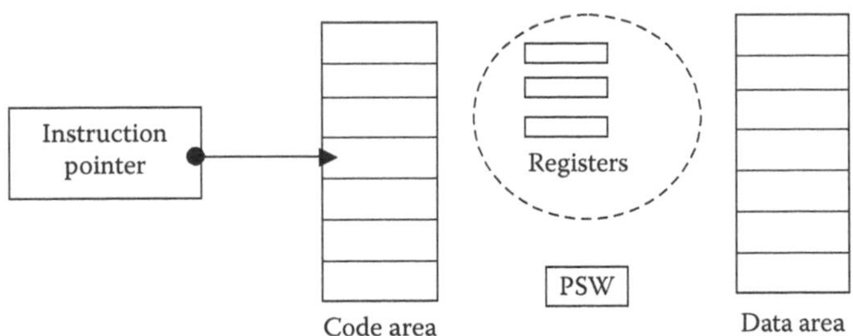

FIGURE 5.1 A schema of an abstract machine for implementation.

Figure 5.1 illustrates a low-level abstract machine for the implementation of programming languages. After compilation, store is translated to data area, and control abstractions are translated to code area. Both the code area and the data area are modeled as single-dimension arrays. A memory location in the code area is denoted as *c[code-index]*, where "code-index" is an expression when evaluated gives the offset from the first instruction in the code area. A memory location in the data area is denoted as *d[data-index]*, where the data index is an expression when evaluated gives the offset from the base address of the currently visible data area.

Depending upon the programming paradigm, the data area and the code area can be further refined to define a specific abstract machine. Most of the abstract machines need at least *control-stack*, *heap*, and directly accessible memory area for global and static variables.

In static allocation schemes, the base address after compilation is 0 and is omitted. However, in the stack-based allocation scheme, *frames* are placed at different memory locations on the stack, and data elements are accessed using (1) the base address of the frame using a pointer called *frame-pointer* and (2) the offset with respect to the base-address of the frame.

The simplest abstract machine for functional programming paradigm is called a *SECD machine*, which has four stacks: S, *a stack for expression evaluation*; E, a *stack to hold environment*; C, *a stack to hold command string*; and D, *a stack to hold the dump*—a sequence of environment of calling procedures in last-in-first-out order.

The logic programming paradigm also supports *backtracking*—a way to go back and undo some of the computations and try out alternate paths to search for solutions. Implementation of backtracking needs an additional stack called *trail-stack* as well as a regular control stack. The trail-stack keeps track of control points to go back and undo the actions for trying out alternate actions.

Object-oriented languages use a stack-based virtual machine such as *Java Virtual Machine*, and use heap to store the objects. Imperative languages support one stack- and one heap-based implementation.

In this chapter, we discuss the von Neumann–based abstract machine for imperative programming paradigm. Other abstract machines are discussed in the corresponding chapters. For example, SECD machine is discussed in Chapter 9, and Warren Abstract Machine (WAM)—an abstract machine for logic programming—is discussed in Chapter 10. A schematic of implementation of object-oriented languages is discussed in Chapter 11.

5.2 TRANSLATING CONTROL ABSTRACTIONS

The translation of control abstraction can be grouped as follows: (1) translation of evaluation of expressions, (2) translation of assignment statement, (3) translation of conditional constructs, 4) translation of iterative constructs, and (5) translation of subprogram call. In this section, we discuss all other control abstractions except the call to a subprogram, which is be discussed separately in Sections 5.3 and 5.4.

5.2.1 Translating Expressions

Evaluation of an expression is done using a combination of processor registers and memory locations. To reduce the overhead of memory access, the intermediate values after the evaluations of subexpressions are retained in the processor registers. Common subexpressions are evaluated once and retained in the registers for future use; after the first evaluation of common subexpressions, register allocation is done at compile time analysis using various optimization techniques. For example, evaluation of an expression (X + Y + 5) + 2 * (X + Y) will be translated as the following:

```
load X, R1 % load value stored in memory-loc(X) into register R1
add Y, R1, R2 % add value stored in memory-loc(Y) into register R2
add #5, R2, R3 % add constant 5 to R3 and store in the register R1
multiply #2, R2, R4 % value-in(register R4) = 2 * value-in(register R2)
add R3, R4, R4 % add the values stored in the registers R3 and R4
```

The above example illustrates the breaking of the evaluation of expressions to a sequence of low-level abstract instruction to load the variable values from the corresponding memory locations and the storing of the partial results temporarily in the registers. Note the common subexpression (X + Y) is evaluated only once and stored in the register R2 for future reuse. In the future, to discuss the translation of other control abstractions, we will avoid translation of an expression and denote it by *evaluate* (*expression*).

5.2.2 Translating Assignment Statement

Assignment statement *<variable>* = *<expression>* is equivalent to a sequence of low-level abstract instructions that evaluate the expression first and then store the resulting value into the memory location of the left-hand-side variable. For example, an assignment statement such as X = Y + 5 is equivalent to the following:

```
load Y, R0 % load the value in Y-location into register R0
load # 5, R1 % load constant 5 into register R1
add R0, R1, R0 % add value of R0 and R1, and store the result in R0
store R0, X % store value-in(R0) into location of variable X
```

Knowing that assignment operation is a combination of load, evaluate, and store operations, we denote the assignment operation by a simple command *assign*(*<variable>*, *<expression>*).

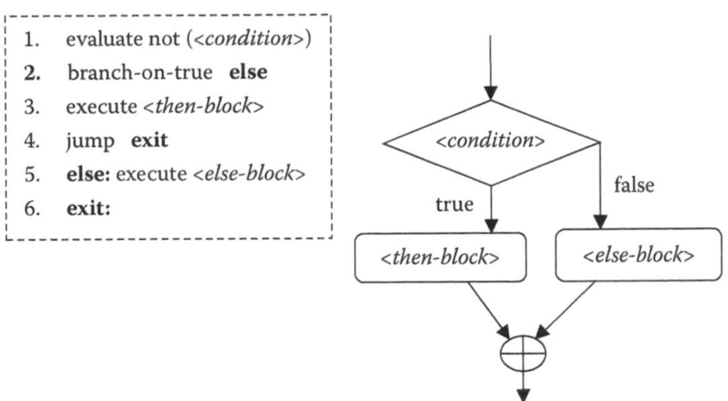

FIGURE 5.2 A schematics of translating control flow diagram to abstract machine.

5.2.3 Translating If-Then-Else Construct

The control flow of the if-then-else statement has three steps: (1) evaluate the Boolean condition; (2) based upon the evaluation of the Boolean condition, set status flags at hardware level; and (3) based upon the value of the status flags' branch conditions, make the jumps to execute either the block of statements in then-part or the block of statements in the else-part. Figure 5.2 shows a control flow diagram of an if-then-else statement and the corresponding schematics for low-level translation.

A control flow diagram is a two-dimensional planar figure. It has to be translated to code area, which is single dimensional. To ensure mutual exclusion of *<then-block>* and *<else-block>* and to avoid stepping into the *<else-block>*, a jump statement is needed after the execution of *<then-block>*.

After the control flow passes through *<then-block>*, a jump statement is executed to bypass the *<else-block>* and to take the control to the location after the *<else-block>*. The control jumps to *<else-block>* if the conditions are false. Since the control flow moves in the forward direction by default, jump is needed only to go to *<else-block>*, as the control will automatically flow to *<then-block>* if the conditions are true. To facilitate this, negation of the condition is checked. If the negation of the condition is true, then control jumps to the *<else-block>*. Otherwise, the control passes through the *<then-block>* by default.

5.2.4 Translating Case Statement

Case statements can be simulated by using a sequence of if-then-*<then-block>* statements, as shown below, to execute one of the mutually exclusive possibilities as follows:

```
1. result = evaluate(<expression>);
2. if (result == value_1) then {<block_1>; jump-to exit;}
3. if (result == value_2) then {<block_2>; jump-to exit;}
   ...
   If (result == value_N) then {<block_N>; jump-to exit;}
   <default-block>;
   <exit>:
```

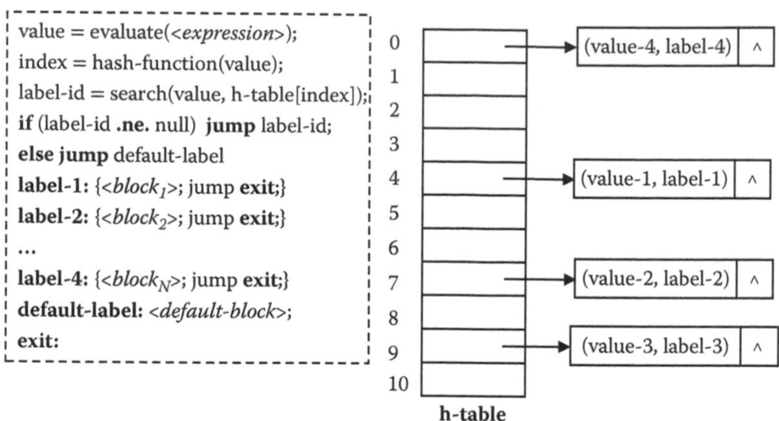

FIGURE 5.3 Translating case statement using a hash table.

Alternately, it can be implemented using a hash table that stores the array of locations where the control can jump based upon the evaluation of the *<expression>*, as shown in Figure 5.3. After evaluating the expression, the result is fed into a hash table. The hash table contains a triple of the form (*expected-value, label-number, pointer to next triple*). If the expected value is the same as the derived value, then the corresponding label is returned; otherwise "null" value is returned, and control jumps to the default label. After executing a block of statement corresponding to the expected value, the control bypasses other blocks by jumping to label "exit."

5.2.5 Translating Iterative Constructs

Iterative constructs are translated using the translation of if-then-else construct. If we translate while-loop using if-then-else statement, then the statements looks like the following:

```
1. Initialize the variables;
2. loop: if (evaluate(not <expression>) = true) {
               3. jump-to exit;
         4. else {execute <while-block>
                5. jump-to loop;}
         }
6. exit:
```

if we translate {**if** *<expression>* **then** {*<while-block>*; jump-to **loop**} using prior knowledge of translating if-then-else statement, then the translation would be as given in Figure 5.4. First *not*(*<expression>*) is evaluated. If *not*(*<expression>*) is true, then the control exits out the while-loop. Otherwise, the control falls through the *<while-block>*, executes it, and jumps back to the start of the while-loop.

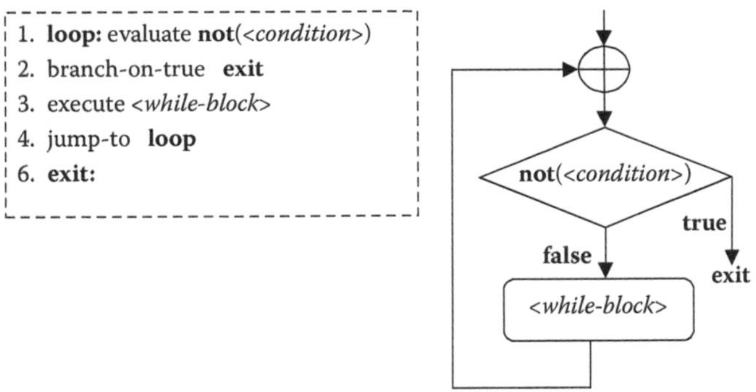

FIGURE 5.4 A schematics to translate while-loop to low-level abstract instructions.

5.2.5.1 Translating for-loop

A for-loop construct is of the following form:

```
for (i = <initial-expression>';'<final-expression>';'<step-expression>)
         <for-loop-block>';',
```

For-loop is simulated using while-loop, where the final condition is (index > evaluate(<*final-expression*>)). The translation of the for-loop using while-loop would be as follows:

```
1) assign(i, evaluate(<initial-expression>);
2) while (not evaluate(<final-expression>))
       {3.execute <for-loop-block>;
        4. assign(i, i + evaluate(<step-expression>));
       }
```

Using the knowledge of iterative constructs, the for-loop, is translated as shown in Figure 5.5.

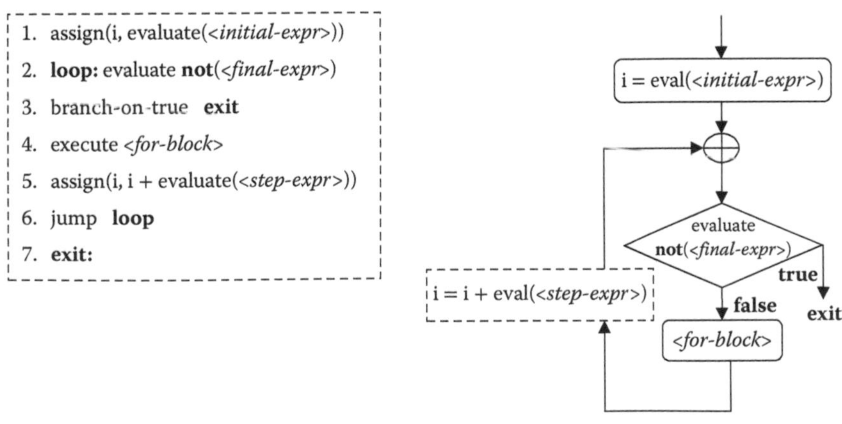

FIGURE 5.5 A schematics for translating for-loop to low-level abstract machine.

Everything in Figure 5.5 is similar to the translation of while-loop, except two statements: (1) assignment statement to initialize the index-variable and (2) update of the index-variable in a user-transparent way. The box to increment the index has been shown using dashed lines to indicate that the user has no control over incrementing the index-variable within the statement block other than specifying an expression for the step-size at the beginning of the for-loop.

A similar translation can be done for the do-while loop and has been left as an exercise to the students. Iterators are very similar to the translation of while-loop and for-loop. The only difference is that iterators step through the collection of elements one by one, and at the beginning of the loop they check if they have reached the end of the collection.

5.3 STATIC ALLOCATION

Modern-day languages support recursive procedures, memory reuse, recursive data structures, and dynamic data objects. Hence, a hybrid model is preferred. However, some of the concepts of static memory allocations are helpful in understanding stack-based implementation and the hybrid approach. We take an example from Fortran-66, since later versions of Fortran support stack-based allocation.

The data area is represented as an array, and the notation d[i] is used to denote direct access to *i*th memory location in the data area. The shared memory locations are allocated first, followed by a memory location storing the return address of the next instruction after the instruction that called the current subprogram. The return address is stored in the data area, because its value changes every time the subprogram is called from a different location. All local variables that are not shared are allocated locally in the activation record of the subprogram.

Different program units can be compiled separately. The compiled codes are *linked* together in a user-specified order. The data areas and code areas are linked in the same user-specified order, and the information available in the symbol tables of different compiled codes is merged. The calling program will not have any information about the data areas and code area of the called subprogram if they were compiled separately; the information would be available only after merging the symbol table. The information derived from symbol tables includes (1) information about imported and exported entities, (2) information about where to jump when a subprogram is called, and (3) information about where to return after the current subprogram is over. During compilation, each subprogram assumes that the set of missing information would be made available to them through the symbol table of the corresponding program units during linking period.

During linking, the data areas of different compiled codes are merged such that common block declarations are joined together, followed by return address, followed by local variables. Let us assume that the data area of a compiled code is (C, $R_{I\,(I>1)}$, $L_{I\,(I>1)}$), where C denotes the common block, R_I denotes the return address memory location of the *i*th compiled code area, and L_I stands for set of local variables in the *i*th data area, which then, after linking the data area would look like (C, L_1, R_2, L_2, …, R_I, L_I, … R_N, L_N) in the same order as specified by the user. All the common blocks having the same name in different compiled codes are joined together as one shared block and are placed at the beginning of the joint data area. The memory address of the return addresses and local declarations gets shifted during the linking process due to the merging of the data areas. Linking is a two-pass process: (1) in the first pass, all the data areas are merged, jump to subroutines and return from subroutines are computed, and offsets for data areas of different program units are computed and (2) in the second pass, the corresponding offsets are incorporated in the code area.

Call to a subprogram involves (1) storing the memory location of the next executable instruction of the calling subprogram in the return address, (2) exchanging parameter information, and (3) jumping to the first instruction of the called subprogram.

Example 5.1

Figure 5.6 illustrates an implementation using the static allocation scheme. We assume that the main program and the subroutine *FIND_MAX* have been compiled separately. Both the main program and subprogram share a common block consisting of a variable *MAX* and a fixed array *M* of size 4. In the main program, the local unshared variables are *I* and *J*. The program *MAIN* reads the data in the common array *M* and then calls the subroutine *FIND_MAX*. The subroutine *FIND_MAX* finds the maximum of the array *M* and uses the common block to share the value of the variable *MAX*. The low-level code uses direct memory access as shown by d[<*m*>] and uses the offset method to access an element of an array.

In the main program, after the compilation, the address of the variables *M*[1] to *M*[4] is respectively d[0] to d[3], the variable *MAX* maps to memory location d[4], and the local variables *I* and *J* in the main program map to the memory locations d[5] and d[6]. In the subroutine *FIND_MAX*, the address of variable data elements *M*[1] to *M*[4] is respectively d[0] to d[3]; the variable *MAX* maps to memory location d[4]; and the return address in the subroutine occupies the memory location d[5], followed by the local variable *I* that occupies memory location d[6].

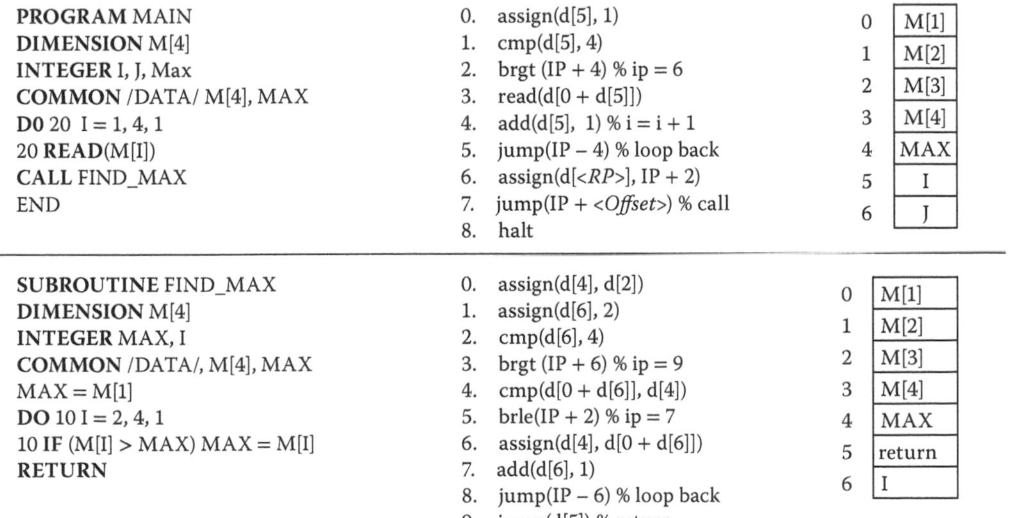

FIGURE 5.6 Code area and data area for static implementation after compilation.

This program shows low-level translation for call to the subroutine *FIND_MAX* and two major control abstractions: *do-loop* (a variation of for-loop) and *if-then-else* statements. The translation of do-loop in the main program starts from code area location c[0] and ends at c[5]. The statement at c[0] assigns the value 1 to memory location d[5] that is the memory location of the variable *I*. The statement c[1] compares the value of d[5] with the constant 4. The statement c[2] branches to IP + 4 if the value of d[5] is greater than 4. Note the use of "IP + 4" to take the control out of the loop. The technique to add offset to the instruction pointer makes the low-level code independent of memory locations where a program is loaded in the computer for execution. The instruction at c[3] uses the offset method to access the memory location of *M*[*I*]: 0 is the base-address of *M*[0], and d[5] stores the value of the variable *I*. The instruction at c[4] increments the value of d[5] by 1, and the instruction at c[5] takes the control back to c[1], which starts the next cycle of the iteration. The instructions c[6] and c[7] are related to the call of the subroutine *FIND_MAX*. The instruction at c[6] stores "IP + 2"—the address of the next high-level instruction after subroutine call in the currently unknown return address of the subroutine *FIND_MAX*. Since the information is unknown, a new symbol is created and is entered into the symbol table. During linking, the memory address of the return pointer is looked up from the symbol table. The instruction c[7] uses the jump offset from the current instruction to jump to the first executable instruction in the called subroutine *FIND_MAX*. However, the offset is also unknown before linking and is computed at the linking time.

The compiled low-level code for the DO-loop in the subroutine *FIND_MAX* occupies the memory locations between c[1] to c[8], and the translation of the if-then-else statement, nested inside the DO-loop, occupies the locations between c[4] to c[6]. The instruction at c[1] is equivalent to initializing the index variable *I* to 2. The instruction at c[2] compares the value of the index variable *I* with the upper-bound 4. The instruction at c[3] takes the control out of the DO-loop if the index variable is greater than 4. The instruction at c[4] compares the value of the subscripted variable *M*[*I*] with the value of the variable *MAX* and jumps to c[7] if *M*[*I*] is less than or equal to the value of the variable *MAX*. Otherwise, c[6] assigns the value of *M*[*I*] to the variable *MAX*. The instruction at c[7] increments the value of d[6] (local variable *I*) by 1. The instruction at c[8] subtracts 6 from the instruction pointer *ip* to jump back to the instruction c[2] to start the next iteration cycle. The last instruction at c[9] picks up the address stored in the return-address memory location and returns the control to the next instruction in the calling procedure. Since a subroutine may be called multiple times from different parts of the code area, the address stored in the return-address memory location would be different every time.

As shown in Figure 5.7, after linking, the common-block variables M[1] to M[4] and the variable MAX are merged and placed in the memory locations d[0] to d[4]. The memory locations of the return address and the local variable *I* in the procedure *FIND_MAX* are placed after the last memory location allocated to the main program in the data area. The return address is mapped on d[7], and the local variable I_{sub} is mapped on d[8]. The unshared memory locations of the subroutine *FIND_MAX* have been shifted by size of (activation record of program main)—size-of(common block in Subroutine *FIND_MAX*) = 7 − 5 = 2. This offset is added to the memory locations in the compiled code part of the subroutine

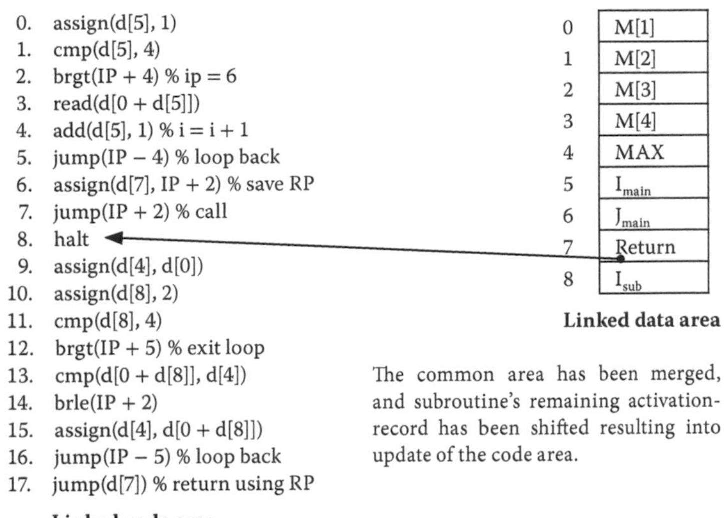

0. assign(d[5], 1)
1. cmp(d[5], 4)
2. brgt(IP + 4) % ip = 6
3. read(d[0 + d[5]])
4. add(d[5], 1) % i = i + 1
5. jump(IP − 4) % loop back
6. assign(d[7], IP + 2) % save RP
7. jump(IP + 2) % call
8. halt
9. assign(d[4], d[0])
10. assign(d[8], 2)
11. cmp(d[8], 4)
12. brgt(IP + 5) % exit loop
13. cmp(d[0 + d[8]], d[4])
14. brle(IP + 2)
15. assign(d[4], d[0 + d[8]])
16. jump(IP − 5) % loop back
17. jump(d[7]) % return using RP
Linked code area
0 M[1]
1 M[2]
2 M[3]
3 M[4]
4 MAX
5 I_{main}
6 J_{main}
7 Return
8 I_{sub}
Linked data area
The common area has been merged, and subroutine's remaining activation-record has been shifted resulting into update of the code area.

FIGURE 5.7 Linking the code areas and data areas.

FIND_MAX during the linking to generate the linked code. The symbol <*R*> in C[6] is substituted by 7—the memory address of the return pointer—and the symbol <*S*> is substituted for the offset value 2 to jump to the first instruction of the subroutine.

5.4 HYBRID ALLOCATION

Block structured languages and languages that support recursive procedures or dynamic objects use a hybrid allocation that includes control stack, heap, and static allocation, as shown in Figure 5.8. In an integrated model, a data area is a triple of the form (*static area, control-stack, heap*). The static memory allocation is fixed at the time of compilation and is used for static variables and global variables. The control-stack and heap change dynamically with the program execution. The heap and the control-stack grow in the opposite directions, since, depending upon the input data, different invocation of the same program may need different sizes of the heap and the control-stack. Many times the size of the control-stack is larger, and at other times heap needs more memory to expand. By keeping the heap and control-stack at two ends growing toward each other, the ends of heap and control-stacks are determined by the usage, and the memory space is utilized to the maximum.

The first pointer to an object allocated in the heap resides either in the frame of a procedure or in a register. The heap allocation is discussed in detail in the next chapter. However, in this chapter heap has been mentioned to allow for discussion of parameter passing when objects are allocated in the heap.

Each frame of a called subprogram contains multiple information: (1) memory locations for the incoming formal parameters, (2) saved state for the calling subprogram, (3) various pointers to access the control-stack (including frame of the calling program and nonlocal variables) and to recover the computational state of the calling subprogram, (4) memory allocation for local variables, (5) memory locations for scratch pad computations, and (6) memory locations of the outgoing actual parameters.

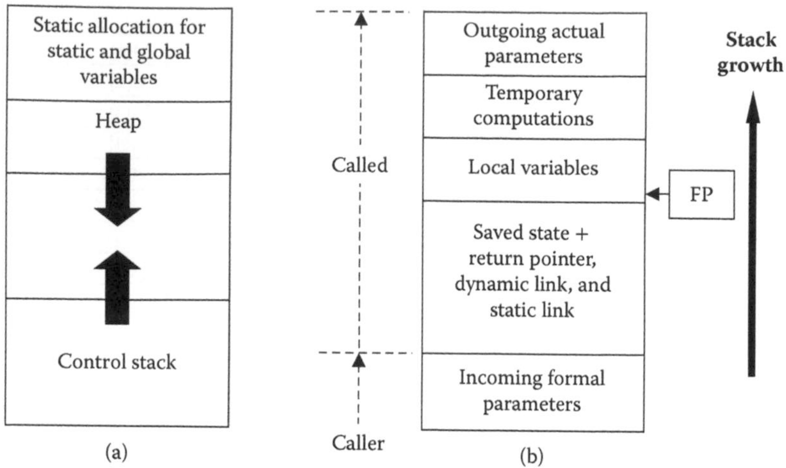

FIGURE 5.8 Data area for a block-structured language. (a) Overall data area (b) a frame.

When a subprogram is called, the part of the computational state that would be altered in called subprogram is saved. The saved computational state includes (1) the values for those registers that get altered in the called subprograms, (2) PSW, and (3) various pointers used to access the code and data area of the calling subprogram. Various pointers are set up to access: (1) the code and data area of the called subprogram, (2) the memory locations of the base-addresses of the actual-parameters in the frame of the calling program if parameters are passed by call-by-reference, and (3) the frames of the procedures under which the current procedure is embedded. A memory location is allocated for each local variable. Variable may be mapped to data objects of varying size. For example, a character may occupy just one byte; and an integer occupies 4 bytes on a 32-bit machine. On top of the frame, there are some memory locations that are used to store the results of partial computations while evaluating expressions, and finally it holds actual parameters to be passed to the called subprogram.

The expressions in the actual parameter are evaluated, and the resulting value is stored in a new temporary memory location in outgoing parameter area. Dynamic local variables that have nonoverlapping scope can be mapped on the same memory locations for better memory utilization. After a subprogram terminates, the frame of the called subprogram is discarded, pointers are set to access the frame of the calling program, and the top of the stack pointer is reset to point to the memory location right after the frame of the calling program. This way the whole memory used by the called program is reclaimed for future reuse.

5.4.1 Roles of Various Pointers

There are five major pointers needed for handling a called subprogram. After saving the computational state of the calling program and access links to the frame of the calling subprogram, the control jumps to the first instruction of the called programs. Five pointers are needed during the execution of a subprogram in the stack-based allocation as follows:

Frame pointer, denoted by FP, allows access to: (1) data area of the currently executing subprogram, (2) incoming parameter locations, and (3) saved state information of the calling subprogram. The frame-pointer points to the first memory location after the saved state; and all other memory locations are accessed by adding an offset to the base address stored in the frame pointer. *Top-of-the-stack pointer*, denoted as TOS-pointer, points to the next free memory location on the control-stack. *Dynamic link*:, denoted by DL, stores the frame pointer of the calling program and is used to restore the frame pointer of the calling program. *Return-pointer*, denoted by RP, stores the address of the next instruction in the calling program after the instruction that invokes a subprogram. *Static link*, denoted by SL, stores an address to access the frame of the subprogram under which the currently executing subprogram is nested. It is used to access nonlocal variables. Many of the pointers such as FP and TOS-pointer are stored in registers for faster access.

5.4.2 Calling Subprograms

When a subprogram is called, pointer changes occur in the following order:

1. The outgoing parameter space is set based upon the parameter passing type and the number of parameters. During this time TOS-pointer keeps getting updated due to values of the actual parameters being pushed.
2. The memory location of next instruction in the calling subprogram is stored in the return pointer of the called subprogram.
3. The dynamic link is set equal to the frame-pointer of the calling subprogram.
4. The static link is set to point to the frame of the subprogram under which the called subprogram is nested.
5. The current value of the TOS-pointer is copied in the frame-pointer (FP) to set up the FP to point to the frame of the called subprogram.
6. The instruction pointer is set to the memory location of the first instruction in the called subprogram.
7. In the called subprogram, TOS-pointer is incremented by the size of the local environment to allocate local dynamic variables on the stack.

The order in which these pointers are set is important. For example, if we change the FP before copying the old value of FP into the dynamic link, then the access to the frame of the calling subprogram would be lost. Similarly, if we set up the TOS-pointer first, then new FP cannot be set.

The order in which the pointers are arranged is dependent upon the implementation. Here we assume that RP has offset –3, dynamic link has offset –2, and static link (if present) has offset –1. In case the static link is missing, because the language does not support nesting of procedures, then RP has the offset –2, and DL has offset –1. Thus, the return pointer is accessed by d[FP – 3] (or D[FP – 2] if SL is missing); dynamic link is accessed by d[FP – 2] (or d[FP – 1] if SL is missing); and static link, if present, is accessed by d[FP – 1]. We also assume that FP and TOS-pointer are stored in a processor register for efficient memory

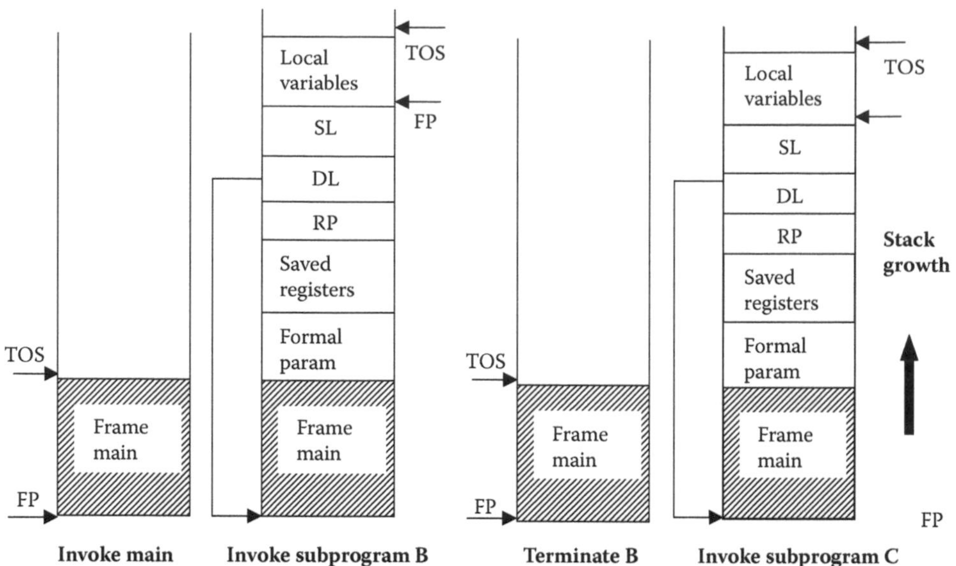

FIGURE 5.9 Frame movement during procedure calls.

accesses. In understanding most of the concepts, we ignore the saved registers and PSW for our convenience, without losing any information.

Example 5.2

Let us consider a scenario. Program *main* calls subprogram A. The subprogram A calls subprogram B. The subprogram B executes and terminates. Then the main program calls subprogram C. Figure 5.9a shows a snapshot of the control-stack when the control is within the main program. Figure 5.9b shows the control stack when the control is in the subprogram B. Figure 5.9c shows the control-stack when the subprogram B has terminated, and the control is back in the main program. Figure 5.9d shows the control stack when program C has been called by the main program. Note the reuse of the memory space that was originally occupied by the frame of the subprogram B.

5.4.3 Generating Data and Code Areas

The store of the called subprogram includes (1) the store created by global variables, (2) the store created by the nonlocal variables, (3) the store created by the local variables, (3) the store created by parameter passing, and (4) the store created by the local variables. The global variables and static variables are allocated statically at the beginning of the data area and are accessed using a single memory access, without the use of any pointer. The local variables are accessed by adding an offset to the frame pointer in the control stack. For example, d[FP + 3] gives access to memory location that is three locations away from the address pointed to by the FP.

Static link (SL) stores the base address of the frame of the next outer level of the subprogram, a nonlocal variable can be declared N ($N \geq 1$) level above the nested procedure. Two techniques are used to access nonlocal variables: (1) chain of static links and (2) display registers.

The chain-of-static links scheme uses the knowledge that the static link is stored in a memory location that has fixed offset: d[FP – 1] in our case. This technique uses successive jumps using a static links stored in the frame. If a nonlocal variable is declared "*m*" nesting level away, then the low-level data access can be written as d(*m*, *offset*), and the algorithm to access the nonlocal variable is as follows:

```
while (m > 0) {
         m = m - 1; frame-base = d[SL];
         SL = frame-base - 1;}
         value = d[frame-base + offset]
```

The chain-of-static-links is a naïve technique, and suffers from the overhead of going through multiple memory accesses if the nesting level is large. The second technique uses *display registers. Display registers* are processor registers or cache memory that holds the direct address of the nonlocal variables. The display register is populated when the called procedure is invoked, and the nonlocal variables are accessed efficiently without going through a chain of pointers.

The variables in the frame of calling procedures are accessed when parameters are passed as reference. In call-by-reference, the formal parameter is a pointer to the first memory location of the data entity being referenced. If the data abstraction is a composite data entity, then different attributes are accessed by adding the offset of the attribute to the first memory location of the data entity. Similarly, if the abstraction is a collection of data entities, then offset is calculated by multiplying the index by the size of the individual data element, and the memory location of the *i*th data entity is calculated by d[reference-link] + i * size-of(individual data-element), and the value inside the memory location is accessed as d[d[reference-link] + i * size-of(individual data-element)].

Called subprograms accesses memory location in local frames using only its frame pointer and offset. The frame of the called subprogram is pushed on the control stack when a subprogram is invoked, and the memory occupied by the frame is reclaimed after the subprogram terminates. The sequence of instructions to call a subprogram is as follows:

1. Evaluate the expressions in the actual parameters, and copy the values in the outgoing parameter area.
2. Save the PSW of the calling program and the registers in the saved state area of the called programs frame. Only those registers that get modified during the called subprogram are saved.
3. Update the return-pointer (RP) to point to the next instruction in the calling program after the control returns from the called subprogram.
4. Copy the FP into the dynamic link of the called subprogram.
5. Update the static link if nonlocal variables are supported by the language.
6. Update the frame pointer by copying the TOS as new FP value.

7. Update the return-pointer to point to the next instruction in the calling program.
8. Make the jump to the first instruction in the called subprogram.
9. In the code area of the called subprogram, the first instruction allocates the local environment for the called subprogram by adding the size of the frame of the called program to the top-of-the-stack pointer (TOS-pointer).

The return from the called subprogram performs the following operations to recover the environment and the store of the calling subprogram:

1. The stored registers and PSW are copied back to corresponding registers to regain the computational state of the calling subprogram.
2. The TOS of the calling subprogram is recovered to reclaim the memory used by the called subprogram. TOS-pointer of the calling subprogram is recovered by subtracting the size of the incoming parameters, the saved state and various-pointers from the frame-pointer of the called subprogram.
3. The dynamic link DL is copied into FP to recover the frame of the calling subprogram.
4. The return-pointer RP is copied into IP to take the control back to the calling program.

After the return from the called subprogram and before executing the next instruction, the results of the formal parameters are copied from the memory locations of the outgoing parameter area to the memory locations of the actual parameters.

In our remaining discussions, we omit saving of the PSW and registers without losing any relevant information related to the implementation of call to the subprograms.

Example 5.3

Figure 5.10 illustrates the concept using a block structured program. Instruction pointer (IP), frame pointer (FP), and top-of-the-stack pointer (TOS-pointer) are stored in processor registers for efficient access. The program has been tailored to show different control abstractions, data abstractions, and static allocations. The program main reads four data elements into m[4] and calls the subroutine *find_max*. The snapshot shows the scenario when the control is inside the subprogram *find_max*.

The instructions *brgt* (branch on greater-than), *jump* (unconditional branch), *brle* (branch on less-than-equal-to), and *cmp* (compare) are intuitive. The instruction *add* updates the first argument by adding the value stored in the second argument. For example, the instruction "add(TOS, 1)" is equivalent to "TOS = TOS + 1." Many abstract instructions when translated to assembly-level instructions map to a sequence of instructions. For example, the instruction "assign(d[FP], IP + 2)" is equivalent to a pair of statements "add(IP, 2, R1), assign(d[FP], R1)." Similarly, the read statement would invoke an operating system call to read the data. The instruction "push" pushes the argument in the control stack and increments the TOS by one.

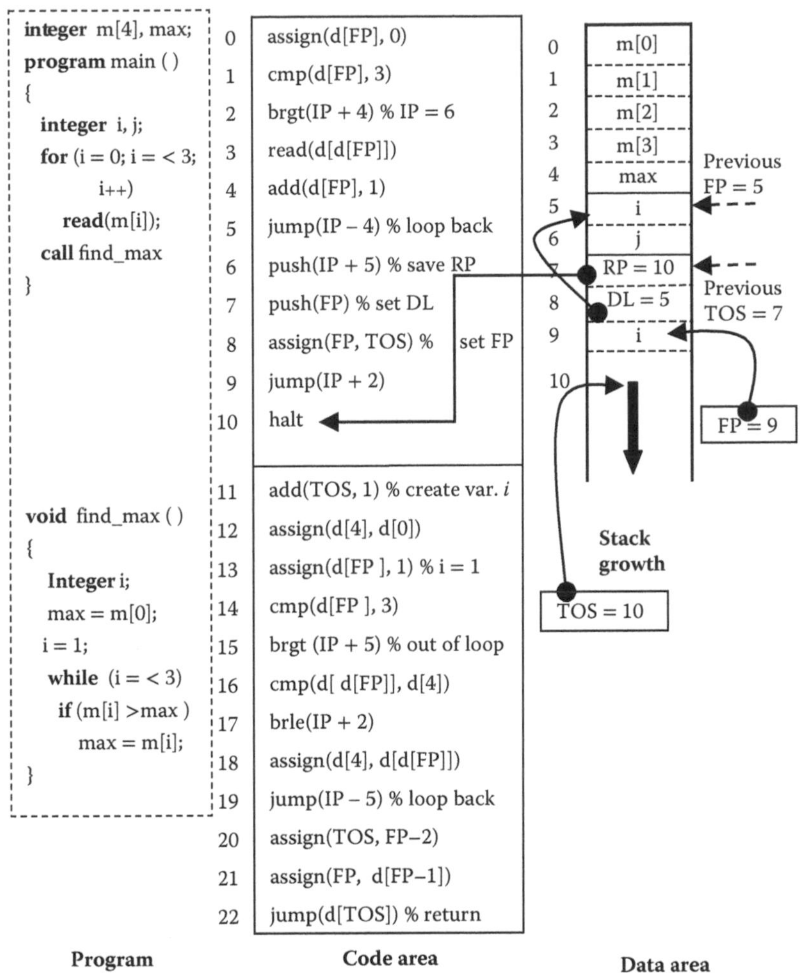

FIGURE 5.10 Code and data generation in a stack-based implementation.

The data area has three partitions: static area for the global variables, frame of the main program, and the frame of the called subroutine *find_max*. The global variables consist of the array m[4] and the variable *Max*. Global variables are allocated statically: *m*[0] ↦ d[0], m[1] ↦ d[1], m[2] ↦ d[2], m[3] ↦ d[3], and max ↦ d[4]. The frame of the main program lies between d[5] to d[6], and the frame of the called procedure *swap* is between d[7] and d[9]. The frame for the main program does not contain the return pointer and the dynamic link. The locations d[7] and d[8] save the return pointer and the dynamic link to revert the control back to the main program.

The frame of the program *main* contains memory locations for two local variables: *i* ↦ d[5] and *j* ↦ d[6]. The frame of the subprogram *find_max* contains the return pointer (RP) ↦ d[7], dynamic link (DL) ↦ d[8], and the local variable *i* ↦ d[9]. FP points to the location d[9], TOS-pointer to the location d[10], and DL points to the base of the frame of the main program; that is, d[5]. The return-pointer (RP) points to the location c[11] in the code area that corresponds to executing "halt" instruction.

The code area uses the frame pointer (FP) to access the memory locations within the frame of currently executing procedure. The instructions between c[0] and c[5] correspond to the execution of the for-loop. The instruction at c[0] assigns the value 0 to the memory location pointed by FP and is the low-level translation of the statement "$i = 0$"; the instruction at c[1] compares the value d[FP] with the constant 3. The execution of this statement sets the PSW accordingly, and the instruction at c[2] uses the flags in PSW to exit out of the while-loop by jumping to the statement c[6]. The instruction at c[3] reads the value of the subscripted variable *m*[*i*] by using two memory accesses: first reading the value of d[FP] (value of *i*) and then reading the indexed memory. The reading invokes a system-level call. The instruction at c[5] subtracts 4 from the instruction pointer to jump back to c[1]—the start point of the for-loop.

The instructions at code area locations c[6] to c[9] are used to set up the various pointers and jump to the first instruction of the called subroutine. The instruction located at c[6] sets up the return pointer by pushing "IP + 5" to the top of the stack (at d[7]), and TOS is incremented by 1. The instruction located at c[7] sets up the dynamic link by pushing the value of FP on top of the stack (at d[8]), and TOS is incremented by 1. Register values, if modified, are also pushed on the stack. The instruction located at c[8] sets up the FP to access frame of the subroutine *find_max* by assigning FP to TOS. The offset is controlled by the memory locations needed to save the computational state of the calling subprogram; this would change if there were saved registers, PSW, and a static link to be saved. The instruction at c[9] passes the control to the first instruction of the subroutine *find_max* by looking up the offset from the symbol table: the offset is 2 in this case. The last instruction c[10] corresponds to the high-level instruction "halt."

In the subroutine *find_max*, the instruction at c[11] sets up TOS-pointer by adding the number of memory locations allocated to the local variables: the size is 1 in our case. The instruction at c[12] corresponds to the high-level instruction "max = m[0]": max ↦ d[4] and m[0] ↦ d[0]. The instruction at c[13] corresponds to the high-level instruction "i = 1": i ↦ d[FP]. The instructions between the locations c[14] and c[19] correspond to the while-loop, and the instructions between the locations c[20] and c[22] correspond to the return from the subroutine *find_max*. The instruction at the location c[14] compares the variable *i* with the constant 3. The instruction at the location c[15] gets out of the while-loop if the value of the variable *i* is greater than 3. The instruction at the location c[16] corresponds to the high-level instruction *m*[*i*] > *max*: *i* ↦ d[FP], *max* ↦ d[4], and d[d[FP]] ↦ *m*[*i*]. The instruction at the location c[17] corresponds to the vacuous else-part of the if-then-else statement and takes the control to the instruction at the location c[19]. The instruction at the location c[18] is equivalent to the high-level instruction *max* = *m*[*i*]. The instruction at the location c[19] subtracts 5 from "IP" to jump back to the start of the while-loop.

The return from the subroutine resets the pointers in the following order:

1. The instruction located at c[20] copies the value of the FP – 2 into TOS to reclaim the memory used by the frame of the subroutine *find_max*. The value 2 is the sum of the size of the saved-state information that is 2 in our case: return pointer and dynamic link.
2. The instruction located at c[21] recovers the frame pointer of the main program by copying the dynamic link (d[FP – 1]) into FP.

3. The instruction located at c[22] passes the control back to the main program by copying the address stored in RP (d[TOS]) into IP. After changing TOS-pointer in c[20], TOS-pointer is pointing to the same memory location as the return pointer of the called subprogram.

Accessing array variables requires two memory accesses: one to get the value of the index variable, and the second to get the value of the memory location corresponding to the array element. This additional memory access is reduced by compilers by storing the index variables in the registers. Use of cache memory also improves the memory access overhead.

5.5 IMPLEMENTING PARAMETER PASSING

Depending upon the type of parameter passing mechanisms, the implementation part corresponding to information exchange is different. For example, call-by-value requires copying of the evaluated value of the expression to the memory location of the formal parameter. Call-by-reference needs to store the base address of the actual parameter in the memory location of the formal parameter.

The memory space for the actual parameters in the calling subprogram is called *outgoing parameter space*, and the memory space for the formal parameters in the frame of the called subprogram is called *incoming parameter space.* Outgoing parameter space belongs to the frame of the calling subprogram, and the incoming parameter space belongs to the frame of the called subprogram. However, the *outgoing parameters space* and *incoming parameters space* are superimposed into one, so that both the calling subprogram and the called subprogram can access it. The called subprogram saves the computational state into the saved-state information right after the incoming parameter space. FP points right after the saved-state information. The offset for a formal parameter is negative with respect to the FP of the called subprogram. However, offset of the remaining local variables is positive with respect to FP.

5.5.1 Implementing Call-by-Value

During call-by-value, the value of the actual parameter expressions are evaluated and stored in the outgoing parameters space. The called subprogram saves the PSW, registers modified in the called subprogram, and various pointers. The incoming parameters are accessed as d[FP – offset] for single data entities and d[FP – base-address-offset + d[FP + index-offset]] to access an array element where base-address-offset is the offset of the first element in the array, and index-offset is the offset of the index variable i to access an array element $a[i]$. Complex objects allocated in heap are also accessed by copying the reference to the object into the memory locations of the formal parameter.

Example 5.4

Figure 5.11 illustrates the abstract implementation of call-by-value. The main program copies values of two variables, i and j, to the corresponding formal parameters x and y. The subprogram *swap* swaps the values of formal parameters x and y. The results of the computations in the subroutine *swap* are not passed back.

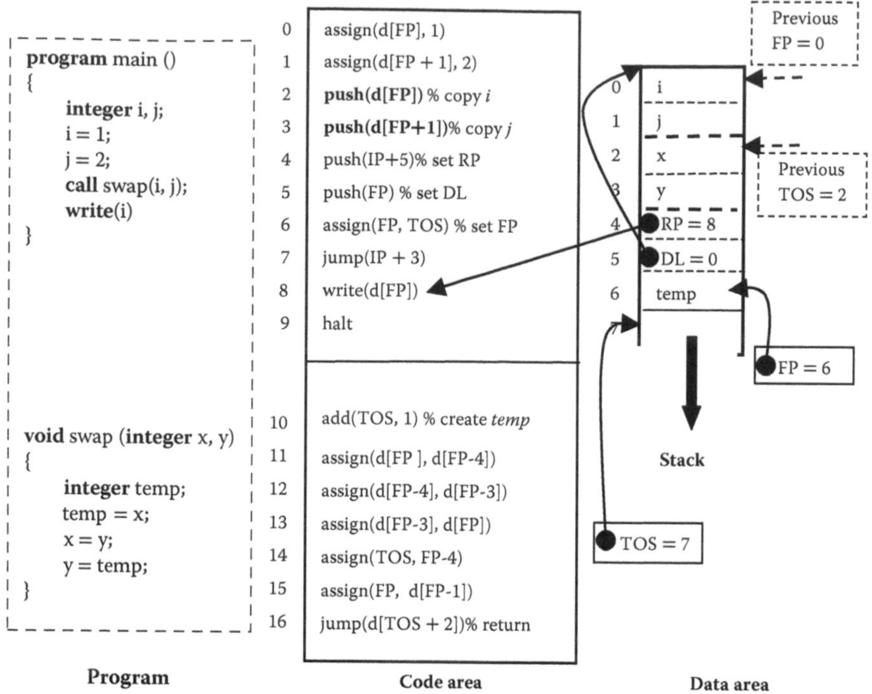

FIGURE 5.11 Schematics of implementing call-by-value.

The frame of the main program consists of location d[0] to d[3], and the frame of the called subprogram consists of d[2] to d[6]. The overlapping locations d[2] and d[3] correspond to the formal parameters *x* and *y*. The locations d[2] and d[3] are treated as outgoing parameter space by the main program and incoming parameter space by the called procedure *swap*. The locations d[4] and d[5] are saved pointers RP (return-pointer) and DL (dynamic link): the return-pointer location d[4] stores the address of the next instruction write(i); and the location d[5] stores the dynamic link. The variable *temp* maps to the location d[6].

Before calling the procedure *swap*, the control is in the main program, and the FP points to d[0], and TOS-pointer points to d[2]. The FP of the subroutine *swap* points to the location d[6]. D[FP – 1] is the dynamic link location, and d[FP – 2] stores the return address of the next instruction to be executed in the program main. Static link is missing.

The parameter passing uses the pointer FP to copy the values in the outgoing parameter space. The instructions located at code locations c[2] and c[3] copy the value of actual parameters in d[FP] and d[FP + 1] into the outgoing parameter locations using the instructions *push(d[FP])* and *push(d[FP + 1])*. Remaining instructions in the main program are used to set various pointers and jump to the subroutine *swap*, and have already been discussed in Example 5.3.

The notations d[FP – 4] and d[FP – 3] are used to access the value of the formal parameters *x* and y, respectively. The offsets *4* and *3* are calculated by knowing the size of the saved-state information and size of the incoming parameters. The instruction at c[11] copies the value of the variable *x* into the memory location of the variable *temp*. The instruction at code-location c[12] copies the value of the variable *y* into *x*, and the instructions at code-location c[13] copies the value of variable *temp* into the memory location *y*. The remaining three instructions located between code locations c[14] and c[16] are the instructions to pass the control back to the calling program, and have been discussed in Example 5.3. In instruction c[16], the location of the return pointer can be determined at compile-time by knowing the size of the outgoing parameter locations. The size is 2 in our case: actual parameters *i* and *j*.

5.5.2 Implementing Call-by-Reference

During call-by-reference, a calling subprogram stores the base-address of a data entity in the outgoing parameter space that becomes the incoming formal parameter space for the called subprogram. Referenced objects are accessed by d[d[FP – parameter-offset] + index-offset], where *parameter-offset* is the offset of the base-address of the incoming formal parameter in the called subprogram, and the *index-offset* is the offset of the local index variable in the frame of the called subprogram. The second term can be replaced by appropriate addressing mechanism if the index variable is other than the local variable. Nothing special is done during the return from the called subroutine, as call-by-reference allows continuous mutation of the actual variable, and there is no need to explicitly pass back any result.

Example 5.5

In Figure 5.12, the memory locations for the formal pointers *x* and *y* store the address of the memory locations of the variables *i* and *j*, respectively. Since *x* and *y* store the addresses of the actual parameters *i* and *j*, parameter passing in location c[2] pushes the memory address FP instead of the r-value d[FP], and the location c[3] pushes the memory address FP + 1 instead of the r-value d[FP + 1].

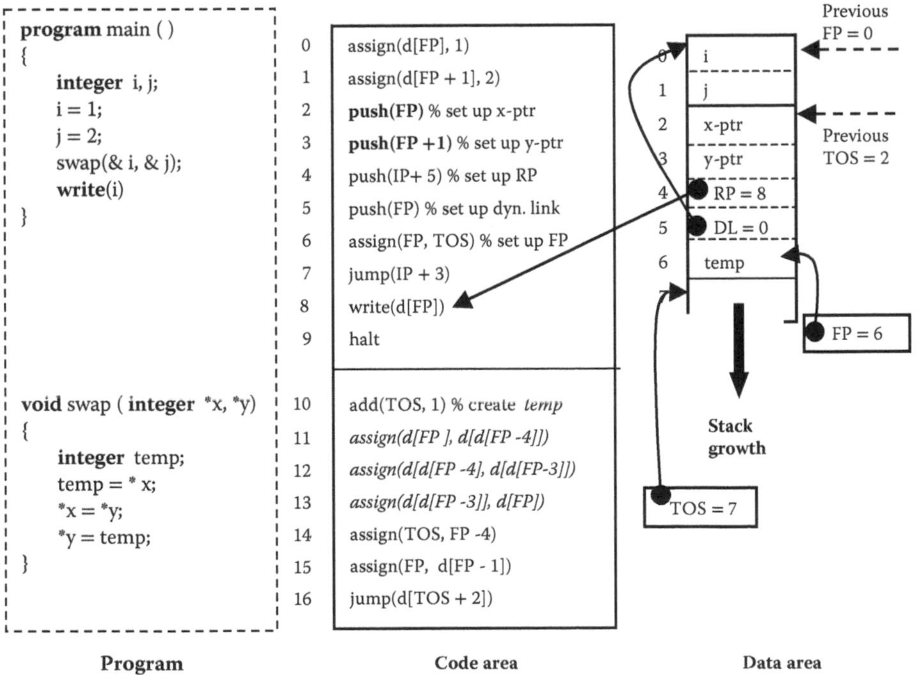

FIGURE 5.12 A schematics of implementing call-by-reference.

The instructions at locations c[11], c[12], and c[13] that use the formal parameters *x* and *y* to indirectly access the actual parameter have two memory accesses denoted by d[d[FP – 4]] to access the actual parameter value for the variable *i* and d[d[FP – 3]] to access the actual parameter value for the variable *j*. Note that d[FP – 4] gives the memory address of the actual parameter *i*, and d[FP – 3] gives the memory address of the actual parameter *j*. Remaining instructions are similar to the call-by-value and have been discussed before.

5.5.3 Implementing Call-by-Value-Result

During call-by-value-result, the calling subprogram evaluates the value of the expressions for the actual parameters and stores the resulting value in the corresponding memory locations in the outgoing parameters' space. After the termination of the called subprogram, the result is copied back by the calling subprogram from the outgoing parameter space to the corresponding variable locations, if needed. Since outgoing parameter space of the calling program and the incoming parameter space of the called subprogram are superimposed, the result is automatically passed back to the calling subprogram.

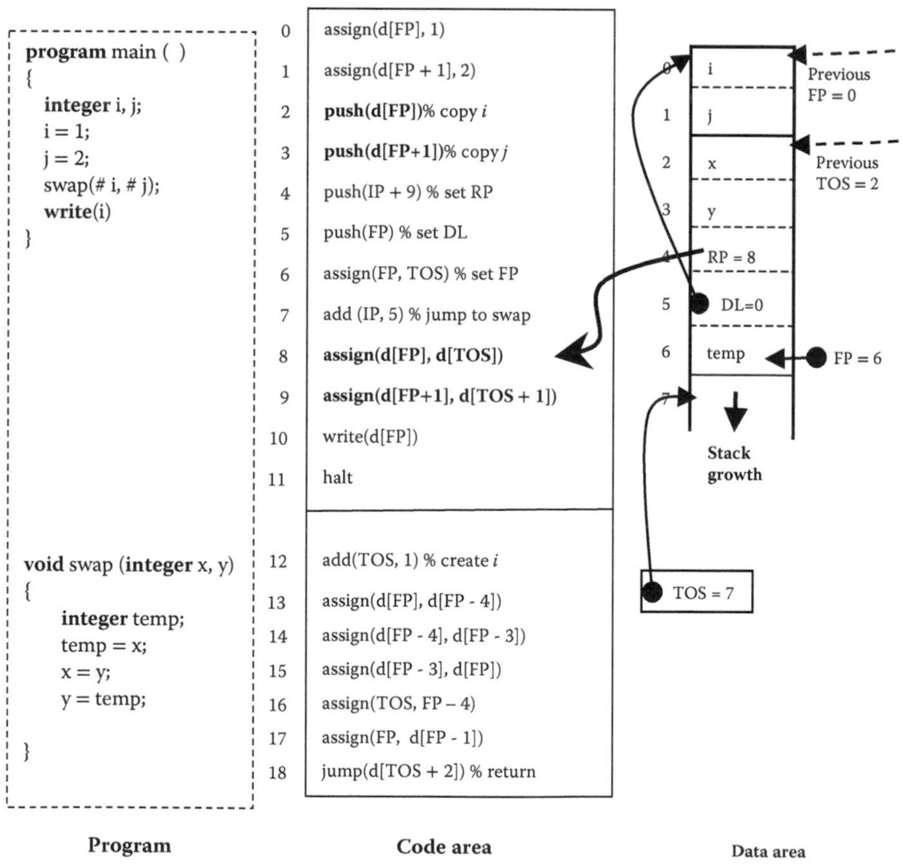

FIGURE 5.13 Schematics of implementing call-by-value-result.

Like call-by-value, formal parameters are also treated as local variables and are accessed as d[FP – offset] for single data entities and d[FP – base-address-offset + d[FP + index-offset]] to access an array element where base-address-offset is the offset of the first data element in the array, and index-offset is the offset of the index variable i to access an array element $a[i]$.

Example 5.6

Figure 5.13 shows the same program, with call-by-value-result parameter passing. The call-by-value-result actual parameters have been tagged using symbol '#'. The data area is the same as in call-by-value. There are two additional instructions in the code area: the instruction at c[8] and c[9] copies the result back from the outgoing parameter space to the actual memory location of the variables x and y, respectively. After the control comes back to the code area, the result of the formal parameter is copied back to the actual memory locations before executing the next high-level instruction. Note that outgoing parameter space is reusable, is used by multiple calls to different subprograms, and cannot be used as an image of specific variables. All other instructions are similar to call-by-value and have been discussed in Section 5.5.1.

5.6 LOW-LEVEL BEHAVIOR OF RECURSIVE PROCEDURES

Recursive procedures invoke a new activation record for every recursive call. The return address in the frames of the each instance points to the same code area location, and the static link of the frames of each instance of the recursive procedure points to the frame of the same procedure under which the recursive procedure is nested. However, the dynamic links of the m_{th} ($m > 2$) instance points to the base address of the frame of the $(m - 1)_{th}$ instance, and the first instance of the recursive procedure points to the frame of the procedure that invoked the recursive procedure.

Example 5.7

The difference between static links and dynamic links in the invocation of recursive procedures has been explained in Figure 5.14, showing the calling pattern of the procedures for *factorial* subprogram program invoked by the main program.

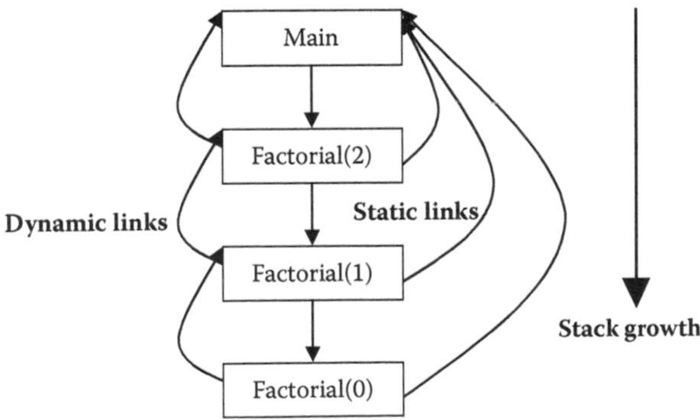

FIGURE 5.14 Dynamic and static links in recursive procedures.

The main program invokes *factorial*(2). *Factorial*(2) calls *factorial*(1) and suspends itself waiting for the value from *factorial*(1). The dynamic link of *factorial*(1) points to the base of the frame of the invocation *factorial*(2). Similarly, *factorial*(1) calls *factorial*(0). The dynamic link of the frame of the procedure invocation for *factorial*(0) points to the base of the frame of *factorial*(1). However, the static links in the frames of all the invocations of *factorial*(2), *factorial*(1), and *factorial*(0) point to the base of the frame of the Main program under which the *factorial* program is nested.

5.7 IMPLEMENTING EXCEPTION HANDLER

Exception handlers are implemented using jump statements. After the program returns from executing a statement with exception handlers or from the called routine, a flag for "successful termination" is checked. If the statement has executed successfully without any exception, the control jumps to the instruction after the exception handlers using a branch statement. Otherwise, the control falls through the exception handlers. It checks the conditions in every exception handler and picks up the instruction location from the corresponding location in the frame of the procedure if the condition is true. If an exception is declared, then the activation record has an entry in the frame of the procedure that stores the location of the routine to be executed. After the exception handler has been executed, the control returns back to check the next exception condition.

If the exception handler is successfully executed, then the control remains in the currently executing subprogram. Otherwise, if none of the exception handlers match the exception condition, a branch instruction takes the control to the return sequence, and the control returns to the calling subprogram, which checks the exception flag, and repeats the process.

Another implementation approach is to keep a separate stack of exceptions called *exception stack*. The address of the first instruction to the exception handler and the topmost activation frame is pushed on the execution stack. At the end of the successful execution of the exception handler, all the frames above the frame of the successful exception handler are removed from the control stack.

5.8 SUMMARY

In this chapter, we studied the abstract implementation model of imperative programming languages. Exact implementation is much more detailed. However, this discussion gives us a better understanding of the implementation and access of various data and control abstractions at an intermediate level.

This chapter has discussed the concept of an abstract machine and its behavior in terms of data area, control area, instruction pointer, and registers. Data area is continuously modified using fixed code, except for the languages that support first-class objects. Data area consists of three parts: static area, control-stack, and heap. Static area is fixed at compile time and does not grow or shrink at run time. Stack and heap can grow and shrink at run time. Stack grows linearly. Heap is a common memory area visible to all subprograms, and data cells can be allocated on demand at any time. Physical cells of a logical data structure in a heap are connected through chains of pointers.

Static allocation is a fixed memory allocation scheme that allocates memory at compile time, such that memory locations are accessed directly without the use of any pointer. Static allocation is used for the data entities that need fixed memory mapping and do not change memory location at run time. Global variables and static variables are allocated using static allocation. Dynamic local variables and data objects, whose lifetime is limited to the program units or block where they are declared, are allocated in the control stack. Recursive data structures and dynamic data objects that have lifetime beyond the program unit they are declared in are allocated in heap.

The control flow diagram is used to translate high level control abstractions to abstract-level instructions. The negation of conditional expression, followed by the conditional branch statements such as *brlt*, *breq*, *brne*, *brgt*, *brge*, and *brne*, are used to take the control to *<else-statement>* in the if-then-else constructs and to get out of the loop in the iterative constructs.

Static implementation uses static allocation and direct data access. Procedure calls require storing the memory address of the next instruction to be executed and making a jump using the branch instruction to pass the control to the called subprogram. After finishing all the instructions, the called subprogram assigns the return-pointer value to the instruction pointer to pass the control back to the next instruction in the calling subprogram.

The stack-based implementation uses multiple pointers and separate frames for each subprogram invocation. The frame-pointer along with the offset is used to access the memory location of a local variable or a data object. A dynamic link is used to get back the frame of the calling subprogram after the called subprogram is over. The static link or display-registers are used to access the nonlocal variables. Before calling a subprogram, parameters are passed to the outgoing parameter space that becomes the incoming parameters space for the called subprogram. The registers, pointers, and the status word PSW of the calling subprogram are saved; the pointers are altered to access the frame of the called procedure, and the control is passed to the first instruction of the called subprogram. After the called subprogram is over, the reverse process is used to restore the environment of the calling program, pass the parameter values depending upon the parameter passing mechanism, and pass back the control to the next instruction in the calling subprogram.

To share a complex object allocated in a heap, call-by-value is used that allows access to the data object allocated in a heap. During call-by-reference, a pointer to the actual parameter is saved in a formal parameter location. The retreival of r-value requires two memory accesses: one to access the memory location of the actual parameter and the second one to access the r-value. Call-by-value-result copies the actual parameters to the outgoing area that becomes the incoming parameter area in the called subprogram. After the return from the called subprogram, the results from the outgoing parameter area are copied back to the actual memory locations of the variables.

Recursive functions create a new frame for every invocation. The static link and dynamic link of a recursive function point differently: the dynamic link points to the base of the frame of the calling procedure, and the static link of all the invocations points to the base of the frame of the procedure under which the recursive function is nested.

Exception handling is implemented using one of the two techniques: (1) keeping the location of exception-handling routines in the frame of the subprogram or (2) a separate exception-stack. In the first technique, if all the exception handlers fail, then the control returns to the calling program, and the process is repeated. In the second technique, if the exception handler succeeds, then all the frames above the successful frame are removed from the exception-stack.

5.9 ASSESSMENT

5.9.1 Concepts and Definitions

Abstract instructions; abstract machine; active space; call-by-value; call-by-reference; call-by-value-result; code area; control-stack; data area; display registers, dynamic link; exception handling; exception stack, frame; frame-pointer; heap; idle space; recursive data structures; recursive procedure; return-pointer; stack based allocation; static allocation; SECD machine; static link; top-of-the-stack pointer; trail stack, WAM (Warren Abstract Machine).

5.9.2 Problem Solving

1. Write a sequence of low-level codes to implement the statement X = Y + 2 + 3 + 5.

2. Draw a control flow diagram for the case statement, and then write a low-level abstract code to implement case statement.

3. Write the code area and data area for the following Fortran 66–like program using static allocation.

PROGRAM MAIN
INTEGER I, J, K[20]
COMMON/B1/K[20]
DO 20 I = 1, 20, 1
20 READ(K[I])
CALL SORT
DO 30 I = 1, 20, 1
30 WRITE(K[I])
END
	SUBROUTINE SORT
	INTEGER M[20], I, J
	COMMON/B1/M[20]
	DO 10 I = 1, 20, 1
	J = 20 - I + 1
	10 CALL MAX
	RETURN
		SUBROUTINE MAX
		INTEGER M[20], I, J, K
		COMMON/B1/M[20]
		DO 30 I = 1, J, 1
		IF (M[I].GT. M[I + 1])
		    K = M[I]
		    M[I] = M[I + 1]
		    M[I + 1] = K
		ENDIF
		RETURN

4. Write a function that finds out the *factorial*(n), and show the control stack for computing *factorial*(2) when the control is in the invocation that is computing *factorial*(0).

5. Write the code area and data area for the following program using stack-based allocation for the following block-structured language. Assume that the block-structured language supports three types of parameter passing: call-by-value, call-by-reference, and call-by-value-result. Call-by-reference is denoted by a "&" tag in the actual parameter, and call-by-value-result is denoted by an "#" sign in front of the actual parameter. In the called subprogram, the value of the actual parameter is accessed by putting "*" before the formal parameter.

```
program main ()
integer maxvalue;
{   integer i, j, a[20]
    for (i = 0; i = < 19, i++)
       read(a[i]);
       sort(ref &a);
       maxvalue=d[19];
       i = 0;
       do
         write(a[i])
       while (i++=<19);
       write(max);
}
```

```
void sort (integer *d)
integer i,j
j = 20;
while(j > 1) {
 max(d, # j);
return
```

```
void max(integer *d, j)
{ integer i, temp;
  i = 0;
while (i = < j) {
 if (*d[i] < *d[i+1])  {
        temp = *d[i];
        *d[i] = *d[i+1];
        *d[i+1] = temp;
  }
 }
 j = j - 1;
}
```

6. Write a simple program for bubble sort that calls a procedure *find-max* using exchanging adjacent values instead of keeping temporary max. It passes the array using call-by-reference. Show the data and the code area, and highlight the parameter passing code. Clearly mark the pointers when the control is in the main program and when it is in the procedure *find_max*. Assume that the size of the array is 5.

5.9.3 Extended Response

7. What are the major drawbacks of static allocation? How does a hybrid allocation scheme consisting of static allocation, stack-based allocation, and heap-based allocation solve the problem?

8. Explain the role of frames in stack-based implementation. Why is it necessary to keep the visibility restricted mainly to local frame? Explain.

9. What are the advantages of superimposing outgoing parameters and incoming parameters area? Explain using schematics.

10. Assuming that you have both control stack and heap available to you, design a parameter-passing mechanism that acts like a read-only call-by-reference during the execution of the called subprogram if the value is being read, and passes any changes to the actual parameter at the end of the called procedure to update the portion of the actual parameter that has been modified during the execution of the called procedure. Discuss the advantages and drawbacks of such a scheme compared to call-by-reference and call-by-value-result.

11. Do a literature search to implement iterators, and write various techniques to implement iterators and the issues with implementing iterators.

12. What are exception handlers, and how are they handled in a stack-based implementation?

13. Draw a control flow diagram for selection construct with multiple conditional exits in ADA, and write a low-level code for the construct.

FURTHER READING

Brent R. "Efficient implementation of first fit strategy for dynamic storage allocation." *ACM Transactions on Programming Language and Systems*. 11(3). July 1989. 388–403.

Diehl, Stephan, Hartel, Pieter, and Sestoft, Peter. "Abstract machines for programming language implementation." *Future Generation Computer Systems*, 16.2000. 739–751.

Hanson, David R. "Fast allocation and deallocation of memory based on lifetimes." *Software Practice and Experience*, 20(1). January 1990. 5–12.

Jones, Richard, Hosking, Antony, and Moss, Eliot. *The Garbage Collection Handbook*. CRC Press/Taylor & Francis Group. 2012. 481.

Jones, Richard and Lins, Rafael D. *Garbage Collection: Algorithms for Automatic Dynamic Memory Management*. New York: John Wiley. 1996.

Wilson, Paul R., Johnstone, Mark S., Neely, Michael , and Boles, David. "Dynamic storage allocation: A survey and critical review." *Proceeding IWMM '95 Proceedings of the International Workshop on Memory Management*. Springer Verlag. 1–116.

CHAPTER 6

Dynamic Memory Management

BACKGROUND CONCEPTS

Abstract concepts in computation (Section 2.4); Abstract implementation (Chapter 5); Data abstraction (Section 4.1); Data structures concepts (Section 2.3); Principle of locality (Section 2.4.8).

Memory reuse is an important requirement for executing large-scale programs. Different program subunits and dynamically created objects have a lifetime. Allocated memory gets released when the corresponding program units have finished execution, or when the recursive data structures or the dynamic data object are manually released by a programmer action. The memory is allocated in the *control stack* or the *heap*. The memory from the control stack is released when the frame of the called procedure is discarded, and the memory location can be reused for allocating another frame. *Heap* is a common area used mainly for recursive data structures and dynamically created data objects as in object-oriented programming languages such as C++, Java, and C#. The data structures allocated in heap are extended dynamically based upon run-time programmer request, and outlive the procedures in which they were created. Heap is visible throughout the life of a program. The allocated memory in heap is recycled only after the data entity has been released.

Dynamic memory management is concerned about allocation, deallocation, and recycling of memory locations needed for data structures allocated in heap at runtime. If the objects are extended piecemeal, then it is difficult to access them in the frame of a procedure, as different allocations need to be chained and accessed sequentially, while data objects in a frame are accessed using index and offsets. The data structures and objects that are constructed once and deleted before the lifetime of the subprogram that created them can be allocated in the control stack for efficient access and better memory reclaim. A language that supports allocating complex dynamic objects in the control stack has to support *call-by-reference* for sharing the object with the called subprogram. If the dynamic data objects are allocated in the heap, then the control stack stores only the pointer to the first memory location of the dynamic object, and *call-by-value* (or *call-by-sharing*) is used to pass the pointer for accessing the object in the called subprogram.

6.1 HEAP ORGANIZATION

Heap is organized as a linear array, where dynamic data entities are allocated. Heap has three types of memory blocks: *allocated, released*, and *free. Allocated blocks* are active blocks being used by a process. *Released blocks* are not being used by a process and need to be recycled to be in the free pool. *Free memory blocks* can be allocated to a process based upon a run-time request. Each type of memory block contains the size information along with it. Heap can be modeled as a sequence of quadruples, where an element of the sequence can be (*allocated, <block-size>, <start-address> <end-address>*), or (*released, block-size, <start-address> <end-address>*), or (*free, block-size, <start-address> <end-address>*).

Free blocks are grouped together either using a chain of pointers or using an indexible structure that groups memory blocks of similar sizes together using chains of pointers. The right chain can be searched using an efficient search scheme such as binary search or hash-functions. After the chain connecting the similar-sized blocks is identified, the optimum block can be identified by sequentially following the chain.

The first pointers for heap objects are stored in the processor registers, control stack, or handles—a tuple of references to resources used by a process. The data cells of the same data structure scattered within a heap are accessed through a chain of pointers connecting scattered memory blocks belonging to the same data structure. During parameter passing, by copying the first pointer to a dynamic object, the data object can be accessed in the called subprograms. Only call-by-value to copy the reference to objects is required to access the heap objects.

Example 6.1

In Figure 6.1, the heap starts at location 13,500 and ends at location 0. It stores two data entities. Data entity 1 is spread over three memory blocks: 13,500–11,501, 8000–6001, and 5000–3001; data entity 2 is spread over two memory blocks: 10,500–8001

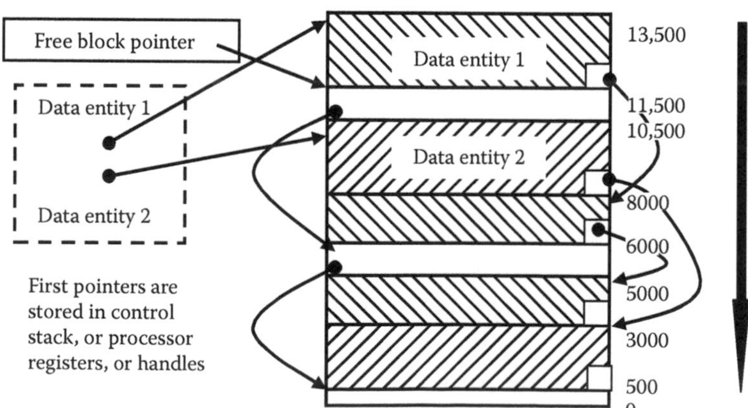

FIGURE 6.1 Schematics of heap organization.

and 3000–501. The first pointer for the data entity 1 points to the base address 13,500, and the first pointer for the data entity 2 points to the base address 10,500. The free blocks are also interspersed into three memory ranges: 11,500–10,501, 6000–5001, and 500–0, and the free block pointer points to the base of the first free block located at 11,500.

The heap blocks are allocated on the basis of the run-time request by a *process*—an active version of a program. A heap block is a collection of data cells that has three types of fields: *header field*, *information fields*, and *fields pointing to other data cells*. A *header field* carries multiple information such as flags to mark the cells as allocated, number of fields and their memory offsets from the start of the cell, number of memory locations in a cell, and number of pointers pointing to a cell. An information field carries the type of information and the values. For example, a linked list is a sequence of pair (header, info-field, pointer to next cell), and a cell in a binary tree is a 4-tuple of the form (header, info-field, left-pointer, right-pointer), as shown in Figure 6.2. Cells in a logical data structure are traversed using the chain of pointers connecting the heap blocks in the same data structure.

Free space in a heap can be organized in four different ways: (1) as a chain of free blocks, as shown in Figure 6.1; (2) as an indexed chain of pointers linking the blocks of similar sizes together, as shown in Figure 6.3; (3) as a group of indexible stacks, as shown in Figure 6.4; or (4) as bitmapped allocation markers. The third

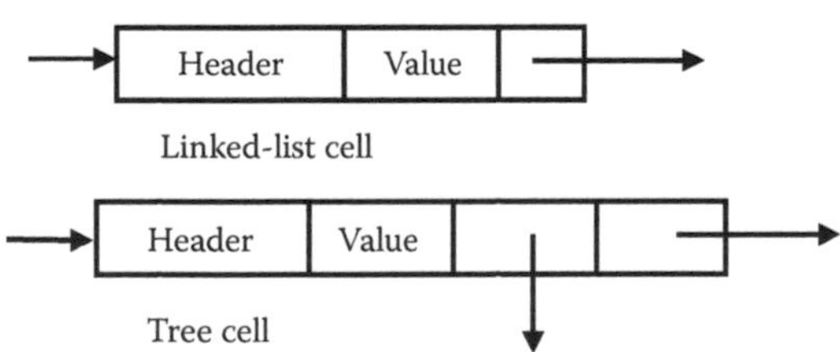

FIGURE 6.2 Unit cells in allocation of linked lists and trees.

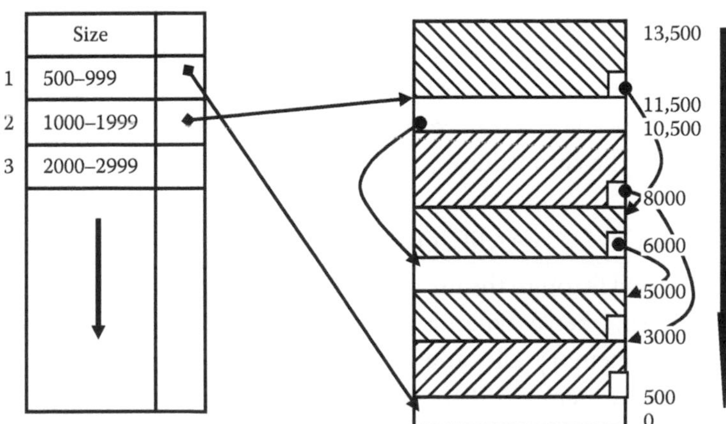

FIGURE 6.3 Size-based organization of heap for fast search of optimum block.

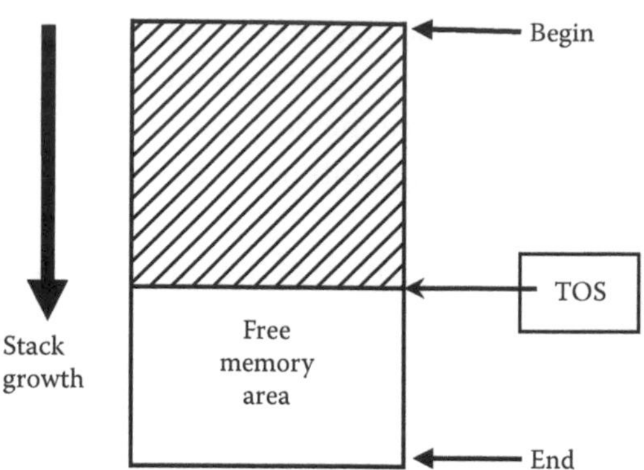

FIGURE 6.4 Stack-based allocation in a heap.

scheme is similar to the control stack studied in Chapter 5. The fourth scheme is popular in system programming to map files on physical blocks in secondary memory.

Figure 6.3 shows schematics of an organization of the heap for efficient search of the optimum size free memory block. An indexed table is organized by the similarly sized blocks. The corresponding chain of free blocks is identified by efficient search schemes.

In the stack-based allocation, a heap is represented as a chunk of memory locations that are marked by a *begin-address* and an *end-address*. The *top-of-the-stack pointer* (TOS-pointer) starts from the begin-address and moves toward the end-address after every run-time allocation of the requested memory block. A TOS-pointer is used to allocate the next requested memory block and to start the garbage collection—recycling process for released memory blocks that returns them to a pool of free memory. When the value of TOS-pointer goes past the end-address of the stack, garbage collection starts.

In the *bitmapped scheme*, the heap space is divided into fixed block sizes that are mapped on to a bitmap block such that the address offset of a bit from the start of the bitmap block represents the index of the data block in the heap. If the bit is set to "1," then the block having the same index as the address offset of the bit is allocated; otherwise it is free. The list of free blocks can be identified by (1) traversing the bitmap block to identify the bitmaps with zero value or (2) by keeping a sequence of address offsets of bits having the zero value.

6.2 ALLOCATION OF DYNAMIC DATA OBJECTS

Data objects are allocated through the use of constructors in object-oriented programming, such as C++, C#, or Java, when a dynamic data object is created or programmatically using an explicit request for memory allocation as in C. When a request is made, the heap is checked for the available memory. If the memory is available, it is allocated. Otherwise, released memory blocks are recycled using garbage collection. There are three popular memory allocation schemes: *first-fit*, *next-fit*, and *best fit* (or *optimal-fit*).

In *first-fit* algorithm, the first free block that is bigger than the requested memory block is allocated, and the header of the newly allocated block and chain of pointers connecting free block are adjusted accordingly. The pointer in the previous block is updated to point to the next memory address after the last memory address of the newly allocated memory block.

Next-fit scheme is a variation of the first-fit scheme. It is based upon memorization of the position of the last free memory block during the last search. The search continues from that position in the following request. After hitting the end of the heap, it starts again from the beginning of the heap. The advantage of this scheme is that allocated blocks are more evenly distributed. By evenly distributing the memory blocks, the probability and amount of *fragmentation* is reduced. *Fragmentation* is caused by the creation of leftover small memory blocks that do not have sufficient contiguous memory space to satisfy a request for memory blocks. However, the advantage is not that straightforward, because different subprograms have different lifetimes, and different programs have different amounts of allocations and deallocations of objects in a heap. A subprogram that has many allocations and deallocations of objects in a heap contributes more to fragmentation. The allocation can be done either using fixed-sized blocks or using a variable-sized block. The idea is to avoid fragmentation. If the size of the memory block left after the allocation is very small, then instead of saving the fragment as a free block, the whole block is allocated to avoid fragmentation. After allocation of the memory, a pointer is set from the extensible data entity to point to the newly allocated data cells. If the whole free block is allocated upon the request, then the free block is deleted from the linked list of free blocks. Otherwise, the chain of free block is adjusted accordingly to point to the remaining smaller free block.

The *best-fit scheme* finds out the most optimum size that is closest to the requested size. The best-fit scheme is more suited to the indexed structure of the heap, where the heap is organized in a group of similar-sized free blocks connected through linked lists. Efficient search schemes identify the chain of pointers with the nearest range and traverse the chain of pointers to find the best fit of the size that matches the allocation request. Again, if the computation shows that the fragment size of the remaining block is very small, then the whole block is allocated, and the chain of pointers is adjusted accordingly. If the remaining block-size is bigger than the threshold of a very small fragment, the remnant free block's size is computed, and the new free-block is deleted from the current linked list of pointers and inserted into the linked list of free-blocks that appropriately reflects the size of the smaller free block. In case the current chain of pointers being searched is empty, the search is carried to the next higher size range, until a nonempty chain of pointers is found. If no nonempty chain of pointers is identified, then the memory recycling process starts using *garbage collection*.

6.3 DEALLOCATION OF DYNAMIC DATA OBJECTS

Deallocation is done in two ways: (1) manual deletion of the data structure by a programmer action; or (2) automated deletion of the objects when a called subprogram is over, and lifetime of the dynamically created object is limited to the lifetime of the

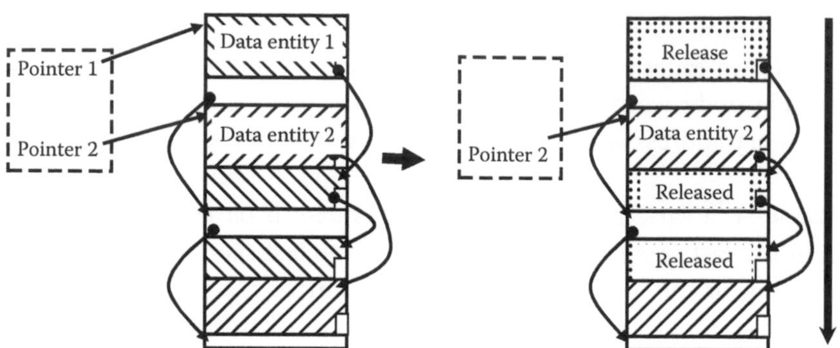

FIGURE 6.5 Deallocation of an object from the heap.

called subprogram. After a dynamic data object is deleted, the occupied memory block is *released* and is ready to be returned to the pool of free blocks. Deallocation of an object does not require traversal to every memory location occupied by the dynamic data entity. Instead, the first pointer that pointed to the base-address of the first memory block of dynamic data entity is reset to *null*. During the garbage collection process, the traversal starts from the collection of the first pointers to the stored dynamic data entities in the heap. The memory locations that are unreachable through the chain of pointers starting from the collection of first pointers are treated as garbage and are collected as free-blocks. Hence, the deletion of a pointer from the set of the first pointers is sufficient to release the whole dynamic data entity. Deallocation is illustrated in Figure 6.5.

6.4 FRAGMENTATION

Fragmentation is the process of formation of interspersed, smaller-sized memory blocks that cannot be allocated to a bigger-sized memory block request. As the process of allocation and deallocation continues, the block-size of the free blocks keeps decreasing, and the number of very small-sized free blocks keeps increasing. These fragments combined together can be used to allocate requests for bigger memory blocks. However, individually these free-blocks are too small to be of any use.

Example 6.2

Let us consider the example of a heap consisting of the following blocks: {(allocated, 200, 0, 199), (free, 400, 200, 599), (allocated, 300, 600, 899), (allocated, 500, 900, 1399), (free, 200, 1400, 1599)}. The total sum of the combined free memory space scattered across fragments is 600 bytes. A request comes for 500 bytes. None of the free-blocks individually can handle the request.

One could argue that the requests could be split into smaller requests and be chained together to accommodate a bigger request. However, there is another problem with fragmentation. If the data structure is scattered all over the heap, then it may be split into different pages in the virtual memory area. Principal of locality states that a program accessing a data or control entity works in the physical vicinity of the program. The computer hardware exploits the principal of locality to (1) populate the cache memory with a block of physically contiguous memory locations and (2) to load physically contiguous pages from hard disks into RAM when a page-fault occurs. If a data structure is scattered across the heap, the cache memory will not have the next cell of the data entity due to physical remoteness, and page-fault will not bring the adjacent data cells of the data entity due to significant distance between the physical addresses. This means that the overhead of the populating cache or page-faults would increase as the fragmentation in the heap increases.

There are two approaches to solve the problem of *external fragmentation*: (1) avoid the external fragmentation; or (2) join the external fragments during garbage collection to form bigger free memory blocks. *External fragments* can be avoided if we use fixed-sized blocks and allocate multiple fixed-sized blocks when a request is made. However, using the fixed-sized blocks causes *internal fragmentation*, since $M \times$ *fixed-block-size* $-$ *requested memory* ≥ 0. This wasted memory in the last allocated memory block is known as *internal fragments*. External fragments can be joined if (1) the adjacent blocks are identified as free-blocks during garbage collection; or (2) all the occupied state blocks are moved to a new memory space. Using copying one logical data structure at a time such that logically contiguous data structures also become physically contiguous.

6.5 GARBAGE COLLECTION—RECYCLING HEAP MEMORY

Memory locations in heap go through three states, as shown in Figure 6.6. A memory location becomes *active* after it is allocated to a data entity in a process. A process uses the active memory block and eventually deallocates the block. After the data entity occupying the memory is released, the memory location becomes *released*. The released memory is still not available for allocation to another data entity, because it is not a part of the chain of free memory blocks. Memory is recycled using a software process, called *garbage collection*, that takes the released memory and makes it part of the free memory blocks. A memory location is *free* if it becomes part of pool of free memory and can be allocated to a process.

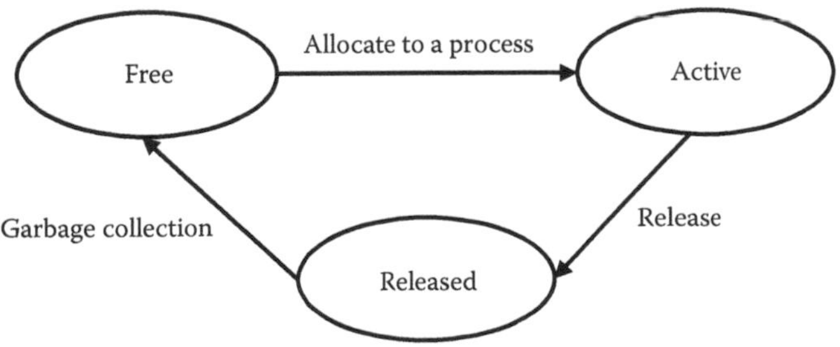

FIGURE 6.6 States of memory location in a heap.

After an object is released from the heap, there can be many actions, depending upon the memory recycling techniques as follows:

1. The released object is collected and recycled immediately. Such garbage collection is called *continuous garbage collection.*
2. The released object is marked for recycling. However, memory recycling is done at some time in the future when the heap runs out of free memory blocks.

The garbage collection process starts from the set of first pointers (root pointers) and traverses, using the chains of pointers, to mark the live data entities. After identifying live data entities, two approaches can be used:

1. Live data entities are relocated to new contiguous locations, freeing the old memory space.
2. The nonactive memory space in the heap is chained together as free space and reused.

Garbage collection has run-time and memory overheads that slow down program execution. The execution time overheads of modern garbage collection implementation are somewhere between 20–30% of the total execution time. In the garbage collection schemes that suspend the program execution during garbage collection, important real-time events or data are lost due to the delay caused by temporary program suspension. For example, a nuclear plant monitoring system may miss a major radioactive leak; or a spacecraft may miss an important sensor reading; or a painting robot on the shop floor of a car manufacturing company may suddenly stop painting, causing uneven delays in the assembly line. In interactive programs, this delay causes perceptible performance deterioration. For example, a video clip may not be rendered in real time.

6.5.1 Garbage Collection Approaches

The garbage collection process can be (1) *start-and-stop* or (2) *real-time.* A start-and-stop algorithm suspends the execution of the program temporarily when the garbage collection starts. Real-time garbage collectors allow interleaving of program execution with garbage collection. *Real-time garbage collection* does not suspend the program completely during garbage collection and performs better for handling real-time events. There have been multiple approaches to achieve real-time garbage collection such as (1) *incremental garbage collection,* (2) *continuous garbage collection,* (3) *concurrent garbage collection,* and (4) *hard real-time garbage collection.*

Incremental garbage collection allows program execution during garbage collection by splitting garbage collection into multiple smaller collections interleaved with program execution. However, during this smaller garbage collection period, program execution is still suspended. The interleaving gracefully degrades the execution of programs during the garbage collection process and allows interaction with the outside world in a limited way.

Continuous garbage collection collects the released memory immediately and puts it in the chain of free space. However, there is overhead of releasing and putting in the free space. Reference-count garbage collection is a popular continuous garbage collection scheme.

Concurrent garbage collection allows garbage collection along with the program execution using multiple threads or using multiple processing units: separate processing units running the garbage collector and program execution. The processes of recycling the released memory and the process of allocating memory are possible if the following conditions are met:

1. The memory space of the collection process and the memory allocation process are separated.
2. The two processes are synchronized when working on the same memory location.

With the use of multicore computers and multiprocessors, concurrent garbage collection is becoming quite popular. Interleaving program execution and garbage collection means that new memory locations are also allocated during garbage collection time. While incremental garbage collection and concurrent garbage collection schemes help in reducing the time delay degradation of performance graceful, there are still many real-time jobs—such as space–shuttle control, automated flight control, engine control in modern cars, robot control, and many mission-critical tasks—that have hard deadlines that must be met. Hard real-time garbage collection schemes try to handle this problem at the scheduler level by temporarily suspending the garbage collection if a high-priority interrupt or process needs to be processed.

The garbage collection has two types of processes: *mutator* and *collector.* A *mutator* is the program execution part that may allocate new memory locations or releases a data structure during garbage collection. A *collector* is the garbage collection process that converts the released memory blocks to free memory blocks. The data structures in the heap can be modeled as a *directed cyclic graph* that contains three types of nodes: *active, released,* and *free.* Active nodes are the union of all the nodes in multiple data structures. Each data structure is a graph of nodes. These data structures are chained together and start from the root pointer. Since there are multiple data structures, there are multiple root pointers. The set of root pointers is the starting point of garbage collection. The process of garbage collection involves collecting the set of active nodes using a chain of pointers. Once the active nodes are marked, the remaining nodes are treated as garbage nodes and are collected. Many data structures share a part of their data structures. Thus, there may be more than one incoming edge to a node.

6.6 START-AND-STOP GARBAGE COLLECTION

There are two major approaches to stop-and-start garbage collection: (1) mark all the active cells and then collect all the remaining cells as free cells, and chain them together; (2) relocate the active cells into another memory space, and free the old memory space.

Both approaches have been used: *mark-and-scan* algorithms use the first approach, and the *copying garbage collection* uses the second approach. This section describes three popular stop-and-start techniques used in modern uniprocessor machines.

6.6.1 Mark-and-Scan Algorithm

Mark-and-scan algorithm, also known as mark-and-sweep algorithm, has two phases: *mark phase* and *scan phase.* In the mark phase, starting from the first pointers stored in processor registers, control stacks, or handles, a recursive descent algorithm traverses all the data cells of a data entity by tracking the chain of pointers. The scan phase searches the entire heap array sequentially, one memory location at a time, and collects all the memory locations for which the corresponding mark bit is not set. After traversing an active cell, it resets the mark bit to facilitate the collection of the memory location released in the next cycle.

The traversal utilizes the depth-first search. For an information node containing more than one unexplored pointer, the address of the information node is pushed on the stack, and the unexplored left-most pointer is traversed in a depth-first manner until a leaf node—a cell with no outgoing pointer—is found. After traversing the leaf node, the top-most pointer is popped from the stack, and the node is traversed again using the depth-first search for the remaining unexplored part of the graph, until there are no more nodes to be traversed. The process is repeated for every data structure. During the traversal, the header of every active cell sets its *mark-bit*—a single-bit flag in the header of every data cell of the data structure. The *mark-bit* tells the scan phase that the cell is active and should not be collected.

During the scan phase, the chunks of contiguous memory are grouped together into one big memory block. These free blocks are either chained in a single chain of free blocks or as indexible multiple shorter chains arranged by similar size free blocks, as explained in section 6.1.

Mark-and-scan algorithm has many disadvantages, as described below

1. Mark-and-scan algorithm traverses the active cells twice: once during the mark phase to mark the active cells, and then during the scan phase to collect the cells that were not marked during the mark phase.

2. Mark-and-scan algorithm uses the depth-first search algorithm to traverse the data structures. Depth-first search is recursive by nature. Any recursive algorithm has memory overhead of using a stack. The size of the stack is at least as deep as the depth of the data structure being traversed. Linked lists are linear and do not need a stack to traverse the nodes. However, tree and graph traversal requires stack. The depth is log(N) for balanced binary trees, where N is the number of data cells in the tree. However, most of the times, due to contiguous insertion and deletion, trees are not balanced, and the depth is between log(N) and N.

3. Traditional mark-and-scan algorithm is a stop-and-start algorithm and is not suitable for languages supporting real-time events.

4. Mark-and-scan algorithm coalesces only adjacent free blocks. However, the free blocks are interspersed with active blocks embedded in. As the time progresses, the size of the free blocks gets smaller, and fragmentation increases. The fragmentation causes (1) increase in the number of page faults and (2) increase in the frequency of invocation of the garbage collector; as memory requests cannot be satisfied in fragmented memory space. Fragmentation results in slow execution speed due to (1) the lack of cache hit; (2) excessive page faults; and (3) frequent invocation of garbage collector due to lack of utilization of very small memory blocks.
5. Mark-and-scan algorithm's scan phase is dependent upon the heap size and not on the number of active cells, since it sequentially scans every memory cell.

6.6.2 Copying Garbage Collection

Copying garbage collection gathers all the scattered free spaces into one big space and does not suffer from the fragmentation and heap-size dependence problem present in mark-and-scan algorithm. The major advantages of the copying garbage collection are threefold as follows:

1. Copying garbage collection compacts all the free memory blocks into one contiguous memory space removing fragmentation completely.
2. Copying garbage collection copies one data structure at a time, keeping them in physically contiguous memory space improving the cache hit ratio and reducing the page faults that may be caused due to interspersed data structures.
3. Only active cells are copied.

Figure 6.7 illustrates the basic scheme of copying garbage collection. In copying garbage collection, the heap is divided into two spaces: *active space* (also called *from-space*) and *inactive space* (also called *to-space* or *idle space*). At any time, only the active space is used to allocate the data entities. The allocation is done using stack-based heap organization.

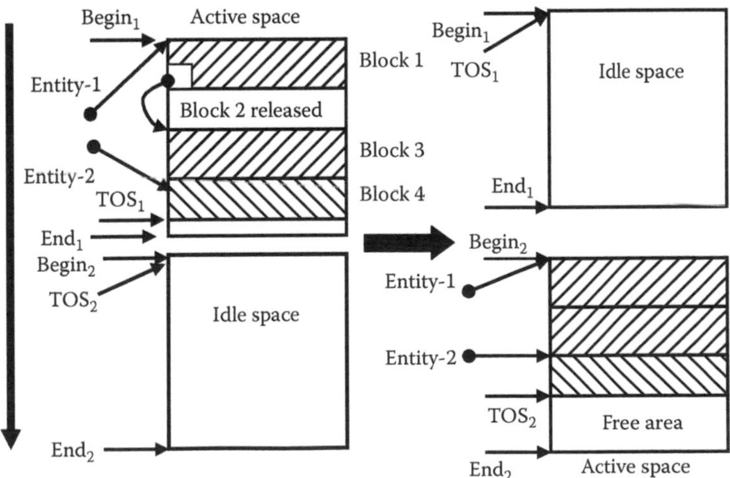

FIGURE 6.7 Copying garbage collection.

When the *top-of-the-stack pointer* hits the end marker, garbage collection is started. During garbage collection, active data structures are traversed one at a time, starting from the first pointer stored in registers, control stack, or the handles. All the interspersed memory blocks corresponding to single logical data structures are copied sequentially in a physically contiguous space to utilize efficiency due to the principle of locality. After all the data structures are copied, the spaces are switched: active space (from-space) becomes inactive space (to-space), inactive space (to-space) becomes active space (from-space), and the program execution resumes.

The copying garbage collection has the following pointers: start-address, end-address, TOS pointer, and forwarding pointer. The role of start-address, end-address, and TOS pointer is the same as used in the stack allocation. The forwarding pointer is used to copy shared data. Conventionally, a heap grows in the reverse direction: from higher address to lower address.

The left-hand side of Figure 6.7 shows a snapshot of a heap before the garbage collection, and the right-hand side shows the snapshot of the heap right after the garbage collection. The active space before the garbage collection shows two data entities: data-entity$_1$ and data-entity$_2$. Data-entity$_1$ occupies memory block 1 and 3, and data-entity$_2$ occupies memory block 4. Block 2 is a released block that needs recycling, along with the small memory fragment at the top of the stack. TOS pointer is very close to the end marker of the active space.

A request for memory block allocation such that TOS−size-of (requested memory-block) < end-marker initiates the next garbage collection. During the garbage collection, blocks 1 and 3 belonging to data-entity$_1$ are copied and placed contiguously, followed by the block 4 belonging to data-entity$_2$. The free space gets coalesced at the other end.

A forward-pointer is used to point from the from-space to the new memory location in the to-space, where the data has been transferred. Each memory location in the from-space that has been copied into the to-space is overwritten with (1) a tag that the cell has already been copied and 2) the forwarding pointer points to the corresponding memory locations into to-space. The purpose of creating a forwarding pointer is to avoid copying data cells shared between two data structures in two different memory locations.

Many data structures share part of the memory locations. Once the shared part has been copied during the copying of the first data structure, the shared space cannot be duplicated again while copying the second data structure. In order to avoid duplication of the shared data structure during copying, a forwarding pointer is stored in the old memory location of the from-space. During traversal, after seeing the forward-pointer tag, the forwarding pointer is copied to the pointer field of the last visited cell of the second data structure, and the copying process for the second data structure stops. The use of forwarding pointer is illustrated in Figure 6.8.

Example 6.3

Figure 6.8 has two data entities that share the data cells: "s-1" and "s-2." After data-entity$_1$ has been copied into to-space, there is no need to copy the shared data cells "s-1" and "s-2." The copied memory locations carry a forwarding pointer to the

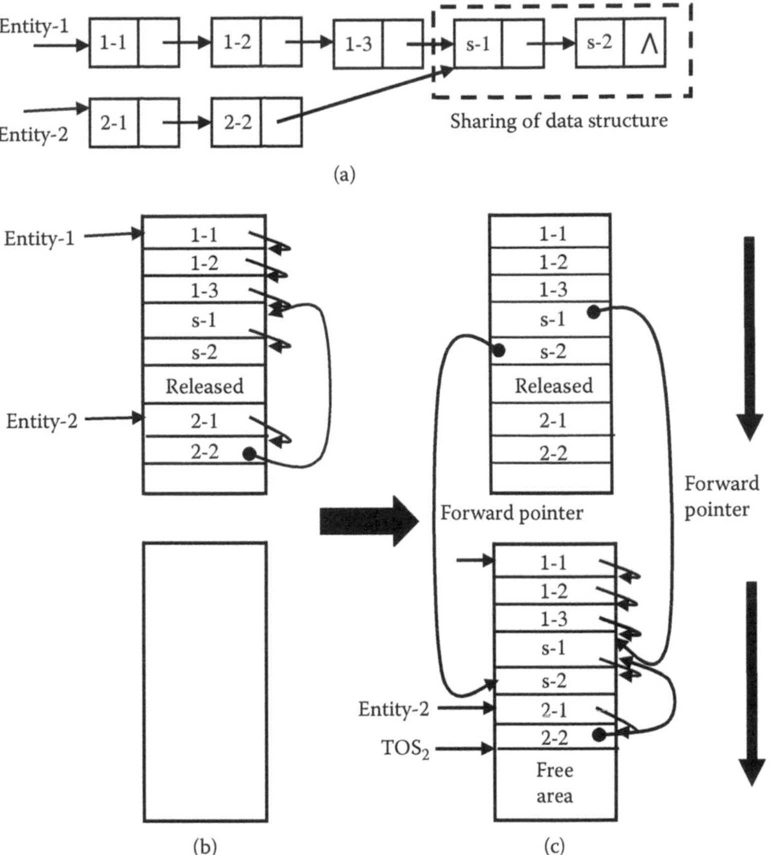

FIGURE 6.8 Role of forwarding pointer during copying garbage collection. (a) Two linked-lists sharing data structure; (b) snapshot before copying garbage collection; (c) snapshot during copying garbage collection.

memory locations they have been copied to. After copying cell "2-2" of data-entity$_2$ the third cell "s-1" is a shared cell. This forwarding pointer stored in the cell "s-1" in the from-space is copied into the pointer field of the cell "2-2" into the to-space. There is no need to copy any additional cells after setting up the pointer, as shared cells are already in to-space.

Naïve copying garbage collections have following disadvantages:

1. It is a start-and-stop algorithm and is not suitable for programs handling real-time events.
2. It uses a depth-first search that has a memory overhead of stack space.
3. Only 50% of the heap space is utilized at a time.
4. The principal of locality states that older data structures are less likely to be used. These older data structures keep getting copied between the from-space and the to-space every time garbage collection is done, contributing to the execution time overhead of excessive copying.

There are two major approaches to solve the problems in naive copying garbage collection:

1. The first approach is to replace the memory overhead of stack by using *breadth-first search* and by treating the full to-space as a queue. The use of the breadth-first search removes the overhead of stack, and the use of to-space as a queue removes the inherent memory overhead of using queue in breadth-first search.
2. The second approach uses multiple version spaces instead of two semispaces, while keeping only one space inactive at a time. This approach improves the memory utilization to 100 * $(n - 1)/n$%. With three version spaces, the memory utilization efficiency is 67%, and with four version spaces, the memory utilization efficiency improves to 75%.

6.6.3 Cheney's Modified Copying Garbage Collection

Figure 6.9 illustrates an algorithm, first developed by Cheney, that removes the memory overhead of stack by using breadth-first search and utilizing the full to-space as a queue. The use of to-space as queue avoids any inherent memory overhead of a separate queue. The data cells in to-space have three states: *black cells*, *grey cells*, and *white cells*. *Black cells* are active cells that have been visited and copied into to-space. *Grey cells* have been copied into the to-space, but their children are in the from-space. *White cells* are unreachable cells that cannot be reached starting from any of the first pointers. The scheme uses an additional pointer called *scan-pointer* that starts from the pointer $begin_2$ in to-space and chases the TOS-pointer during the garbage collection period. The scan-pointer acts like a pointer that is pointing to the next cell to traverse in the queue, and the TOS-pointer acts as a rear-pointer of the queue, where new data elements are to be inserted. The memory location pointed by the scan-pointer is checked to see if it holds a data value or a pointer.

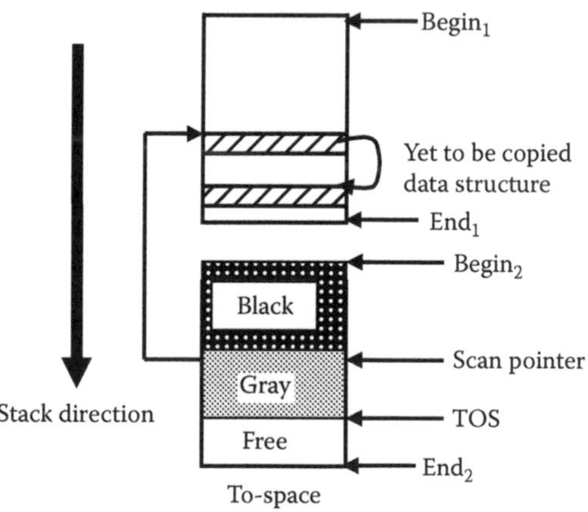

FIGURE 6.9 Cheney's garbage collection scheme using breadth-first search.

Depending upon whether the memory cell holds a data or a pointer, the following three actions are possible:

1. If the memory location holds a pointer to a data cell in the from-space, then a new cell is allocated on top of the to-space stack, the value of the memory cell in from-space is copied in this new cell, and the forward-pointer in from-space is set to point to this new cell. The scan-pointer is incremented by one location.

2. If the memory location holds a data element, then the scan-pointer is incremented by 1.

3. If the memory location holds a pointer to a memory location in to-cell, then the scan pointer is incremented by 1.

After the scan-pointer catches up with the TOS-pointer, all the cells of one data structure have been copied from the from-space to the to-space. The process is repeated for every data structure starting from the first pointer. When all the data structures have been copied and the scan-pointer has caught up with the TOS-pointer, the garbage collection process terminates.

6.6.4 Generational Garbage Collection

Generational garbage collection keeps more than two spaces. However, only one space is kept idle. The space that gets full is copied into the idle space. The major advantage of this scheme is (1) The memory utilization increases; (2) copying cost into the idle space gets reduced, as the size of the buckets is smaller; (3) the frequency of garbage collection for the younger generation of data structures increases due to the principle of locality and the fact that most recently used spaces are accessed more often and get full first; and (4) the overhead of copying older data structures that are not being used is minimized by keeping the older data structures in older buckets.

The principle of locality states that the data accessed in the recent past are accessed more frequently unless the locality changes due to called subprograms. Splitting the heap into multiple smaller version spaces exploits this property of program execution: different version spaces (subunits of heap) tend to collect data structures according to their age. Each version space is marked with a version number according to its age. The overall scheme is illustrated in Figure 6.10.

New data structures are mostly short lived and are collected before the next garbage collection cycle. A smaller percent go through more than one cycle, and the percentage becomes progressively smaller with age. This means that very old data structures occupy fewer version spaces and grow very slowly. The garbage collection part does two operations: (1) copies the newer data structures into the idle space and (2) if a new cell has been allocated for a very old data structure, it is transferred to the older version space; where the rest of the old data structure is residing. When the old version space is full, it is also collected like any other version space.

One of the problems that multiple version spaces scheme has is that, unlike copying garbage collection with two semispaces, where all the data structures are stored in one semispace, a single logically contiguous data structure is scattered over multiple version spaces. Large percentages of such data structures are older. There is a need to change the pointers when the partial data structure from one version space is copied to idle version space.

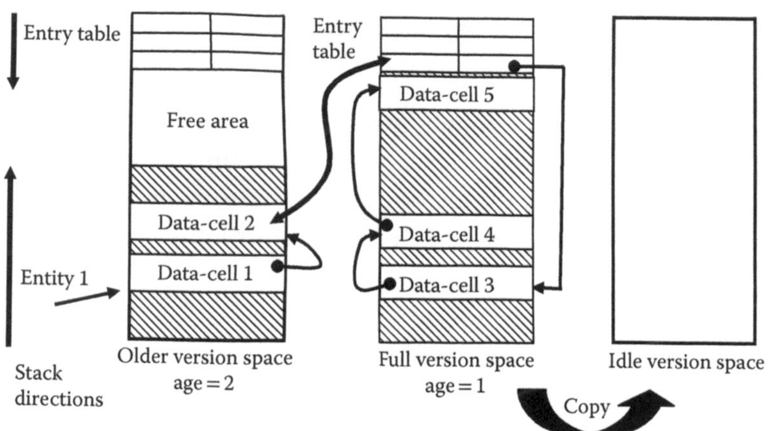

FIGURE 6.10 Generation garbage collection and use of entry table.

In order to remove the problem of changing pointers in the scattered data structures in multiple version space, a table of pointers is used in every version space. The table of pointers has two functions:

1. It carries backward pointers to the previous cell of the same data structure that points to the next cell of the data structure stored in the corresponding version space.
2. The table points to the next cell stored in the version space. The advantage of this scheme is that when the data cells are compacted during copying garbage collection, the address is changed only in the entry tables, and not in different version spaces.

In addition to all the pointers used in copying garbage collection, there are three additional pointers in generational garbage collection: (1) forward-pointer to entry table, (2) backward-pointers from the entry table, and (3) TOS for the entry table. The forward-pointer is used to point from the last cell of a data structure in a version space to the entry table location that stores the address of the next cell in another version space. The backward-pointer points back to the previous cell from the entry table that stores the address of the next cell. If more than one data structures are sharing the next cell, then there is a linked list of back pointers pointing to all the previous cells that point to the next cell in the version space. The table of back pointers and front pointers is toward the top end of the version spaces, as shown in Figure 6.10. The table is extensible, grows like a stack in the reverse direction to the growth of the heap, and needs another top-of-the-stack.

During garbage collection, when the entry table of the version space is copied into an inactive version space, two actions have to be taken:

1. The forward-pointer from the previous cell has to be modified to point the new table entry into the to-space. This is done by using the backward pointer to go back to the previous cell, then storing the address of the new location into the to-space where the table entry has been copied.

2. The compaction is done for each table entry, because each table entry points to the first cell of the partial data structure stored in the version space.

Example 6.4

Figure 6.10 shows a heap consisting of three version spaces. The left two version spaces are active, and the right-most version space is inactive. The left-most version space is older with age = 2, and the middle-version space is the newer one with age = 1. The data-entry$_1$ spans across both the version spaces. The version space with age = 1 fills up faster, according to the principle of locality, and is copied frequently into the idle space. The second data cell in the older version space (age = 2) has a forward-pointer to a table entry that points to the third data cell of the same data structure in the newer version space (age = 1). The same table entry in new-version space has a back pointer that points to the second data cell in the older version space (age = 2).

When the middle-version space becomes full, garbage collection starts, and the middle-version space is compacted and copied in the rightmost version space. The address of the table entries and the data cells changes. The forward-pointer from the second cell is changed by first using the backward-pointer to get the address of the second cell, and then updating the pointer field of the second cell.

While generational garbage collection improves naive copying garbage collection by improving the memory space efficiency and by not copying older data structures, it has an additional overhead of maintaining the table of entries to handle data structures scattered across the version spaces. As the number of version spaces increases, the overhead also increases.

6.7 INCREMENTAL GARBAGE COLLECTION

An *incremental* garbage collector (see Figure 6.11) intertwines garbage collection and program execution. After the garbage collection starts, the garbage collector does partial garbage collection followed by some program execution. Garbage collection process once started goes into the following sequence: *pgc*, *pe*, *pgc*, *pe*, *pgc*, *pe* ... where *pgc* stands for *partial garbage collection*, and *pe* stands for *program execution*. The garbage collection period is quite small. The garbage collection is done quite frequently. More released memory

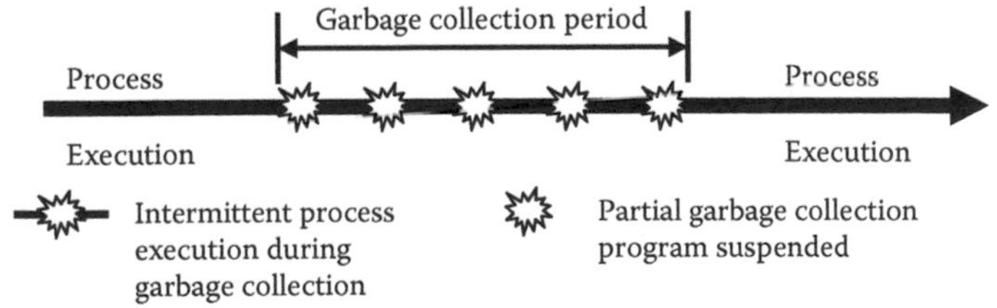

FIGURE 6.11 Incremental garbage collection.

is recycled than the new memory allocation during program execution during garbage collection period. After the garbage collection is over, regular computation resumes.

6.7.1 Baker's Algorithm

Baker improved Cheney's algorithm by periodically collecting K ($K >> 1$) cells from the from-space to the to-space before allocating one requested memory location in to-cell during garbage collection. As soon as the garbage collection starts, the from-space is sealed, and the new memory allocation requested by the executing process during garbage collection period is done only in the to-space. If a memory block of size m is requested, then the first $m * K$ cells are copied from the from-space to the to-space before allocating m cells in the to-space. The new cells are allocated from the other end of the semispace to ensure the physical contiguity of the data structure being copied from the from-space to the to-space, as shown in Figure 6.12.

As shown in Figure 6.12, there are two pointers—top-pointer and bottom-pointer—in addition to the scan-pointer. The bottom pointer points to next free space in the free area for copying the data structures from the from-space, and the top-pointer points to the next free area for new allocations. Bottom-pointer and top-pointer move in opposite directions. After copying the data entities from the from-space, the bottom-pointers gets fixed, and only the top-pointer moves, until it hits the bottom-pointer when the new garbage collection cycle starts.

There are three types of cells: *black*, *grey*, and *white*. *Black cells* are the cells that have been copied from the from-space to the to-space along with their children nodes. In the to-space stack, they lie between the begin marker and the scan-pointer. *Grey cells* are those cells that have been copied from the from-space to the to-space. However, they still have possible pointers to other cells in the from-space that have not been copied to the to-space. In the to-space stack, they lie between the scan-pointer and the bottom-pointer. *White cells* are cells in the from-space that have been released and need not be copied to the to-space. Newly allocated cells are treated as black cells.

Since the data to be traversed is the same as the copied data from the from-space, no additional memory space is needed. The total allocated memory is given by $N + N/K$, where

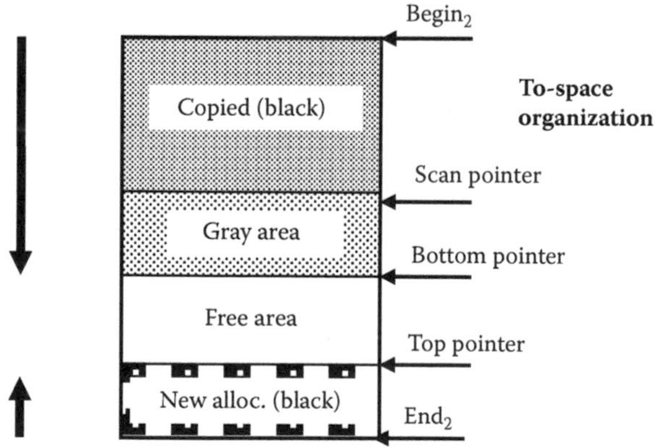

FIGURE 6.12 To-space during incremental copying garbage collection.

N is the active memory locations in the from-space and N/K is the new memory allocated in the to-space during garbage collection. The remaining memory in the to-space is $M - (N + N/K)$. Assuming a steady state—that the rate of allocation and deallocation are equal—the remaining memory in the to-space should be the same as the active memory copied from the from-space. Hence, $N = M - (N + N/K)$, which gives $M = 2N + N/K$. Using this equation, one can derive the heap size, given the size of active memory location and the ratio K.

6.8 CONTINUOUS REFERENCE-COUNT GARBAGE COLLECTION

Another popular scheme for garbage collection is reference-count garbage collection, which is a continuous garbage collection scheme that handles the problem of dangling pointers in the presence of shared data structures. As described earlier, part of a data structure may be shared with other data structures. The shared part of the data structure cannot be recycled, unless all the data structures that share are logically deleted. Reference-count garbage collection solves this problem by keeping a reference-count in the header of every data cell. The reference-count is incremented upon the creation of a new reference and decremented upon the removal of a reference. Reference can be created by creating a new data structure that shares the part of the data structure. Reference can be removed by deleting a data structure.

After a data structure is deleted, all the data cells reachable from the first pointer (root pointer) of the deleted data structure are decremented by 1. Decrementing the reference-count by one makes the reference-count zero if the reference-count was equal to one. All the data cells having reference-count = 0 are collected and chained together as free memory block. The process of decrementing the reference-count stops after decrementing the reference-count of a cell with reference count >1 because the following structure is being used by at least one more data structure.

Example 6.5

Figure 6.13 shows two data entities represented as linked lists. Data-entity$_1$ consists of a chain of data cells: #1, #2, #3, #6, and #7. Data-entity$_2$ consists of a chain of data cells #4, #5, #6, and #7. Data cells #6 and #7 are shared, and the data cell #6 is pointed by two pointers: pointers from data cells #3 and #5. The reference-count of cells #1, #2, #3, #4, #5, and #7 is 1, and the reference-count of the cell # 6 is 2.

Consider a scenario when Data-entity$_1$ is deleted. The reference-count of cells #1, #2, and #3 becomes 0, and the reference-count of the cell #6 becomes 1. The three

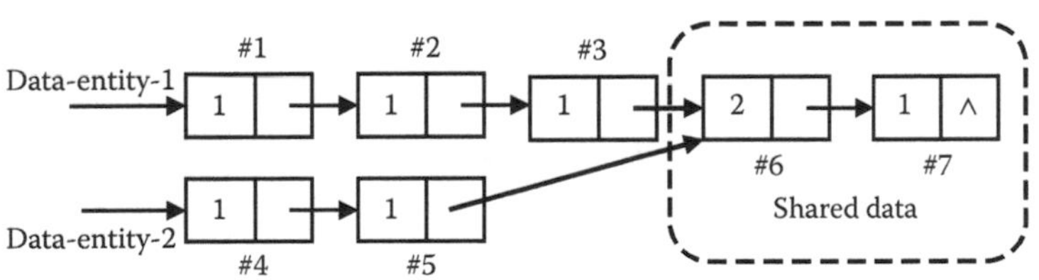

FIGURE 6.13 Reference-counts for shared data-structures.

cells—cells #1, #2, and #3—are collected and recycled. However, cells #6 and #7 have a reference count of 1 and are not collected.

There are many problems with naive reference-count garbage collection as follows:

1. Reference-count garbage collection makes no effort to compact the free space, as it is a continuous garbage collection scheme and thus suffers from fragmentation.
2. Reference-count garbage collection cannot recycle cyclic data structures efficiently. In case of cyclic data structure, as shown in Figure 6.14, the reference-count of the first cell is equal to 2 and not 1, as in the case of the acyclic data structures. When the data structure is deleted, the reference-count is decremented by 1 for the first data cell, and the decrementing process stops. Since none of the cell has reference-count = 0, the released data structure cannot be collected unless the cycle is detected. Cycle detection has significant memory and execution time overheads and is avoided in realistic garbage collectors. The phenomenon of inability to recycle the released memory cells is called *memory leak*, because these memory cells can never be allocated to other data structures during the lifetime of the program. Since, empirically, only a small percentage of data structures are cyclic, the memory leak is allowed to maintain the execution time efficiency of the garbage collection.
3. Decrementing the reference-count of a large data structure has significant execution time overhead. To reduce this overhead, the reference-count decrementing is deferred until there are requests for memory allocation. Instead of eagerly decrementing the reference-count, the address of the first cell of the released data structure is pushed on the top of a stack called *to-be-decremented stack* (TBD). When an allocation request for *m* memory locations is made, an address is popped from the TBD stack, and "*m*" memory locations are collected from the corresponding released data structure. The collection starts, if after decrementing, the reference-count is equal to 0, and ends when the reference-count after decrementing is greater than or equal to 1.
4. Storing the reference-count has significant memory overhead. Using just 4 bits for at most 16 references will cause a significant memory overhead. Many times 4 bits are not sufficient, and there can be more references. Multiple variations of reference-count garbage collection have been proposed to handle the overflow caused by excessive references.

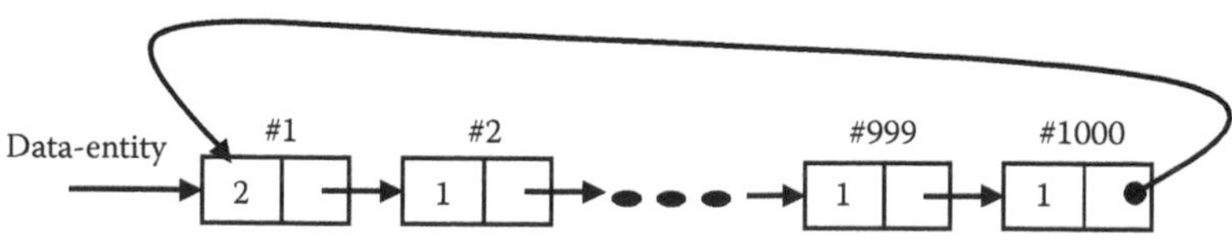

FIGURE 6.14 Cyclic data structure problem in reference-count garbage collection.

6.9 CONCURRENT GARBAGE COLLECTION

Incremental garbage collection alleviates the problem of program suspension to a large extent. However, incremental garbage collection still suffers from smaller program suspensions. Concurrent garbage collection allows much finer grain garbage collection by running the garbage collector and the program execution concurrently on two separate processors or using two different threads on a uniprocessor machine. While one processor performs program execution, the second processor performs the garbage collection. There are two issues in concurrent garbage collection: how to handle the shared data structures, because the garbage collector (also called the collector) and the program execution (also called the mutator) cannot simultaneously perform operations on the same data cell. The process of releasing or allocating a data cell is a multi-instruction operation. Once started, the operation must be completed without relinquishing the control. This property is called *atomicity*, and the operation is called *atomic operation*. Both atomicity and fine-grain synchronization are crucial for concurrent garbage collection. These restrictions impose additional overheads compared to sequential garbage collection schemes.

6.9.1 Concurrent Copying Garbage Collection

Concurrent copying garbage collection allows copying and compaction of the live objects from the from-space to the to-space without ever suspending program execution. Both collector thread and mutator thread keep executing concurrently. The overall scheme is illustrated in Figure 6.15.

It uses two special data abstractions: *relocation map* and *mutation log. Relocation map* stores entries of the form (*from-space memory location, to-space memory location*) to mark the memory locations that have been copied from the from-space to the to-space, and the mutation log keeps all the changes made in the objects in the from-space during garbage collection that have to be incorporated into the to-space before switching the semispaces.

Unlike start-and-stop copying garbage collection, no forward-pointer is created in the from-space. Relocation map is used to avoid copying the shared part of data structures. Relocation map is implemented using a hash table with a from-space address as the key. Each entry in the relocation map is equivalent to a nondestructive version of the

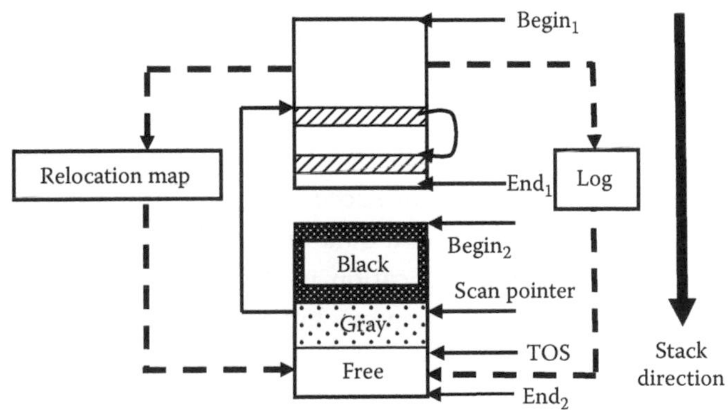

FIGURE 6.15 Concurrent copying garbage collection.

forward pointer in the incremental garbage collection. The purpose of the relocation map is the same as the forward-pointer in the start-and-stop copying garbage collection. Before copying a memory location from the from-space to the to-space, the relocation map is checked. An entry in the relocation map indicates that the memory location is shared between two data structures, and the memory location is not copied.

The concurrent garbage collection does not allow a mutator (program execution) to use the replicated objects in the to-space until all the live objects have been completely copied. The new modifications made by the mutator are collected in a *mutation log*. The collector keeps checking the log and updates the corresponding replicated image in the to-space.

The overall algorithm is that the from-space is copied one data structure at a time to the to-space using Cheney's improvement. When a memory location is copied from the from-space to the to-space, an entry of the form (*from-space-address, to-space-address*) is entered in the relocation map. Any new modification in the from-space during garbage collection is also inserted in the log. Under the assumption that the collector runs much faster than the mutator, both copying of the live objects and the log-based updates will be copied to the to-space in realistic time.

One major issue in concurrent garbage collection is handling the conflict of simultaneous modification of the data structure by the mutator, which is also being copied at the same time by the collector. One scheme is to use *atomic operation* and copy the data structures one cell at a time. If it is essential to pass the control to an operating system routine immediately, then all the effects are undone, and atomic operation starts again from the beginning to get the correct effect. To insure atomic operations, locks are used. Locks are special purpose hardware-supported memory locations that provide exclusive control of a computer resource to a process. The control is released only after resetting the lock.

6.9.2 Concurrent Real-Time Garbage Collection

As described in the previous section, concurrent garbage collection has to go through atomic operation at the information cell level and synchronization to provide correct interleaving of the mutator and the collector. For real-time concurrent copying garbage collection, two types of schemes have been used: (1) use of information cell level locks to provide the cell-level atomic operations and (2) lockless garbage collection by realizing that the frequency of the collectors and mutator performing operations on the same cell is rare. In case of the collision of the mutator and the collector, copying is aborted. The avoidance of the use of locks improves the performance.

6.10 ISSUES IN GARBAGE COLLECTION

Garbage collection requires information for successful recycling of the memory. It needs the information whether a memory cell is a data or a pointer. Similarly, each cell needs a mark-bit to mark whether a memory location is allocated or released. The reference-count garbage collection requires extra memory bits to keep the reference counts. There is a significant execution time overhead of garbage collection. It takes around 20–30% of the total execution time. Stop-and-start garbage collection techniques suspend the program execution, which can result in missing real-time events and sudden deterioration in performance, especially in interactive programs. To support the real-time computation and to avoid sudden deterioration in program performance, collection of released memory and program execution are interleaved.

Incremental garbage collections solve the problem of long-time suspensions. However, there are still patches of short-time suspension delays that cause delay in processing of hard real-time events. To handle this problem, collection and program execution are executed concurrently on multiprocessor systems and multicore computers. Concurrent execution relieves the problem of smaller suspensions in traditional incremental garbage collections to a large extent. However, concurrent execution has its own problem of handling: (1) *atomic operations*—multiple instructions treated as one big single instruction; (2) handling shared memory locations that are accessed both by the garbage collector and program execution; and (3) and the handling of real hard-time events. In order to handle real hard-time events, modern systems allow for the suspension of the garbage collection process by the task scheduler in the operating system when the real-time events need to be handled.

The process of garbage collection starts with traversing the data cells starting from the root-pointer. The key is to identify a pointer, as both numbers and addresses are combinations of zeros and ones. Many programming languages such as C allow for pointer arithmetic treating addresses as integers to step through memory locations. This makes the separation between integers and pointers very difficult. Integers are data and pointers are addresses of other data cells to be collected after the data structure is released. Generally, pointers can be identified by knowing the structure of the data cell. However, if a pointer is missed and treated as data during garbage collection, then the following released data cells are not reachable, are not marked active, and are collected as garbage cells causing incorrect program behavior. If a data cell is treated as pointer, then the garbage cells may be marked as active cells and may not be collected during garbage collection causing that cell not to be used in the future, which results in *memory leak*. Memory leak is a situation when a released data cell is not available for reallocation. Garbage collectors are conservative and prefer memory leaks over the incorrect behavior of the program. Many other precautions are taken to distinguish between pointers and data. For example, any data cell that has a value not pointing to within the heap area is treated as a data cell, and the data cells are word-aligned and initialized to proper null value to separate them from pointers.

Another major issue is the allocation of data on a heap or stack. Those dynamic data, whose lifetime is limited to the scope of a invoked procedure are allocated in stack for faster memory access and recovery.

6.11 SUMMARY

In this chapter, we discussed the memory management of dynamic data structures and recursive data structures using *heap*. The *heap* is a common area visible to all the subprograms, and all the subprograms can allocate a dynamic data in heap. This heap is different from the heap you studied in data structures. The free memory area in a heap can be organized in many ways: (1) using a single chain of pointers of free memory blocks; (2) using chained group of free blocks sorted by similar sizes; and (3) a collection of stacks. Data can be allocated automatically either when a new dynamic object is created or requested programmatically.

Allocation strategy can be *first-fit, next-fit,* or *best-fit.* The memory cells in a heap are in three states: *active, released,* and *free.* Garbage collection recycles memory from the *released state* to the *free state.* Dynamic data cells can be released automatically or programmatically. During the process of continuous allocation and deallocation, the free space keeps getting interspersed into isolated smaller memory chunks. Many of the memory blocks are so small that they individually cannot be used for effective memory allocation. This kind of formation of interspersed small blocks is called *fragmentation,* and it negatively affects the execution time performance of garbage collectors and program execution by decreasing the hit ratio in cache and increasing the frequency of page faults and garbage collections.

Garbage collection can be done continuously as in reference-count garbage collection or periodically after no more memory can be allocated in the heap space. Periodic garbage collectors can be start-and-stop or incremental. Start-and-stop garbage collectors suspend the program execution completely during garbage collection and are unsuitable for handling real-time events, since garbage collection causes significant delay due to memory and execution-time overhead. To avoid this problem, many approaches have been tried such as incremental garbage collection, concurrent garbage collection, and continuous garbage collection. In incremental garbage collection, one big garbage collection period is divided into multiple smaller garbage collection periods interleaved with small periods of program execution, such that the collection rate is faster than the memory allocation rate. Concurrent garbage collection runs program execution and garbage collection simultaneously. However, they have to handle the issues of atomicity of multiple instructions involved in a common atomic operations, and provide synchronization while sharing the memory location between the garbage collector and program execution. Both incremental garbage-collection schemes and concurrent garbage-collection schemes alleviate the problem of soft real-time problem such as video display by graceful degradation. However, they are unable to solve the problem of hard real-time events involving sensors or mission-critical operations. In order to handle hard real-time events, the process scheduler in the operating system has to defer and suspend the garbage collection process when high-priority real-time events occur.

Periodic garbage collectors work in two ways: (1) marking active spaces and collecting the remaining space in the heap as free memory area and (2) collecting active memory area and copying it in another memory space to remove fragmentation. Marking active spaces suffers from many drawbacks such as overhead of stack, overhead of traversal of active space twice, and fragmentation. Copying garbage collection suffers from wastage of memory. In order to alleviate this problem, multiple-version spaces are used. Each version space has an age associated with it. Only one version space is inactive, and all other version spaces are active. Owing to principle of locality, older data structures are located in the older version spaces, and younger data structures are located in younger version spaces. The younger-version space gets full faster than older version space. Increasing the number of version space also avoids excessive copying of older, less-accessed data structures. The drawback of the version space is the overhead of keeping a table of forward and backward pointers to connect the data structures interspersed across multiple version spaces.

Reference count garbage collection is a continuous garbage-collection technique that is useful in handling shared data structure by keeping a reference-count. When a data structure is released, the reference-counts of the corresponding cells are decremented, and all the data cells with reference count = 0 are collected. It faces the problem of handling cyclic data structures that make 2–3% of the total allocation. To maintain efficiency, reference-count garbage collection schemes prefer *memory leaks* over the costly garbage collection of cyclic data structures.

In real-time concurrent garbage collection mechanisms, the smallest unit of the collection is one information unit or an object at a time. This can be done using atomic operations. Collector uses a lock to copy one information node or one object at a time.

There are many issues with garbage collection. It is difficult at times to distinguish between data cells and pointers. If pointers are treated as data cells during garbage collection, then the following memory cells are not marked, resulting into incorrect collection of active cells as garbage. If the data cells are treated pointers, then additional cells are marked as active cells and are never collected, resulting into *memory leak*. Garbage collection programs use a conservative approach and prefer memory leak in case of doubt. Empirical study shows that 2–3% of memory is lost to memory leak.

There are many variations of the class of garbage collections discussed in this chapter. In recent years, a lot of effort has gone in developing concurrent versions of garbage collectors for various parallel architectures. There are also distributed versions of garbage collectors for Internet-based languages.

Java uses a concurrent version of garbage collection in its implementation. Different implementations of C++ use variations of mark-and-scan garbage collection as well as copy garbage collection. C# uses generational garbage collection with three version spaces. Scala is built on top of Java and uses JVM's garbage collector. ADA does not have a default garbage collector. Haskell uses a parallel generational garbage collection. Current implementations of Ruby use a variant of mark-and-scan garbage collection.

6.12 ASSESSMENT

6.12.1 Concepts and Definitions

Active memory blocks; backward pointer; Baker's algorithm; best-fit allocation; black cell; Cheney's algorithm; concurrent garbage collection; concurrent copying garbage collection; continuous garbage collection; copying garbage collection; entry table; first-fit allocation; forward-pointer; fragmentation; free memory block; from-space; garbage collection; generational garbage collection; grey cell; hard real-time garbage collection; heap; inactive space; incremental garbage collection; mark-bit; mark-and-scan algorithm; memory allocation; memory deallocation; mutation log; real-time garbage collection; reference-count garbage collection; relocation map; scan-pointer; semispaces; TBD-stack; to-space; version space; white cell.

6.12.2 Problem Solving

1. Consider an indexed chain organization of free spaces, where each range of memory location is 1000 bytes. For example, the ranges are 0–999, 1000–1999, 2000–2999, 3000–3999, 4000–4999, and so on. Show the indexed heap organization with the free space sizes as 1024, 512, 1024, 2048, 1536, 4096, 32, 16, and 128. A request for memory allocation for two blocks of size 480 bytes and 1000 bytes is made. Show the

indexed heap organization for the heap after the allocation of requested blocks using an allocation algorithm that picks up the first free blocks in the chain of free blocks in the best matching range.

2. Describe a data structure representation for indexed free space organization of free space in a heap, and write a high-level algorithm to allocate requested data using the indexed chain for optimum allocation. Show clearly how the table gets readjusted after the allocation.

3. A heap has the following free blocks: 1200, 3000, 340, 560, 790, and 4002 bytes. Assume that the free-blocks are chained using a single chain of pointers. Allocate the three memory block requests of 2000, 450, and 740 using first-fit and next-fit strategy, and show the remaining free blocks after each allocation.

4. Write a high-level algorithm for naive mark-and-scan algorithm.

5. Write a high-level algorithm for naive stop-and-start copying garbage collection.

6. Write a high-level algorithm for Cheney's copying garbage collection.

7. Write a high-level algorithm for incremental copying algorithm with Baker's improvement.

8. In Baker's algorithms, assume that the size of a semispace is 200 kb, and the ratio of the cells collected to cells allocated during garbage collection is 8. What would be the number of memory locations allocated during the garbage collection process, assuming that the percentage of live data entities in a semispace is 50% at the time garbage collection starts?

9. Two data entities: data-entity$_1$ (size 2000 bytes) pointed by the pointer P and data-entity$_2$ (size 1000 bytes) pointed by the pointer Q share 400 bytes of common space. Assuming that the deletion uses TBD stack, show the entries in TBD stack for the following actions: delete(P), delete(Q), new(A1), new(A2), new(A3). Where A1 is an object that requires 200 bytes, A2 requires 60 bytes, and A3 requires 1440 bytes. Explain your answer using a figure and various snapshots of the TBD stack.

6.12.3 Conceptual Type

10. What are the problems with start-and-stop garbage collections? Explain.

11. What are the problems of naive mark-and-scan algorithm? How can they be corrected?

12. Explain the problems with naive copying garbage collection. How can they be corrected?

13. Explain the problems with reference-count garbage collections. How can they be corrected?

14. Explain in detail Cheney's improvement to traditional start-and-stop copying garbage collection with a clear explanation of the role of breadth-first search and scan pointer to reduce the memory overhead of a stack.

15. Explain the role of forwarding and backward pointers in pointer tables stored in generational garbage collection.

16. Explain the Baker's improvement in Cheney's algorithm to make it incremental garbage collector.

17. What are the various steps taken to handle real-time events during garbage collection? Explain.

18. Explain the differences between Baker's algorithm and concurrent copying garbage collections.

19. How does the age and table-of-pointers affect the overhead in generational garbage collection?

FURTHER READING

Abdullah, Saleh E. and Ringwood, Graem A. "Garbage collecting the Internet: A survey of distributed garbage collection." *ACM Computing Surveys*, 30(3). 1998. 330–373.

Boehm, Hans. "Space-efficient conservative garbage collection." In *Proceedings of the ACM SIGPLAN '93 Conference on Programming Language Design and Implementation, SIGPLAN Notices* 28(6). June 1993. 197–206.

Brent, Richard P. "Efficient implementation of the first-fit strategy for dynamic storage allocation," *ACM Transactions on Programming Languages and Systems (TOPLAS)*, 11(3). 1989. 388–403.

Dijkstra, Edgar W., Lamport, Leslie, Martin, Alain J., Scholetn, Carel S., and Steffens, Elisabeth F. M. "On-the-fly garbage collection: An exercise in cooperation." *Communications of the ACM*, 21(11). 1978. 966–975.

Hanson, David R. "Fast allocation and deallocation of memory based on lifetimes." *Software Practice and Experience*, 20(1). 1990. 5–12.

Hosking, Antony L. "portable, mostly-concurrent, mostly-copying garbage collection for multiprocessors." In *Proceedings of the 5th International Symposium on Memory Management*. 2006. 40–51.

Jones, Richard, Hosking, Antony, and Moss, Eliot. *The Garbage Collection Handbook*. Chapman Hall/ CRC Press/Taylor & Francis Group. 2012.

Jones, Richard and R. Lins. *Garbage Collection: Algorithms for Automatic Dynamic Memory Management*. John Wiley. 1996.

Marlow, Simon, Harris, Tim, James, Roshan P., Jones, Simon P. "Parallel generational-copying garbage collection with a block-structured Heap." *Proceedings of the 7th International Symposium on Memory Management*. 2008. 11–20.

Nettles, Scott M. and O'Toole, James W. "Real-time replication garbage collection." In *SIGPLAN Symposium on Programming Language Design and Implementation*. 1993. 217–226.

Pizlo, Filip, Petrank, Erez, and Steensgaard, Bjarne. "A study of concurrent real-time garbage collectors." In *Proceedings of the SIGPLAN Conference on Programming Languages Design and Implementation*. 2008. 33–44.

Wilson, Paul R., Johnstone, Mark S., Neely, Michael, and Boles, David. "Dynamic storage allocation: A survey and critical Review." In *Proceeding IWMM '95 Proceedings of the International Workshop on Memory Management*. Springer Verlag. 1995. 1–116.

CHAPTER 7

Type Theory

BACKGROUND CONCEPTS

Abstract concepts in computation (Section 2.4); Data structure concepts (Section 2.3); Discrete structures concepts (Section 2.2); Boolean logic and predicate calculus (Section 2.2.2), data abstractions (Section 4.1).

Type system is an important part of a program. Type may be explicitly declared, may be derived by compile time program analysis, or may be derived at runtime after a variable gets the value. The basic question arises, what are types, and what is their relationship with program execution and program abstractions? Why should we have type at all? Do we get sufficient advantages of declaring objects of one type or another? With the increase in software size, the number of type declarations increase, which makes it difficult for programmers to keep track of all the type declarations. So another set of questions is as follows: Is type declaration user-friendly? Is it good for better productivity?

These questions have been debated in programming language community against the backdrop of efficiency, correctness of program behavior, better memory allocation, and avoiding runtime crash. On the one hand, inclusion of type declaration makes it difficult for a programmer to keep track of types of data objects. On the other hand, type declaration provides better execution efficiency, better precision, low memory overheads, much smaller runtime error, and much smaller number of program crashes. Languages that do not explicitly declare types suffer from the disadvantages of lower execution efficiency, memory allocation overheads, and unpredictable runtime crashes due to type mismatches.

Statically-typed languages bind specific types to identifiers at compile time and check all the detectable type-mismatch errors at compile time. The type of a variable once bound to a type in statically-typed language cannot be altered. The type of an identifier can either be explicitly declared or it can be derived using type-inference. For example, languages such as C, Pascal, ADA, Scala, C++, C#, and Java are statically-typed languages. A language is *strongly-typed* if it fixes the type of every variable at compile time, or the compiler can infer the type of every variable at compile time. In contrast, dynamically-typed languages allow the variables to be bound to different types of objects at runtime depending upon the values they are assigned to.

The type of a variable keeps changing dynamically depending upon the assigned value. Languages such as Lisp, Prolog, and Ruby are dynamically-typed languages. Dynamically-typed languages carry the type information at runtime, and are generally less efficient in execution time and memory space compared to statically-typed language.

7.1 ADVANTAGES OF TYPE DECLARATION

Type declaration provides many advantages at compile time during code generation. Some advantages are as follows:

1. Optimum memory allocation;
2. Achieving better accuracy that can handle some of the exception handling problems due to runtime error conditions;
3. Compile time coercion—converting an object bound to one data type to another data type without any information loss;
4. Compile time disambiguation of overloaded operators—operators with multiple meaning;
5. Finding out unsafe operations on a specific type of object;
6. Better long-term evolution and maintenance of software through the declaration of user-defined types and abstract data types;
7. Code optimization for faster execution of the compiled programs;
8. Use of generic procedures that can handle different types of data objects; and
9. Declaring semaphores and monitors to avoid deadlocks and facilitating concurrent execution.

The program in Figure 7.1 illustrates many of these advantages that can be achieved at compile time due to type declaration. The advantages of types in handling generic procedures will become clear when we discuss polymorphism in Subsection 7.5. The advantages of types in facilitating concurrent execution and deadlock avoidance will become clear when we discuss concurrency in Chapter 8.

Example 7.1

The variables x and y occupy only four bytes on the control stack. In the absence of type declaration, the information about the type of the variables x and y is known only after the first assignment. In that case, it may not be possible to compute the frame size and the offset of following variables, and the variables will have to be treated as pointer to objects in a heap that stores the value of the variables x and y. Stack-based allocation is faster to access than heap-based allocation and the allocated memory can be easily reclaimed and reused after the called subprogram is over. The declaration of type in this case facilitates optimum memory allocation on stack and improves execution efficiency of the program.

```
(1)  program main                          %
(2)        struct galaxy {                 %
(3)              integer starCount;        %
(4)              double float distance;    %
(5)              }                         %
(6)  {                                     %
(7)        integer x, y; % integer takes 4 bytes
(8)        float w, z;  % float takes 8 bytes
(9)        double integer a, b;  % takes 8 bytes
(10)       string  c, d;                   %
(11)       galaxy neighbors[10];           %
(12)       x = 4; y = 6;                   %
(13)       w = x + y; % the evaluation is coerced to float
(14)       z = w + y; % 'y' is coerced to float
(15)       c = "Milky Way"                 %
(16)       d = z + c  % type mismatch error
(17)       neighbors[1].starCount = 32567823418; % accuracy
(18)       neighbors[1].distance = 4.5 E**12;    %
     }
```

FIGURE 7.1 Illustrating some advantages of use of types.

The declaration of the struct *galaxy* provides data abstraction capability that can be easily modified at high level if certain additional attributes characterizing *galaxy* would be needed in future. It also makes the program comprehensible and easy to maintain. For example, suppose we want to extend the abstraction *galaxy* by adding an additional attribute such as *known planets in the galaxy*; it can be easily added.

In line 13, the statement "w = x + y" adds two integer values, and the result is assigned to a floating point variable. Since the operands of the addition are of the type integer, the overloaded symbol gets disambiguated at compile time to *integer-addition*, and the appropriate low-level code is generated. After the expression evaluation, the integer value is transformed to the equivalent floating point value without loss of any information. This process is called *compile-time coercion*, and the translated code is generated at compile time avoiding runtime overhead. The notion of coercion will discussed in detail in Section 7.5 under polymorphism.

In line 14, the statement "z = w + y" mixes up two different types of operands: the variable *w* is of the type floating point and the variable *y* is of the type integer. The value of the variable *y* is coerced to floating point representation at compile time, and the overloaded symbol '+' is disambiguated as floating point addition during the code generation.

In line 16, the addition operation is not possible due to the type mismatch error. In lines 17 and 18, user-defined type information has been used to pick up appropriate fields of an element of the collection *galaxy.* In the absence of extra accuracy, the language would not be able to represent numbers bigger (or smaller) than the representation possible on one word size that is 32-bit or 64-bit for modern-day computers. While 32 bit is an extremely big number: 32-bit is around 4 billion. Yet, 32-bit word is not suitable for very large numbers such as planets and stars in observable galaxies that may have around 170 billion stars and even more planets.

If an object is of integer type, and is used frequently in a computation, then it can be temporarily stored in the processor register during the computation to reduce the overhead of memory access and to improve the execution efficiency. However, without type declaration such code optimization would not be possible at compile time.

In summary, programming domains that require efficient execution, better memory management, and additional accuracy will choose statically typed languages. In contrast, the domains that are not time critical may prefer specific dynamically typed languages for programming constructs and the ease of program development.

7.2 NOTION OF TYPE

We have seen from the intuitive understanding of types during program development that type information can be used for many purposes. Now we ask the question formally: Is there any mathematical basis of type theory, and how is type theory related to data abstractions?

Mathematical types such as integers, floating points, and Booleans, define a domain that has well-defined properties and operations. For example, integers are made of digits; floating points are made of digits, a period, and possible exponentiation; and Boolean has two values: *true* or *false*. In all of these domains, the operations are well defined that take as input one or more operands and generate a value. There are two types of operations on mathematical types:

1. The operations that generate the value in the same domain as the input operands. This is called *closure property*.

2. The operations generate a truth value. Such operations that map an input domain to a truth domain are called *predicates*.

Data entities are of the same type if they belong to the same domain having well-defined properties. Type theory is deep rooted in the notion of sets because we can take a set, associate well-defined attributes and operations with the set, and all the elements in the set will automatically inherit the operations and attributes associated with the corresponding set.

So the first notion of type is that types are sets with well-defined properties and well defined operations. If types are sets, then the natural question arises: What happens to the definitions if we derive new sets using operations on sets? There are many operations that can be performed on sets such as union, intersection, Cartesian product, mapping of one set onto another set, disjoint union, power set of a set, and ordering a set. We know that these set operations derive yet another set. If types are sets, then these derived sets are *user-defined types* and correspond very closely to *aggregation* in the data abstraction, as we discussed in Chapter 4.

Abstract data types are sets that have additional user-defined restrictions and well-defined abstract operations on them. However, the implementations of these operations are kept private for ease of modification. In most languages, these abstract operations are made public to be accessed by other subprograms. For example, complex numbers, rational numbers, stacks, and queues are abstract data types. Complex numbers are modeled as a pair of floating point numbers with four abstract operations: *complex-add, complex-subtract, complex-multiply,* and *complex-divide.* Similarly, rational numbers are modeled as a pair of integers with four abstract operations: *rational-add, rational-subtract, rational-multiply,* and *rational-divide.* Stack is a sequence with an additional restriction that data is inserted and deleted from only one end, and it has four abstract operations on sequences to simulate *push, pop, is_empty,* and *top.*

In compiled statically typed languages, the information about types is not carried at runtime. After a type is declared statically, declaring a variable of that type means that the variable can be assigned one or more values that are elements of the set corresponding to the declared type. For example, a declaration such as "integer x" says that the variable *x* can be bound to only an integer value and will support only integer operations. Any attempt to associate a value that is not an integer will result in type-mismatch condition. This property is used to test for type mismatch of operations and coercing the type of an object to another type.

7.2.1 Basic Data Types

A programmer can use mathematical sets such as integers, floating points, define their own sets that are called *enumeration types*, or use set operation on previously derived sets to derive new sets. On the basis of how the sets are derived and what operations are associated with them, one can broadly classify the types into three categories: *basic types, structured types*, and *abstract data types. Basic types* can be the following:

1. Mathematical domains such as integers, Boolean, floating point, etc.;
2. Mathematical domains extended for additional accuracy such as double float, quad integer, and long integer;
3. Scalar types, such as *fixed type*, where accuracy of the representation is user defined. ADA uses the notion of fixed type using *delta*, where *delta* is the minimal accuracy for a number. For example, delta for the U.S. currency dollar is 0.01;
4. Text representation domains such as characters;
5. Information representation domains such as bits, bytes, and words; and
6. Synchronization and concurrency domains such as semaphores.

Basic types are built in as part of the language and need not be defined by the user unless a user wants to override the default definition of an operation. Object-oriented languages such as C++ and C# support such an override. Some languages such as Ada, support complex numbers and rational numbers as part of built-in types, and call them *numeric types.* However, most of the languages treat complex numbers and rational numbers as user-defined abstract data types.

Strings have a dubious classification. Strings are sequences of characters. However, many languages have a separate declaration for string. Similarly, files are sequences of characters or sequences of data entities. Depending upon the nature of the file, these sequences can be indexible. Ruby supports strings as indexible sequence.

7.3 SET OPERATIONS AND STRUCTURED TYPES

The simplest form of user-defined type is to define your own set that can be enumerated. For example, *days-of-the-week* is a set of seven data-entities, and *months-in-a-year* is a set of 12 data-entities.

Example 7.2

Let us consider the following declaration:

```
summer_month = (May, June, July, August);
...
summer_month    vacation_month;
```

The above declaration says that *summer_month* is an enumerable set containing four elements. The variable *vacation_month* can hold one of the four values. Any attempt to assign a variable that is not one of the four data values will result in an error condition.

7.3.1 Ordinal Types

An ordinal type is a user-defined, enumerated set where every element is associated with a position. Ordinal types are ordered bags where each element is associated with the corresponding position in the bag. Most of the programming languages such as Pascal, C, C++, C#, Java, and ADA take this approach.

Two operations, *successor* and *predecessor*, can be performed on a variable that is declared of the ordinal type. The operation *predecessor(<variable>)* gets the data element that occurs immediately before the current value of the variable, and the operation *successor(<variable>)* retrieves the data element in the enumerated set that occurs right after the current value of the variable. The operation *predecessor(<variable>)* is undefined if the current value of the variable is the first element of the ordinal type, and should raise an exception. Similarly, the operation *successor(<variable>)* is undefined if the current value of the variable is the last element of the ordinal type.

Example 7.3

The following example shows the declaration of an enumeration set *school_year* that is also an ordinal type. The variables *senior_student_year*, *student_year*, and *junior_student_year* are declared of the type *school_year*. The variable *senior_student* is a successor of the variable *student_year*, and the variable *junior_student* is the predecessor of the variable *student_year*.

```
{
...
school-_year = (freshman, sophomore, junior, senior)
...
```

```
school_year senior_student_year, student_year, junior_student_year;
...
student_year = junior;
if (student.ne. senior)
     senior_student = successor(student_year);
if (student.ne. freshman)
     junior_student = predecessor(student_year);
...
}
```

Some languages such as Ada, also use enumeration type to model Boolean values: *false* precedes *true* because internally *false* maps to 0, and *true* maps to 1. Ordinal type is also expressed as a subrange by just mentioning the lower bound of the range and the upper bound of the range. For example, *date* can be expressed as ordinal type 1..31; and *month* can be expressed as ordinal type 1..12.

7.3.2 Cartesian Product and Tuples

As discussed in Chapter 2, Cartesian product of two sets S_1 and S_2 produces exhaustively a set of pairs of the form $(a, b) \in S_1 \times S_2$, where $a \in S_1$ and $b \in S_2$. The Cartesian product of n sets $S_1, S_2, \ldots, S_n$ produces a set of n-tuples of the form $(a_{i,}^{1}\ a_{i,}^{2}\ \cdots,\ a_{i}^{n}) \in S_1 \times S_2 \times \cdots \times S_n$, and the j_{th} $(1 \leq j \leq n)$ field of an element in an n-tuple comes from the set S_j. The number of elements in the set is given by multiplying the size of the individual sets.

Tuple is an abstraction used to represent a composite data entity where each field may be an aggregation. The composite data entities are written using syntactic constructs "struct" or "record" in different languages. Any language supporting tuples using any syntactic sugar can model composite data entities. For example, a complex number corresponds to a set *real* × *real*, and a rational number corresponds to a set *integer* × *integer*. The Cartesian product *real* × *real* is a set of pairs such that both fields are real numbers, and the Cartesian product *integer* × *integer* is a set of pairs such that both fields are integers.

7.3.3 Finite Mapping and Arrays

In Chapter 4, we saw that collections of data entities can be implemented using arrays, linked lists, vectors, and trees. Arrays are popular data structures used for implementing a collection of data entities. An array is an indexible sequence such that the index is a set of natural numbers that maps using many-to-one into-mapping to a set declared by the programmer. For example, a declaration integer $m[10]$ has two sets: *0..9* and *integer*. Every element of the set *0..9* maps on an *integer*. The subrange *0..9* forms a finite domain, and *integer* forms the range. It is called finite mapping because the domain is a finite set.

Finite mapping in the simplest form that maps a set of index to any data type, and is equivalent to arrays. However, the concept is easily generalized to *association lists*. In association lists, the domain is an enumerable set of keys that are associated with a value in the range. For example, a domain could be the set {*world-war II, 1967, Earth*} and the range is {*1939, man-on-moon, water*}, and the three domain elements can be mapped on the range values such as {*world-war-II* ↦ *1939; 1967* ↦ *man-on-moon;* and *Earth* ↦ *water;*}

7.3.4 Power Set and Set Constructs

Given a set of the form $S = \{x_1, x_2, \ldots, x_n\}$, a power-set is defined as a set of all subsets of S. Given an enumerable set, a variable can be declared of the type power-set of the enumerable set. This means that a variable can be bound to any subset of the enumerable set S. This becomes the basis of set-based programming as one can define all the set operations on these subsets. Let us understand set-based programming using Example 7.4. The example uses Pascal's "set of" construct that is equivalent to the defining power-set.

Example 7.4

The declarations shown below declare an enumerable set *student*, and a variable *regular_students* that is a power-set of the enumerable set. The variable *regular_student* can be bound to any of the eight subsets: {{}, {*Tom*}, {*Phil*}, {*Jean*}, {*Tom, Phil*}, {*Tom, Jean*}, {*Phil, Jean*}, {*Tom, Phil, Jean*}}.

```
type student = (Tom, Phil, Jean) % enumerated set
var regular_students : set of students;
```

7.3.5 Disjoint Union and Variant Record

Many programming languages support the notion of disjoint union—union of two disjoint sets S_1 and S_2 such that individual elements retain the identity about the original sets they belong to. As described in Chapter 2, two sets S_1 and S_2 are disjoint if $S_1 \cap S_2 = \emptyset$. The elements in the disjoint sets are colored with distinct colors to maintain the set identity. For example, let us consider two sets: *{Mary, Nina, Ambika, Susan}* and the second set *{Tom, Rubin, Mark}*. As is evident by the names, the first set is made up of girls, and the second set is made of boys. If we join them in a classroom setting for the learning purpose, the union will be a disjoint union. However, to maintain the distinction between elements of the two sets, we can color them separately: the first set is colored by the value *girl*, and the second set is colored by the value *boy*. The disjoint union would be *{(girl, Mary), (girl, Nina), (girl, Ambika), (girl, Susan), (boy, Tom), (boy, Rubin), (boy, Mark)}*. The disjoint union of sets can be modeled as the union of the Cartesian product of color and the corresponding sets. Given two sets S_1 and S_2, the disjoint union is modeled as $\{Color_1\} \times S_1 \cup \{Color_2\} \times S_2$. For example, the above disjoint union set is modeled as *{girl}* × *{Mary, Nina, Ambika, Susan}* ∪ *{boy}* × *{Tom, Rubin, Mark}*.

In programming languages, disjoint unions are used to model variant records. Variant records have two parts: *fixed part* that is common to different tuples, and the *variant part* that corresponds to disjoint information from different sets. The variant part occupies the same memory location for elements from the different sets. However, based upon the color value, the location is interpreted differently as shown in Example 7.5. The simplest form of color is the binary color given by the set {*true, false*}.

Example 7.5

Let us consider an example of modeling an assignment. An assignment has a number of questions, and the date assignment was given to a student. This is the fixed part.

The variable part is that a student may have either turned in the homework or may not have turned in the homework. If a student has turned in the homework, then the variable part just contains the *score*; if a student has not turned in the homework, the variable part contains the *date_expected*. Note that the two fields *score* and *date_expected* are interpreted on the basis of the value of the Boolean variable *turned_in*. If by mistake at runtime, the Boolean variable *turned_in* gets altered, then the variant part will be interpreted differently and may cause erroneous program behavior.

```
possible_score = 1..100;
possible_date = 1..31;
type assignment = record
     questions : integer;
     assignment_date: possible_date;
     case (turned_in : Boolean) of
       true: (score: possible_score);
       false: (date_expected: possible_date)
     end {case}
end
```

The set operations in the above example can be expressed as *integer* × *(1..31)* × *({true}* × *(1..100)* ∪ *{false}* × *(1..31)})*. In the Cartesian product, the first term "integer" corresponds to the number of questions, the second term "(1..31)" corresponds to the date of assignment, and the third term corresponds to the variant part. The Boolean values *true* and *false* act as mutually exclusive colors.

Another problem with disjoint union is that two disjoint sets may be of different types with different internal format to represent values. If the Boolean variable is set incorrectly at runtime, the operations on the memory locations can corrupt the values.

We denote a disjoint union using a symbol "⊎." For example, a disjoint union of two sets *integer* and *character* is represented as *integer* ⊎ *character.*

7.3.6 Set Operations for Recursive Data Types

In Chapter 4, we modeled extensible data abstractions using linked lists, trees, and vectors. These data structures are modeled recursively for extension, as shown previously. Recursive data types are modeled using a combination of *Cartesian product* and *disjoint union*.

As we have discussed earlier, that recursive data structure are defined as multiple definition involving *recursive-part* and *base-condition*.

```
<linked-list> ::= <data-item> <linked-list> | nil
<binary-tree> ::= <binary-tree> <data-item> <binary-tree> | nil
```

The above definition of *<linked-list>* states that *<linked-list>* could be any sequence of *<data-item>*. Alternately, a given linked list is an element of a set that contains all possible sequences of varying sizes containing *<data-item>*. Similarly, a binary tree is an element of the set that contains binary trees of all possible depths.

If we substitute the "|" by *disjoint union*, the concatenation by the *Cartesian product*, and *<data-item>* by the *<data-type>* of the *<data-item>* then we get the following equivalent recursive data types:

```
<linked-list-type> ::= <data-type> × <linked-list-type> ⊎ nil
<binary-tree-type> ::= <binary-tree-type> × <data-type> ×
                       <binary-tree-type> ⊎ nil
```

The recursive type *<linked-list-type>* represents the set of all possible linked lists of different sizes (including an empty list) that contains data elements of the type *<data-type>*. Similarly, the recursive type *<binary-tree-type>* represents the set of all possible binary trees of different depths (including an empty tree) that contains data elements of the type *<data-type>*. A variable of the recursive data type gets bound to one of the elements in the set.

It can be shown using the definition of disjoint union that the set-based definition of *<linked-list-type>* generates the set of all possible sequences of the *<data-item>* by unfolding the Cartesian product and applying the definition of disjoint union. The definition *<linked-list-type>* can be written as {{(*false, nil*)} ∪ {(*true, <data-type> <linked-list-type>*)} using the first unfolding. The second unfolding of the definition *<linked-list-type>* embedded inside the first unfolded version gives {(*false, nil*)} ∪ {(*true, <data-type>* {(*false, nil*) ∪}(*true, <data-type> <linked-list-type>*)}. By removing the colors in the unfolded part, the expression becomes {*nil*} ∪ {*<data-type>*} ∪ {(*<data-type> <data-type>*}∪ .... This expression can be further unfolded to get an indefinite expression that is a union of sets of all possible sequences of differing sizes. This set is the same as the set derived earlier by recursive definition of linked lists. We can reason the same way about unfolding *<binary-tree-type>*.

Recursive data-type implementation uses pointers because recursive objects extend indefinitely and cannot be allocated at compile time due to indeterminate size. Pointers are used to extend recursive data items on demand at runtime. For example, a linked list is represented as

```
struct mylist {   integer mydata;
                  mylist *list _ pointer
              }
```

The above structure represents a tuple of the form (*mydata, list_pointer*) where the field *list_pointer* points to the next tuple. Each tuple needs memory locations, and tuples are connected through a chain of pointers. The last tuple contains a null pointer that can be extended by creating more tuples in the heap and linking the newly created tuple with the last cell by updating the pointer field of the last tuple.

7.4 LIMITATIONS OF TYPE THEORY

The static declaration of type for variables provides many advantages as described earlier. However, the explicit information about the types is lost after the compilation. Part of the information is implicitly embedded in the code and data area in terms of allocated optimized memory and the disambiguated low-level operations. Since most of the type-related information is lost, it is difficult to reason about the *runtime property violation* without incurring additional execution time and memory overhead. Some of the runtime errors are as follows:

1. Accessing ith element of an array can lead to array-bound violation. Since the index variable can be modified at runtime, checking that *lower-bound* ≤ *value of the index variable* ≤ *upper bound* is a runtime property. Not checking the boundaries will lead to incorrect program behavior because index may fall out of range of the memory locations where the array is located. Allowing runtime checks leads to serious runtime overheads.

2. Dynamically computing a substring can lead to string-bound violation. String is a sequence of characters, and the operation *substring(<string>, <substring-start>, <substring-length>)* picks up a substring starting at position *<substring-start>* of the length *<substringlength>*. If the value of *<substring-start>* is greater than the size of the string, or the sum of the values of *<substring-start>* and *<substring-length>* is greater than the size of the string, then it will be a case of string-size violation and will lead to incorrect program behavior. Since all three arguments of the operation *substring* can be computed at runtime, no compile time declaration can check this type of bound error.

3. Pointer arithmetic on a collection of data items can easily violate the boundaries of the collection of data items during runtime.

4. Erroneous runtime modification of Boolean variables controlling the interpretation of the variant part in variant records may lead to incorrect interpretation of mutually exclusive variant parts.

Example 7.6

Let us take the following example. The program declares an array a of size 50 data elements. The for-loop computes the value of an index variable $j = 64$. The statement $a[j] = 120.2$ tries to assign the value to memory location of $a[64]$ that is memory address of $a[0] + 64$ * *size-of*(one-data-element). The program will write in some memory location that does not correspond to array a and corrupts that memory location. To fix this problem, it has to be checked that $j \geq 0$ and $j \leq 49$ before the assignment statement. However, that would mean additional execution overhead before accessing an array element.

```
program main
integer i, j;
real a[50];
      ...
{   j = 1;
```

```
        for (i = 1; i < = 6; i++) j = 2 * j; % finally j = 64.
        a[j] = 120.2; % A non-existent a[64] is assigned a value
    ...
    }
```

Example 7.7

Let us take the following example that calls a built-in function *substring* with three arguments as described earlier. It picks up the substring starting from position *j* up to the position *j* + 4. The value of *j* is computed using a for-loop. The loop is executed four times, giving the value of *j* as 16. The length of the string "Arvind" is only six. However, the function substring tries to pick up the substring from position 16 to position 20. It will pick up some other memory locations not belonging to the variable *my_name* and return some arbitrary value. In order to fix this problem, it has to be ensured that the start position *j* and the end position *j* + 4 are within the range 0 to 6. This information cannot be checked at compile time.

```
program main
   {    string my_name, short_name;
        integer i, j, k;
        my_name = "Arvind";
        j = 1;
        for (i = 0; i < = 3; I++) j = 2 * j;
        short_name = substring(my_name, j, 4);
   }
```

In summary, the properties that are computed at runtime cannot be checked by compile-time declarations unless the boundary information is carried at runtime and checked before accessing the data items. This causes significant runtime overhead and slows down the execution. Many compilers allow a compile-time switch that performs the range check during the program debugging phase, and the program is recompiled without the switch after debugging to improve the execution speed.

7.5 POLYMORPHISM

Till now we have studied types as sets of values with well-defined operations. Sets could be basic sets, user-defined sets formed using well-defined set operations, or sets with abstract user-defined operations. The sets could be built-in such as *integer* or *real* or they could be user defined *enumerated sets*. Infinite sets have infinite subsets. We associate certain properties with the original set, then these properties can be inherited by the subsets, and programmers do not have to duplicate the definition of operations already defined for the original set.

Well-defined operations can be associated with handling the data items or performing operations on structure-related properties that have nothing to do with the types of data items in the set. For example, integer addition is an operation that is specific to data items. However, counting the number of elements in a linked list is a generic operation and is applicable to any linked list irrespective of the type of the data items stored in the linked list. These generic operations can be applied on potentially indefinite types of data objects. A function or a procedure written to implement such generic operations need not be tied to any specific type of data item.

Earlier in the history of the development of programming languages, language designers did not separate the operations specific to data items and the generic operations associated with structures. Functions were associated with a specific data type. For example, there was a separate function to count the number of elements in a list of integers, and a different function to count the number of elements in a list of real numbers, and yet another function to count the number of elements in a list of complex numbers. Languages that support a separate function for the same generic operation on different types of data objects are called *monomorphic languages* as they required programmers to write the same code multiple times for different types of data objects.

In all of these functions, the generic part is counting the number of elements that needs to be coded once and types of data items on which the operation occurs may be passed as parameter to specialize the function behavior for the specific call. In 1978, Robin Milner developed a language called ML that separated the generic operations on structures from the operations on data items, and started a new class of languages called *polymorphic languages*—languages where generic functions are written once, and the type of data items is passed as parameter to specialize the function for specific type of data items. Passing the types as parameter to a generic function is called *parametric polymorphism.*

The polymorphism is defined as support of operations that work uniformly on possibly indefinite types of data objects without redefining the operations every time the type of data items changes. Polymorphism in languages supports reusability of code for different types of objects while preserving static typing—the ability to infer and verify potential computational errors by dividing objects in different classes with well-defined properties and operations. Static typing along with reusability are important concepts for structured and robust software development.

Two major classes of polymorphism are supported: *universal polymorphism* and *ad hoc polymorphism. Universal polymorphism* is defined as supporting the same operation on possibly indefinite number of data types, and *ad hoc polymorphism* is limited to finite number of data types. Universal polymorphism is further divided into two categories: *parametric polymorphism* and *inclusion polymorphism. Ad hoc polymorphism* is divided into two categories: *coercion* and *overloading.* Universal polymorphism is different from ad hoc polymorphism because universal polymorphism supports operations that act on possible indefinite number of data types, while ad hoc polymorphism supports only finite number of data types.

7.5.1 Parametric Polymorphism

Parametric polymorphism allows the use of generic functions to perform the same operation on different (possibly indefinite) types of data objects because the operation is associated with the structure of the data objects rather than the property of individual data elements. Some of the examples of such operations are counting the number of elements in a list, duplicating the elements, appending two lists, finding the *i*th element of a list. None of these operations are specific to the type of the data item. Another situation is adding a sequence of elements. If the data type of elements in the list is an integer, then the addition works as integer-addition; if the data type of elements in the list is floating point, then the addition works as floating-point-addition. On the basis of the value of the type variable passed as a parameter, the interpretation of the addition operator changes at runtime.

In all these cases, a generic function is written, and the type is expressed in the form of *input-mode → output-mode*, and the type of arguments are passed as parameters involving type variables in the type domain. There are two types of variables: *concrete variables*, associated with values or the objects in the actual call to the subprogram; and *type variables*, associated with type information associated with specific data-items in each function or generic methods in object-oriented languages. The input mode is the set-operations involving the incoming arguments in type domain, and the output mode is the set-operations involving the outgoing arguments in type domain.

Example 7.8

For example, the type information for a function such as *counting a list of integers* that takes as input a list of data elements and generates a scalar value of the type *integer* is written as follows:

```
list(τ) → integer where list(τ)::= τ × list(τ) ⊎ nil
```

The Greek symbol τ is a type variable passed as parameter, the input mode is list(τ), and the output mode is *integer.* We have already seen the definition of the recursive data type list: list(τ) denotes a set of all possible lists of different lengths including an empty list that have data items of a generic type passed as parameter. If the value of the type variable passed as parameter is *integer* then the input mode becomes *list(integer)*; and if the value of type variable passed as parameter is *real* the input mode becomes *list(real).*

Example 7.9

The type information for a function that appends two lists of the data elements of the same type as input and generates a list of data elements as output can be written as follows:

```
list(τ) × list(τ) → list(τ) where list(τ)::= τ × list(τ)
                                              ⊎ nil
```

The type of the first input argument is list(τ); the type of the second input argument is list(τ); and the Cartesian product of the two sets gives a set of pairs such that each field is a list of a generic type passed as parameter, and the output mode gives a set of lists of the same generic type. If the type passed as parameter is *integer* then the input mode is *list(integer)* × *list(integer)*, and the output mode is *list(integer).*

Parametric polymorphism can be explicitly declared before the procedure; alternately, it could be inferred at compile time. Polymorphism where parametric polymorphism of a function or procedure is explicitly declared is called *explicit polymorphism*, and where parametric polymorphism is not admitted explicitly can be derived at compile time using type-inference is called *implicit polymorphism*.

7.5.2 Inclusion Polymorphism and Subtypes

As we have discussed that types are sets with well-defined operations working on the data items in the set, these same well-defined operations can be associated with subsets of the original set. If the original set is of indefinite size such as *integer* or *real*, it will have indefinite number of subsets that will inherit the same set of well-defined operations. A subset can be declared as subtype of the original type, and the subtype will automatically inherit the well-defined operations on the original type; the operations already defined on the original type need not be redefined on the subtypes. This form of polymorphism is called *inclusion polymorphism*.

For example, probability is a subtype of real numbers, and natural numbers are subtypes of integers, probability will inherit all the floating-point operations of the real numbers, and natural numbers will inherit all the operations of integers. There is one problem: original type may be closed under an operation. However, a subtype may not be closed under the operation and may need to have error-handling mechanism. For example, a real number is closed under addition, subtraction, multiplication, and division (except divide by zero). However, probability is not closed under addition, subtraction, and division: 0.6 + 0.6 gives 1.2; 0.2 – 0.5 gives –0.3; 0.6/0.2 gives 3.0. All such cases of nonclosures need to be handled using exception handlers.

Example 7.10

The following example taken from language Ada describes subtyping. The first example tells that *month* is a subtype of integer with range between 1 and 12. Any value beyond this range will generate error. Subtype *age* is an integer between the range 0 and 150. The declaration *Workingdays* is a subtype of the enumeration type *Weekday.*

```
subtype Month is INTEGER range 1..12
subtype age is INTEGER range 0..150
type Weekday is (Sun, Mon, Tue, Wed, Thu, Fri, Sat);
subtype Workingdays is Weekday range Mon..Fri
```

Subclasses and inheritance

Inclusion polymorphism is also present in object-oriented programming as a subclass inherits all the properties and declarations in a class unless the declarations are sealed within the class. Inclusion polymorphism for object-oriented programming is given by *Liskov's substitution principle*, which states that objects of a type may be replaced with objects of the corresponding subtype without altering the program's extrinsic public behavior with respect to other objects. We defer the topic of polymorphism in object-oriented languages until Chapter 11.

7.5.3 Overloading

Many operators such as '+', and '*', '/', '–', reserved words such as function names in the Pascal family of languages can have multiple meanings. For example, the symbol '+' can be interpreted as integer addition, floating-point addition, rational-number addition, or complex-number addition. The meaning of an operator can be disambiguated if the type of the operands is known. In statically-typed languages, the overloaded operator can be disambiguated at compile time because the information about the type of the operands is available during compilation. However, in dynamic languages the meaning of an operator is disambiguated only at runtime, based upon the type of the arguments.

Example 7.11

Let us consider a simple program segment as follows:

```
integer x, y;
float a, b;
...
x = 3; a = 5.3;
y = x + 6;
b = a + 7.4;
```

In this example, the operator '+' is overloaded. The first occurrence of the addition operator '+' is an integer-addition because both the operands are integers. However, the second occurrence of the same operator is a floating-point-addition because both operands are floating point numbers.

7.5.4 Coercion

Coercion is an automatic conversion of a type of data element to another type of data element for computation such that the information is not lost. For example, integer value 1 can be coerced to floating point value 1.0 without loss of any information. However, if we try to convert a floating point number to an integer, then information may be lost, and it will be *casting* and not coercion. An interesting confusion even among computer scientists is to confuse three different concepts: *coercion, casting*, and the *same low-level representation* for different types of implementation. Coercion is information preserving conversion to allow for mixed-type operands so that the meaning of overloaded operators can be disambiguated. Casting causes loss of information, and most statically-typed languages provide a specific library function to provide programmer directed casting. Without the use of the programmer directed library functions, casting is unsafe. However, many dynamic, specifically web-programming, languages mix up types and provide type conversion in an unsafe way to provide user-friendliness. For example, language Javascript can treat number as a string and concatenate a number to a string when resolving the meaning of "+" operator that mixes a string with a number. Many times, low-level abstract machines such as JVM use the same abstract instruction for different types of high-level instructions due to the availability of limited abstract instruction set. It is neither coercion nor casting.

Coercion is supported by almost all the modern programming languages to provide natural interface with the human way of handling the computations in the mathematical world. Coercion is a antisymmetric and transitive relationship. If we represent coercion by an arrow symbol "→" then lower-precision type → higher-precision type. For example, integer → floating-point, floating-point → double-float, integer → long-integer. Using transitivity we can also infer integer → double-float. However, we cannot infer that long-integer can be coerced to double-float.

Coercion of the consumer occurrence of variables works well for both statically and dynamically-typed languages. However, for the producer occurrence of variable, coercion acts differently in statically and dynamically-typed languages. Statically-typed languages do not allow the type of a variable to be altered, while dynamically-typed language change the type of the variable based upon the value.

Example 7.12

Let us take the following C++ like code that has three types of variable declarations: int, float, and double. The statement $y = m + x$ mixes an integer to a floating point number. The value of the variable m will be coerced to floating point value 4.0 and y will be assigned the value 7.4. Since float → double, the statement d1 $= n + y$ will coerce the value of the variable y to double float version of 7.4 and add it to double float version of the value 5 to derive the double float version of 13.4. The last statement converts the integer value 5 to double float version and adds to get a double float version of the value 18.9.

```
int m, n;
float x, y;
double d1, d2;
{m = 4; n = 6; x = 3.4; y = m + x; d1 = n + y; d2 = d1 + 5;}
```

7.6 TYPE SYSTEM IN MODERN PROGRAMMING LANGUAGES

Modern programming languages support monomorphic type, polymorphic type, and pointers (or reference type). Monomorphic type is further categorized as scalar, structure, and reference type. Scalar type consists of integer, real, Boolean, char, semaphores, byte, word, and ordinal type. Ordinal type can be an enumerated set or a subrange. Many modern languages such as C++, ADA and Java support all four forms of polymorphism: *parametric, inclusion, coercion*, and *overloading*. Figure 7.2 shows that overall type structure of modern programming languages. Reference type or pointers are used to access dynamic objects.

A *structured type* is derived using set operations on previously declared sets. The set operations are *Cartesian-product* to model tuples; *finite-mapping* to model arrays, association lists (or maps), and indexed sequences; *power-set* to model set of the subsets for set-based programming; *disjoint-union* to model variant records, combination of *Cartesian-product* and *disjoint-union* to model recursive data types. *Strings* are sequence of characters. However, they have been treated in many different ways. Strings have been modeled as packed arrays of characters as in Pascal and C, or through a built-in class declaration of the type "string" as in Java.

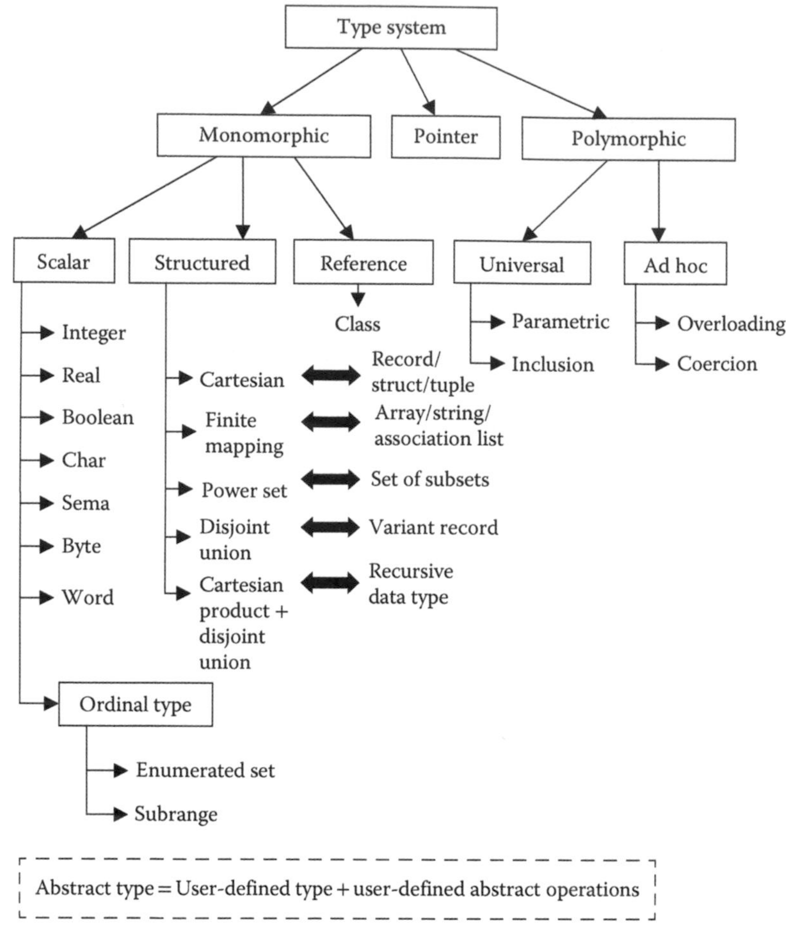

FIGURE 7.2 An overall type structure of modern programming languages.

Pointers are addresses that point to a memory location. Independent pointers can be potentially unsafe if they allow arithmetic operations to step through memory locations. Pointers have been treated differently in different programming languages depending upon the languages' problem domains and philosophy. For example, Pascal does not allow an independent status of pointer. Rather, pointers are associated only with structured types such as recursive data types to prevent unintended runtime program-crashes.

7.6.1 Universal Reference Type

Object-oriented languages such as Java, C#, CLU, Modula-3, and Ruby use a reference-type that is an internal representation to access an object in the heap. Reference-types do not permit pointer arithmetic or independent status to references. In Java, this type is called *object type*; in C++ and C#, it is called *void* *; in CLU it is called *any*; and in Modula-3 it is called *refany*. The type of the object referred to by the universal type can be dynamically altered on the basis of the data type of the assigned object. Compiling universal reference is generally not type safe because, at runtime, a universal type can be associated with different incompatible data objects. There are two approaches to handle this problem of type compatibility at the runtime: (1) casting and (2) dynamic-type tags. Each data entity in an object has a tag that keeps its type, and the objects are checked for type compatibility before an operation is performed on the object. If the types are different, then type casting is used dynamically to make the data entity compatible with the data type used by a method. As explained in Section 11.4.2, there are two types of casting: *upcasting* and *downcasting*. In upcasting, the information is not lost, and is type safe. In downcasting, type information is lost, and is potentially unsafe. Upcasting can be done automatically. However, downcasting must be explicitly initiated by a program action.

7.7 TYPE EQUIVALENCE

One important question: Can two variables of seemingly different types be equated? Ideally, corresponding variables of two types carrying same information should be equated with each other. A natural approach is to look at two sets that contain the same information. Mathematically speaking, we can impose a condition for type equivalence that a bijective mapping should be defined between the two sets, which means that for every element in one set, there is a corresponding element in the other set that carries the same information. However, there are two problems with structured types: (1) many set operations such as Cartesian-product of sets can shuffle the fields while preserving the information and (2) structuring can be nested to any depth. If we model the structured types using tree representation, then the basic types are at the leaf nodes. If we collect all the leaf nodes of two trees carrying the same number of fields having same basic types, then the information they carry is the same. However, it is computationally difficult to infer that two complex structured types are equivalent due to the following reasons:

1. Cartesian product and disjoint union cause permutation of the fields while retaining the same information.
2. Different fields may be grouped differently at different nesting levels.
3. In case of languages that allow flexible base index in the domain in finite mapping, some computation has to be done to match the types. For example, integer a[1..10] and integer y[0..9] can be equated since a[i] corresponds to y[i − 1]. However, the relationship needs to be inferred.
4. There may be more than one occurrence of same basic types at the leaf nodes carrying semantically different information.
5. The basic types may be the same, but the semantic entity they represent may be completely different, because data abstractions may be using the same structure and basic types to model semantically different entities.

Let us look at two tree-based representations of equivalent structured types in Figure 7.3. Clearly both the representations potentially represent the same information with different structures due to permutation and grouping of fields. The two integer fields are difficult to equate due to multiple occurrences. Humans can disambiguate some of this information by

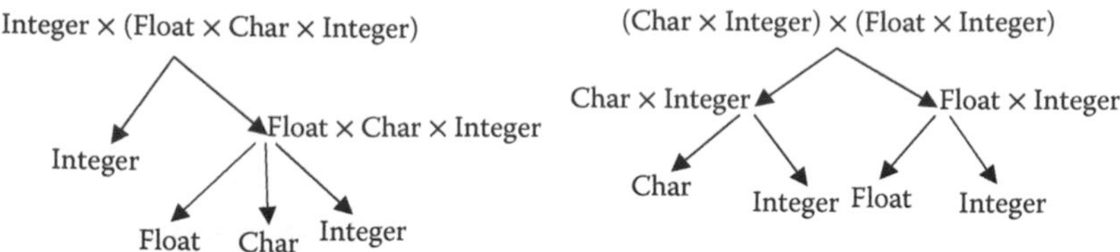

FIGURE 7.3 Two types carrying same basic types with different structure.

associating similar meaning to the name of the fields. In order for the computer to equate two variables that carry possibly the same information but different structure, individual fields are programmatically equated to avoid ambiguity of multiple occurrences of the same basic type.

Example 7.13

Given the two type declarations for *student1* and *student2*, the corresponding set for *student1* is given as *integer* × *string* × *float*, and the set for *student2* is given by (*string* × *integer*) × *float* where the parenthesis shows a nested structure. Although two representations have a clear bijective mapping and carry the same information, the order of the field names is not the same.

```
typedef struct {integer age; string name;
                float assignment_score;} student1;
typedef struct {string name; integer age;} person;
typedef struct {person individual;
                float assignment_score;} student2;
```

7.7.1 Structure versus Name Equivalence

Programming languages support two types of type equivalences: *structural-equivalence* and *name-equivalence*. *Structural equivalence* is based upon structural matching, and *name equivalence* is based upon two types having the same name in addition to carrying the same information.

Strictly speaking, *structural-equivalence* means that two structures are equivalent if they carry the same information even if the individual fields are permuted or if the individual fields have different names. Even if two named tuples (record or structs) have different names but have same basic types in permuted order they are structurally equivalent. However, such a definition of structural-equivalence is not practical for equating two structures because of (1) ambiguity caused by multiple occurrence of the same basic type in the structure and (2) permutation of the basic types in Cartesian-product and disjoint-union that can cause excessive computation in matching the nested structure.

Programming languages take a conservative approach to define structural-equivalence. If the individual fields of two types have the same name and the same basic types, then they are called *structurally equivalent*. Some languages put an additional restriction on structural-equivalence by disallowing the permutation in the field arrangement at the basic type level (leaf nodes of the tree in Figure 7.3). With this definition, two types given in Figure 7.3 will not be considered structurally equivalent, although they may carry the same information and are bijective. Languages such as ALGOL-68, Modula-3, C, and ML support restricted definitions of structural-equivalence. In ML, permuted order is allowed if the field names and their basic types match. Modula-3 supports a restricted form of structural-equivalence. The equivalence relationship is defined using subtypes. Two type declarations are structurally equivalent where both are subtypes of each other. The definition of Modula-3 subtypes is described in Section 7.8 under the case study of Modula.

Another problem with structural-equivalence is that it does not distinguish between two *semantically different entities*. For example, a point in a two-dimensional plane can be modeled as a pair of integers, and a complex number also can be modeled as a pair of integers. However, a complex number is semantically very different than a point in a two-dimensional plane, and the two types cannot be equated despite having same basic type and field names.

Example 7.14

The following structs are structural-equivalent because they carry the same information, have same basic types, and the same field names specified in the same order.

```
struct {integer a, b} record1;
struct {integer a, b} record2;
```

However, if we change the order of fields *a* and *b* in the type declaration *record2*, then two type declarations will not be structurally equivalent in a language such as C or Modula-3.

Example 7.15

The following Modula-3 syntax will treat the two structures to be structural-equivalent. However, the two types represent two different semantic entities. The first type represents a point in a two-dimensional plane, and the second type represents a complex number.

```
Type Coordinate = Record x, y : INTEGER;
Type Complex = Record x, y : INTEGER;
```

Name equivalence states that two types may carry the same information. Yet they are not type-equivalent unless they have the same name. This definition that two types are name-equivalent if they carry the same information and have the same name is more restrictive, yet it protects the programmer's intention of giving a different name to two different types that may model two different entities. Most modern languages support name-equivalence for (1) the ease of type matching during compilation and (2) to preserve the intention of the programmer to keep the semantics associated with the names. Languages such as Ada, Java, and C# support name-equivalence.

7.8 IMPLEMENTATION OF TYPES

Type information and various attributes are carried in the symbol table. In statically-typed languages, most of the descriptors that are not associated with runtime range checking are not carried forward after the compilation. Some of the type-related information—such as memory allocation, coercion, disambiguation of overloaded operator, and additional accuracy—are implicitly embedded in the code area of the program.

The various attributes of a type declaration are collectively called *type-descriptor.* Type information can be grouped as a pair of the form (*type-descriptor, memory allocation*). The information about memory allocation is given as a *memory-offset* from the base-address of an object. The information about type-descriptor includes (1) the name of the type declaration; (2) classification of the type such as record, array, and ordinal type, (3) domain-related information; (4) codomain-related information; (5) memory size of each data element; (6) number of elements in the type declaration; and (7) the offset of the memory location where the a specific data element or the field in a composite data-entity starts.

The number of bytes to be allocated for the built-in basic types is known to the compiler and does not need any additional information. The type-descriptor information for user-defined types has to be built. Tuples carry the information: *(name of the tuple, number of fields in the tuple, information about each specific field of the tuple, number of bytes to be allocated to individual fields, offset of the individual field where tuple will be allocated).* Arrays carry the information: *(name of the array; number of data elements; domain related information such as type of the domain, lower index value, upper index value; codomain related information such as type of the individual data element, number of bytes used by individual data element, range of allowed values, if any).* In case a user-defined type uses another user-defined type in its definition, then the name and reference of the corresponding user-defined type is stored. The exact information allocation is compiler dependent.

After the compilation is over, information needed for type-mismatch checking is discarded, and the memory allocation and memory access information for individual variables of specific type gets implicitly embedded in the code area of the program. However, all the information related to runtime error checking such as range-check error in arrays can be either kept dynamically for runtime check or can be embedded in the code area as conditional checks depending upon the compiler.

The following examples show an abstract representation of various type descriptors such as record, array and nested structures.

Example 7.16

Let us take a simple declaration of the record *student* described earlier. A student is abstracted as a triple of the form (*name: string, age: integer, major: string*). It can be represented as a named tuple *student* with the three named fields: *name, age*, and *major.* Let us also assume that the maximum size of the string is fixed to 256 characters. The type-descriptor is as follows:

```
(record, student, 3, 514,
    (string, name, 256, 0),
    (integer, age, 2, 256),
    (string, major, 256, 258)
).
```

The type-descriptor states that the type is a record. The name of the record is *student*. It has three fields, and the total size of the data element is 514 bytes. The information about the first field is as follows: it is of the type *string*, has name, size of the field is 256 bytes, and its offset is 0. The information about the second field is as follows: it is an *integer* with name *age*, occupies 2 bytes, and its offset is 256. The information about the third field is as follows: it is of the type *string* with name *major*, occupies 256 bytes, and the offset is 258.

Example 7.17

Let us take an array *class* that is an array *[1..30] of student*. The domain type is an *integer* with lower index value as 1 and upper index value as 30. The range type is *student*. The description of *student* is given by a reference to another type descriptor *student-descriptor*. The size of the individual element is 514, which is picked up from the *type descriptor student*, and total size of the array is 30 × 514 = 15420. A possible type descriptor for the above type declaration would be as follows:

```
(array, class, 30, 514,
        (integer, 1, 30),
        (student, reference(student-descriptor))
)
```

The implementation could use pointers, index, or hash functions to access the descriptor of the user-defined type *student*. The implementation of the type-descriptor is implementation dependent, and different implementations may use different data structures to implement it. In addition to the above information, the type-descriptor also contains the information about the program subunit where it was declared to check for the scope rule.

A code generator takes this information from type-descriptors, and performs memory allocation based upon the information of the offsets and the order of declaration of variables. Memory locations are allocated only for declared variables. For stack-based implementation, memory allocation for the local variables is done relative to the base-address of the frame. Since the current frame pointer is added to an offset to access memory location in the frame, the relative addressing based on offset makes the variable allocation independent of the absolute memory location. The individual fields of a composite type are accessed by adding the base-address of the allocated variable and offset of the field given by the corresponding type-descriptor. Similarly, the individual elements of an array are accessed by using the equation *base-address of the array + index value * size-of(individual data element)*.

Example 7.18

A variable *cs_class* of the type class is being allocated, and the base-address of the array is 12020. The ending memory location is 12020 + 30 × 514 = 27440. Each individual element is calculated in the code area using the base address 12020 + *value-of(index-variable)* * 514.

7.8.1 Type Inference and Checking

Polymorphic type is inferred if it is not declared explicitly. The process of inferring the parametric polymorphic type is called *type-inference*. Given an explicit polymorphic declaration, the process of checking the declared types with inferred types is called *type-checking*. In statically-typed polymorphic languages such as Scala, parametric type information may not be given, and it is automatically derived at compile time using a robust type checking system. In fact, languages such as Lisp and Prolog are polymorphic languages where the declaration of parametric type information is implicit.

In polymorphic languages, type information can be as follows: (1) type variables such as alpha, beta, and so on; (2) a concrete type such as *integer, Boolean, float*, and so on; (3) union of types; (4) disjoint-union of types; (5) Cartesian-product of two types; and (6) mapping of two types. The parametric polymorphism of a function is expressed in the most general form as $\alpha \rightarrow \gamma$ where α and γ are parametric types that are associated with input and output arguments. If there are N ($N > 1$) input arguments in a function, then the domain is represented as a Cartesian product of the form $\alpha_1 \times \alpha_2 \times \ldots \alpha_N \rightarrow \gamma$ where each α_I ($1 < I \leq N$) represents parametric type for an input argument, and γ represents parametric type for the output argument. A parametric type of a composition of a function f∘g is $(\alpha \rightarrow \beta) \rightarrow \gamma$, where the type variable α denotes the type of the input argument for the function g, the type variable β denotes the parametric type of the output of the function g, and the type variable γ denotes the parametric type of the output of the function f. Some of the parametric types for kernel functions are given in Table 7.1.

In order to infer polymorphic type, each argument is initially treated as most generic form, and it is specialized progressively to specific types based upon the structure information, operand type information, and the operator information.

Example 7.19

```
defun my_sum(DataList)
      (if (null DataList) 0 (+ (first DataList)
                               (my_sum (rest DataList))))
 )
```

The program states that *my_sum* takes an argument *DataList*. If the argument *DataList* is null then return 0. Otherwise add the first element of the *DataList* to the output of recursive application of the function *my_sum* on the rest of the list.

TABLE 7.1 Polymorphic Types of Some Known Functions and Functional Forms

Polymorphic Type		Polymorphic Type	
Function	**Type**	**Function**	**Type**
First	$list(\alpha) \rightarrow \alpha$	Length	$list(\alpha) \rightarrow integer$
Rest	$list(\alpha) \rightarrow list(\alpha)$	Append	$list(\alpha) \times list(\alpha) \rightarrow list(\alpha)$
Cons	$\alpha \times list(\alpha) \rightarrow list(\alpha)$	Insert	$\alpha \times list(\alpha) \rightarrow list(\alpha)$
Null	$list(\alpha) \rightarrow Boolean$	Apply_all	$(\alpha \rightarrow \beta) \times list(\alpha) \rightarrow list(\beta)$

Type inference starts with the generic form $\alpha \rightarrow \beta$. After looking at the recursive definition, we know that the input argument is a list, and output argument is an output on which '+' operator is defined. Thus the input type is refined to *list(α)* $\rightarrow \beta$ where β could be any type on which '+' operator is defined. Depending upon the language, it could be an *integer, float, long integer, double float*, or *string.* Now we look at the base case. It verifies that the input argument is a list, and output value '0' is mapped to the type domain *integer.* Thus the polymorphic type of the function is further specialized to *list(α) → integer.*

Many polymorphic languages, such as Scala, ML, and Haskell, although statically-typed languages, also have the option of inferring the type of the data entities based upon the known types of the literals and explicitly declared types of data entities.

7.8.2 Implementing Polymorphic Type Languages

Since polymorphic type procedures containing type variables can be bound to any type of data object at runtime, it is not possible to allocate memory at compile time since the actual type is not known. The type variable is specialized at runtime. Owing to runtime specialization, polymorphic languages have been implemented differently based upon the mapping of source code to machine code and the way data objects are stored. If the source code and machine code both exhibit polymorphism, then the implementation is called *uniform polymorphism.* For example, the functional programming language ML supports *uniform polymorphism.* If polymorphism exists only at the source code level, then it is called *textual polymorphism.* In object-oriented languages, the data is represented differently for polymorphic code, and it is called *tagged polymorphism.*

Textual polymorphism utilizes multiple possible specialized codes. An appropriate code is picked at runtime based upon the specialization. The major issue with this technique is the excessive requirement of memory for multiple specialized codes. This technique has another problem: it does not allow functions as first-class objects, because low level code is fixed to a set of specialized codes that can not be altered or expanded. The implementation of generics in ADA uses a variation of this approach.

Uniform polymorphism uses a uniform machine code for different types of specialization. This means that memory allocation for different types of data objects should also be uniform. In this approach, optimum memory allocation for different type of data objects is not possible, and data is excessed using a pointer. This means that there is an overhead of accessing the data in heap due to indirect addressing. This type of implementation is found in ML.

Tagged polymorphism uses the uniform code representation both at the source level and at the machine level. However, data is tagged by their types, and the generated code uses this tag to determine how to process the data. The operator overloading is handled using this technique. Object-oriented methods use this technique to handle same-name methods for different objects at runtime. Different dynamic sequences of code are executed for different objects.

7.9 CASE STUDY

In this section, we discuss the type system of Ada and C++. The type system supported by most of the languages is very similar for basic types and structured types. Recent object-oriented languages and multiparadigm languages also support polymorphism through the use of generics, template-based programming, subtyping, and inheritance. All the modern programming languages support overloading and coercion. However, the definition of coercion is loosely defined by many languages. Many languages do not support set-based programming mainly due to the lack of usage by the programmers. String is treated as a sequence of characters. However, many languages allow string declaration through the use of library. Subrange is also not supported by many languages such as C++.

7.9.1 Type System in Ada

Ada is a strongly typed language that supports objects, reference, structured types, and all the basic types including extensive numeric types. The overall type system of ADA is described in Figure 7.4.

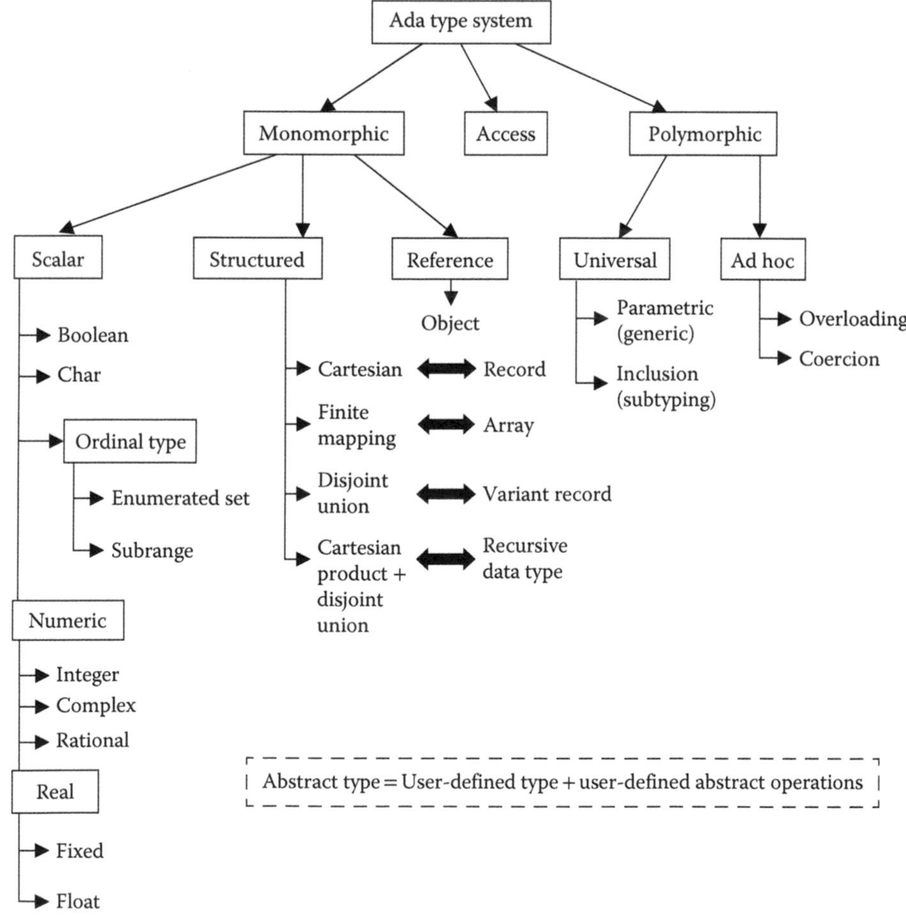

FIGURE 7.4 Types supported in Ada 2012 programming language.

Complex and rational types are built in the numeric package. Ada supports ordinal types including enumerations and subrange. Under universal polymorphism, it supports generics and subtyping, and like any other language, it supports overloading and coercion. Under scalar type, it supports Boolean, char, ordinal types, and numeric. The numeric package supports integer, real, complex, rational, and real numbers. Real numbers support fixed type in addition to traditional float type.

As described in Chapter 4, fixed type allows the decimal number using a *delta increment* that makes a quantum jump by delta amount, as additional bits do not provide additional accuracy. For example, the delta for U.S. dollar is one penny; that is 0.01. Ada also supports two additional declarations, *Task and Protected*, for concurrent programming that has been omitted in type system discussion. These two declarations will be discussed in Chapter 8 on concurrent programming. Note that Ada does not support set-based declarations.

7.9.2 Type System in C++

C++ is a strongly-typed and statically-typed object-oriented language. C++ supports basic types; structured types such as arrays, struct, union, and recursive data types; pointers; and object class. It also supports parametric polymorphism through the use of *template-based programming*, and inclusion polymorphism through *inheritance*. Template-based programming declares a generic method using a type variable that is instantiated at run time. It also supports overloading and coercion. However, C++ does not support subrange. Built-in basic types include only integer, float, char, Boolean, and string. Unlike Ada, complex numbers and rational number declarations are not built-in. C++ supports string as a class library. It also supports extra precision integers and floating point declarations such as double.

7.9.3 Type System in Modula-3

Modula-3 is a strongly and statically typed programming language that supports objects and modules. Modula-3 uses structural equivalence instead of name equivalence. Structural equivalence is defined as the leafs of the expanded tree matching in two equivalent types. It supports ordinal types: both subrange and enumeration types. It supports double precision, such as LONGREAL. It supports all the structured types including set-based constructs. It supports pointers as independent type. The pointer can have a value either nil or any valid address. It also supports procedure type, and procedure can be passed as parameter. It supports objects that are implemented using a reference type. Under polymorphism, it supports subtyping, coercion, and overloading.

7.10 SUMMARY

Type system is classifying the values based upon well-defined properties and operations on them. Most of the primitive types come from the mathematical world or text processing or information representation inside the computer. The definition of types is deep rooted in sets, and new types can be created by set operations on the constituent sets. The major set operations are as follows: *ordered bags, Cartesian-product, finite mapping, disjoint-union,* and *power-set*. *Cartesian-product* corresponds to a set of tuples and corresponds to struct (or record) in programming languages. *Finite mapping* corresponds to arrays or associative maps. An array can be viewed as a subset of integer domain mapping to another domain. *Disjoint-union* takes union of two disjoint sets and corresponds to *variant records*. A combination of disjoint-union and Cartesian-product is used to model recursive data types where recursive definition and base case are connected using disjoint-union. *Power-sets* correspond to a set of all subsets of a given set, and has been used in set-based programming.

Ordered bags have played a major role in defining various types. For example, *enumerated types* are ordered sets, *ordinal types* are ordered bags, and *sequence* of data elements are ordered bags. Sequence is an important abstract data type. For example, strings are sequence of characters. Lists are sequence of data elements. Stacks and queues are abstract data types that can be modeled using sequence.

There are many advantages of declaring types. Type declaration is used for type-mismatch error, efficient memory allocation, precision declaration, accuracy declaration, ease of modeling of real-world entities through abstraction, compile-time disambiguation of overloaded operators, and compile-time coercion. The major advantage of type declaration is to remove a major class of errors in large programs that may save programs from crashing specially in mission critical applications. However, type declaration cannot capture runtime errors such as range check, finding out a substring, runtime violation of types in variant records, and type violation due to pointer arithmetic.

An important concept in type theory is the notion of type equivalence. If two declared types are equivalent, then the variables of the equivalent types can be equated. Theoretically, two types are equivalent if they carry the same information and there is a bijective mapping between the corresponding sets. However, set operations such as Cartesian-product and disjoint-union produce sets that carry the same information yet permute the fields. It is difficult to find such equivalent sets due to ambiguity caused by multiple occurrences of the same basic type, change in name of the fields, and permutation of the fields. Programming languages support two types of equivalence: *structural-equivalence* and *name-equivalence*. Structural-equivalence imposes additional restrictions in identifying the sets carrying the same information. Many languages impose the restriction that, when expanded, the field names and basic types should be in the same order. Some languages such as ML allow permutation of the fields. One of the major problems in supporting structural-equivalence is that two structurally equivalent types may represent two different real-world entities that can never be equated. Name equivalence puts additional restriction that two types that have been named differently can never be treated equivalent. Most of the modern languages support name-equivalence. However, there are exceptions such as ML and Modula-3 that use structure-equivalence.

A major limitation in early days was the tight integration of functions with the data type. This tight integration made them unusable for different types of data objects, although the functions performed operations that were not related to specific type of the data objects. Some of the examples of such operations are *appending two lists*, *counting the number of elements in a list*, *stack operations*, *queue operations*, and *swapping two elements*. This led to the development of polymorphism that is closely related to the *reusability* of the functions and operations for multiple types.

Polymorphic types are classified into two broad categories: *universal polymorphism* and *ad hoc polymorphism*. Universal polymorphism allows indefinite number of data types to be associated with an operation or function, while ad hoc polymorphism allows limited number of data types to be associated with an operation. Universal polymorphism again can be divided into two classes: *parametric polymorphism* and *inclusion polymorphism*. Ad hoc polymorphism is also divided into two classes: *overloading* and *coercion*.

Parametric polymorphism allows types to be passed as parameters using type variables. A function instead of being associated with a concrete type is associated with a definition possibly containing a type variable that can be bound to different specific concrete types at runtime during procedure calls. The parametric polymorphism of a function is given in the form *generic-type(input-arguments) → generic-type(output-arguments) or concrete-type*. The generic-type definition includes *type variables*. The major advantage of parametric polymorphism is *reusability of code*, and indirectly it supports robustness of software development because the reusable code can be debugged and included in the library to be imported; polymorphic function does not have to be duplicated for different type of data objects. Parametric polymorphism was first implemented in the functional programming language ML, and since then it has been incorporated in many languages, specifically modern languages supporting functional programming paradigm and objected-oriented programming paradigm. Some of the modern popular languages that support explicit parametric polymorphism are ADA, C++, C#, Java, Scala, Ruby, and Haskell. ADA, Java, and C# use generic functions and type variables, and C++ calls generic functions as templates using type variables.

Inclusion polymorphism allows functions and operations on a type to be reused on a subset of the original set. Again the major advantage is that abstract operations defined on the original type can also be used on subtypes. If the original set is an infinite set, then it can have infinite possible subsets, and the operations on the original set can be used on indefinite number of possible subtypes. Inclusion polymorphism is used extensively in ADA and Modula-3 as *subtype* and is used in object-oriented languages through the definition of subclass. Subclass supports inheritance—a common trait in inclusion polymorphism.

An overloaded operator has multiple meanings. However, the actual meaning can be derived by looking at the type of the operands. If the type of the operands is available at compile time through type declaration, then the overloaded operator can be disambiguated at compile time. In dynamic languages, the meaning of overloaded operator changes dynamically, depending upon the dynamic type of the operands.

Coercion is related to type conversion of values to facilitate computation without loss of information. For example, integer value can be coerced to a floating point value. Any type conversion that loses information is not suitable for automatic conversion, because it can derive erroneous results.

For statically-typed information, most of the type-related information is kept in type-descriptor and is not carried during runtime unless it is needed to derive runtime errors. Type-descriptor contains information of basic types or reference to user-defined types, offset of the various fields, the information about domain and the type and size of data elements. All this information is needed to verify the type compatibility at compile time and to allocate memory for various variables in the frame of a procedure.

Polymorphic types can be either declared explicitly or they may be implicitly present. In statically-typed polymorphic languages, type-inference can be used to infer and verify the polymorphic type of a function. The technique is to start with the most generic form and progressively make it more specific it by looking at the structure, operators, and type of operands.

Implementation of polymorphic languages is classified into three major categories: *textual polymorphism, uniform polymorphism,* and *tagged polymorphism*. Many modern languages combine different implementation techniques. Textual polymorphism has polymorphism only at the level of high-level source code, while the lower-level code is split into different cases that can be selected dynamically based upon the specific type information. It has extra memory overhead. Uniform polymorphism has polymorphism both at the high-level source code and low-level code. It has an overhead of representing all different types of objects using a uniform memory space. Tagged polymorphism keeps a tag of data type with the object and selects the appropriate method by looking at the tag. Tagged polymorphism is used to implement object based languages.

7.11 ASSESSMENT

7.11.1 Concepts and Definitions

Abstract type; ad hoc polymorphism; basic type; built-in type; Cartesian-product; closure property; coercion; concrete type; delta disjoint union; dynamically-typed; enumerated set; enumeration types; explicit polymorphism; finite mapping; fixed type; implicit polymorphism; inclusion polymorphism; limitations of types; name-equivalence; named tuples; ordinal types; parametric polymorphism; power set; predicate; recursive type; reference type; set operations; statically-typed; structural-equivalence; strongly typed; structured type; subtype; tagged polymorphism; textual polymorphism; type checking; type-descriptor; type equivalence; type implementation; type inference; type variable; uniform polymorphism; union type; universal polymorphism; user-defined type.

7.11.2 Problem Solving

1. Model a student with the following fields: *student_id* of the type array of characters, *name* as string, *age* as integer, and *major* as enumeration type. Write the corresponding type declaration in C++ and ADA and the set operations to derive the corresponding set.

2. For the following set operations, give a realistic type declaration that models a real-world entity.

   a. $[0..30] \rightarrow \text{real} \times \text{real}$

   b. $(\text{real} \times \text{real} \times \text{real}) \times (\text{integer} \times \text{integer} \times \text{integer})$

   c. $\wp(1..12)$ where $\wp$ stands for power set

   d. $T = \text{nil} \uplus T \times (\text{string} \times ([0..30] \rightarrow (\text{string} \times \text{integer} \times \text{real})) \times T$

3. Using type variables, write a parametric polymorphic declaration for a function that takes two input arguments that are lists of any type and the output is a list of integers
4. Write a polymorphic type declaration for the following functions, and explain the logic:
   a. Sorting a sequence of data items
   b. Searching an element in a sequence of data items
   c. Searching an element in a tree
   d. Merging two sorted lists into a single list
5. Write a C++ program using template to merge two sorted sequences.
6. Write data structures for the following real-world abstractions, and then derive the corresponding set operations:
   a. A class of students where each student is a tuple (*name, age, department, year*)
   b. A university modeled as a collection of colleges, where each college is a tuple of the form (*area, dean-name, collection of departments-names*). You can model name as a string.
   c. A galaxy as a collection of stars where each star is a tuple of the form (*name, brightness-type, number of known planets, distance from earth*).
7. Write a simple program in Lisp or Haskell to delete an element from a list, and use the type inference mechanism to show that polymorphic type of the delete function is given by list(α) → list(α).
8. Write a simple program in Lisp or Haskell to count the number of elements in a list, and use type inference mechanism to show that the polymorphic type of the count function is given by list(α) → integer.
9. Derive the polymorphic type for all the stack operations and all the queue operations, and explain your reasoning.

7.11.3 Extended Response

10. Explain the difference between monomorphic type and polymorphic type using an example.
11. What is the purpose of incorporating polymorphism in programming languages? Explain.
12. Briefly describe each type of polymorphism in programming languages using at least one simple example.
13. Read about the numeric package from ADA 2012 official report, and explain the difference between fixed point and float using a simple example.

14. What are the different set operations and the corresponding structured types? Explain using simple examples.

15. Compare all three implementation mechanism for the implementation of polymorphism.

FURTHER READING

American National Standard Institute. *Programming Language—C++, DS/ISO/IEC 14882*. 2011. http://www.ansi.org

American National Standard Institute, *Programming Language Ada*, CSA *ISO/IEC 8652:201z*. 2012. Available at http://www.adaic.org/ada-resources/standards/ada05/

Burstall, Rod M., MacQueen, David B., and Sannella, Donald T. "HOPE: An experimental applicative language." In *Conference Record of the 1980 LISP Conference*. 1980. 136–143.

Cardelli, Luca. "Basic polymorphic type checking" *Science of Computer Programming. (SCP)*, 8(2). 1987. 147–172.

Cardelli, Luca, Donahue, Jim, Jordan, Mick, Kalsow, Bill, Nelson, Greg. "Modula 3 type system" *International Conference on Principles of Programming Languages (POPL)*. 1989. 202–21. Cardelli, Luca and Wegner, Peter. "On understanding types, data abstraction, and polymorphism." *ACM Computing Survey*, 17(4). 1985. 471–522.

ECMA International. *C# Language Specifications*, 4th edition ECMA-334, *ISO/IEC 23270*. 2006. Available on www.ecma-international.org/publications/files/ECMA-ST/Ecma-334.pdf.

Ellis, Margret A. and Stroustrup, Bjarne. *The Annotated C++ Reference Manual*. Addison-Wesley. 1990.

Liskov, Barbara and Wing, Jeannette. "A behavioral notion of subtyping" *ACM Transactions on Programming Languages and Systems (TOPLAS)*, 16(6). 1994. 1811–1841.

Milner, Robin. "A theory of type polymorphism in programming" *Journal of Computer and System Sciences*, 17(3). 1978. 348–375.

Morrison, Ron, Dearle, Alan A., Connor, Richard C. H., and Brown, Alfred L. "An ad-hoc approach to the implementation of polymorphism." *ACM Transactions on Programming Languages and Systems*, 13(3). 1991. 342–371.

Reynolds, John C. "Types, abstraction and parametric polymorphism" In *Proceedings of the IFIP 9th World Computer Congress*, edited by R. E. A. Mason, Elsevier Science Publishers B. V. (North-Holland) 1983. 513–523.

CHAPTER 8

Concurrent Programming Paradigm

BACKGROUND CONCEPTS

Abstract concepts in computation (Section 2.4); Abstractions and information exchange (Chapter 4); Control abstractions (Section 4.2); Discrete structure concepts (Section 2.2); Grammar (Section 3.2); Graphs (Section 2.3.6), Principle of locality (Section 2.4.8), Nondeterministic computation (Section 4.7); Operating system concepts (Section 2.5); Program and components (Section 1.4).

Concurrency is concerned about dividing a task into multiple subtasks and executing each subtask as independently as possible. There are two potential advantages of exploiting concurrency: (1) efficient execution of programs; and (2) efficient utilization of multiple resources, since each subtask can potentially use a different resource. With the available multiprocessor and multicore technology, concurrent execution of programs has tremendous potential of speeding up the execution.

The goal of exploiting concurrency is in the speedup of large grand challenge software, such as weather modeling genome sequencing, designing aircrafts with minimal drag, reasoning about nuclear particles, and air-traffic control. In recent years, because of the availability of multicore processors, concurrent execution is also available on personal computers. With the availability of multiple processors, it is natural for processors to map multiple tasks on different processors for efficient execution.

Parallelization can be incorporated at many levels in solving a problem by (1) designing a new algorithm more suitable for parallel execution of a task, (2) taking an existing algorithm and identifying subtasks that can be done concurrently, (3) taking an existing sequential program and developing a smart compilation process for incorporating parallelism, and (4) writing a parallel program with concurrency constructs. Parallelism can also be incorporated by speculatively executing conditional computations in anticipation that at least one of them will be needed in the future. In speculative computation, the input values should be known. Otherwise, the computation gets suspended waiting for the input values.

Executing multiple subtasks requires sharing of the resources and sharing of the information. Resources could be shared devices, shared memory locations, shared data structures, or shared codes. Sharing introduces sequentiality, and there have to be criteria for executing the different parts of the programs, so that the outcome of the program is the same whether they are executed sequentially on a uniprocessor machine, a multiprocessor shared address space machine, a multiprocessor with distributed address space, or on distributed processors with distributed address spaces. This property is broadly called *sequential consistency*. Sequential consistency is the foundation of automatic parallelization effort in programming languages. Automatic parallelization transforms a sequential program to a concurrent version of the program, without violating sequential consistency. Parallelism is exploited by distributing data and code effectively for maximum utilization of a multiprocessor time while minimizing the communication overhead with the constraint of sequential consistency. The programming constructs and compiler technology have been developed to achieve this final goal.

8.1 CONCURRENT EXECUTION AND ABSTRACTIONS

There are multiple approaches to exploit concurrency: (1) develop parallelizing compilers that transform sequential programs to concurrent programs and (2) develop programming constructs that allow multiple threads—sequence of actions—to run concurrently. Both the approaches have been used to exploit concurrency with success. However, the approaches are not free of problems as we will see in the following discussions.

Any effort to exploit concurrency is inherently limited by the *dependency* between the actions. *Dependency* means that an action depends upon another action. Dependency is caused because of (1) causality of actions, (2) sequentiality imposed by the control abstractions in a programming language, (3) sequentiality due to flow of data between the statements, and (4) sequentiality caused on uniprocessors because of the order of statements imposed by the programmer. An example of *causality-based dependency* is turning in the solution to a homework that is dependent upon the action of getting the homework. A student cannot turn in the solution to a homework problem unless a teacher has given him the homework to solve. An example of *control dependency* is the iterative-loop, where the sentences in for-loop, and while-loop are executed sequentially after the execution of the conditional expression. An example of data dependency is the execution of sequence of statements $x = 4; y = 5; z = x + y$. Here the third statement is dependent upon the successful termination of the first and second statement. However, the first two statements are independent of each other under the assumption that the variables x and y are not aliases and are not modifying the same memory location. *Programmer-induced sequentiality* is clear in the same example. Although there is no restriction on the order of execution of statements $x = 4$ or $y = 5$, a uniprocessor machine executes them sequentially because of programmer-specified order.

Dependency causes "sequentiality of execution," and one of the tasks is to minimize the dependency. The causality-based dependency is inherent in the solution of the problem and cannot be avoided. However, the effort is to minimize the control and data dependency that has been introduced in a program because of the programmer's action or due to the control and data abstractions. For example, if we take a for-loop that adds 4 to every element of a sequence, then there is a control dependency introduced by for-loop: the statement in for-loop cannot be executed unless the conditional expression is checked. However, if we unroll the for-loop, and replace for-loop by a set of statement individually adding 4 to different elements of the array, then the program is highly parallelizable.

Exploiting concurrency is a general problem. Sequentiality is also caused (1) if the number of available resources at a time is less than the resources needed by the subtasks at any given time; (2) to avoid racing condition (see Section 8.1.1) to maintain sequential consistency; (3) by the need for the shared resources at a very fine grain level that is not explicitly reflected in a high-level instruction; or (4) by limited availability of hardware components such as a limited number of processors, memory ports, limited bandwidth, and available memory banks. Shared resources could be memory locations or hard disks or processors or i/o devices or communication devices.

The subtasks can be executed independently if (1) they do not communicate with each other to share the information, (2) they do not share a common resource, and (3) they are not dependent upon each other through data and control dependencies. This means that given a program with n independent statements and *m* processors, the execution will take *ceiling(n/m)* unit time if the execution of each statement takes one unit time. However, in reality, different subtasks communicate with each other; share information; and share resources such as memory locations and I/O devices; and have statement-level dependencies. While one subtask is using a shared resource, other may have to wait, depending upon the action and the resource being shared. Sharing of resources involved in destructive update of the information such as write statements, causes inherent sequentiality.

8.1.1 Race Conditions

There are many high-level actions that require more than one low-level instruction to complete while using a shared resource. If the high-level action is done partially by one subtask and then control is passed on to another subtask, then the overall program behavior may be corrupted. This kind of corruption of the shared resource such as memory space is called *data-race*, and needs to be avoided. A sequence high-level actions should be completed before the shared resource is passed on to other subtasks. The shared resource has to be *locked*, so that other subtasks do not have access to it while the subtask currently holding a shared resource is using the shared resource. This notion of treating a sequence of actions as a unified single action is called *atomicity*. Enforcing atomicity using a "memory lock" is essential if we do not want the concurrent programs to get into data-race condition. This atomicity of action also enforces sequentiality among subtasks as other subtasks have to wait for the shared resources until the current subtask is finished. If the current subtask holds the resource more than the minimal time needed to ensure atomicity, then concurrency is reduced.

Example 8.1

Let us take an example of race-condition. Assume that we have a set of sequence of statements such that the variables *x* and *w* are aliases.

```
x = 4; y = 8; z = x + y; w = 5; y = 2 * w
```

Sequential execution of the above sequence of statements will give the final values as $x = w = 5$; $y = 10$; $z = 12$. However, if we execute the statements concurrently, without enforcing sequential consistency, we may get a different inconsistent set of values, since the order of executions may be different. If the statements terminate in the order of $x = 4$; $w = 5$; $y = 8$; $z = x + y$, $y = 2 * w$, then the final values would be $x = w = 5$; $y = 10$, and $z = 13$, which is different from the value we get after sequential execution.

8.1.2 Threads and Dependencies

Modern concurrent languages such as Java, C#, and C++, support the notion of *threads*. *Threads* are sequence of activities. A process can spawn multiple threads that merge back to the process after the corresponding subtasks are over. However, a program cannot be arbitrarily split into multiple concurrent threads and executed, because there are dependencies due to shared variables that cannot be split across threads without enforcing sequentiality between the statements. Such relaxation of condition would cause race-condition violating sequential consistency.

Example 8.2

Let us take the following sequence of statements: $x = y = 2$; $z = y + 4$; $x = 4$; $w = x + 2$; $y = 8$. Let us split the statements into two threads executing concurrently, as shown in Figure 8.1. The statement $x = y = 2$ is executed before spawning two threads. Thread_1 executes the statements $z = y + 4$; $x = 4$, and Thread_2 executes the statements $w = x + 2$; $y = 8$.

The two threads share two variables x and y. Thread_1 produces a new value of x, and Thread_2 consumes the value of x. Thread_2 produces a new value of y, and Thread_1 consumes the value of y. One could argue that the two threads can be executed concurrently. However, the presence of shared variables and destructive update of the shared variables inherently imposes a sequential consistency restriction.

Assuming that within the threads statements are executed sequentially, there are six possible outcomes as follows:

1. First statement of Thread_1 → second statement of Thread_1 → first statement of Thread_2 → second statement of Thread_2. In this sequence, the final values are $x = 4$, $y = 8$, $w = 6$, and $z = 6$.

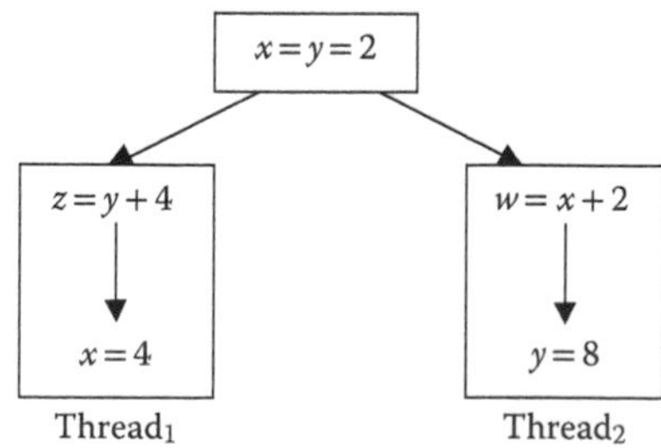

FIGURE 8.1 Race condition because of shared variables in threads.

2. First statement of Thread_1 → First statement of Thread_2 → second statement of Thread_1 → second statement of Thread_2. In this sequence, the final values are $x = 4$, $y = 8$, $w = 4$, and $z = 6$.
3. First statement of Thread_1 → First statement of Thread_2 → second statement of Thread_2 → second statement of Thread_1. In this sequence, the final values are $x = 4$, $y = 8$, $w = 4$, and $z = 6$.
4. First statement of Thread_2 → second statement of Thread_2 → First statement of Thread_1 → second statement of Thread_1. In this execution order, the final values are $x = 4$, $y = 8$, $w = 4$, and $z = 12$.
5. First statement of Thread_2 → First statement of Thread_1 → second statement of Thread_2 → second statement of Thread_1. In this order of execution, the final values are $x = 4$, $y = 8$, $w = 4$, and $z = 6$.
6. First statement of Thread_2 → First statement of Thread_1 → second statement of Thread_1 → second statement of Thread_2. In this order of execution, the final values are $x = 4$, $y = 8$, $w = 4$, and $z = 6$.

None of the solutions other than the sequential execution of Thread_1 followed by Thread_2 generates values consistent with sequential execution. If we look closely, then first statement of Thread_2 is dependent on the second statement of Thread_1 because of the consumption of the value of x and can be executed only sequentially.

8.1.3 Synchronization and Mutual Exclusion

A *lock* is associated with a shared resource. The purpose of locks is to enforce mutual exclusion of the subtasks competing for resources. The memory-locks, alternately also called *semaphores* or simply *locks*, have two states: *released* or *occupied*. When a process or a thread tries to use a shared resource, they have to check for the state of the lock. If the lock's state is *occupied*, then the subtask waits. Otherwise, it changes the state of the lock as *occupied* to block other subtasks from using the shared resource and starts using the shared resource. After finishing the use of shared resource, the subtask releases the shared resource and turns back the state of the lock as *released* to allow other subtasks to use the shared resource. The amount of time other processes have to wait while the shared resource is occupied inherently leads to sequentiality.

There are many problems with the use of locks as follows:

1. If the locks are allocated aggressively to one process, then other subtasks may starve—may not get a chance to use the shared resource.

2. Locks have to be declared and managed. Somehow humans are not very good in complex concurrent reasoning and make mistakes in lock management, causing the shared resource to be used unsafely by more than one subtask or blocking out all the subtasks by keeping the lock in *occupied* state.

3. A locking mechanism also causes overhead of waiting if the current subtask holds the lock longer than needed.

Programming languages provide a high-level construct called *monitor* that ensures only one of the competing subtask is using a resource at a time. A *monitor* is a passive high-level construct that includes many mutually exclusive subtasks sharing a common resource. The small chunk of code where a subtask uses the shared resource is called *critical section* and is treated as an *atomic operation*. During the execution of an atomic operation, the execution of the currently executing subtask cannot be broken; the shared resource cannot be allocated to another subtask. Other subtasks have to suspend their execution and wait until the current subtask is out of the critical section.

Example 8.3

Let us take an example of presidential opinion poll conducted by a newspaper company. It uses a central computer with an array that notes the vote of the person polled over the Internet. There is a global variable storing the maximum number of voters to be surveyed. The program also has a global shared variable that is incremented every time a person votes "yes" or "no" for a candidate. Let us assume the code of a single thread spawned to poll a candidate.

```
integer counter; max-to-be-surveyed = 3000;
string vote_array[max_to_be_surveyed]; vote;
Thread vote-count:
{
   ask; read(Vote);
   if (response(Vote) = "yes" ) counter = counter + 1'
   if (counter < max-to-be-surveyed) vote_array[counter] = vote;
}
```

All these threads are sharing the shared variables *counter* and the shared array *vote_array*. Let us assume that two people are voting at the same time: two threads are executed simultaneously. Let us also assume that the counter value before two threads start running is 10. Assume that the first thread reads a vote and increments the counter to 11. At the same time, the second thread read another vote and increments the counter to 12, before the first thread records its vote. Now the first thread records its vote at *vote_array*[12], and *vote_array*[*11*] is skipped. After that, the second thread writes its vote on the same memory location *vote_array*[12], destructively updating the location. As a result, one vote is lost completely and cannot be recovered. This problem could have been avoided using a lock associated with the array and the counter, because the locks would force the second thread to wait until the locks for the shared objects are released. The update of the variables *counter* and *vote_array* should be done as an atomic operation in a critical section that cannot be suspended by another thread before the locks are released.

8.1.3.1 Synchronization

The process of waiting for other subtasks or threads before executing the next instruction is called *synchronization*. This can happen if one of the subtasks is using a shared resource, and others are waiting. This kind of waiting is very common in the execution of the

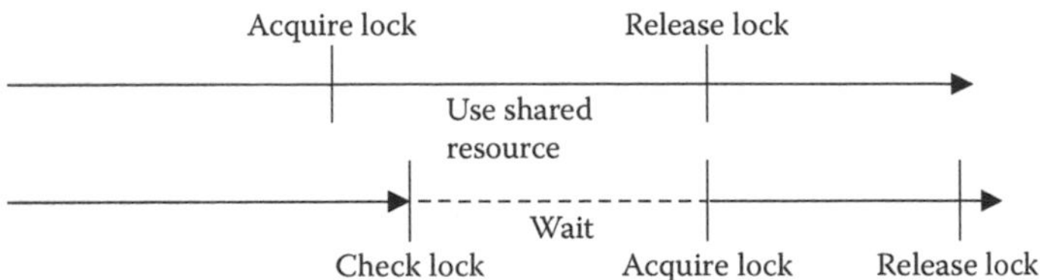

FIGURE 8.2 Lock and synchronization.

programs when (1) one subtask is writing on a shared memory location, and others are waiting; (2) when a subtask is reading a shared variable, and an other subtask is waiting to write; and (3) when a task is waiting for one or more subtasks to complete, because its resumption is dependent upon their completion. The waiting subtasks are awoken only after the executing subtask has released the shared resource. Synchronization is needed to avoid race conditions and introduces inherent sequentiality for the results to be consistent.

Waiting by the processes and mutual exclusion is achieved by associating a lock with a shared resource. When multiple processes are competing for a resource, one of the processes acquires the associated lock in a single operation, uses the resource in a critical section as one atomic operation, and releases the shared resource after the use. After releasing the shared resource, it releases the corresponding lock. Other processes compete to acquire the lock as soon as the lock is released. The process of synchronization between competing processes using a lock is illustrated in Figure 8.2.

8.1.4 Sequential Consistency

We have seen in Chapter 2 that store is altered when a command is executed. A group of commands C_1; C_2, …, C_N (N > 1) when executed with respect to a store *S* gives a new store *S'*. If all the permutations of the group of statements give the same final store *S'*, then changing the order of execution of the statements does not alter the final store, and the group of statements can be executed concurrently. Abstractly, if *permute*(C_1; C_2, …, *CN*) (*S*) ≡ (C_1; C_2; …; *CN*) (*S*) → *S'*, then the sequence of statements can be executed concurrently. This restrictions is satisfied under two conditions: (1) evaluation of expressions under commutative operations such as addition, multiplication, logical-OR, and logical-AND and (2) the store can be modeled as N mutually exclusive partitions that do not share memory locations, and each command C_I ($1 < I \leq M$, $M \leq N$) can alter at most one of the different mutually exclusive partitions at the same time using an atomic operation.

The reading action does not introduce sequentiality, as the store remains unaltered. Thus evaluation of an expression cannot cause sequentiality. It is only when a variable value is updated and later used in other statements or updated by multiple statements, then sequentiality has to be introduced in a concurrent version of a program to avoid race condition. One of the conditions of automatic parallelization is that *the sequential consistency must be maintained during the exploitation of concurrency.* In order to ensure this sequential consistency, dependency analysis is done, and sequential flow is weakened only if the final state remains the same as that obtained from the sequential execution. We will see dependency analysis in the next section.

8.2 PROGRAM DEPENDENCY AND AUTOMATIC PARALLELIZATION

This section focuses on the dependency among the program statements that regulates the exploitation of concurrency between the statements. If the statement order is relaxed too much without honoring the dependency imposed by the control and data abstractions, then the execution may become unsafe and inconsistent with the sequential consistency constraint.

We denote dependency by the binary symbol "≻." If the statement S_j is dependent upon a statement S_i, then we represent as $S_i \succ S_j$ ($i \neq j$). Dependency relationship is *transitive, antisymmetric,* and *antireflexive. Transitive* means that if a statement S_2 is dependent upon S_1 and the statement S_3 is dependent upon the statement S_2, then the statement S_3 is dependent upon S_1. *Antisymmetry* means that if a statement S_2 is dependent upon a statement S_1, then S_1 can never be dependent upon S_2. Otherwise, the program will be *deadlocked*—statement S_1 and S_2 will never be executed waiting for each other. *Antireflexive* means that a statement cannot be dependent upon itself.

In order to exploit concurrency between the statements, the program is modeled as a graph, such that each statement is a node, and the dependency between the statements is modeled as an edge between the corresponding nodes. The edge could be because of *control dependency* or *data dependency. Control dependency* is caused by the control abstractions, and *data dependency* is caused because of the information flow between the program statements. Control and data dependency are discussed in detail in Sections 8.2.1 and 8.2.2.

Dependency graph is a "directed acyclic graph," which means there does not exist a sequence of statement S_1, S_2, … S_N ($N \geq 2$) such that $S_1 \succ S_2 \succ \ldots \succ S_N \succ S_1$. Otherwise, every statement in the cycle can be dependent upon other statement including itself causing *deadlock*. Owing to the transitivity property of connectivity, if there is a directed edge between S_1 and S_2, and there is a directed edge between S_2 and S_3 then there is no need to show an edge between S_1 and S_3; the dependency between S_1 and S_3 is implicit. The dependency graph is analyzed, independent statements are identified, and synchronization points are set up to honor the sequentiality among the dependent statements. There are many types of control and data dependencies.

Control dependency and data dependency together are known as *program dependency.* Some of the control dependencies are artificially imposed because of the control abstractions and can be removed by appropriate techniques. The process of automatic parallelization is as follows:

1. Transform the program using various techniques so that artificially imposed control-dependencies is transformed to a program with minimal control-dependencies. Some of these techniques are *loop-unfolding* and *loop-lifting*, as described in Section 8.2.

2. Use data dependency analysis to build a data dependency graph for the program.

3. Superimpose control dependency graph and data dependency graph to make a program dependency graph, and execute those parts of the graph concurrently that are not connected to each other through dependency edges.

Partitioning of dependent statements on multiple processors causes time-overhead due to information exchange between statements involved in dependencies. The overhead can be significant, nullifying the advantage gained by the exploitation of concurrency. How this problem effects the overall execution, and how to reduce the overhead of information exchange between statements involved in dependencies is explained in Sections 8.2.5 and 8.2.6.

8.2.1 Control Dependency

Control dependency is caused by the control abstractions and has nothing to do with the flow of data between the statements. Control dependency imposes sequentiality because of the dependence of other statements on conditional expressions in control abstractions and because of the single-entry point restriction in the subprograms. Control dependency is imposed on a program instruction S_2 by an instruction S_1 if one of the following conditions is satisfied:

1. The execution of S_1 dominates the execution of S_2, which means the statement S_2 cannot be allowed without the execution of S_1. A statement S_2 is dominated by a statement S_1 if all the execution paths from start point of the control abstraction to S_2 include S_1.
2. S_1 is postdominated by S_2, which means all paths from S_1 to the end of the control structure go through the statement S_2.
3. Any statement between S_1 and S_2 on any path is postdominated by S_2.

In if-then-else statements, then-part or else-part can be executed only after executing the conditional expression. Thus, there is a control dependency between the conditional expression and the then-part; and conditional expression and the else-part. If there is more than one statement in then-block or else-block, then each of the statements is dependent upon the conditional expression. Similarly, in while-loop, there is control dependency between the conditional expression and every statement inside the while-loop (see Figure 8.3). That means no statement in while-loop can be executed without successfully executing the conditional expression. Similarly, there is control dependency between the entry-point of a subprogram and all other statements in the subprogram.

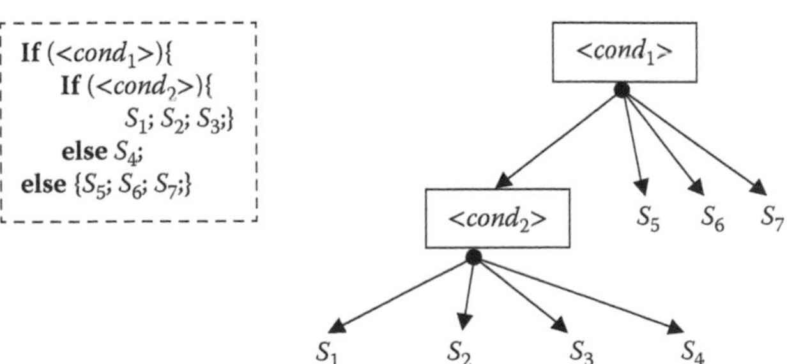

FIGURE 8.3 An illustration of control-dependency graph.

Control dependency is different from control flow diagrams, because control dependency is due to sequentiality caused by control abstractions only. A control flow diagram also includes programmer-imposed order of statement in addition to sequentiality caused by control abstractions. Thus, control dependency has a more relaxed condition of sequentiality. Control dependencies are derived by the control flow diagram of a program using the three conditions for control dependency stated earlier.

A graph representing control dependencies is called CDG. Control flow diagram describes programmer-imposed execution behavior of a control structure or a subprogram, whereas CDG describes the graph with essential sequentiality imposed by conditional expressions in the if-then-else loop or while-loops or single-entry point restriction in subprograms. In a CDG, each statement or conditional expression of the control abstraction is shown as a node, and there is a directed edge between the nodes if there is a control dependency. There is an edge between the entry-point and other statements in a control abstraction.

Example 8.4

Let us take the nested if-then-else statement given in Figure 8.3. The nested then-block has three statements S_1, S_2, and S_3; the nested else-part has a single statement S_4, and the outer else-block has three statements S_5, S_6, and S_7. The CDG for the nested if-then-else statement shows two levels: the top level for the outer if-then-else statement, and the second level for the inner if-then-else block. The statements within the outer block are dependent on the condition "Cond_1," and there is a dependency edge between the node "Cond_1" and S_5, S_6, and S_7, respectively. Similarly, there is an edge between the predicate node "Cond_2" and the statements S_1, S_2, S_3, and S_4. Dependency is transitive that means the statements S_1, S_2, S_3, and S_4 are also dependent on the condition Cond_1.

Example 8.5

Let us understand CDG involving while-loop and subprogram call using the example given in Figure 8.4. For better comprehension, we take the program of Chapter 5 that calls the subprogram *find_max* to find out the maximum value in a sequence. The statement numbers on the left side have been used in the CDG nodes to show the correspondence between the statements and the CDG nodes. Declarations have been left out, as they are not part of the execution.

The main program has six nodes related to the control dependence, and the subprogram *find_max* has seven statements. The beginning of the program and end of the subprograms are important, as they represent the start and end point of the corresponding subgraph. The for-loop in the main program has three statements built into one: (a) "i = 0"; (b) the condition "i =< 3"; and (c) "i++." We will label these high-level instructions as statement #2a, statement #2b and statement #2c. The if-then-else statement in the subprogram *find_max* is nested inside the while-loop and is dominated by the conditional expression of the while-loop. The nodes have been labeled according to their statement number. The start-node of the subprogram

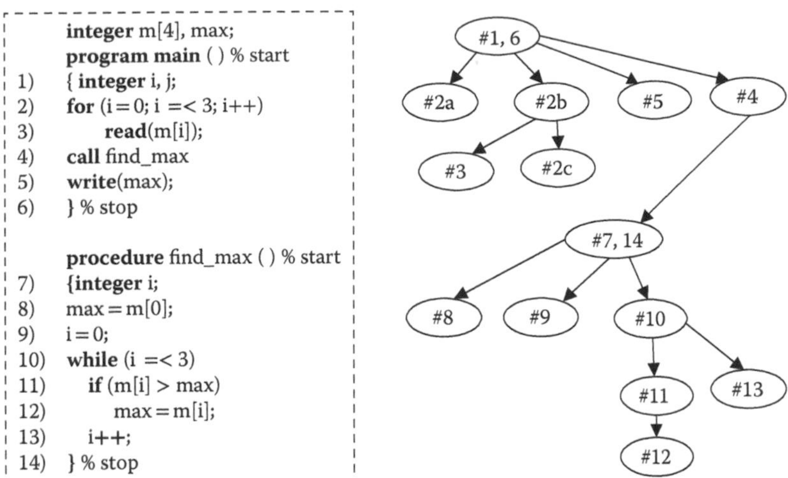

integer m[4], max;
program main ( ) % start
1) { integer i, j;
2) for (i = 0; i =< 3; i++)
3) read(m[i]);
4) call find_max
5) write(max);
6) } % stop
procedure find_max ( ) % start
7) {integer i;
8) max = m[0];
9) i = 0;
10) while (i =< 3)
11) if (m[i] > max)
12) max = m[i];
13) i++;
14) } % stop
#1, 6
#2a
#2b
#5
#4
#3
#2c
#7, 14
#8
#9
#10
#11
#13
#12

FIGURE 8.4 An illustration of control-dependency graph.

contains the scope of the subprograms: "nodes #1 and #6" correspond to the main program, and "nodes #7 and #14" correspond to the subprogram *find_max*.

There are control-dependency edges between the entry-point of the main program (Node #1, #6) and the nodes #2a, #2b, #4, and #5, because none of the statements can be executed unless the main program is started. There are control-dependency edges between nodes #2b, #3, and #2c, because without the evaluation of conditional expression in statement #2b, statements #2c and #3 cannot be executed.

In the subprogram *find_max*, the statements #8, #9, and #10 are dependent upon the statement #7, because these statements cannot be executed before entering the subprogram. Statement #10 has an embedded if-then-else statement on which the statements #11 and #13 are dependent. Statement #11 is another conditional statement on which statement #12 is dependent.

8.2.2 Data Dependency

Data dependency is the sequential execution restriction imposed between the program statements because of the destructive update of the shared memory locations between two statements and the restriction imposed by sequential consistency. If the statements are simply reading the values of the variables, then there is no restriction, as the values of the variables are not being modified. However, if the memory locations are updated with a new value, then the following statements must use the updated value, and statements previous to the statement updating a memory location are restricted to consume the old value stored in the memory location.

There are inherently three types of data dependencies: (1) *Producer–consumer* relationship—the statement reading the value from a memory location cannot be read unless it has been written into; (2) *Antidependence*—all the consumers of the previous value of the variable being updated must be executed before executing a statement that rewrites on the corresponding memory location; and (3) *Output dependence*—maintaining

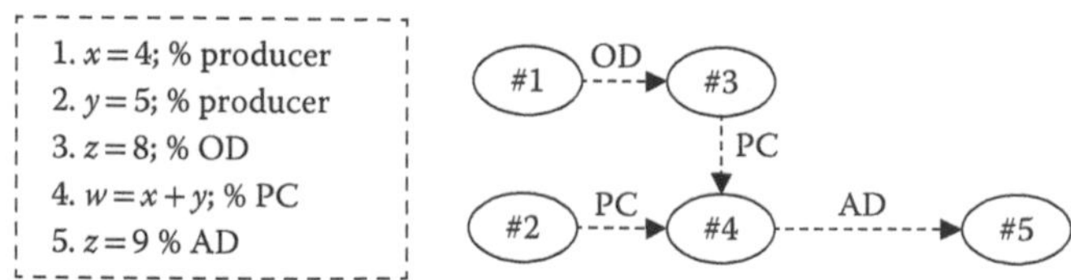
1. x = 4; % producer
2. y = 5; % producer
3. z = 8; % OD
4. w = x + y; % PC
5. z = 9 % AD
#1
OD
#3
PC
#2
PC
#4
AD
#5

FIGURE 8.5 An illustration of data-dependency graph.

the relative sequential execution order of statements involved in destructively updating aliased variables to ascertain that consumers use actual values and not updated values. Except producer–consumer relationship, other two types of dependencies are caused by destructive update of the memory locations. Producer–consumer relationship is also present in declarative languages with assign-once property and can be used to exploit concurrency in declarative languages.

The dependency graph made with statements as nodes and data dependencies as edges is called a *data-dependency graph.* A data-dependency graph is used to exploit concurrency by concurrently executing two *independent* statements. Two statements are *independent* if there is no path between the corresponding nodes in the data-dependency graph.

Example 8.6

Figure 8.5 illustrates data dependencies and the corresponding data-dependency graph using a sequence of statements. We abbreviate producer–consumer relationship as "PC," antidependence as "AD" and output dependency because of aliasing as "OD."

Let us consider the sequence of statements in Figure 8.5. We assume that variables x and z are aliases. The first two statements—$x = 4$ and $y = 5$—are independent, because both x and y map on different memory locations and can write concurrently in the corresponding memory locations under the assumption that computer hardware supports concurrent update in separate memory locations.

Statement #3 is dependent upon statement #1 through OD, because x and z are aliases. Statement #4 is consuming the values of x and y. Hence, statement #4 is dependent upon statement #3 and statement #2 using PC. Statement #5 destructively updates the value of the variable z. The variable z can be updated only after all the consumers of the old value have completed. The variable x is an alias of the variable z. Hence, statement #4 must be executed before statement #5. The dependence is because of combination of aliasing and antidependency.

Example 8.7

Let us take the same example of finding out maximum value of a sequence of numbers as given in Example 8.3 and find out data dependencies in the program. The type of data-dependency edges are given in Table 8.1. Figure 8.6 shows the corresponding

TABLE 8.1 Types of Edges in Dataflow Graph of Figure 8.5

Edge	Type	Edge	Type
(#2a, #2b)	PC	(#9, #11)	PC
(#2a, #3)	PC	(#9, #12)	PC
(#2b, #2c)	AD	(#10, #13)	AD
(#2c, #2b)	PC	(#11, #12)	AD
(#2c, #3)	PC	(#12, #5)	PC
(#3, #2c)	AD	(#12, #13)	AD
(#3, #8)	PC	(#13, #10)	PC
(#3, #11)	PC	(#13, #11)	PC
(#8, #11)	PC	(#13, #12)	PC
(#9, #10)	PC		

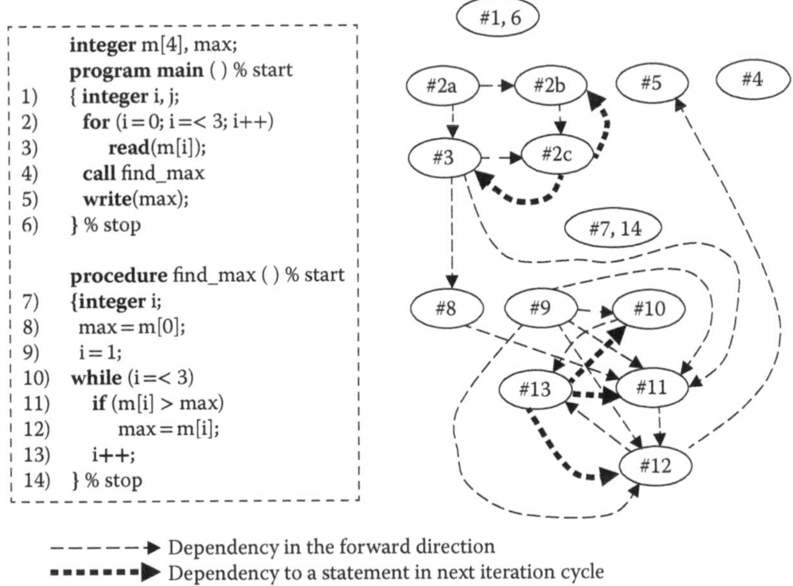

FIGURE 8.6 An illustration of a data-dependency graph.

data-dependency graph. The slim dashed edges show the regular data dependency between the statements. The thick dashed edges show data dependency between statements in the block of the for-loop to the conditional statement. They are special because they seem to show cycle because of iterative nature of for-loop, whereas actually there is no cycle as explained in the following paragraphs.

Producer–consumer (PC) relationship between the statement pairs (#2a, #2b), (#2a, #3), (#8, #11), (#9, #10), (#9, #11), and (#9, 12) are straightforward. The producer–consumer data dependency edges between statement pairs (#3, #8), (#3, #11), and (#12, #5) are because of the writing data into the global variables. The antidependency between the statement pairs (#2b, #2c), (#3, #2c), (#10, #13), (#11, #12), and (#12, #13) is straightforward. There is no aliasing of variables, and there is no OD.

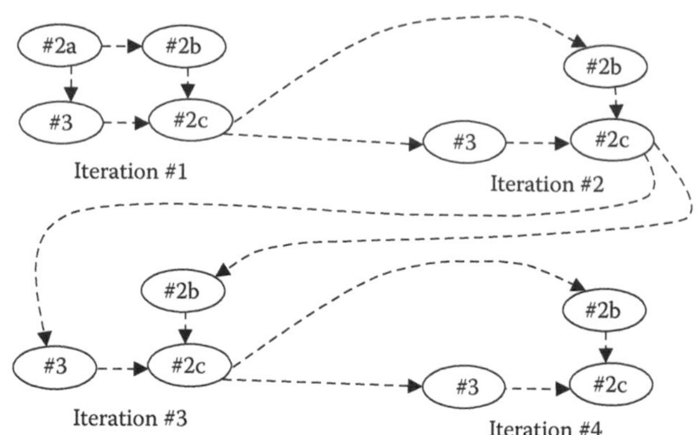

FIGURE 8.7 Data-dependency graph of unrolled for-loop.

Although we understand that there should be no cycles in dependency graph, five pairs of edges are problematic and seem to form cycles: (1) (#2c, #2b) seems to be causing the cycle {(#2b, #2c), (#2c, #2b)}; (2) (#2c, #3) seems to be causing cycle {(#2c, #3), (#3, #2c)}; (3) (#13, #10) seems to causing the cycle {(#13, #10), (#10, #13)}; (4) the pair (#13, #11) seems to be causing the cycle {(#13, #11), (#11, #13)}; and (5) (#13, #12) seems to be causing the cycle {(#13, #12), (#12, #13)}. In reality, there are no cycles if we unroll the iterative loops. In the representation of iteration graph, multiple statements from different iterative cycles performing same actions are superimposed on the same node and need to be unrolled to remove the apparent cycle. There is also a redundant edge (#3, #11) that can be inferred from the transitive relationship (#3, #8) and (#8, #11). However, it has been retained, because different data elements in the array *m* are responsible for these dependency edges.

The unfolded version in Figure 8.7 shows the edges between (#2c, #2b), and (#2c, #3) are edges across different statements in next iteration cycle, and there is no cycle in the data-dependency graph.

8.2.3 Program-Dependency Graph

Program-dependency is the dependency between the statements in a program. A program-dependency graph (PDG) is obtained by superimposing the control-dependency graph and data-dependency graph of a program. PDG shows the sequentiality imposed by the statements. However, to exploit maximum concurrency, PDG has to be transformed to minimize the control dependency caused by iterative loop and index variables. One such technique is "loop-unrolling" as explained in Section 8.2.4.

We superimpose the control-dependency graph in Figure 8.4 and data-dependency graph in Figure 8.6 to derive PDG, shown in Figure 8.8. Some of the dependency edges can be removed using the transitivity property. However, it has been left as an exercise.

8.2.4 Parallelization Techniques

The parallelization techniques take a PDG and find out multiple independent paths between the beginning and end of the program. Multiple threads handling independent

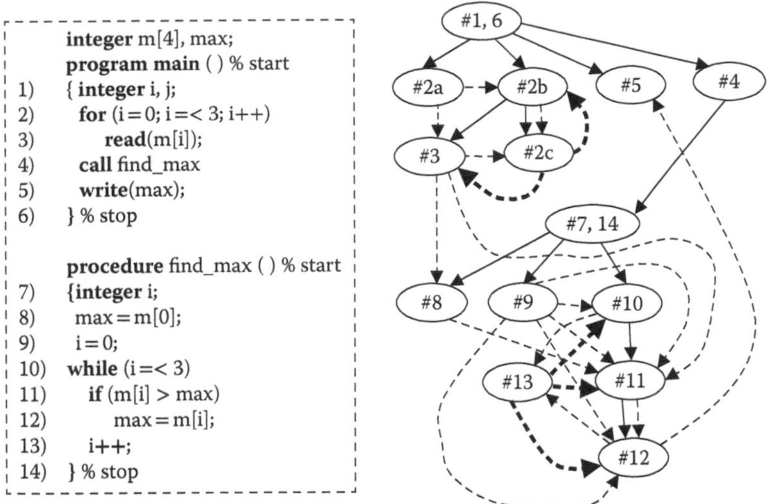

integer m[4], max;
program main ( ) % start
1) { integer i, j;
2) for (i = 0; i =< 3; i++)
3) read(m[i]);
4) call find_max
5) write(max);
6) } % stop
procedure find_max ( ) % start
7) {integer i;
8) max = m[0];
9) i = 0;
10) while (i =< 3)
11) if (m[i] > max)
12) max = m[i];
13) i++;
14) } % stop
#1, 6
#2a
#2b
#5
#4
#3
#2c
#7, 14
#8
#9
#10
#13
#11
#12

FIGURE 8.8 An illustration of program-dependence graph (PDG).

paths can be mapped on different processors. However, there are three problem areas that need to be sorted out: (1) allocation for nodes sharing data-dependency edges, (2) removing redundant dependency caused by control abstractions, and (3) keeping the data transfer overhead between processors minimal. The simplest form of automated parallelization is to map the PDG on multiple processors at the statement level. The mapping of transformed PGD is done in such a way to minimize the data transfer overhead between processors and to avoid unnecessary idling processor.

Example 8.8

Figure 8.9 illustrates processor allocation to exploit concurrency under the assumption that execution of each statement takes one unit time and data transfer across processor is instantaneous (a big unrealistic and simplified assumption). Under this assumption, the concurrent execution will take minimum of four unit time and a maximum of two processors. Putting extra number of processors will not improve the execution time, because of dependencies. Note that 4 units of time is not a linear speedup. There is expensive interprocessor data transfer overhead between statement #2 that is mapped on processor #2 (abbreviated as proc. 2 in Figure 8.9) and statement #4 that is mapped on processor #1.

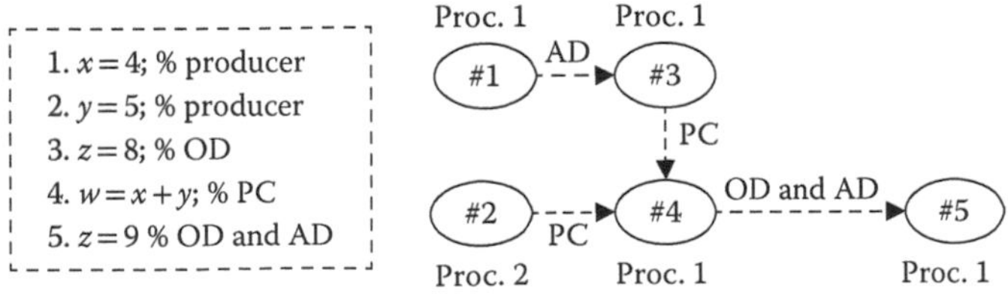

1. x = 4; % producer
2. y = 5; % producer

5. z = 9 % OD and AD

FIGURE 8.9 Processor allocations for data-dependency graph.

Data transfer overhead across processors are significant because of (1) repopulation of cache memory of other processors, (2) data-transfer between random access memories of the associated processors, (3) slowness of I/O bus in case of distributed processors, and (4) packing and unpacking cost of information or object being transferred across distributed processors. Packing and unpacking cost are serious in the case of distributed processors.

8.2.4.1 Loop-Unrolling

One of the problems in writing single-thread program is the use of the control abstractions for iterations. Although indefinite iterations are difficult to unroll, definite iterations can be unrolled to reduce sequentiality caused by the use of index-variables. The advantage of loop-unrolling is that redundant dependencies introduced by index-variables are reduced or removed, and unrolled blocks are executed concurrently on a multiprocessor machine. Let us take the example of for-loop

```
for (i = 0; i =< 3; i++) read(m[i]);
```

which is equivalent to *read(m[0])*; *read(m[1])*; *read(m[2])*; *read(m[3])*. All four statements are independent of each other without any data dependency, and hence can be executed concurrently. If there were four processors, then each one of them may be working concurrently, with no overhead of sequentiality induced by index-variables.

Let us consider another for-loop:

```
for (i = 0; i =< n, i++)   a[i] = b[i] + 4
```

Such dependence is caused only by the index-variables in the iterative loop and can be removed by unrolling the iterative loop. Many parallel programming languages replace these for-loop by a data-parallel construct: *a[1: n] = b[1:n] + 4.*

Let us assume that we are executing a block of statement 1000 times as follows:

```
for (i = 0; i =< 1000; i++)
{a[i] = 10; b[i] = a[i] + 4; c[i] = 2 * b[i];}
```

The for-loop can be unrolled four times. The new transformed block would be

```
for (i = 0; i =< 250; i++)
{ a[i] = 10; b[i] = a[i] + 4; c[i] = 2 * b[i]; % block 1
  a[i+1] = 10; b[i+1] = a[i+1] + 4; c[i+1] = 2 *b[i+1]; % block 2
  a[i+2] = 10; b[i+2] = a[i+2] + 4; c[i+2] = 2 * b[i+2];  % block 3
  a[i+3] = 10; b[i+3] = a[i+3] + 4; c[i+3] = 2 * b[i+3]; % block 4
}
```

There are four blocks of statements: each block contains three statements. There is no dependency between these blocks, and they can be executed concurrently on a four processor machine. In addition, data can also be partitioned in the corresponding address spaces at compile time for efficient execution. Concurrent execution of independent statements without any data transfer overhead achieves near linear speedup as follows:

```
n = 10000;
for (i = 0; i =< 1000, i++){
  a[i, 1:n] = 10;
  b[i, 1:n] = a[i, 1:n] + 4;
  c[i, 1:n] = 2 * b[i, 1:n];
}
```

While-loops can also be unrolled like for-loop. However, because of indefinite size in the while-loop, there has to be a provision to exit out of the loop if the condition is not satisfied after the execution of an unrolled block. Let us consider the following while-loop:

```
while (<cond>) <block>;
```

After unrolling four times, it would look like

```
while(<cond>)
{      <block>; if (<cond>) exit;
       <block>; if (<cond>) exit;
       <block>; if (<cond>) exit;
       <block>;
}
```

Loop-unrolling causes the compiled code to become quite large. In a vector computer, many other techniques are used to remove data-dependencies present due to nested iterations. Some of these techniques are node-splitting to remove the data-dependency cycle, and sorting the dependency graphs to rearrange the vector operations.

8.2.5 Granularity and Execution Efficiency

Exploitation of concurrency does not necessarily mean the improvement of execution efficiency, because of program dependency. Program dependence causes sublinear speedup, and the mapping of fine-grain statements on distributed memory space on distributed processors cause data transfer overhead for transferring the values of the variables that are produced on one processor and consumed on other processors.

In multiprocessor machines there are also many architectural limitations. The memory structure in a multiprocessor could be either shared among a group of processors, or it could be distributed to individual processors. In either scheme, there is significant overhead of data communication, partly because of much slower data transfer in IO bus compared to memory bus. Besides, the shared memory has limited number of memory banks, and each memory bank has limited number of address ports. Just increasing the number of processors does not scale up the execution efficiency; increasing the number of processor has to be matched with the increase in memory banks and address ports, which is either technologically not feasible or is cost-prohibitive. This problem is enhanced by multiple folds if the data-dependency graph at the statement level is mapped across distributed computers that may not share the same addressing mechanism and communicate using local area network or wide area network, and the data has to linearized and packed with extra information at one end and unpacked at the other end before it can be used. The data transfer between two computers may introduce an overhead that is 100 times slower than sequential computation on the same processor, when executing a PC like $x = 4; y = 5; z = x + 7$ that will take three units of time on the same processor. However, it would take 102 units of time on a distributed processor assuming that the overhead of data transfer between two processor takes 100 units of time.

By *ganularity* we mean the number of statements that are grouped together and executed sequentially on the same processor to avoid overhead of data transfer. A naive PDG is at the statement level. This kind of concurrency is called *fine-grain concurrency* and has significant overhead of data communication and *packing-unpacking* among processors. In order to achieve efficiency, statements have to be grouped together and mapped on multiple processors in such a way that there is minimal overhead of data communication, while concurrently executing the groups of statements. However, statements within each group are executed sequentially within the same processor. This kind of concurrency is called *coarse-grain concurrency.* The advantage of coarse-grain parallelism is to improve the efficiency of the program execution by reducing the data-communication overhead between the processors by (1) grouping excessive dependency edges across processors and (2) mapping the corresponding nodes on a single processor.

The question is, What kind of program will have coarse granularity where the overhead of information transfer is significantly less than the efficiency gained by concurrent execution? If multiple coarse-grain blocks of programs are identified, such that they have minimal communication between them, then each block can be mapped separately on a separate processor to exploit maximal concurrency without communication overhead.

Example 8.9

Let us consider the same sequence of statements as given in Figure 8.9. The data-dependency graph in Figure 8.9 shows that there is an interprocessor communication between statement #2 and statement #4. Assuming that there is an overhead of m unit time to transfer the data across processors, the total time taken to execute the concurrent version of the code is $4 + m$, compared to 5 units of time for sequential execution. In the equation $4 + m < 5$, the data-transfer overhead should be less than one unit for concurrent version to be useful. The overhead of data transfer is

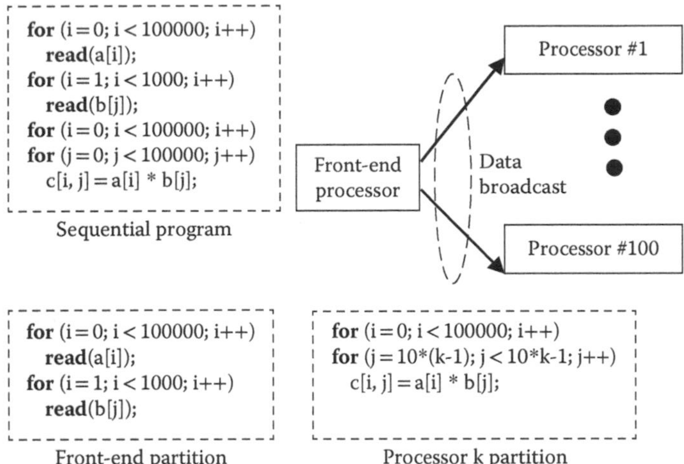

FIGURE 8.10 Partitioning a program for concurrent execution of independent data sets.

generally a lot more, depending on the architecture and the overhead of data-object migration from one processor to another.

Example 8.10

Let us assume that we want to map the program in Figure 8.10 for maximum concurrency without loss of efficiency. Let us also assume that we have a 100 processor distributed machine, and the overhead of transferring a packet of data is around *m* unit time (*m* >> 1). Looking at the program, we clearly see that each inner loop can be split into 100 different partitions such that i_{th} processor works b[10*(i–1)] to b[10*i – 1] data elements. Since the operation c[i, j] = a[i] * b[j] is independent of elements computed on other processors, these partitions can be executed concurrently on the distributed processors getting near linear speedup.

Exploitation of statement level concurrency based upon statement level PDG causes fine-grain parallelism and causes data dependencies that cause data-communication overhead after processor allocation. To improve the execution efficiency, fine-grain parallelism has to be transformed to coarse-grain parallelism. In the following section, we will study how programs can be split into multiple concurrently executing coarse-blocks that map on multiple processors, such that there is minimal communication overhead between the processors.

8.2.6 Program Slicing

Program slicing is a compile-time program analysis technique that analyzes PDG to optimize program properties. It has been used in matching programs, identifying duplicated code in programs, debugging the programs, software maintenance, and automated parallelization. In this section, we will discuss the application of program slicing for automated coarse-grain parallelism.

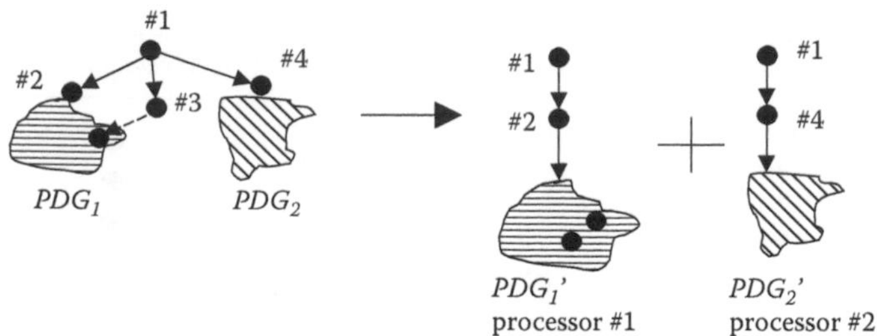

FIGURE 8.11 Program slicing for coarse-grain concurrency.

Program-slicing is possible if the PDG can be split into a set of smaller PDGs, such that there is minimal number of dependency edges across the sets. Figure 8.11 illustrates the concept of program slice for coarse-grain parallelism. The node #1 is connected to nodes #2, #3, and #4 through control-dependency edge. Node #2 is the start node of PDG_1, and Node #4 is the start node of PDG_2. Node #3 is a single node with a data-dependency edge to a node inside PDG_1. If we want to map the PDG on left-side to exploit maximum concurrency.

To improve the execution efficiency, we can take three actions: (1) merge node #3 in PDG_1 to give a new enhanced PDG_1'; (2) duplicate node #1; and (3) map (node #1, node #2, PDG_1') on processor 1 and (node #1 copy, node #4, PDG_2) on processor 2 and execute them sequentially. Because PDG_1' and PDG_2'v have no communication overhead, the program will run efficiently while exploiting concurrency.

If we look at the PDG, there are two cases of dependency edges that are suitable for program slicing: (1) nodes with multiple outgoing edges and (2) data-dependency edges that connect two different processors. Nodes with multiple outgoing edges are replicated to split PDG into two separate PDGs that can be mapped on separate processors without communication overhead. The second case requires joining the two subgraphs by allocating them in the same set and thus on the same processor to remove interprocessor communication. The start nodes of the subgraph for the control abstractions and the nodes corresponding to producers are potential nodes with multiple outgoing edges. Let us understand program slicing for coarse-grain parallelism using the following example.

Example 8.11

Figure 8.12 illustrates the improvement of execution efficiency by transforming fine-grain concurrency to coarse-grain concurrency by reducing the processor-to-processor data transfer. Let us consider the following code: $\{x = 4; y = 5; w = 8; m = x + y; n = x + w;\}$.

The data-dependency graph and a naive processor allocation of the code is given on the left side of the Figure 8.12, and the sliced program is given on the right side of the figure.

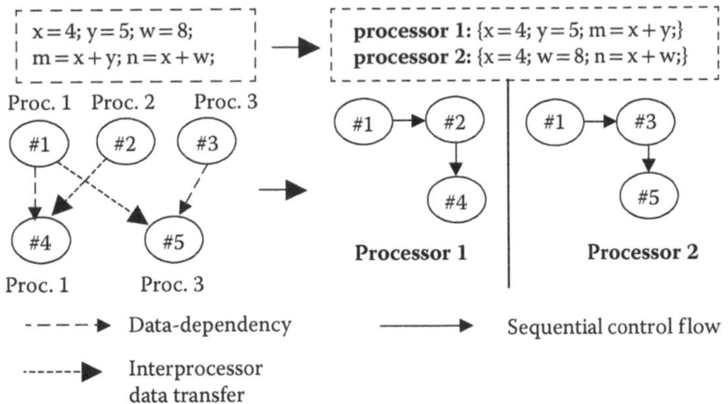

FIGURE 8.12 Program slicing to remove data-dependency overhead.

There are interprocessor data-transfer edges between the statement #2 (node #2) and statement #4, (node #4) and statement #1 (node #1) and statement #5 (node #5). If we assume that the data transfer overhead is 10 unit time, then fine-grain concurrency will take at least 12 unit time, which is much worse than sequential execution, which is 5 unit time. To improve the situation, we can use program slicing. The edge (#1, #5) is a producer–consumer dependency, and the producer node #1 has two outgoing edges. The node #1 can be replicated to avoid the interprocessor communication. Similarly, the node #2 can be grouped with nodes #1 and #4 to remove interprocessor communication.

The resulting program slice codes are given below

```
Processor 1: {x = 4; y = 5; m = x + y;}
Processor 2: {x = 4; w = 8; n = x + w;}
```

The overall execution, if the program is concurrently executing program slices, is 3 units of time, which is better than 5 units of time in sequential execution and 12 units of time, in fine-grain concurrency that includes 10 units of data transfer overhead.

The problems with program slicing at compile time are (1) it is not easy to get even near-optimum program slices with minimal overhead and (2) all parts of programs do not execute equally. Those parts that execute more frequently need to be analyzed and profiled for better performance. Only frequently executed part of the programs are speculatively transformed to exploit program slicing.

8.3 TASK AND DATA PARALLELISM

Task parallelism is concerned about executing different commands concurrently using multiple threads running on multiprocessor machines. *Data parallelism* executes the same commands on multiple data items on multiple processors. The major difference between task parallelism and data parallelism is the ability of task parallelism to execute different commands concurrently on multiple processors. Many problems that involve executing the same set of instructions on a large sequence of data elements can exploit data parallelism, as there are massive parallel single instruction multiple data (SIMD) computers with large numbers of simple processors. In this section, we will discuss task parallelism constructs, data parallelism constructs, and integration of task and data parallelism.

8.3.1 Task Parallelism

Task parallelism is based upon spawning and managing multiple processes or threads to perform different subtasks, possibly on different processors, and joining the results with the parent-processes. Task parallelism has been implemented in most of the traditionally single-thread languages such as C and C++, using an external thread library such as Posix, and Windows' thread library. However, Ada and Java have explicit constructs to exploit task parallelism. Java uses *synchronized methods* to access shared resources, and Ada uses *task construct* to start another subtasks.

There are four important concepts in handling multiple concurrent threads: (1) synchronization and handling of shared resources, (2) communication between the threads and the parent-processes and communication among threads, (3) resource allocations to avoid starvation and deadlocks, and (4) resource deallocation when the process or threads are terminated or aborted. All these concepts are interlinked. For example, if a thread has some shared resource and is suspended or aborted prematurely without appropriately releasing the shared resource, then other processes and threads may get *starved*—indefinitely suspended, waiting for the resources to be released so that they resume execution.

The processes or concurrently executing threads behave in two ways: (1) without mutual communication that does not use any shared resources (including memory locations) and (2) the use of shared resources. Shared resources can be accessed only sequentially. To enforce this sequentiality, locks (traditionally called semaphores) are introduced. These locks protect the associated shared resource by enforcing mutual exclusion among the processes that try to access the shared resource. Java uses *synchronized methods* to access the shared resources. Synchronized methods can be invoked by only one process at a time, which forces mutual exclusion. Shared resources can also be handled using a program abstraction called *monitor. Monitors* are high level program declarations that include all the mutually exclusive routines. Every corresponding process that wants to access the shared resource has to go through the monitor that enforces mutual exclusion. The constructs to use shared resources have been discussed in detail in Section 8.7.

The manipulation of shared resource puts on additional restrictions. Although the shared resource is being updated, it is in *transient state*, and is disturbed during this transient state to avoid data corruption. A high-level instruction or action is equivalent to many low-level instructions. Unless restricted, a process scheduler in an operating system or a hardware interrupt can suspend a thread (or a process). If another process accesses the same shared resource during transient state then the shared resource such as memory location would be corrupted, the resulting in incorrect program behavior. For program to behave correctly, high-level instruction using shared resources must be executed as an *atomic operation.* This part of the code inside a thread that is protected by a lock is called *critical section.* The general mechanism for enforcing the critical section is explained in Table 8.2.

TABLE 8.2 Lock and Critical Sections in Multithreads

Thread$_1$:	Thread$_2$:
…	…
wait_and_acquire(my_lock)	**wait_and_acquire**(my_lock)
<*critical section block$_1$*>	<*critical section block$_2$*>
release(my_lock)	**release**(my_lock)
…	…

The operation *wait_and_acquire(lock)* waits for the lock to be in a *released* state. If the thread is in suspended mode, then it is notified by a system routine to be awoken. The next thread scheduled on the processor that was waiting for the lock grabs the locks, puts the lock in the *occupied* state, and enters the critical section. After executing the critical section, it releases the lock, and then a system routine simultaneously notifies all the suspended threads to wake up and be brought into CPU by the scheduler.

The major problems in enforcing atomicity are as follows:

Problem 1: If the operating system has to abort or suspend the currently executing thread, then the critical section has to be restarted from the beginning. To ensure that it starts from the same state, all the actions in critical sections have to be recorded in a log, and the computational state has to be recovered before restarting the critical section again. This action has considerable execution time overhead.

Problem 2: If too many instruction are included within the critical section, then mutual exclusion delays the resumption of execution of other processes, making the execution slow. During a longer critical section, other threads have to wait for the *release-signal* even after a shared resource is not needed any more.

Problem 3: If the locked object is a large collection of data items, then accesses to those member elements that are not being updated during the critical section are also being blocked causing unnecessary sequentiality.

Problem 4: Multiple processes may be performing read operations on a shared object. Multiple concurrent reads are allowed on the shared objects. The only condition is not to allow writing when the information is being read.

Problem 5: There is an execution time overhead of checking and notifying the threads about the status of the locks.

Problem 6: Programmers cannot comprehend complex concurrent programming situations, resulting into incorrect placement of the locks, which may result in incorrect program behavior such as deadlocks or starvation.

Multiple solutions have been proposed to solve the sequentiality caused by atomic operations. Problem #4 has been solved using two different types of locks: *shared-read locks* and *exclusive locks. Shared-read locks* allow simultaneous reading of the shared data entities. However, it blocks those threads and processes that try to update the shared data entities. *Exclusive locks* disallow an access to any other threads or processes. A process that is updating a shared data entity will set an exclusive lock, whereas a process that is reading a shared data entity will set a shared-read lock.

Recently, there have been efforts to address Problems #2 and #3 using the concept of *transactional memory* both at the hardware level using cache memories and at the software level using *shadow copies* of the data entities being updated. The concept of transactional memory comes from database transactions. A *transaction* is a sequence of instructions that may read and write a set of data entities and may either commit to changes at the end of transaction or undo the changes completely in the case of transaction abort. In transactional memory schemes, there are four relevant operations: (1) start transaction, (2) end transaction, (3) read data entity, and (4) update data entity. During a transaction, process creates a copy of the updated data entities that is copied into actual data entities at the end of transaction, when the transaction is *committed.* During a transaction, the actual data elements in the collection of data elements that are not being updated during the transaction can be read. The data elements that are updated during the transaction are blocked automatically from further write by other threads, unless the transaction is successfully committed. Updated data entities can be read by other threads even before the transaction-commitment, provided there is no read–write conflict in transactions, and threads updating data items do not abort and are definitely going to commit. However, finding out conflicts with other threads/processes that would be affected because of memory update after a transaction started and before the transaction ended is a complex issue. Each transaction keeps its own *read set* and *write set* for conflict resolution. *Read set* is the set of references to the data entities that are read during a specific transaction, and *write set* is the set of references of data entities that are updated during that transaction. These read sets and write sets are visible to other transactions for conflict resolution. Conflict resolution can be done eagerly at the "transaction start" or done lazily at the "transaction commitment" time. Conflict resolution is done by checking the operations in the read sets and write sets. Eager conflict resolution is generally pessimistic, finds more conflicts, and introduces more sequentiality. Lazy conflict resolution is generally optimistic, finds less number of conflicts, and permits more concurrency. However, it may cause late detection of conflicts, rolling back the whole transaction and causing excessive overhead. There is also problem with exception-handling that causes transactions to roll back. There is also significant overhead of matching read sets and write sets.

Problem #6 of programmers' lack of comprehension of handling locks in complex situations has been partially solved using the concept of *monitors* as discussed in Section 8.7.1.

One purpose of the shared variables is to exchange information between two threads. Information can also be exchanged asynchronously between two threads or two processes. In an asynchronous information exchange, the producer thread writes the information into a predesignated memory buffer, and the consumer thread picks up the information from the buffer whenever the information is available. This requires help from the operating system and uses *message passing* and *mailbox*—a buffer space for the corresponding process. Mailbox is generally stored in the operating system area and has a copying overhead.

Task parallelism can be realized on multi-processor architectures generically called *multiple instruction multiple data* (MIMD), which means the processors can concurrently execute different instructions on different data elements. MIMD architectures are further categorized as (1) shared-memory multiprocessors and (2) distributed-memory multiprocessors. Shared-memory multiprocessors share the memory among multiple processors. They have an advantage of faster memory access and keeping only one copy of the data. However, they have the bottleneck of accessing the shared resources and face traffic congestion in the data bus. Distributed-memory multiprocessors have the advantage of local memory. However, the data transfer across physically separated memory without a common high-speed bus is slow. The constructs used to implement task-level parallelism include *fork-and-join, cobegin-coend* or *parbegin-parend*, use of *locks*, use of *monitors*, and *spawning of multiple threads.*

The problems in distributed computing are somewhat different because of object migration. Object migration involves overhead of *linearization-delinearization* (also called *packing-unpacking*) of data structures. Distributed processors and distributed memory supercomputers use message passing to communicate between the processes and use a client-server model of execution that uses *remote procedure calls* (RPCs), as discussed in Section 8.5.

8.3.2 Data Parallelism

Data parallelism is exploited by applying the same instruction on multiple data-elements. Instruction is broadcasted to different processing elements. The main restriction is that the data-parallel operations on the processing elements should not show dependency on one an other. Data parallelism works well on flat data structures such as sequences, arrays, and sets. Let us illustrate this concept using the following example.

Example 8.12

Let us take the following two for-loops. The first for-loop when unrolled is equivalent to 1000 independent statements that perform the same operation: add 4 to the previous value. Hence, they are data parallel. However, the second for-loop is not data parallel, because each element is dependent upon the element occurring before it.

```
1.   for (i = 0;i =< 1000;i++) a[i] = a[i] + 4 % data-parallel
2.   for(i = 1;i =< 999;i++) a[i] = a[i - 1] + a[i + 1]
```

Data parallelism is exploited if the region where the same instruction can be executed is identified. As shown in the first for-loop, data parallelism performs the same operation on a large sequence of elements. Some of these elements are masked to work on a subsequence of elements. Let us consider the following example:

```
for (I = 0; i =< 1000, i++)
            if (a[i] < 0) a[i] = -1 * a[i];
```

This is also a data-parallel operation, as all the elements are compared simultaneously with 0 to create a *mask* that will filter only those numbers that are < 0 and then multiply them with −1 in a data-parallel manner.

Another property of data parallelism is to *reduce* a sequence of values to a smaller set of values using a user written function. *Map-reduce model* of exploiting data parallelism, as the name suggests, is based upon two techniques: (1) map and (2) reduce. In the first stage, the user written map function takes a collection of (key, value) pairs, repartitions them, and sorts them, such that identical keys can be passed to reduce function. The user-written reduce function groups the key-value pairs into multiple groups, such that each group is a subset that can be uniquely processed using a user-defined function. The output is the union of the values derived by applying the function.

8.3.3 Integrating Task and Data Parallelism

As discussed, task parallelism and data parallelism are quite different from each other. Task parallelism is bothered about spawning multiple processes on different data sets involved in a computation and managing shared resources, whereas data parallelism is bothered about performing the same computation on a large data set. Technically, task parallelism is synonymous to MIMD, and data parallelism is involved with single program multiple data (SPMD). There are many problems that require the integration of data parallelism and task parallelism. In recent years, efforts have been made to integrate task and data parallelism. Some of the languages that integrate both types of parallelism are "Fx", "Opus", "Orca", "Braid", and "Fortran-M."

Different languages have taken different approaches to perform the integration. Fx allows for spawning multiple tasks, such that each subtask is capable of performing non-interfering data-parallel computation. One class of supported data-parallel operations is *map-reduce operation*, as explained in Section 8.3.2. Fx is good for multiple-data parallel computations that are performed concurrently.

Opus is another integrated programming language that uses a "shared abstract data type" called SDA. SDA is controlled by a coordinator process that spawns multiple subtasks, such that each subtask concurrently writes on the SDA in a data-parallel manner. The advantage of Opus is in spawning multiple-data parallel tasks concurrently. A simple example of SDA is a common buffer being shared between a producer and a consumer process.

Orca is a programming language that uses a "distributed shared data structure." Multiple subtasks are spawned concurrently that work in a data-parallel manner on different subsets of processors. A data-parallel operation operates the same operation, so that each processor writes on its partition of the shared-data structure. A task can access partitions in other processors. Objects that are not shared are stored in one processor and can be replicated into multiple processors. An instruction can be broadcast on all the processors for data-parallel computation.

8.4 DISTRIBUTED COMPUTING

With the advent of computer connectivity, it has become quite common to map a computation on multiple distributed processors. These processors can (1) execute procedures concurrently in a distributed fashion, (2) transfer data on separate processors that can perform computation and transmit the result back, or (3) transfer the code to another remote processor that computes locally. Depending upon the application, all three approaches or their combinations are used. The major issue is to minimize the overhead of information exchange between the calling subprogram and the called subprogram on a remote processor. *Data migration* is useful if the amount of information exchange is not too large, and the load has to be distributed. *Code migration* is preferred if the overhead of code transfer is smaller than the overhead of data transfer. Code migration is also preferred if the server is overloaded, and the load has to be shed by placing the subtasks on remote processors.

The communication between processes on distributed processors passes through multiple layers of network protocols that cause significant overhead. A complex data structure is linearized, padded with parity information to protect against corruption of data during communication, made into packets, and then transmitted. On the receiving end, the reverse process is done. The information transfer can be both synchronous and asynchronous. In synchronous communication, the sender is blocked, unless the receiver receives the message and sends an acknowledgment. In asynchronous communication, the sender deposits the message in a *mailbox* handled by the middleware and keeps executing. Similarly, a communication can be "persistent" or "transient." In persistent communication, the message is stored until transmitted, whereas in transient communication, the message is stored as long as the sending and receiving application is executing.

There are many issues in invoking a distributed computation: (1) the remotely executed procedure should have access to resources, such as files and IO devices; (2) the address of the resources should be uniform; and (3) the communication port between calling subprogram and called subprogram should be uniformly accessed by the processes.

Generally, distributed programming provides another layer of transparency than simply sending the information and receiving the information. The transparency is provided using the parameter passing mechanism present in programming languages. The mechanism is called *remote procedure call (RPC)* and is discussed in Section 8.4.1.

8.4.1 Executing Remote Procedures

Remote procedure call (RPC) means executing the called subprogram on a remote processor. The issues executing an RPC are as follows: (1) there is more than one address space that has to be accessed; (2) information has to be physically transferred across machines using the communication network layer; which means setting up the channel between the calling and the called subprograms, (3) the result may be passed back across the processors.

There are two ways to access the data objects in the calling subprograms: (1) by passing the reference and (2) by passing the value. If the reference is passed, then there are issues about how to access the address space of another processor using the communication channel that may incur significant overhead. If the value is passed, then the calling subprogram stores the information about the channel, number of bytes, and buffer address where transmitted data is stored/received, along with the return-address. Similarly the remotely called procedure also has to store (1) the channel and mailbox-ID, where transmitted data is stored to retrieve the data and (2) channel and buffer of the outgoing message.

Owing to the difference in address spaces, the calling program packs the information in *stub* and sends it to the operating system that sets up the channel with the operating system on another processor. The operating system on the other side receives the parameter information and invokes the called subprogram. The called subprogram generates the result, packs the result, and asks the remote-operating system to send the data across to the calling subprogram. When the result is received at the processor executing the calling subprogram, a system routine unpacks (also called unmarshalling or delinearizing) the result and sends the signal to the calling subprogram to resume and read the information from the system area.

Packing the parameter values is called *marshaling* (or *linearizing* or *packing*). The process of marshaling (1) puts the unique name of the procedure making the call and (2) linearizes the data to be transferred. The mechanism works fine if the operating system and data representation mechanism is homogenous, which means both the processors are executing compatible operating systems that use the same internal representation format for different data types. However, for a large cluster of distributed computers, machine architectures are different, and operating systems are different. So moving from one processor to another heterogenous processor also involves conversion of the format representation and having a common format representation in Unicode during transmission. Unicode is a universal 16-bit representation. In addition, the calling subprogram and the called subprogram should use the same format presentation for arranging the data during linearization of the complex data structure.

8.4.2 Parameter Passing in Remote Procedure Calls

There are three types of parameter passing mechanisms in RPCs, as discussed briefly in Chapter 4. These mechanisms are similar to already discussed mechanisms: (1) call-by-value, (2) call-by-reference, and (3) call-by-value-result. However, the major difference is that multiple address spaces, communication protocols, communication channels, and buffers are involved in data communication, which causes overhead of communication, specially for accessing small amounts of data multiple times across the address spaces. Passing the reference of the object means that remote call has to access the address space of the objects stored in the address space of the calling subprogram. Copying an object means the called subprogram works only on the local copy of the object. Sometimes this is also called *call-by-move* or *call-by-copy.* If the copied object is modified and the modified object is needed by the calling program, then the resulting object has to be moved back to the calling subprogram. This is called *call-by-copyrestore* and is similar to call-by-value-result in uniprocessor languages.

Parameter passing using a reference parameter is difficult because of distributed address space. The problem with call-by-reference is the communication overhead to access the data from distributed address space. The address space of the other processor is not directly accessible to the remote process. Instead, it has to go through the operating system of the processor where the calling subprogram is located. In order to facilitate call-by-reference, the object is copied into the buffer space of the operating system, and only needed data elements are transmitted as value to the remote procedure. After receipt of the modified data element, the corresponding data element in the complex data structure is updated. If the call-by-reference is read only, then the object is moved to the processor of the remote procedure to avoid the overhead of communication network.

Call-by-copy and *call-by-copy-restore* have to go through marshaling (packing) and unmarshalling (unpacking) of the object. The overhead of marshaling and unmarshalling can be few milliseconds, and hence the granularity of the RPC should be sufficiently large to justify the RPC.

There are problems both for call-by-reference and call-by-copy. If a communication is lost in call-by-reference during the execution of the RPC, then the effect of the changes made to the object have to be undone, which requires keeping a log of changes, and the loss of communication will cause the failure of computation. In call-by-copy-restore, the original data structure and the copy of the data structure may become inconsistent, until the original data structure is restored with the changes made in the remote copy. Call-by-copy-restore also forces the suspension of the calling subprogram to maintain the consistency of the data structure. Consistency can also be maintained using *automatic coherence*, provided the replicated remote data structure does not become unavailable because of communication failure.

An alternative scheme that finds out a middle ground between call-by-reference and call-by-copy has also been implemented. This scheme performs the following operations:

1. It copies the immutable simple objects such as integers, Boolean, and floating point numbers.

2. It avoids call-by-reference if the communication failure is probable.

3. It sends only the object_id before copying. If the object is already in the remote store, then it avoids copying the object.

Emerald is a distributed programming language that supports all three forms of parameter passing. Java supports distributed computing using *remote method invocation* (RMI), which is slightly different than RPC because of dynamic invocation of objects. Distributed objects and RMI will be discussed in detail in Chapter 11.

8.5 COMMUNICATING SEQUENTIAL PROCESSES

Communicating sequential processes (CSP) is a theoretical model of concurrent programming languages that is based upon guarded commands. It was first proposed by C.A.R. Hoare, and has formed the theoretical basis of a major class of concurrent programming languages. Although the model was described using guarded commands, the concepts and the algebra for concurrent processes have been utilized in the development of other concurrent programming languages.

The language uses guards as *input sources* and commands as the *output source. Input source* is a process generating input data, and *output source* is the corresponding process processing the input data. If the information is available in the input process, then the corresponding command of the output process is executed. Two concurrent processes *P* and *Q* are modeled as P || Q. The concurrent processes are disjoint, do not share any variable, and do not communicate using global variables. Hence, they can execute concurrently without any synchronization. All the concurrent processes start simultaneously, and parallel command P || Q terminates when both *P* and *Q* are finished. The set of events seen by a process *P* are called *alphabets of process*, and are denoted by αP. If the sets of events of two processes are the same, then they get engaged together. However, if one of the events belongs only to a specific process *P*, then only *P* gets engaged.

Example 8.13

Let us take two processes: *P* with alphabet {*move_left, move_right, move_back, move_forward*} and process *Q* with alphabet {*move_back, move_up, move_forward*}; then *P* and *Q* will be engaged together if the next action is *move_forward*. However, only *P* can do *move_left*. Similarly, only *Q* can take an action *move_up*.

8.5.1 CSP Algebra

The parallel processes are controlled by simple algebraic rules. Given three processes *P, Q* and *R*, their alphabets would be denoted by αP, αQ, and αR, respectively. Let us assume that $\alpha P = \alpha Q = \alpha R$, and $\{c, d\} \subseteq \alpha P$. If each process has the same set of actions, then the processes are controlled by the following rules:

Rule 1:	commutativity:	$P \parallel Q \equiv Q \parallel P$
Rule 2:	associativity:	$P \parallel (Q \parallel R) \equiv (P \parallel Q) \parallel R$
Rule 3a:	deadlock:	$P \parallel Stop_{\alpha P} \equiv Stop_{\alpha P}$
Rule 3b:	running:	$P \parallel Run_{\alpha P} \equiv P$
Rule 4a	agreement:	$(c \rightarrow P) \parallel (c \rightarrow Q) \equiv (c \rightarrow (P \parallel Q))$
Rule 4b	disagreement:	$(c \rightarrow P) \parallel (d \rightarrow Q) \equiv Stop$

The commutativity and associativity rules tell that the order in which the concurrent processes are executed does not matter. Rule 3 states if one of the parallel processes is deadlocked, then it can deadlock the whole system. However, if a process is running, then it does not alter the outcome of other process *P*. Rule 4 states two parallel processes are simultaneously engaged if the first action of the two processes is the same. Two processes with the same alphabet get deadlocked if their first actions differ.

These rules have been generalized for the processes if their alphabets are different. Given two processes *P* and *Q* their alphabets αP and αQ are not equal, and $\alpha\ (P \parallel Q) = \alpha P \cup \alpha Q$. Let us assume that $a \in (\alpha P - \alpha Q)$, $b \in (\alpha Q - \alpha P)$, and $\{c, d\} \subseteq (\alpha P \cap \alpha Q)$. Then the rules (1)–(4) would be the same, because the rule for simultaneous engagement is the same as the initial events for two processes are the same. However, two additional rules will be added to take care of events in the alphabets that are not shared. These rules state that if the events are not common to the alphabets of the process, then there is no need for simultaneous engagement. The symbol "|" in Rule 6 denotes logical-OR.

Rule 5a: $(a \rightarrow P) \parallel (c \rightarrow Q) \equiv (a \rightarrow (P \parallel (c \rightarrow Q))$

Rule 5b: $(c \rightarrow P) \parallel (b \rightarrow Q) \equiv (b \rightarrow (Q \parallel (c \rightarrow P))$

Rule 6: $(a \rightarrow P) \parallel (b \rightarrow Q) \equiv (a \rightarrow (P \parallel (b \rightarrow Q)) \mid (b \rightarrow (Q \parallel (a \rightarrow P))$

8.5.1.1 Sequential Composition Rules

The rules can also be extended for sequential composition of processes. Rule 7 shows the property of associativity in sequential composition. Rule 8 states that if the first process does not perform any computation, then the sequential execution is equivalent to running the other process alone. Rule 9 states if the first process is "abort" then the following process is not executed. The Rule 10 states that if the process is executed after the event "a" followed by the process *Q*, then it is equivalent to say that processes *P* and *Q* are executed only after the event "a" has occurred.

Rule 7: associativity $(P; Q); R \equiv P; (P; R)$

Rule 8: with_unit skip; $P \equiv P$

Rule 9: with_zero Abort; $P \equiv$ Abort

Rule 10: distributivity $(a \rightarrow P); Q \equiv a \rightarrow (P; Q)$

8.5.1.2 Guarded Command Rules

There are three guarded command rules. Rule 11 says that guarded commands are associative, which means they can be grouped in any fashion and tried. Rule 12 shows that changing the order of guarded commands does not alter the outcome. Rule 13 states that sequential process coming after guarded commands can be distributed to execute sequentially after every guarded command.

Rule 11: associativity $(P \square Q) \square R \equiv P \square (Q \square R)$

Rule 12: commutativity $P \square Q \equiv Q \square P$

Rule 13: distributivity $(P \square Q); R \equiv (P; R) \square (Q; R)$

There are some other rules that can be derived using these basic rules; rules about recursion and input and output commands have been omitted here.

8.5.2 Communicating Sequential Process Language

CSP uses these rules of concurrency, sequential composition, and the guarded commands. The statements of CSP are *input statement → output commands*, where *input statement* is analogous to guard, and *output command* is analogous to command part of a guarded command. The input statement can be a declaration, a Boolean condition, or a process with input data. The output command can be skip, assignment, alternative command, parallel command, iterative command, or a process that outputs data. The syntax for the input process supplying the data to the output command is of the form *<process>?<input-data>*, where the symbol "?" denotes that input data follows. Similarly, the syntax for the output process is of the form *<process>!<output-data>*. The output process stops when all the input processes terminate or no guard condition is true. Termination of an input process in the guard part is treated as failure of the guard part.

The language supports single-data entities, arrays, structured data, array of processes, assignment, alternative commands, iterative commands, parallel commands, recursive processes, and subroutines. The semantics of iterative loop is that all the input processes must terminate for the output processes to terminate. The iterative loop will continue till the time there is a single-input process that is supplying data to the output command. The abstract syntax for the guarded commands is the same as the guarded commands discussed in Chapter 4, with some syntactic differences. The process-array syntax has been omitted for simplicity. A simplified abstract syntax (in extended Backus-Naur form) for the parallel commands is given below

```
<parallel-command> ::= '['(<process> {|| <process>}*']'
<process> ::= <identifier> {(<command> | <declaration>}+
<command> ::= skip | <assignment> | <input-statement> |
                <output-command> |<alternative-command> |
                <iterative-command> | <parallel-command>
<assignment> ::= <variable> = <expression>
<input-statement> ::= <process-name>?<variable>
<output-command> ::= <process-name>!<variable>
<iterative-command> ::= '*'<alternative-command>
<alternative-command> ::= '['<guarded-command>
                             {'□'  <guarded-command>}* ']'
```

Example 8.14

The following example shows a producer–consumer process on a buffer using CSP syntax. After a "producer" process gets an input character, the character is inserted in the buffer location pointed out by the front-pointer, and the pointer value is incremented by one using modulo arithmetic to handle the circular queue; the count is incremented by one; and the buffer is marked "not empty." The delimiter ';' shows sequentiality, and the symbol '|' separates the guarded commands. The symbol '*' in the beginning denotes that the command is an iterative construct and will terminate only when all the guards fail and the input processes terminate. The symbol '∧' denotes logical-AND. Because all the conditions are covered in the guards and at least one condition is always true, the iteration will continue forever.

```
iobuffer::
        m = 80; buffer(0..m - 1) character; c character;
        rear, front: integer; rear = 0; front = 0; count = 0;
        full, empty: Boolean; full = false; empty = true;
      * [         count == m - 1 → full = true ▯
                   count == 0 → empty = true ▯
                   not(full); producer?c →
                     % read the value of c from the producer
                            buffer[front] = c;
                            front = (front + 1) mod m;
                            count = count + 1;
                            empty = false ▯
                   not(empty) →   consumer ! buffer(rear);
                     % consumer removes the element
                            rear = (rear + 1) mod m;
                            count = count - 1;
                            full = false
        ]
```

8.6 MEMORY MODELS FOR CONCURRENCY

Before we develop concurrency constructs in programming languages, the memory models about how the memory behaves under different concurrent programming abstractions should be clearly specified for the following reasons:

1. The concurrency behavior of a program must be explained with respect to a sound memory model that explains how the shared resources are managed abstractly in the language.
2. Compiler optimizer reorders the statements, as they would do in a single-thread program, for optimization, which may violate the sequential consistency in multi-threaded programs.
3. The memory model should facilitate execution that is free of data race and supports sequential consistency. The model should not cause unnecessary sequentiality while supporting causality and safety.
4. In the absence of safe models, Internet languages such as Java and .NET languages such as C# may introduce loopholes that may be exploited by the attackers.

The earlier memory models were flawed, and most of the languages, such as C++, do not have appropriate memory model, and use external thread libraries. In recent years, memory models are being refined as more and more concurrent programs are being written to exploit the available multicore processing. A memory model should be simple and robust; otherwise, program development becomes error prone, and software maintenance becomes difficult. There are some synchronization properties that should be followed to maintain sequential consistency. A program should (1) be data-race free; (2) follow a synchronization order that is consistent with the program order; (3) a lock (or release) action should be followed by the release (or lock) action on the shared resource; (4) once a lock has been placed on a shared resource, no process should be able to violate it, unless the resource is released; (5) a read operation reads the last written value; and (6) a write operation cannot be performed unless all the read operations on the previous write in the program order have been executed. Rule 5 is the producer–consumer relationship. Rule 6 is the antidependency property. Rule 4 is a difficult one to maintain, because most of the languages except Ada use synchronized and unsynchronized methods: synchronized methods follow the restrictions, but unsynchronized methods may write on the shared resources, causing errors in program behavior.

Multithreaded models of programming languages have to treat other variables differently than the locks and shared resources, because these variables follow the synchronization order consistent with program order and cannot be rearranged by the optimizing compilers, which may result in the violation of one of the above memory model rules, resulting into the violation of the fundamental restriction of sequential consistency. The latest memory model (after Java 5) uses the declaration *volatile variable*. Volatile variables can not be touched or reordered by the compilers to maintain the program order.

One of the problems in multithreaded programs is the use of shared variables that are updated in one thread and consumed in another. Although a data-dependency analysis within a single thread may not show a sequential dependency, if both the threads are analyzed together, the dependency becomes clear (see Figure 8.1). Many times the use of the shared variables in two threads gives rise to cyclic reasoning and breaks up the sequential consistency. In such cases, a global analysis of the threads has to be done to break up the cycles. Such an analysis should preserve *causality of actions* even if it may seem to violate the property of sequential consistency. The reason is that sequential consistency is used to enforce causality and it is the causality of actions that is the most important property to be preserved.

A thread may be blocked to perform input, output, interaction with an external event, or because of an operating system action. Owing to synchronization order, all other threads that are waiting for *locked resource* or *volatile variables* will also be blocked indefinitely, unless the resources are released.

One of the limitations of the Java memory model is that Java threads have no limit on how long they are going to hold a CPU, unless they are forced by the process scheduler of the operating system. Another problem with concurrent execution is the cyclic reasoning in multiple threads that may be difficult to reason simply by global dependency analysis, and compiler transformation may yield incorrect behavior inconsistent with causality restriction.

8.6.1 Memory Model of C++

Current executions of C++ use a single-thread memory model for C++ interleaved with multithread memory model of the libraries such as "Pthread" and "Microsoft word thread library." Single-thread execution without any formal interface definitions with multithread libraries does not prevent compilers from performing code reorganization such as register optimization or code reordering, that may violate the fundamental sequential consistency property in concurrent programs. Multiple-thread libraries use lock primitives to restrict access to normal variables. Currently, there are efforts to develop a similar cohesive memory model for C++, based on the lines of Java memory model. The basic approach is to define what are atomic operations and how to prevent data race by compiler actions. The model is still evolving.

8.7 CONCURRENT PROGRAMMING CONSTRUCTS

Concurrent programming constructs are divided into the following categories: (1) constructs for data parallel programming, (2) constructs for parallel spawning of subtasks, (3) constructs for spawning multiple threads, (4) synchronization constructs to handle shared resources, and (5) constructs for invoking remote procedures. This section describes generic constructs present in various concurrent programming languages.

8.7.1 Coroutines

Coroutines are used to execute two processes alternately, such that control switches between two processes. Each process saves its state before passing the control to another process. When signaled by the other process to resume, it resumes from the last suspended state. The control keeps switching between two or more processes, such that each process resumes execution where it left off in the past. Coroutines are much faster than process and thread, because there is no operating system involvement. Coroutines have been implemented in Simula, Modula-2, Ruby, Lua, and Go. They can be used to implement threads and iterators. Coroutines can also be used to model discrete event simulation, where multiple events are being generated by separate threads such that they interact with each other.

The coroutines in Ruby are implemented using *fibers*. Fibers are code blocks that can be paused and resumed later. Fibers are not preempted, and unlike threads, the scheduling has to be provided by the programmer inside a program. Syntax for coroutines in Ruby using fibers is given below:

```
fact = Fiber.new
     m, n = 1, 1
     loop do
            Fiber.yield n
            m = m + 1
            n = n * m
     end
end
5.times {puts fact.resume}
```

There are two coroutines: function *fact* that generates a factorial value, and the external loop that prints out the value of the factorial. The external loop starts the coroutines fact, which returns the next value through the call *Fiber.yield* and suspends itself. The control is passed back to the external loop, which resumes the fact coroutines again using the method *fact.resume*. The coroutine *fact* increments the value of the variable m, generates a new value of $n = n * m$, and returns the new value of the variable n again using the method *Fiber.yield n*. The process continues until the external loop is executed five times. The generated values are 1, 2, 6, 24, and 120.

8.7.2 Cobegin–Coend

The construct *Cobegin-Coend* is used in the programming languages Concurrent Pascal, Algol-68, and C. It is also used in the Internet language SMIL. SMIL can display multiple media streams concurrently using parallel spawning of subtasks. These subtasks do not share any variable. After all the subtasks are over, the main control flow resumes. A generic Concurrent Pascal syntax is as follows:

```
Cobegin
        <statement-block1>;
        <statement-block2>;
        …,
        <statement-blockN>
Coend
```

The following code illustrates a simple example of cobegin–coend.

```
x : = 0; z : = 4;
cobegin
      begin x := 1; x := x + 1 end; % concurrent activity 1
      begin z := 2; z := z - 1 end; % concurrent activity 2
Coend;
print(x, z)
```

The syntax for parallel spawning of subtasks in SMIL is given below:

```
<par>
     <text src = "my_resume.html" region = "text_area"
        dur = "60s"/>
     <video src = "my_presentation.mpg" region = "Video_area"/>
</par>
```

The above code starts two streams: a text stream that is displayed for 60 seconds, and a video presentation that starts concurrently along with the text.

8.7.3 Fork-and-Join

Fork-and-join operations spawn multiple subtasks that execute concurrently and independently. The parent process is suspended until all the spawned threads have terminated successfully. After all the threads terminate, the parent process is signaled, and it resumes its execution again. Languages such as Ada, Java, JRuby (a variation of the language Ruby), C++, C#, Modula-3, and Scala support multiple-thread based programming. Although Ada uses the notion of a task described in Section 8.8.1, other languages use "spawning of threads." A thread can be created, linked to a runnable procedure, started, run concurrently with other threads, while providing mutual exclusion on shared variables and joining with other threads. It can be suspended using the "wait" command, or sleep for specified time. A suspended thread can be awoken using a "notify signal." After a signal is broadcast, then all the threads waiting for that signal become eligible for the resumption of activities. Some of the thread-based operations that various languages support are new **thread**(<*thread-name*>), <*thread-name*>.**start**() and <*thread-name*>.**join**().

8.7.3.1 Synchronization Constructs for Shared Resources

There are two major constructs to handle shared resources: *locks* and *monitors*. Object-oriented languages like Java use *synchronized methods* to access the shared resources. Locks (or semaphores) are low-level constructs that are associated with shared objects. Monitors are associated with a group of mutually exclusive procedures. Both locks and monitors are declared, along with other declarations. Both locks and monitors allow one process at a time to access the shared resource. There should be total synchronization order that is imposed on lock-release operation: *lock(v)* must be followed by *release(v)* to avoid indefinite blocking of threads or processes; sequences like lock(v), ... lock(v) or release(v), ... release(v) are not allowed. However, it is very difficult for programmers to follow the low-level restrictions on locks.

8.7.4 Monitors

Monitors are high-level passive declarations in languages, which support mutual exclusion of procedures described within the monitor. The embedded entities in a monitor are shared resources and mutually exclusive procedures that work on shared resources in a mutually exclusive manner. A monitor also has an initial body that is executed when a monitor is called. A monitor can access its own variables. However, other processes can read only the public variables of the monitor. Any process that needs to access the shared resource has to use a monitor and the corresponding procedures to access the shared resource; there is no direct access to the shared resource. The general abstract syntax for a monitor is as follows:

```
type <identifier> =
monitor (<parameter-list>)
<shared variable declarations>
procedure <identifier> <procedure-body>
procedure <identifier> <procedure-body>
...
```

```
procedure <identifier> <procedure-body>
<initial-body of the monitor>
```

System routines ensure that only one process corresponding to one of the monitor procedure can access a shared resource at a particular instance of time. If a process is using a shared variable, then all other processes requesting to access the same shared resources have to wait until the currently executing mutually exclusive procedure within the monitor releases the shared resources. The waiting can be done in two ways: *spinlock* or by *suspension of the process.* In *spinlock* a process keeps looping in a "busy-wait loop" and checking for the availability of the shared resource. In the *suspension mode*, the process requesting to use the shared resource gets blocked until the shared resource is released by the currently executing procedure. Upon release of the shared resource, all the blocked processes are notified and made ready to be rescheduled by the CPU scheduler.

Example 8.15

An example of the use of a monitor is the buffer management problem, where one process writes in a buffer and other processes read from the buffer. An intuitive code has been given in Figure 8.13. The shared buffer *buffer[0..m-1]*, the rear pointer *rear*, and the front pointer *front* are shared resources. There are two procedures in the monitor:

```
monitor iobuffer;
m = 80;
character buffer[0..m - 1];
integer rear, front, count;
condition nonempty, nonfull;
procedure insert(character element)
{  if (count == m - 1) then await(nonfull);
       buffer[rear] = element;
       rear = (rear + 1) modulo  m;
       count = count + 1;
       signal(nonempty)
}
procedure remove (char element)
{   if (count == 0) then await(nonempty)
          result = buffer[front];
          front = (front + 1) modulo n;
          count = count - 1;
          signal(nonfull)
}
% initial body of the monitor
{ rear = 0; front = 0; count = 0;
nonempty = false; nonfull = true }
...
iobuffer  console;
...
{
...
console.insert('a');
...
}
```

FIGURE 8.13 An example of a monitor.

write-buffer and *read-buffer*. Operations on shared resources are atomic, and the shared resources are not released until the operations are complete. The procedures *insert* and *remove* are mutually exclusive.

Any process can access the buffer only through the use of the user-defined type *iobuffer*. Different periphery buffers can be declared of the type iobuffer. There is also notion of *condition-variables*, which are Boolean variables and can be set to true by *signal* operation. A signal can be read by the corresponding suspended processes and is reset after it has been read by the *await* operation. The await operation waits for the condition to become *true*, and turns the condition *false* again before entering in the critical section. The *false* condition prohibits other processes to enter the critical region at the same time.

The process of buffer update is the same as described in Section 8.2.5 except now we have a notion of transparently delaying the insert and removing operations instead of simply terminating, as was the case in the single-thread description in Section 8.2.5.

8.8 CASE STUDY

In this section, we discuss various constructs used in three representative programming languages Ada, Java, and Emerald. Both Ada and Java have built-in support for concurrency.

8.8.1 Concurrent Programming in Ada

Ada supports concurrent programming using *tasks*. *Tasks* are equivalent to *Java threads*. Tasks are embedded inside a master program. When a master program is invoked, then the embedded tasks are also invoked. The master program does not terminate until all the tasks terminate. Task is like a module that exports the declared *entry points* to communicate with other tasks. For every entry point, there is a corresponding *Accept* statement inside the statement block of the corresponding *task body* that pulls the information into the task body. A pair of "ENTRY-POINT" and "ACCEPT" is used for parameter passing from other tasks. Parameters are of the form *<formal-parameter-name>*: *<formal-parameter-type>*. All the local variables and statements are inside the task body. The general syntax for tasks is given below.

```
Procedure <proc_identifier>
task <task_name1> is <entry_points1> end <task_name1>
task body <task_name1> is <block1> end <task_name1>
...
task <task_nameN> > is <entry-pointN> end <task_nameN>
task body <task_nameN> is <blockN> end <task_nameN>
begin null end <proc_identifier>
```

Example 8.16

The program given below gives an example of the use of the tasks and the use of *entry point* and *accept* feature. The main body of the procedure has a loop that spawns the task assignment with a different index value. Tasks are spawned like objects with the *Problem_index* being passed as parameter. The task *SolveProblem* puts a delay of 240 time-units and terminates. The main body terminates when all the tasks terminate.

```
WITH Ada.Text_IO;
USE Ada.Text_IO;
PROCEDURE Assignment IS
  TASK SolveProblem IS
  ENTRY start_thinking(Problem_index: INTEGER); END SolveProblem;
  TASK BODY SolveProblem IS

BEGIN
      ACCEPT start_thinking(Problem_index: INTEGER) DO
       delay 240.0; -- delay to solve a problem
       Put_Line("write answer");
      END start_thinking;
END SolveProblem;

BEGIN
  FOR Index IN 1..5 LOOP
   SolveProblem.start_thinking(Index)
   Put_Line("Solving the next problem");
  END loop;
  Put_Line("Assignment done");
END Assignment
```

Tasks can also be declared as type declaration, and later multiple-task objects having the same template can be created.

```
task type my_type is … end my_type;
task body my_type is … end my_type;
…
Task_1, Task_2 : my_type;
```

8.8.2 Concurrent Programming in Java

Java uses a combination of *thread primitives* and *synchronized methods* to handle concurrency. A thread is created using the construct **new Thread**(<*thread-name*>). There are many primitive thread operations such as <*thread-name*>.**start** (), <*thread-name*>.**yield** (), <*thread-name*>.**sleep**(<*duration*>), and <*thread-name*>.**setPriority** (<*priorityValue*>). The names are intuitive. In addition, threads can raise an interrupt to wake up the blocked threads and exit after waiting for a predefined sleep time.

Example 8.17

The following code gives a simple class definition in Java to start two threads that execute the method *run*. The main thread creates two new threads—$Thread_1$ and $Thread_2$. Thread_1 and Thread_2 start executing using the commands Thread_1.**start** ()

and $Thread_2$. **start**, and the main thread waits for 1000 milliseconds. By that time, two threads have terminated, and the program terminates. The main thread can communicate to other threads using static variables, and set the conditions for threads to stop if needed. The threads can also be made to yield to higher priority threads by setting up the priority using the primitive command $Thread_1$.**setPriority**( ) or $Thread_2$.**setPriority**( ) and using the primitive operations $Thread_1$.**yield** or $Thread_2$.**yield** primitive.

```
class myThread extends java.lang.Thread {
  public static void main(String args[]) throws
     InterruptedException {
  myThread thread1 = new myThread(); //Create Thread 1
  myThread thread2 = new myThread(); //Create Thread 2
  thread1.start();    //start Thread 1
  thread2.start();    //start Thread 2
  Thread.sleep(1000); //sleep for one second
}
//Created threads execute this method
public void run() {System.out.println("Created thread");}
```

All the methods that use a shared variable have to be declared as *synchronized methods* to provide mutual-exclusion among the threads that may access the shared object through the use of any of the methods. The shared objects are accessed by only one of the synchronized methods providing mutual exclusion. Synchronized methods are used only for shared objects that are visible to other threads. However, if an instance of a class is confined, then there is no need to synchronize.

Example 8.18

The following code gives an example and syntax for the use of a Java synchronized method that handles the shared class-variable *value*. Only one synchronized method can access the shared variable. Since the variable *value* cannot be read while it is being updated and vice-versa, the two methods *getValue* and *setValue* have to be synchronized.

```
class Cell {
          protected int value;
          public Cell(int v) { value = v; }
          public synchronized int getValue() { return value; }
          public synchronized void setValue(int v) { value = v;}
```

There are some problems with the mutual exclusion realized using synchronized methods as follows:

1. Locks cannot be set in one method and released in another.
2. Locks are not at the variable level but at the method level.

3. An unsynchronized method can also update the shared variables. Thus, synchronized method declaration should be done carefully.

4. Synchronized method causes sequentiality and thus slows down the execution.

8.8.3 Distributed Computing in Emerald

Emerald is a distributed object-based language. Emerald supports *object mobility* that includes both "methods" as well as "data." Emerald does not support class hierarchy. An Emerald object has four components: (1) *unique network id* that can be generated by concatenating *host-name, process-name,* and *local identifier*; (2) data representation local to the object; (3) methods working on the local data; and (4) an optional process that may invoke other objects. Emerald has three types of objects: (1) *global objects* that can be moved directly to any node; (2) *local objects* that are embedded within another object and cannot be moved directly; and (3) *direct objects,* such as basic types that are used to build other objects. Arrays, records, and single entities all are objects in Emerald.

Emerald supports *fine grained mobility.* There are many advantages of process migration in Emerald: (1) it enhances load balancing, (2) active objects can be moved to remote processors for better performance, (3) objects' mobility facilitates robustness against processor failure, and (4) it facilitates the utilization of special software available at different processors. Emerald supports abstract data type that can be implemented differently on different processors. Emerald uses five operations for object mobility. The operations are (1) locate an object; (2) move an object; (3) fix an object at a particular node; (4) unfix an object to make it mobile again; and (5) refix an object that is a combination of unfix, move, and fix at a node.

Emerald supports *call-by-object-reference, call-by visit,* and *call-by move.* The advantage of object-reference is that a remote object's reference can be passed to distributed nodes easily. Arguments are moved if they can be copied easily without much overhead. As we discussed earlier, within the same address space, an object can be addressed directly without intervention of the operating system. However, remote objects can be accessed only by going through the operating system.

8.9 SUMMARY

Concurrent programming is concerned about dividing a task into multiple subtasks that execute concurrently. Concurrency can be exploited using task parallelism, data parallelism, or an integration of task and data parallelism. Data parallelism is concerned about executing the same operation on a large data set. Task parallelism is concerned about spawning multiple processes and threads that work on different data sets. There have been approaches to integrate data and task parallelism, so that complex problems that need both data and task parallelism can exploit maximum parallelism. The integration has taken following approaches: (1) spawning multiple-data parallel tasks concurrently, (2) distributing data structures across processors to exploit data-parallel computation on distributed data structures, and (3) remote access to data structures on other processors.

Two subtasks can be executed concurrently if they are independent of each other. There are many ways to exploit concurrent execution of subtasks (1) using multiple threads that share information using a channel or a shared variable synchronously or asynchronously. The presence of shared resources cause sequentiality. Concurrent execution of the subtasks has to follow the sequential consistency criteria: a program should produce consistently the same outcome as the sequential execution. In order to maintain sequential consistency during transforming a sequential program to its concurrent counterpart, three rules have to be followed: (1) producer–consumer relationship, (2) antidependency, and (3) output dependency. Producer–consumer relationship states that a value cannot be consumed unless it has been produced. Antidependency rule states that the statements consuming of the old values of a shared variable have to be executed before the statement producing a new value of the shared variable can be executed. The third rule states that an order should be maintained to update the aliased variables, because they update the same memory location. Not following these rules will cause "race condition," which means the program may generate inconsistent values each time it is executed.

Concurrency can be exploited in programs either by automatically parallelizing a sequential program or by explicitly writing multiple thread-based programs using concurrent constructs and managing the shared variables, so that only one process or thread can write into the shared resource at a time. Both the approaches have been used.

For the automatic parallelization of programs, dependency between the program statements has to be identified. Theses dependencies can be handled in two ways: (1) transforming programs to reduce the dependencies and (2) incorporating sequentiality to maintain the sequential consistency. A program has two major types of dependencies: data dependency and control dependency. Data dependency is present because of the data flow due to shared variables between the statements, and control dependency is present because of the evaluation of predicates in control abstractions such as if-then-else statement, case-statements, while-loops, and do-while loops. By analyzing data and control dependencies, a program dependency graph can be derived that is a combination of a data-dependency graph and control-dependency graph. Iterative loops can be transformed by loop-unrolling to reduce the dependencies introduced by control abstractions.

A naive way is to exploit fine-grain concurrency by mapping the program at statement level on different processors. The exploitation of fine-grain concurrency introduces excessive communication and data serialization-deserialization overheads between the processors or between distributed memories attached to various processors. This communication overhead is significant in a distributed environment. In order to circumvent the inefficiency introduced because of this overhead in distributed computing or bandwidth congestion overhead on multiprocessor machine, a large number of dependent statements are executed sequentially on the same processor. This form of concurrency is called *coarse-grain concurrency*. In order to reduce the data-transfer overhead between the processors or between the memories associated with different processors, *program slicing* is used. Program slicing reduces the data-transfer overhead by (1) replicating the producers on multiple processors and (2) grouping the block of statements if it reduces the data transfer overhead.

Shared resources can be allocated to only one process at a time. This enforces mutual exclusion between the processes. Mutual exclusion can be forced either using a *lock* or a *monitor.* A lock has two states: *occupied* and *released.* In the occupied state, the shared resource is unavailable to other processes. The portion of the code where a process is in the possession of the lock is called *critical section.* The critical section introduces execution time overhead because of sequential execution and should be as small as possible.

A process changes the state from *released* to *occupied* before entering the critical section to manage the shared resource. However, testing and setting the lock is an *atomic operation.* Unless the lock is released the shared resource is under exclusive control of the process that set the lock-state as *occupied.* All the operations within critical sections are treated as an atomic operations to maintain consistency of results. If the operations are broken because of high priority interrupt, then the changes have to be undone, and the atomic operation has to be redone.

Owing to low-level lock and signal operations, any mistake may cause deadlock of processes, causing the shared resources to be blocked indefinitely. Monitor is a high-level abstraction to ensure mutual exclusion of processes. Processes can access the shared resource only through the monitor. Monitors are declared, like any other declaration, and contain the shared resource and the associated processes. Processes can either wait for the shared resource by blocking themselves while waiting, or they can signal the release of the shared resource by waking up other blocked processes.

Explicit concurrent programming is achieved by providing language constructs or interfacing with operating systems library to spawn multiple threads that do various subtasks concurrently while sharing the address space with the parent process for information sharing. The parent process suspends while threads are executing concurrently and resumes when last of the spawned thread has successfully terminated.

Distributed computing involves *remote procedure calls (RPC)* and *remote method invocations (RMI)* where the called subprogram or the method is executed on a remote processor with different address space. In distributed computing code, data or the whole object can be migrated to remote processors to exploit concurrent execution. The major problems in distributed computing are the multiplicity of the address spaces, failure caused by communication breakdown, overheads due to the use of system utilities for communication and the overhead of packing and unpacking of data. Information exchange between the calling program and remote called procedure involves operating system, setting up the channel between the two processors, and packing and unpacking of the data between the client and the server. This causes significant overhead. There are three major parameter passing mechanisms in remote procedure calls: *call-by-reference*, *call-by-copy*, and *call-by-copy-restore.* They are analogous to call-by-reference, call-by-value, and call-by-value-result in uniprocessor languages. Call-by-copy copies the object to the remote processor that makes future access in the same address space. Call-by-copy-restore copies back the result to the address space of the calling procedure after the remote procedure call is over. The advantage of call-by-reference is that it can be easily passed to other processors to access the remote objects. However, the major disadvantages of call-by-reference are: loss of the object in case of communication failure and delay caused by communication to access the object as it is not in the same address space as the remote procedure call.

Communicating sequential processes is a theoretical framework that forms the basis of concurrent programming languages. It has algebra for concurrent programming. The algebra states the rules for "associativity," "commutativity," "deadlock," "abortion" "agreement," and "disagreement" of guards. CSP languages are based upon stating the input statements and output commands. Input statements are analogous to guards, and output commands are analogous to the command part of guarded commands. A parallel command is a conjunction of statements of the form *input statement → output command.* Input statement can include declaration of the data types, input processes that provide the data, and Boolean expressions.

Programming languages that support multithreaded programming should have a sound memory model to ascertain sequential consistency. Sequential consistency is based upon the causality of action that should be preserved. In order to preserve causality, a program should be (1) data-race free, (2) should follow synchronization order that is consistent with the program order, (3) a lock action on a shared resource should be followed by the release action on the same shared resource, (4) a read operation reads the last written value, and (5) a write operation cannot be performed unless all the read operations on the previous write in the program order have been executed. In recent years, a robust memory model has been developed for Java, and a similar model is being developed for C++.

Programming languages support many abstractions (both data and control abstractions) for concurrent programming such as coroutines, parallel iterative loops, semaphores, monitors, thread based constructs, synchronized methods, subtasks, and remote procedure calls. Coroutines spawn two or more routines that keep passing the control to each other, and suspend the control and computational state of other routines when one of them is executing. Programming languages provide low-level lock mechanism, or higher-level monitor mechanism to provide mutual exclusion of the processing working on a shared resource. Object-oriented programming such as Java provides mutual exclusion using synchronized methods that update the shared object. Any process has to use the synchronized method for accessing the shared object.

There are multiple popular programming languages that support extensive integration with concurrency. Notable languages are Ada, C, C++, Emerald, parallel variants of Fortran, Java, Concurrent Pascal, parallel versions of Haskell, Ruby, Lua, Emerald, Scala, web programming language SMIL, multiple logic programming languages, and languages for massive parallel computing such as HPF, Chapel, and X10. Concurrency in functional programming languages, such as Lisp, Haskell, Ruby, and Scala, is discussed in Chapter 9. The discussion on concurrency in logic programming languages is given in Chapter 10. High-productivity massive parallel languages such as X10 and Chapel are discussed in Chapter 13.

Ada uses a high-level structure *task* to spawn concurrent tasks. Multiple tasks can be embedded in the body of a main program. A task uses *entry-point* and *accept* combination for information exchange. The main program with embedded tasks terminates only after all the tasks terminate.

Java uses multiple threads and synchronized methods to achieve concurrency and mutual exclusion of the shared objects. All the methods that access the shared objects have to be declared as synchronized methods to provide mutual exclusion among the methods. There are some limitations of providing mutual exclusion at the method level: (1) locks are associated not with a variable but a method, (2) a variable can still be updated by an unsynchronized method, (3) lock cannot be set in one method and released in another method, and (4) use of synchronized method slows down the execution by introducing sequentiality in the execution of synchronized methods.

8.10 ASSESSMENT

8.10.1 Concepts and Definitions

Alphabet of process, antidependence; atomicity; atomic operation; cal-by-copying; call-by-copy restore; causality; coarse-grain concurrency; code migration; CSP; CSP algebra; cobegin-coend; control dependence; control-dependency graph; coroutines, critical section; data dependence; data-dependency graph; data migration; data-parallel for-loop; data parallelism; data-transfer overhead; deadlock; delinearization; distributed computing; domination; fiber; fine-grain concurrency; fork-and-join; granularity; input source; interdependencies; linearization; lock; loop lifting; loop-unrolling; marshaling; mailbox; memory model; message passing; monitor; multithreading; mutual exclusion; object migration; output dependency; output source; packing; pipelining; postdomination; process; PDG; producer–consumer relationship; program slicing; race condition; remote procedure cal; sequential composition; sequential consistency; sequentiality; semaphore; shared resource; starvation; subtask; synchronization; synchronized method; task; task parallelism; threads; transaction-abort; transaction commitment; transactional memory; unmarshalling; unpacking; volatile variable.

8.10.2 Problem Solving

1. Draw a data-dependency graph for the following program, and perform statement level processor allocation under the assumption that there is no overhead of transferring data between two processors. Also calculate the unit time taken to execute the code assuming that the execution of each statement takes one unit time. Also assume that variables X and W are aliases.

```
X = 4; Y = 5; Z = X + 5; W = Y + 9; A = Y + W; B = X + Y;
   C = 2 * Z; Y = B + C;
```

2. Make a data-dependency graph for the following program. Show minimum time taken to execute the program, assuming a time delay of 5 microseconds (5 milliseconds) data transfer and packing/unpacking cost when the data is transferred from one

processor to the second processor. Also assume that the time of execution of an instruction in the same processor is 100 nanoseconds. Note: 1000 nanoseconds are equal to one microsecond.

```
X = 4; Y = 9; Z = X + Y; M = Y + 5; N = X + 7; W = Z + M;
  Y = 2 * W;
```

3. Solve the data-dependency graph with maximum fine-grain concurrency. Show the processor allocation and the minimum time taken to execute. Assume that the data transfer cost (including packing and unpacking cost) between two processors is 50 nanoseconds, and the cost of executing a statement is 2 nanoseconds. Also assume that variables *X* and *M* are aliases.

```
X = 4; Y = 5; M = X + 4; W = 2 * Y; N = M + Y; U = 2 * M +
  4; X = U;
```

4. Perform program slicing by replicating the producers and grouping statements to reduce the data transfer overhead and improve the execution efficiency of the fine-grain parallelism for problem 2 above.
5. Write a Java program for producer–consumer operation using synchronized method on a buffer that reads from a buffer of 80 characters.
6. Write a CSP program that filters the negative integers from a list of integers.
7. Write a merge-sort program using CSP, a thread-based model in Java, and Ada's tasking, and compare.

8.10.3 Extended Response

8. What do you understand by program slicing? How can program slicing be used to improve the execution time? Explain using an example.
9. Explain the difference between task parallelism and data parallelism using an example.
10. What are semaphores (locks), and how are they used in concurrent programming with shared resources? What are the issues with semaphores?
11. What are monitors, and how are they different from semaphores (locks)? How are they used to solve the problems with shared resources? Explain.
12. What are remote procedure calls? Explain. How is information exchanged between the calling subprogram and the called remote procedure? Compare various parameter passing mechanisms for remote procedure calls.
13. Discuss the design of an integrated language that supports both data and task parallelism.

14. Discuss the algebra for communicating sequential processes.

15. Discuss the various programming constructs in CSP language.

16. What do you understand by sequential consistency? Explain. What role does sequential consistency play in the multithread-based execution? Explain.

17. What is race condition, and how is it related to sequential consistency? Explain using a simple example.

18. Read about Java memory model for concurrency using articles suggested in Further Readings, and discuss it in detail.

19. Read about the actor-based message passing model; channel-based CSP model, and thread-based model using Internet and articles suggested in the bibliography below, and compare.

20. Read more about Java threads and Ada's tasking using Internet-based literature, and compare Ada's tasking and Java's thread-based concurrency.

FURTHER READING

Agrawal, Hiralal. "On slicing program with jump statements." In *Proceedings of the ACM SIGPLAN '94 Conference on Programming Language Design and Implementation*. 1994. 302–312.

Andrews, Greg. *Foundations of Multithreaded, Parallel, and Distributed Programming*. Reading, MA: Addison-Wesley. 2000.

Ben-Ari, Mordechai. *Principle of Concurrent and Distributed Programming,* 2nd edition. Addison Wesley. 2006.

Benton, Nick, Cardelli, Luca, and Fournet, Cedric. "Modern concurrency abstractions for *C#*." *ACM Transactions of Programming Languages and Systems*, 26(5). 2004. 269–804.

Birrell, Andrew D. *An Introduction to Programming with C# Threads*, Revised 2005. Available at http://research.microsoft.com/pubs/70177/tr-2005-68.pdf

Birrell, Andrew D. and Nelson, Bruce J. "Implementing remote procedure calls." *ACM Transaction of Computer Systems*. 2(1). 1984. 39–59.

Boehm, Hans-J and Adve, Sarita V. "Foundations of the C++ concurrency memory model." In *Proceedings of 2008 ACM Conference on Programming Language Design and Implementation.* 2008. 1–12.

Burns, Alan and Wellings, Andy. *Concurrent and Real Time Programming in Ada.* Cambridge, UK: Cambridge University Press. 2007.

Cytron, Ron, Ferrante, Jeanne, and Sarkar, Vivek. "Compact representations for control dependence." In *Proceedings of the ACM SIGPLAN'90 Conference on Programming Language Design and Implementation. 1990*. 337–351.

Hansen, Per B. "Monitors and concurrent pascal—A personal history." In *Proceeding HOPL-II: The Second ACM SIGPLAN Conference on History of Programming Languages*. 1993. 1–35.

Harris, Tim, Cristal, Adrian, Unsal, Osman S., Ayguade, Eduard, Gagliardi, Fabrizio, Sm*ith, Burton, and Valero, Marteo.* "Transactional memory: An overview." *IEEE Micro*. 27(3). 2007. 8–29.

Hoare, Charles. A. R. "Communicating sequential processes." *Communications of the ACM*, 21(8). 1978. 666–677.

Israd, Michael, Budiu, Mihai, Yu, Yuan, Birrell, Andrew, and Fetterly, Dennis. "Dryad: Distributed data-parallel programs from sequential building blocks." In *Proceedings of Eurosys Conference.* 2007. 59–72.

Lea, Doug. *Concurrent Programming in Java—Design Principles and Patterns*, 2nd edition. Addison-Wesley, MA: Prentice Hall. 1999.

Lee, Edward A. "Problem with threads." *Computer,* 39(5). 2006. 33–42.

Lewis, Bill and Berg, Daniel J. *Threads Primer—A Guide to Multithreaded Programming.* Addison-*Wesley, MA: Prentice Hall. 1996.*

Manson, Jeremy, Pugh, William, and Adve, Sarita V. "The Java memory model." In *Proceedings of the Principles of Programming Languages.* 2005. 378–*391.*

Sottile, Matthew J., Mattson, Timothy G., and Rasmussen, Craig E. *Introduction to Concurrency in Programming Languages.* Boca Raton, FL: Chapman *and Hall/CRC Press. 2009.*

Tanenbaum, Andrew and Steen, Marten van. *Distributed Systems—Principles and Paradigms,* 2nd edition. Boston, MA: Prentice Hall. 2007.

Tip, Frank. "A survey of program slicing techniques." *Journal of Programming Languages,* 3(3). 1995. 121–189.

Wiser, Mark. "Program slicing." *IEEE Transactions on Software Engineering,* 10(4). 1984. 352–357.

Wolfe, Michael. *Optimizing Supercompilers for Supercomputers.* Research Monograph in Parallel and Distributed *Computing.* The MIT Press. 1990.

CHAPTER 9

Functional Programming Paradigm

BACKGROUND CONCEPTS

Abstract computation and information exchange (Chapter 4); Abstract concepts in computation (Section 2.4); Abstract implementation (Chapter 5); Concurrent programming (Chapter 8); Data structure concepts (Section 2.3), Discrete structure concepts (Section 2.2), von Neumann machine (Section 2.1).

Declarative programming hides the control from the programmer. The resulting program is reflective of the underlying logic and is easier to maintain than imperative programs except for the unfamiliar syntax of some declarative languages. There are two major declarative programming paradigms: functional programming and logic programming. Mathematical functions are the basis of functional programming, and predicate calculus is the basis of the logic programming. Both mathematical models have been studied for decades and are sound. In this chapter, we will study functional programming.

Functional programming is important from many aspects: (1) declarative style hides the control from a programmer; program is only logic + abstraction; (2) functional programs are more concise than imperative programs; (3) functional programming is becoming a key paradigm that is being integrated in recent popular languages including many scripting languages such as *Clojure, Ruby, Scala, and Python*; and (4) functions are an important part of almost all programming languages.

The work on functional programming started as early as 1960s with the development of the language Lisp. Although many scientists disagree with the purity of functional programming in Lisp, its contribution to functional programming is undeniable. Lisp contributed many abstractions such as high-level representation of lists, meta-programming, higher-order functions, and recursive style of programming.

Variables in a pure functional program are immutable *assign-once* value holders. *Assign-once* property of variables has caused many programming challenges, as simple forms of index-variable based iterations are disallowed. This limitation gave rise to a new style of recursion-rich programming in Lisp, and the iterative versions of programs were modeled as tail-recursive programs. Lisp developers were quick to realize the advantages of memory reuse and storing the results of previous computations for execution efficiency, and allowing the notion of global variables to support memorization of the previously computed partial results that can be used at later computational steps. New data abstractions such as iterators were also developed for the application of functions on a sequence of data items.

The mathematical theory of functional programming is deep-rooted in λ-calculus that has been well-studied. Later Backus developed and refined a set of kernel functions and a bunch of *functional-forms* that take functions as arguments to form complex functions. *Kernel functions* are basic functions like *arithmetic operations, Boolean operations, data access and update functions,* and *comparison operations.* Declarative programming encourages top-down programming. A complex problem is progressively decomposed to a combination of smaller problems. These smaller problems are progressively reduced to a combination of kernel functions using functional forms.

A λ-expression is an unnamed function with three components: *variable, body* and *parameter.* An example of λ-expression is $\lambda x.\ (x + 4)\ 3$. In this expression, the left-side occurrence of the symbol x is the variable declaration, and the expression within the parenthesis is the body of the λ-expression. The value 3 is the parameter, which is substituted with the variable x, reducing the body of the λ-expression to $(3 + 4)$, which can be further simplified to derive the value 7. If we associate a name with the λ-expression, then it becomes a callable function.

The implementation of functional programs is through an abstract machine that maps the λ-expression to low-level abstract instructions. There are many abstract machines used for different programming languages. The three popular abstract machines are SECD machine, G machine, and ABC machine. The machines differ based on evaluation strategy adopted for the expressions. SECD machine is suitable for both *eager evaluation* as well as *lazy evaluation* of expressions. Eager evaluation means that all the parameter-expressions are evaluated before substituting the variables by the parameter values. The parameter-passing mechanism used in eager evaluation uses call-by-value. Another strategy is to defer the evaluation of expression until expression simplification is needed. This strategy to defer the evaluation until needed is called *lazy evaluation*, which has been described in Section 9.3. Lazy evaluation is used the functional programming language Haskell. Lazy evaluation uses either call-by-name or call-by-need.

Functional programming has been combined with different programming paradigms such as imperative programming in the form of incorporating mutable objects, object-oriented programming paradigm, logic programming paradigm, and concurrent programming paradigm. Multiple multiparadigm languages such as Ruby and Scala have been developed.

The primary data abstraction used in functional programming is *sequence* that is used for strings and collection of data-entities. Sequence was initially implemented using linked list. However, list is a high-level representation at a programmer's level, and programmers do not have to explicitly manage the chain of pointers. That is a big change from imperative languages, where programmer has to explicitly write statements to use chain of pointers.

Functional programming languages support higher-order functions, as it gives them capabilities to (1) treat other functions as data, (2) build a function as data and transform into function, and (3) build complex programs by joining multiple functions in variety of ways.

A higher-order function can treat another function as its argument, and derive new complex function. Higher-order functions can be easily built using *apply* function that converts data into function. For example (*apply* 'square '(2)) will evaluate to give the function (*square* '2) that will evaluate to 4. Another example of a higher-level function is given in function 'foo' written in Lisp. Lisp has been introduced here to give a flavor of the non-imperative programming style employed by functional programming languages that looks quite alien to a student trained in imperative and object-oriented programming style.

```
(defun foo (powerFunction Arg)
       (+ (apply powerFunction Arg) 4)
)
```

The first line shows the function name as the first argument and the data to the function as the second argument. The second line is the body of the function. The body of a higher-level function performs two operations: making the argument 1 as a function with the second argument as argument of the function, and then adds 4 to the outcome. The function itself is written in prefix form like (*function-name* Arg_1, ..., Arg_N). A function call (*foo* 'sqrt *x*) will compute $\sqrt{x} + 4$. Inner level is executed first, followed by the outer level. The use of the single quote before a symbol shows that it is literal and not a function or variable. A list is represented in Lisp as '($element_1$, $element_2$,....$element_N$).

The popular programming languages that support functional programming paradigm are *Lisp, Scheme (Lisp family), ML, Miranda, Haskell, Ruby, Clojure, Lua, Erlang, Python, and Scala.* This chapter discusses some programming styles of the Lisp family, Haskell, Hope, Ruby, and Scala. Chapter 11 again describes Ruby and Scala because of their object-oriented features. Chapter 15 deals with the scripting features of Ruby, and Scala.

9.1 EXPRESSIONS

Many languages such as Haskell and Scheme directly allow user defined λ-expressions in a functional program. A variable declaration in a λ-expression is preceded by the Greek symbol λ to indicate the declaration of the variables and not the consumer occurrence of the variable. Variable has no initial value, as it is assign-once and gets the value only when a parameter is passed. The scope of the variable is in the following expression. If a variable is not declared but occurs in the body of the λ-expression, then it is called a *free-occurrence*; the occurrence of a declared variable is called the *binding occurrence.* Let us take the λ-expression $\lambda x.(x + y + 2)$. Here the variable x is the binding occurrence, and y is the free occurrence. λ-expressions are normally written in prefix-form, as it is easier to evaluate a prefix-form in a stack-based machine. However for better comprehension, we will initially use infix-notation to represent a λ-expression. λ-expressions can be nested. The scope of a variable is limited to the nesting level where it has been declared. The same identifier name may represent two different variables depending upon where it has been declared. A variable declared at a level is visible within that level. λ-expressions can be passed as parameters to another λ-expression.

Example 9.1

```
(λx.λy.(λz. z + 2) x + y) 3 4
```

The above λ-expression has two levels: the outer level has two variables, x and y, and the inner level has z. The outer level expression $x + y$ acts as parameter for the inner level, and the parameter for outer levels 3 and 4 is supplied. The binding of variables to parameters is done in left-to-right order. In this case, the variable x is bound to the value 3, and the variable y is bound to 4. If the outer level expression is evaluated first, then the expression after parameter substitution and simplification of the outer level will reduce to $\lambda z.\ (z + 2)\ 7$. In this case again the variable z is bound to the parameter value 7, and the expression becomes $7 + 2$, which can be simplified to give the value 9.

Example 9.2

```
λx. (x + x) (λz. z + 4) 3
```

In the above example, the λ-expression $\lambda z.\ z + 4$ is a parameter for the left-hand side λ-expression $\lambda x.\ (x + x)$, and the value 3 is a parameter for the λ-expression $\lambda z.\ z + 4$. One way to evaluate this expression will be to bind x to the λ-expression $\lambda z.$ z + 4, and the expression would become $(\lambda z.\ z + 4)\ 3 + (\lambda z.\ z + 4)\ 3$, which would give 7 + 7 = 14. Another way to evaluate this expression would be to bind 3 to the variable z, which would reduce the expression to $(\lambda x.\ x + x)\ 7$ that would reduce to $7 + 7 = 14$.

9.2 EVALUATION OF λ-EXPRESSIONS

The evaluation of a λ-expression requires binding of the variable to a parameter value and simplifying the resulting expression. The binding of the parameter value to a variable simply by textual substitution is called β-*reduction*, and the simplification of the expression that has just constant values is called δ-*reduction*. Let us understand these two reductions using the following example:

Example 9.3

```
λx. λy. (x + y) 3 4
```

In this λ-expression, the leftmost variable x at the same level will be bound to the leftmost parameter-expression 3, and the next variable y is bound to the expression 4. After the first β-reduction, the expression looks like $\lambda y.\ (3 + y)\ 4$, and after the second β-reduction, the expression would reduce to $3 + 4$. Now a δ-reduction would simplify the expression to 7.

In the case of name-conflicts, where a variable has been declared at two places in a nested λ-expression, the inner-level variable has to be renamed to avoid the naming conflict. This renaming of the conflicting variable name is called α-*substitution* or α-conversion. So we have three steps in evaluating a λ-expression: (1) α-substitution to remove the binding conflicts of the separate variables at two different levels by renaming the variable at inner level, (2) β-reduction to substitute the parameters in leftmost-to-right direction one by one, and (3) δ-reduction to simplify the final expression with constant values.

Example 9.4

```
(λx. λy. (λx. x + 4) x + y) 3 4
```

There are two occurrences of declaration of variable *x*. The scope of the variables is limited to their respective levels. The inner level has to be renamed using α-substitution to resolve ambiguity in execution. Let us assume α-substitution renames the inner variable *x* to *z*. The new λ-expression after the substitution is *(λx. λy. (λz. z + 4) x + y) 3 4*. Now the variables are unique and can be reduced using a series of β- and δ-reductions as described before.

9.2.1 Applicative-Order versus Normal-Order Reductions

The nested λ-expressions can be evaluated in two ways: (1) *applicative-order reduction*—first try the innermost rightmost level and progressively move toward the leftmost outermost or (2) *normal-order reduction* (NOR)—evaluate the leftmost outermost level, and then progressively move toward the innermost level. The λ-expressions at the same level are evaluated before moving to the next level. Let us understand applicative-order reduction using a simple example.

Example 9.5

In this example, we will use natural prefix order used in λ-expressions to represent an expression.

```
λx. (+ x x) (λy. + y 4 ) 3
```

The above λ-expression is a combination of two λ-expressions: leftmost λ-expression *λx. (+ x x)* has parameter *(λy. + y 4)*, which has parameter 3. Applicative order reduction (AOR) would follow the below-mentioned reductions to get the final result:

```
      λx. (+ x x) (λy. + y 4) 3
→β    λx. (+ x x) (+ 3 4)
→δ    λx. (+ x x) 7
→β    (+ 7 7)
→δ    14
```

NOR tries from outermost and leftmost and moves toward innermost rightmost. If we try NOR reduction on the same λ-expression, the reduction steps will be as follows:

```
      λx. (+ x x) (λy. + y 4) 3
→β    (+ ((λy. + y 4) 3) ((λy. + y 4) 3))
→β    (+ (3 + 4) ((λy. + y 4) 3))
→δ    (+ 7 ((λy. + y 4) 3))
→β    (+ 7 (+ 3 4))
→δ    (+ 7 7)
→β    14
```

The observations on both reduction techniques are as follows:

1. AOR technique eagerly evaluates the expression of the parameters and then substitutes the evaluated value to the left-hand side λ-expression. The number of the overall reduction steps in AOR is less than in NOR.

2. NOR technique defers the evaluation of parameter-expressions until after the substitution in the body of the left-hand side λ-expression. Deferring the evaluation of expression is called *lazy evaluation*. The number of overall reduction steps in NOR is more than AOR because of the multiple occurrence of bound variables in the expressions. We also observed in the above example that λ-expression becomes bigger after β-reduction because of the lack of eager evaluation of the parameter-expressions. This explosion causes additional reduction steps. This explosion can be optimized if we evaluate the replicated expression once and cache the values for future lookup instead of evaluating replicated expressions. This is the basis of call-by-need and can save significant amounts of computational overhead. In Example 9.5, after the optimization, the expression $\lambda y.\ (y + 4)$ will be evaluated once, and the value 7 will be saved. Next time when the expression is encountered, it is replaced by the value 7. Expression matching is done using graph-based representation. This will decrease the number of reduction steps from six to four, which is comparable to the number of reduction steps used in AOR.

Church Rosser's lemma states that if an expression is reduced either using AOR or NOR technique, then the final normalized form would be the same irrespective of the reduction technique. This lemma allows implementation of programming languages using either technique. Programming languages implementations use both the techniques: Lisp family uses AOR technique, whereas Haskell uses NOR technique with optimizations.

AOR has been implemented using SECD machine with eager evaluation. NOR uses graph-based reduction and has been implemented using G-machines and ABC machines. The implementation technique for lazy evaluation is discussed in Section 9.9.

9.3 FPS—FUNCTIONAL PROGRAMMING SYSTEMS

Pure functional programming, proposed by John Backus, has two parts: *kernel functions* and *functional-forms* that form a complex function out of simpler functions. *Kernel functions* are the low-level basic functions that cannot be decomposed further. Functional forms join two or more functions or provide additional control abstractions to form complex functions. The function-forming abstractions have interesting algebraic properties and model effective control abstractions such as conditionals (if-then-else), iteration (while-loop equivalent) and recursion. The basis of FPS was to identify the minimal set of functions and abstractions that can be used to write functional programs. In this section, we will discuss the various classes of kernel functions and functional forms proposed in FPS by Backus.

FPS separates functions from parameters. A functional program in FPS does not use variables. Rather, it uses an *identity function* to pull the value of the parameters within a function and uses a *constant function* to introduce a constant value within a function. The avoidance of a variable frees this programming style from memory locations and hence from the von Neumann implementation model.

The parameters in FPS are stored in a sequence. For example, the addition of two numbers 3 + 4 is written as +: *<3, 4>*. The major data abstraction in FPS is a sequence of data elements. In addition, a data item could be an *atom*—a single entity that cannot be further split, or a bottom symbol ⊥ to take care of error conditions. An empty sequence is denoted by ϕ or <>. A sequence can contain atoms or a sequence. An atom could be a literal, an integer, or a real number.

9.3.1 Kernel Functions

Kernel functions are divided into the following classes: *arithmetic expressions; logical expressions, comparison functions, selector functions, insertion functions, transpose function, meta-logical predicates, miscellaneous function such as length, reverse, distribute function, rotation function, identity function,* and *constant function.* All functional programming languages support most of these basic capabilities as built-in-functions or as the library functions.

Comparison operations such as '<', '>', '=<', '>=', and '==' check that the first operand satisfies a property compared to the second operand. For example, comparison operation 'operand1 < operand2' is written as < *:<operand1, operand2>.*

The selector functions are *left-selector, right-selector, tl,* and *tlr* to access an element from a nonempty sequence of elements. *Left selector* selects the elements from the left-hand side and is written as *'1l'* for first element from the left, *'2l'* for the second element from the left, and so on. For example, *1l:<a, b, c, d>* will return *a; 2l: <a, b, c, d>* will return *b; 1l:* <> will map to bottom symbol ⊥, because an empty sequence does not have any data-entity, and *1l:* x will map to the bottom symbol ⊥ because x is an atom. The right selector selects the elements of a nonempty sequence starting from the right, and is written as *'1r'* for the last element of a sequence, *2r* for the second last element of a sequence, and so on. The function *tl* will yield the rest of elements of a sequence except the first element.

For example, *tl: <a, b, c, d>* would yield *<b, c, d>; tl :<>* would yield a bottom symbol ⊥ because an empty sequence does not have a data element; and *tl :x* would derive a bottom symbol ⊥ because the argument is not a sequence but an atom. The function *tlr* returns a sequence containing all the elements of an input sequence in the same order except the last element. For example, *tlr : <a, b, c, d>* will return *<a, b, c>; tlr : <a>* will return an empty sequence *<>; tlr : <>* will map to bottom symbol ⊥ because an empty sequence has no element; and *tlr: x* will map to a bottom element ⊥.

The *construction functions* insert an element in the sequence or join two sequences. The main construction functions are *apndl, apndr, insert,* and *append.* The function *apndl* takes a data entity as the first argument and a sequence with n elements as the second argument, and returns a sequence with (n + 1) elements, such that the first argument is the first element in the output sequence. For example, *apndl:<1, <a, b, c>>* will derive the sequence *<1, a, b, c>; apndl:<1, <>>* will derive *<1>; apndl : <<>, <>>* will derive <<>> because the first element is an empty sequence; and *apndl : (1, 2)* will derive the bottom symbol ⊥ because the second argument is not a sequence. Similarly, the function *apndr* takes a data entity as the first argument and a sequence with n elements as the second argument and returns a sequence with (n + 1) elements, such that the first argument is the last element in the output sequence. For example, *apndr:<1, <a, b, c>>* will derive the sequence *<a, b, c, 1>; apndr: <1, <>>* will derive the sequence *<1>; apndr: <<>, <>>* will derive <<>> because the first element is an empty sequence; and *apndr: (1, 2)* will derive the bottom symbol ⊥ because the second argument is not a sequence. The insert function takes three arguments: index, sequence, and element; and insert the element at the position given by the index to derive a new sequence. For example *insert:<3, <a, b, c, d>, x>* would derive *<a, b, x, c, d>*. The append function takes two sequences and joins them, such that the new sequence contains elements of the two sequences in the same order. For example, *append: <<1, 2, 3>, <a, b, c>>* will return *<1, 2, 3, a, b, c>*, and *append: <<>, <a, b, c>>* returns *<a, b, c>.*

The transpose function takes a two-dimensional matrix represented as a sequence of rows such that jth element of the ith row represents the element *a[i, j]* of the matrix. The transpose function derives a new transpose matrix a^{trans} such that *a[i, j]* will become a^{Trans} *[j, i]*. Transposing a sequence of empty sequences derives an empty sequence. If input is not a sequence of sequences, then the output is bottom symbol ⊥. For example, *transpose: <<1, 2, 3>, <4, 5, 6>, <7, 8, 9>>* would give *<<1, 4, 7>, <2, 5, 8>, <3, 6, 9>>.*

The *metalogical predicates* check the type of the objects. For example, *is_float, is_null, is_nonnull, is_atom, is_sequence* are metalogical predicates that check the type or the structure of the data elements. For example, *is_integer*:3 returns *true,* and *is_integer*:a will return *false.* The function *is_atom*:a will return *true.* However, *is_atom:<1, 2, 3>* will return *false.* The function *is_sequence*: <> will return true. However, *is_sequence:a* will return *false.*

The miscellaneous function *length:<a, b, c>* will return 3. There are two distribute functions: *distl* and *distr.* The function *distl* takes an element as the first argument and a sequence as the second argument and generates a sequence of sequences such that the resulting ith element is a sequence of two elements: the first argument and the ith element of the second argument. For example, *distl : <1, <a, b, c>>* will return a sequence *<<1, a>, <1, b>, <1, c>>*. If the second argument is an empty sequence, then the outcome is an empty sequence. The function *distr* takes an element as the first argument and a sequence as the second argument and generates a sequence of sequences such that the ith element of resulting is a sequence of two elements: the ith element of the second argument followed by the first argument. For example, *distr : <1, <a, b, c>>* will return a sequence *<<a, 1>, <b, 1>, <c, 1>>*. If the second argument is an empty sequence, then the outcome is an empty sequence. The *rotation functions* rotate the elements of a sequence by a specified amount. The function *rotl* rotates the elements to the left in a circular way by a specified amount. For example, *rotl : <2, <a, b, c, d, e>>* would return a new sequence *<c, d, e, a, b>*. Similarly, the function *rotr* rotates the element right in a circular way by the specified amount. For example, the function *rotr : <2, <a, b, c, d, e>>* will derive the sequence *<d, e, a, b, c>*. The *reverse function* takes a sequence as input and returns a reversed sequence: the elements of the output sequence are in reverse order. For example, *reverse: <1, 2, 3>* will return *<3, 2, 1>*. The functions *rotl* and *rotr* rotate by one position if the amount of rotation that is not specified. For example, *rotl : <1, 2, 3, 4>* will return *<2, 3, 4, 1>*, and *rotr :<1, 2, 3, 4>* will return the sequence *<4, 1, 2, 3>*.

The constant function takes a constant value function such as $\underline{4}$ and maps every input element to 4, irrespective of the data element, except for the bottom element ⊥. If the input symbol is the bottom element ⊥ then the output will also be the bottom element ⊥. For example, $\underline{4}$ *: a* will return 4. Similarly, $\underline{4}$*: <a, b, c>* will return 4. However, $\underline{4}$*:* ⊥ will return ⊥. Because FPP does not use the notion of variable or literals within a function, constant functions are used to introduce constant values inside a function.

The identity function, denoted by *id*, returns the input value itself. Because FPP does not support variables, the function *id* has the effect of introducing the input parameter inside the function. For example *id: <1, 2, 3>* returns *<1, 2, 3>* within the body of the function.

9.3.2 Functional-Forms for Constructing Complex Functions

There are seven functional-forms that are used to construct complex functions. They are *composition, construction, insert, apply_all, condition, iteration,* and *recursion.*

The *composition functional form* is equivalent to sequence of commands in imperative languages. A composition of two functions denoted as *f · g(x)* is equivalent *f : g :x*, which means first apply the function *g* on the parameter *x*, and then apply the function *f* on the output *g(x)*. The symbol '·' denotes composition. For example, *square · 1l : <4, 5, 6>* will first apply *1l : <4, 5, 6>* to derive the value 4, and then *square : 4* will derive *16*.

The *construction functional form* takes a sequence of function enclosed within square brackets and applies each function to the input parameter to derive an element of the output sequence. For example, *[square · 1l, length, 1r]* : *<4, 5, 6>* would give a sequence *<16, 3, 6>*. The value *16* is derived by the application *square·1l: <4, 5, 6>*; the value 3 is derived by the application *length : <4, 5, 6>*; and the value 6 is derived by the *application 1r: <4, 5, 6>*. If any of the function maps to bottom symbol ⊥, then the overall evaluation is ⊥ because < ..., ⊥, ...> is equivalent to ⊥.

The *insert functional-form*, denoted as */f*, inserts a dyadic operator *f* between every element of the given sequence. For example */+: <1, 2, 3>* is equivalent to *1 + 2 + 3 = 6*. Similarly, */*: <1, 2, 3, 4> = 1 * 2 * 3 * 4 = 24*. The formal definition of insert functional form is recursive as follows: if the parameter is a single-element sequence, then the insert functional form returns the element inside the sequence that is $/f\text{: } <x_1> \equiv x_1$. Otherwise, it applied the function recursively as $/f\text{: } <x_1, \ldots, x_n> \equiv f\text{: } <x_1, /f : <x_2, \ldots, x_{n-1}>>$, which means insert functional form is applied recursively on the rest of the sequence, and then function *f* operates on two arguments: first element of the sequence and the result derived from the recursive evaluation of the rest of the sequence. Another example of the use of insert functional form is /<, which finds out the minimum of a sequence. Similarly, the application of the functional form /> finds out the maximum of a sequence. A functional program for finding out the average of a sequence is *divide· [/+, length]*. The application of '/+' derives the sum of the sequence, and the application of the function *length* derives the number of elements in the sequence. The construction functional form puts these values in a sequence. The division of the total sum by the number of sequence gives the average of the sequence.

The *apply-all functional form*, denoted by α*f*, applies a function *f* to all the elements of a sequence $<x_1, \ldots, x_n>$ and returns a sequence of data elements of the form $<f{:}x_1, \ldots, f{:}x_n>$. For example, α *square : <1, 2, 3>* will return the sequence *<1, 4, 9>*. The functional form *apply-all* is equivalent to iterators in programming languages and *mapcar* in Lisp programming language.

The *condition functional form* is equivalent to the if-then-else statement except the body of then-part and else-part are functions. The syntax of condition functional form is *(if <predicate> <then-function> <else-function>)*. For example, the magnitude of a number can be derived by the conditional function *(if >·[id,* $\underline{0}$*]* $\underline{Id}$*, * · [*$\underline{-1}$*, Id])*. The predicate *> · [Id, 0]* will return true if the given number is > 0. If the predicate returns true, then the then-part function will be executed and will return the number itself. Otherwise, the else-part function ** · [*$\underline{-1}$*, Id]* will be executed, which will multiply –1 with the number deriving the positive magnitude.

The *iterative functional-form* is equivalent to while-loop except the body part is a function. The semantics of the functional form *(**while** <predicate> <function>)* is as follows. The predicate is applied on the input argument. If the predicate returns true, then the function is applied once, and the output of the function becomes the input parameter for the next iteration cycle. The process continues until the predicate returns *false*. After the predicate returns *false*, the last parameter value in that iterative cycle is returned as the result. Let us assume that number of iteration cycles are n before the predicate returns *false*; then output value is *<function> · <function> … n times <input parameter value>*.

Example 9.6

For example, *2l · (**while** > · [1l,* $\underline{0}$*] [– · [1l,* $\underline{1}$*], * · [1l, 2l]]) · [Id,* $\underline{1}$*]* will find out the *factorial(n)* starting from value *n* and progressively reducing the value of *n* and multiplying the current value of n with the previously accumulated value until n becomes < 1 and returning the final value of the accumulator. The accumulator holds the cumulative value of n * (n – 1) …, id at any point of time. The use of accumulators is quite common in the iterative style of programming.

Let us assume that we are trying to find out *factorial*(4); then the input parameter would be 4. The application of the function *[Id, 1]* will return *<4, 1>*. The first argument is the input value, and the second argument is the initial accumulator value 1, which is equivalent to the base case factorial(0). This sequence *<4, 1>* becomes the input for the while-loop. The function > · *[1l, 0] : <4, 1>* checks if *4* > *0*. Because the predicate returns *true*, the function *[– · [1l, 1], * · [Id, 2l]] : <4, 1>* is executed. The application of the function – · *[1l, 1]: <4, 1>* returns 3, because the application of the function *1l:<4, 1>* return the value 4, the application of the constant function *1: <4, 1>* derives *1*, and – : *<4, 1>* returns *3*. The application of the function * · *[1l, 2l]* is equivalent to *4 * 1*, because *1l: <4, 1>* returns 4, *2l: <4, 1>* returns 1, and **: <4, 1>* is equivalent to *4 * 1* = *4*. The input parameters for the while-loop iterations progressively become *<3, 4>* → *<2, 12>* → *<1, 24>* → *<0, 24>*. After this predicate becomes false, the control comes out of the while-loop with final value *<0, 24>*. The application of the function *2l: <0, 24>* returns *24*.

Example 9.7

Let us take the recursive version of writing a factorial program. The recursion functional form is well-known. The definition of a function includes itself such that the next calls progressively move toward the base case.

```
factorial ≅ (if >= ·[Id, 1]*·[Id, factorial· - ·[Id, 1]] 1)
```

The function states that if the input parameter ≥ 1, then call *factorial* (*input parameter* – 1) and multiply the returned value with the value of the input parameter. Otherwise, return 1.

9.3.3 Programming in FPS

Once we have understood the different components of FPS let us write simple programs using FPS to understand the subtle differences between functional programming paradigm that does not allow destructive update. We will understand the use of functions and functional forms used in FPS. We will also see how iterative programming style differs from recursive programming style. Programming concepts have been illustrated using (1) simple sort programs based on finding out maximum value and (2) adding two matrices.

Example 9.8

We find the functional equivalent of sorting. The programming style is different from imperative programming because of the lack of destructive updates and the lack of global variables. The major difference is that, because of the absence of destructive update, operations like in-place swap are missing, and index-based updates are not possible. We use linked list-based representation for sequences instead of indexible sequences. There are many ways to write a program to sort. However, the purpose in this example is to illustrate various functional forms.

In this example, we repeatedly take the maximum, concatenate to the accumulator, delete the maximum value from the sequence, and repeat the process for the remaining subsequence, until the input list becomes empty(null). The examples illustrate the use of composition, insert functional form, construction functional form, condition functional form, iteration, and recursion. It also shows the use of logical expression and arithmetic expression.

```
minimum ≅ /<
delete ≅ (if ==·[2l, 1l·1r]    tl·1r    1r)
                  ·(while   and·[>·[1l, 0], not·==·[2l,
                     1l·1r]] [-·[1l, 1],2l,rotl·1r])
                  ·[length·2l, 1l, 2l]
sort ≅ (if null <> apndl·[minimum, sort·delete·[minimum, Id]])
```

The program has three functions: *minimum, delete*, and *sort*. The function *minimum* finds the minimum of a sequence, and the function *delete* deletes a given element from the sequence. The main function *sort* uses the functions *minimum* and *delete* to sort the sequence. The program says that in order to sort a sequence in the ascending order, there can be two cases: (1) sequence is null or (2) sequence is nonnull. In case the input sequence is null, then return an empty sequence by applying the constant function <>. Otherwise, minimum of the sequence is found using the function *minimum*, and it is deleted from the sequence. The remaining sequence is recursively sorted, and the minimum value is concatenated to the sorted sequence.

The function *minimum* uses *insert* functional form to find out the minimum element, as described earlier. The function *delete* illustrates the use of while functional form to go through the sequence, the use of conditional functional form to extract the value of an accumulator and the use of *accumulator* to store the temporary minimum. The function *delete* has three parts, which are as follows:

1. Construction functional form takes as input *<minimum value, sequence>* and converts into a triple *<length of the sequence, minimum value, sequence)* using the function *[length·2l, 1l, 2l]* that concatenates the length of the sequence to the input parameter. The first element is used as an index value that is decremented by 1 after every iteration cycle.
2. The while-loop keeps repeating until the value of the first element becomes 0. The function inside the while-loop rotates the sequence left in every iteration cycle.

3. The if-then-else statement returns the rest of the sequence except the first element if the first element of the sequence matches the minimum value. Otherwise, it returns the complete sequence.

Let us go through step-by-step about how the functions will be invoked for a sequence *<13, 7, 44, 3>*. First *[minimum, Id] : <13, 7, 44, 3>* derives the pair *<3, <13, 7, 44, 3>>*. The function *delete* is started on this new sequence. The application of construction functional form *[length· 2l, 1l, 2l] : <3, <13, 7, 44, 3>>* returns the sequence *<4, 3, <13, 7, 44, 3>>*. This value becomes the input parameter for the while-loop. Because $4 > 0$ and $13 \neq 3$, while-loop rotates *<13, 7, 44, 3>* left by one position and decrements the counter value to 3; the parameter for next iteration becomes *<3, 3, <7, 44, 3, 13>>*. In the next iteration, again $3 > 0$ and $3 \neq 7$, the operations are repeated again, and the parameter becomes *<2, 3, <44, 3, 13, 7>>*. In the next iteration, $2 > 0$ *and* $3 \neq 44$, the function is applied once more, and the parameter for the next iteration becomes *<1, 3, <3, 13, 7, 44>>*. This time the minimum value 3 is equal to the first element of the sequence, and the control comes out of while-loop with the value *<1, 3, <3, 13, 7, 44>>*. The *if-then-else* again checks if the second element is equal to the first of the rightmost element. Because $3 = 3$, it returns the tail of the sequence, that is *<13, 7, 44>* in the next cycle. After the recursive call of sort *<13, 7, 44>* returns *<7, 13, 44>*. The minimum value 3 is inserted as the first element using *apndl* to return the final sorted sequence *<3, 7, 13, 44>*.

The lack of the use of global variable to store previously computed results forces application of the function *minimum* two times: first to compute minimum value in *sort* function and second in the *delete* function. If we had global variables, the minimum value could be memorized after the first computation and retrieved later using lookup. However, it is shown that nontrivial programs can be written using pure functional programming.

Example 9.9

The next example illustrates the power of *apply_all* to process all the elements of a sequence using matrix addition. The two-dimensional matrix is represented as a sequence of rows, and each row is a sequence of scalar values. Adding two matrices is equivalent to adding a sequence of rows, and adding two rows is equivalent to adding all corresponding elements. The program is given below:

```
group ≅ (if null·1l <> apndl·[[1l·1l, 1l·1r],
                         group·[tl·1l, tl·1r]])
add-row ≅ α+·group
add-matrix ≅ (if and·[==·[length·1l, length·1r],
  =·[length·1l·1l, length·1l·1r] αadd-row·group)
```

The above program adds two matrices using three functions: *group, add-row,* and *add-matrix*. The recursive function *group* takes two sequences and makes a sequence of sequences by pairing the corresponding elements of two sequences. For example, group: *<<1, 2, 3, 4>, <10, 11, 12, 13>>* would return *<<1, 10>, <2, 11>, <3, 12>, <4, 13>>*. The function add-row takes the output of group, and applies add-function '+' on each element of the sequence. For example, add-row: *<<1, 10>, <2, 11>, <3, 12>, <4, 13>>* will return *<11, 13, 15, 17>*. The function *add-matrix* groups the corresponding rows of a matrix together and applies the function *add-row* on each of them using the apply-all functional form.

9.3.4 Comparing λ-Expressions and FPS

The major difference between λ-expression and FPS is (1) use of the variables in λ-expressions, (2) use of functional form abstractions in FPS, and (3) use of naming in FPS that allows callable functions. All of the functional forms in FPS can be programmed using λ-expressions. However, the programs using λ-expressions look lot more complex unless they are associated with names. Most of the high-level functional programming languages associate name with functions and also allow variables. The purpose of lack of variables in FPS demonstrates that programs can be written just as a combination of functions without any need for variables. However, the use of variables in high-level programming languages makes it convenient to access and transform the input parameters. We have seen that lack of variables and constants forces us to use additional functions such as "identity function" and "constant functions" that do not add to better comprehension of the program. Although FPS is a cleaner theoretical tool and clearly marks kernel functions and functional-forms, it needs to be augmented with a more comprehensible style of programming and abstractions.

9.4 ABSTRACTIONS AND PROGRAMMING

Modern programming languages supporting functional programming paradigm are grouped under five major categories: (1) pure functional programming languages, (2) languages that support functional programming along with mutative objects and destructive updates, (3) languages that integrate functional programming with object-oriented programming, (4) function programming languages that support concurrent programming, and (5) multiparadigm languages. Haskell is a pure functional programming language. Lisp, Scheme, and ML mix up the imperative programming paradigm with functional programming paradigm. Ruby, Scala, and Emerald mix up object-oriented programming with the functional programming paradigm. Languages like Haskell, Scala, and Ruby support concurrency. Languages like Scala and Ruby are multiparadigm languages that support functional programming, concurrent programming, and object-oriented programming.

This section deals with various abstractions supported by different languages and how programs can be written in the Lisp family of languages, Haskell, Hope, Scala, and Ruby. Languages such as Ruby, Scala, and Emerald will be discussed again while discussing object-oriented languages to see their multiparadigm programming style.

9.4.1 Abstractions in Functional Programming Languages

Abstractions in functional programming languages, are somewhat different from imperative languages, because the functional programming paradigm is a declarative programming paradigm and does not support destructive update in pure form. The parameters are passed to a function that evaluates an expression in the body of the function after binding the parameter to the declared variables. Owing to the lack of destructive updates, variables cannot be reused to assign a new value. This gave rise to the notion of immutable data structures and creating copies of the data structures when they had to be updated. The main data entity in functional programming is an immutable data sequence. In the early days of functional programming, sequences were implemented using linked lists. Thus the control abstractions were based upon list-based operations. Because list is a recursive data structure, recursive procedures were consider a natural programming style in functional programming. Iterative programming styles that could access and update an array using an index variable were not considered suitable for pure functional programming, because index variable and array locations are frequently destructively updated in traditional index variable based iteration. Instead tail-recursive procedures on sequence-based data abstractions, and later iterators were used for processing collection of data entities. Another approach taken by many functional programming languages is allowing a limited amount of imperative programming with destructive update to (1) store partial computations; (2) memory reuse for iterative constructs such as for-loop, while-loop; and do-while-loop; and (3) allow array-based operations. For example, the Lisp-family and Ruby allow destructive update of variables. Ruby allows mutable dynamic arrays, and Lisp-family, Ruby and Scala allow iterative constructs such as while-loop and for-loop.

9.4.1.1 Data Abstractions

Function programming languages are based upon the use of sequence. In the early days of Lisp, sequences were implemented using linked lists with limited operations such as concatenating in front of the list, appending lists, checking for empty and nonempty lists, association lists, and the use of frames. Lisp also supported the use of imperative programming paradigm by supporting global variables and destructive update of linked lists. Later functional programming languages extended the operation on sequences by treating the sequences as indexible. In addition to sequence in the form of linked lists, the Lisp family of languages also supports arrays, association lists, strings as sequence of characters, and atoms. Haskell and Hope are pure functional programming languages and do not support destructive update. CLOS—a variation of Lisp, Scala, and Ruby—supports object-oriented programming. Every element in Ruby and Scala is treated as an object. Haskell also supports tuples. Ruby and Scala support both mutable and immutable objects. Ruby supports dynamic indexible sequences that can be used as arrays, stacks, queues, and can simulate linked lists.

Functional programming languages inherently support parametric polymorphism. ML is the first language to support declarative parametric polymorphism. *Hope* is another elegant functional programming language that integrated pattern matching like logic programming languages and expression evaluation. Hope is a strongly typed polymorphic language that supports type variables. Haskell supports an hierarchical class of type declarations. Many functional programming languages such as Miranda use union of types to handle the problem of overloaded operators. Some of the functional programming languages like Haskell also support the notion of modules and use export–import to access the functions across modules.

9.4.1.2 Control Abstractions

Almost all the functional programming languages support control abstractions: *composition, apply-all, conditionals, iteration*, and *recursion*. Some of the control abstractions like *insertion* and *construction* can be simulated using other functions. Functional languages support first-class objects, which means that functions can be built as data and can be passed as parameters. In Lisp, the *apply* function converts data into function. For example, (*apply* ‘square ‘(4)) will generate a function call (*square* 4). The Lisp family and Ruby support all three types of iterations: *indefinite iteration*, *definite iteration*, and *iterators*. All functional programming languages support iterators, and many such as Ruby and Scala support while-loop. The Lisp family supports a function *mapcar* that is the same as the *apply-all* functional form.

The evaluation strategy of functional programming languages can be *eager applicative order evaluation* or *lazy normal order evaluation*. For example, the Lisp family of languages uses applicative-order eager evaluation, and Haskell uses lazy evaluation. Lazy evaluation has an advantage of deferring the evaluation until needed. However, it expands the intermediate-reduced expressions that need to be optimized for efficient evaluation. One such technique to optimize is *call-by-need*, which is a variation of call–by-name. The efficiency of call-by-need is between call-by-value and call-by-name. Common subexpressions are evaluated only once in call-by-need, and their value is substituted in multiple occurrence of the same subexpressions instead of evaluating them repeatedly. Call-by-need is slower than call-by-value because of the overhead of bookkeeping to store the value of the evaluated expressions and to delay the evaluation of an expression. The Lisp family and ML use call-by-value for parameter passing because of eager evaluation implementation technique, whereas Haskell uses call-by-need because of optimized lazy evaluation.

Functional programming languages use three techniques to perform I/O operations: (1) stream-based IO, (2) continuation-based IO, and (3) *monads*. Stream-based IO sends a request to the operation system that opens a channel between the file and the program, where the program can perform input and output operations. Continuation-based IO refers to each read and write operation as a transaction. If the operation succeeds, then a success continuation is invoked; if the operation fails, then a failure continuation is invoked. Monads are used exclusively in Haskell. Monads are abstractions for side effect-based IO programming.

9.4.2 Abstractions and Programming in the Lisp Family

Lisp and Scheme use s-expressions (symbolic expression) of the form *(function* Arg_1, …., Arg_N*)* to call a function. For example, adding two numbers is written as (+ 2 3). Data and functions are separated using a quote before the data. They support most of the kernel functions described in FPS and have a rich library of mathematical functions. They use *cons* to insert an element in front of a sequence that is similar in semantics to *apndl* in FPS. However, they have no notion of inserting an element on the right-hand side because of the way linked lists are implemented. The kernel functions like *rotr, rotl, Id,* and constant functions are missing for different reasons. Id and constant functions are not needed, because most of the programming languages use variables instead of identity function or constant functions to pull in the parameter values in the function. In terms of functional forms, they support *composition, condition, apply-all, recursion, iteration,* and *conditionals.* The higher-order function *apply* converts first class objects into functions. The recursive definition shown in Example 9.10 written in Lisp, uses the higher-order function *apply* to build *construction* functional form. A scheme program has similar syntax and constructs as a Lisp program: the reserved word 'defun' would be replaced with 'define.' Scheme language is a member of the Lisp family. The major difference between Lisp and Scheme is that Scheme supports a static scope, whereas Lisp supports both static and dynamic scope.

Example 9.10

The three functions described below illustrate programming style in Lisp. First two functions are straightforward. The function add5 adds the value 5 to the parameter value and returns. The function *square* returns *value * value.* The third function simulates the effect of "construction" functional-form: the first argument is the list of functions, and the second argument is the list of arguments. Data is given in the form of a list because the higher-order function *apply* that expects the arguments in the form of a list. For example, a call like *(construction '(add5 square) '(4))* will give a list *'(9 16).*

```
(defun add5(Value) (+ 5 Value))
(defun square(Value) (* Value Value))
(defun construction(FunctionList Argument)
    (if (null FunctionList) nil
         (cons (apply (first FunctionList) (list Argument))
               (construction (rest FunctionList) Argument))
         )
    )
)
```

The program returns nil (empty list) if the argument FunctionList is an empty list. Otherwise, it converts first element of the argument *FunctionList* into a function, using the higher-order function *apply.* The definition of the functional form *construction* is recursive and calls itself with rest of the *FunctionList* that returns the rest of the output sequence. The function *cons* concatenates first returned value with the rest of the output sequence to derive the final list.

Example 9.11

The following recursive function is written in the programming language Scheme to simulate the functional form *apply-all.* The recursive function uses the higher-order function *apply* to apply the function My*Function* on one argument at a time, and then calls apply-all recursively again to apply the function My*Function* on the rest of the arguments. The output list is built by concatenating the result of the applying the function *MyFunction* on the first element of *ArgsList* and the rest of the sequence derived by recursive invocation of the functional form *apply-all.*

```
(define apply_all(MyFunction ArgsList)
   (if (null ArgsList) nil
      (cons (apply MyFunction (list (first ArgsList)))
         (apply_all MyFunction (rest ArgsList))
      )
   )
)
```

The Lisp family supports multiple types of iterations: *dowhile*, *dolist*, and *dotimes.* The functional *dowhile* is a typical indefinite iteration that keeps applying a function until the predicate becomes false. The functional *dolist* is a typical iterator that keeps selecting the next element out of a list and executing a function until the list becomes empty. The functional *dotimes* executes the iteration a fixed number of times. The application of the functional forms *dolist* and *dotimes* is illustrated in Figure 9.1.

In Figure 9.1 global variables are set by using the *setq* command. A global variable can be bound to a single entity or a complex entity like linked list, or an expression which is evaluated first before binding the resulting value to the global variable. The function *add5_to_all* illustrates the use of the functional form *mapcar* to apply a given function *add_5* to all the elements of the sequence given in *ArgList.* The function *square_and_add* illustrates the nesting of function to simulate the effect of function composition. First the function *square* is called, which becomes the argument for the function *add_5.* The function *hypotenuse* shows the use of passing function as parameter and computes the length of a hypotenuse of a right angle triangle using a built library function *sqrt.* Both *square*(x) and *square*(y) are argument for the addition operator. The output of addition becomes an argument for the library function *sqrt.* The function *factorial* computes the factorial of a given number and illustrates the use of the *if-then-else* construct in implementing a recursive function: then-part is the base case, and the else-part is the recursive definition.

The recursive function *sumlist* adds a sequence of numbers and illustrates the use of the *cond* statement in Lisp. *Cond-statement* is a general form of conditional-statements that can be used for both case statement and if-then-else statement. The function *add_row* illustrates the use of the functional-form *mapcar* to add two sequences. The recursive function *add_matrix* combines the recursive invocation of the functions *add_matrix* and *mapcar* to add two matrices. The program utilizing *dolist* picks up the next row from the sequence matrix and prints out one row at a time, until all the elements of the sequence matrix have been consumed. The program utilizing *dotimes* finds out the number of rows in the sequence matrix using a local variable *size* in the scope of let declaration, and then

```
(defun  greet ( ) (print  "Hello  World")) ; defining a function greet
(defun  add5 (x) (+ x 5)) ; defining a function
(defun  square(x)  (*   x  x))
(setq  m  5)  ; setting the value of a global variable m to 5
(setq  p '(4   "Hello World"))  ; binding a global variable p to a list
(setq  q  (*  (first  p)   m))

; Implementing apply-all using mapcar
(defun  add5 _ to _ all (ArgList) (mapcar  'add _ 5  ArgList))
(defun  square _ and _ add( x ) (add5  (square   x ))) ; composition
(defun  hypotenuse (x y) (sqrt (+  (square  x) (square y))))

; Recursive definition of factorial using if-then-else
(defun factorial(n) (if (= n 0) 1 (* n ( factorial  (-   n   1)))))

; Recursive definition using conditional statement
(defun  my _ sum(DataList)
        (cond    ( (null   DataList)  nil)
                 (t   (+  (first     DataList) (my _ sum  (rest  DataList))))
        )
)

; apply-all  using mapcar to add two sequences
(defun   add _ row (Seq1, Seq2)
  (mapcar   '+   Seq1  Seq2) ; add corresponding elements of Seq1 and Seq2

; recursion with multiple arguments to add two matrices
(defun    add _ matrix [(Matrix1, Matrix2)
   (if  (null Matrix1)  nil  (cons (add _ row   (first Matrix1)  (first  Matrix2))
                                   (add _ matrix   (rest  Seq1)  (rest Seq2)))
        )
)

; printing using dolist
(defun    print-matrix (Matrix)
          (dolist    (V Matrix) (print   V)))   ; Consume all the rows

; printing using dotimes
(defun print-matrix (Matrix)
         (let ((size (length Matrix)))  ;
              (dotimes (Index size) (print (nth  Index Matrix)))
        )
)
```

FIGURE 9.1 Illustrating abstractions and syntax in Lisp.

loops as many times using an index variable *index* that steps through from 1 to *size*. In every iteration cycle, the corresponding row is picked up using the function (*nth index matrix*) and printed using the input–output function *print*.

9.4.3 Abstractions and Programming in Hope

Hope is a polymorphic language that integrates functional programming and pattern matching. Its procedures are a set of rules, like Prolog rules, discussed in Chapter 10. It performs pattern matching between the right-hand side functional call with the left-hand side of a rule and passes the substitutions to right-hand side equation that is evaluated. Hope supports control abstractions such as if-then-else and while-loop and higher order functions that treat functions as data. However, Hope flattens out complex expressions into flat expression like Prolog before calling another function.

Example 9.12

The example shows the declaration of type variable for supporting polymorphism, using a definition for *append* function. The polymorphic-type declaration reads that the function *append* takes a Cartesian product of two lists of any type, denoted by type variable *alpha*, and derives a list of the same type. The symbol '#' denotes Cartesian product. Left- and right-hand side of functional definitions are separated by the symbol '<= ,' and the symbol '::' denotes concatenation of the first element with the rest of the sequence. Here '::' has been used to verify that the list *x::xs* is a nonempty list.

```
typevar alpha
dec append: list(alpha) # list(alpha) → list(alpha)
append(nil, Ys) <= Ys.
append(x::Xs, Ys) <= x::append(Xs, Ys).
```

The above program reads that appending an empty list to another list gives the same second list *Ys*, and appending two nonempty lists is equivalent to concatenating the first element of the first list to the recursive building the list by appending the rest of the first list with the second list. Note that the style of programming is equational.

Another example of supporting higher-order function in Hope is given below:

```
dec apply_all : list(num) # (num → num) → list(num)
apply_all(nil, function) <= nil
apply_all(first:: rest, function) <= function(first):: apply _ all
  (rest, function)
```

The polymorphic-type declaration reads that given a list of numbers and a function that maps a number to number, the functional form *apply_all* derives a list of numbers. The program reads as follows:

1. If the first argument is an empty list, then return nil;
2. If the first argument is a nonempty list and the second argument is a function, then the result is concatenation of applying the function of the first element and the result derived tail recursively by calling the function *apply_all* on the rest of the data elements and the function.

9.4.4 Abstractions and Programming in Haskell

Haskell is a statically typed type-safe functional programming language. Haskell supports parametric polymorphism and functional-forms such as conditionals (case statements and if-then-else in Haskell), iteration, recursion, composition, insert (simulated using *foldl* in Haskell), and apply-all (*map* in Haskell). Haskell implementation uses lazy evaluation. Iteration in Haskell is potentially like indefinite loop. However, because of lazy evaluation, it does not run away. Haskell uses the symbol '.' to compose two functions. For example, (*square.add5*) *x* is equivalent to '*square* · *add5* (*x*). The higher-order functional form *map* in Haskell is used to apply function to all elements of a sequence. The function *map* takes another function and a sequence of data elements as input, and applies the input function to every element of the sequence to generate an output sequence. For example, the function *map (add5)[1, 2, 3]* will generate a sequence *[6, 7, 8]* by adding *5* to every element of the input sequence. The assignment operation is assign-once, and variables are immutable.

Haskell supports sequences in the form of lists. Haskell got rid of the irritating parentheses in Lisp and uses an infix form of expression for better human comprehension. For example, Haskell will evaluate an expression 2 + 3 to 5. A list in Haskell is represented by a pair of square brackets. For example, a Haskell function *length[a, b, c]* would yield a value *3*. A nonempty list is represented as *x* : *xs*, where *x* is the first element, and *xs* is the rest of the list. The elements of a list can be accessed using access function *fst* to pick the first element of a list, and *snd* to pick the second element of the list. Comments are written within {- ... -}.

Haskell programs are divided into modules, and modules can be loaded into other modules. The module name and the file names are generally the same. The functions can be exported out of a module. Unless the functions are exported, they remain private to the module. Exported functions need to be imported by other modules. The functions are imported either by specifying the module or specifying the functions within parentheses after the specification of a module.

A function definition has two sides: the left side contains the name of the function followed by the parameters, and the right side contains the definition of the functions. Left and right sides are separated by '=' symbol. Tuples are put inside parentheses, and strings are put inside a pair of double quotes. Conditionals can be represented as *case-statement* or *if-then-else* statement or multiple mutually exclusive definitions. Haskell also supports guarded commands. Haskell program has a main module and a main function.

Figure 9.2 shows many working programs that compile in GHC (Glasgow Haskell compiler) illustrating many programming features. Although it does not give complete capabilities of Haskell, it illustrates the syntax and programming style of Haskell. For the programs in Figure 9.2, the name of the file is *main.hs*. Once the main function is declared, any number of functions can be declared. In Figure 9.2, there are three different definitions of factorial: *factorial using case-statement*, *factorial1 using if-then-else statement*, and *factorial2 using mutually exclusive definitions.*

The function *square_and_add* illustrates the composition *square* · *add5* of two functions: *square* and *add*. Note that if the type is not declared, then polymorphic type is inferred automatically. The function is polymorphic and can be used for different types of data objects that support overloaded operator '+.' The function *hypotenuse* shows another way to compose a function where the value returned by the first function is treated as an input argument by the second function. In the case of hypotenuse, the expression *square x* + *square y* acts as the input for the library function *sqrt.*

The recursive function *my_sum* shows the use of recursive programming using recursive data structure list. The function *reverse* shows the use of adding an element at the right end

```
module main where
main = putStrLn "Hello World" {- main function -}
add5  x = x + 5  {- Add 5 to parameter and return -}
square  x = x * x  {- return square of a number -}
m = 5  {- assigning a value to a variable    -}
p = (4, "Hello World")  {- assigning tuple to a variable  -}
q = fst p * m
hypotenuse::Float → Float → Float
hypotenuse x y = sqrt(square x + square y)

square_and_add   x = (add5.square)x {- composition  -}
add5_to_all   x  =  map  (add5)x {- apply-all  -}

{- finding factorial using case statement -}
factorial n = case n of
                    { 1 -> 1; {- handle the base case -}
                      _ -> n * factorial(n - 1) {- handle recursive definition -}
                    }
factorial1 1 = 1 {- base case -}
factorial1 n = n * factorial1(n-1) {- recursive definition of factorial -}

{- finding factorial using if-then-else -}
factorial2 n = if n == 0 then 1 else n * factorial(n - 1)

{- Guards -}
my_minimum x y | x <= y = x
               | y <= x = y

{- recursive programming and concatenation at the end -}
my_reverse [ ] = [ ] {- base case -}
my_reverse  (x:xs) = my_reverse xs++[x]  {- '++' adds at the end -}

{- Recursive programming with multiple arguments -}
add_row [ ] [ ] = [ ] {- base case -}
add_row (x:xs) (y:ys) = (x + y:add_row xs ys)

add_matrix [ ] [ ] = [ ] {- base case -}
add_matrix (r:rs) (w:ws) = (add_row r w:add_matrix rs ws)
```

FIGURE 9.2 Abstractions and programming in Haskell.

of a list. The function *add_row* takes the corresponding elements of two sequences, and adds them. It is equivalent to adding two rows of a matrix as the rows are represented as sequences.

The function *add_matrix* takes two matrices and adds them by repeatedly calling *add_rows*. The output matrix is constructed by concatenating the result of adding first corresponding rows with the sequence of rows derived by recursively adding rest of the matrices.

Haskell has influenced the development of many functional programming languages specially Python, later versions of Java, C#, Visual Basic, and Scala. Python is a dynamically typed scripting language that has adopted list notion. Java's generic type has been influenced by type classes in Haskell. C#, Visual Basic, and have adopted ideas from monads.

9.4.5 Abstractions and Functional Programming in Scala

Scala is a multiparadigm language that supports functional programming along with object-oriented programming. Every value is treated as an object in Scala. Every operation in Scala is a method call. Thus the output of a function is an object, and function types are classes that can be inherited by the subclasses. In this section, we will emphasize only the functional programming paradigm. The object-oriented programming part will be discussed in Chapter 11. Scala is also used as a scripting language.

Scala is built on top of Java and has natural interface with Java. Like Java, Scala programs are compiled to JVM byte codes and run like Java programs on any machine. Scala supports both mutable and immutable data structures. Scala stands for "scalable language," and it can grow with user demand by interfacing naturally with Java libraries. The syntax design of Scala has been influenced by Java and C# and looks like a typical dynamically typed scripting language. Its functional programming syntax is influenced by ML, and its class type is influenced by Haskell. Scala supports polymorphism, and if the type of the variable is not given, then it is inferred by the value assigned to a variable.

Scala supports a rich set of data abstractions: arrays, associative maps, lists, tuples, and sets. Arrays are mutable objects, and lists are immutable objects. Lists are used for the functional style of programming, and arrays are used for the imperative style of programming with destructive updates. Sets and associative maps can be used in both mutable and immutable way using *traits*. *Traits* are abstract interfaces that extend the class of the data objects. For example, if we describe a floating point number as a class, then probability class extends the floating point number class by adding three extra traits: (1) probability ≥ 0, (2) probability ≤ 1, and (3) sum of probabilities ≤ 1. In a more real world example, if we have a class vehicle, and a trait 'four wheeler' that extends vehicles, then car is a class with a trait four wheeler. In the case of sets and associative maps, we can associate a trait mutable and immutable with the class set or associative maps to make them mutable or immutable. When we study subclass in object-oriented programming, then we will observe that traits are more like additional methods or properties one can associate with a subclass. The difference between traits and methods specific to subclasses is that traits can be associated with subclasses of different classes to give them the same behavior associated with a trait.

Arrays of integer are declared as *Array[Int](4)*. This means that the object is an array containing four integers. The subscripted variables are kept inside a pair of parentheses. Because every data structure is an object, an array is created using the constructor 'new'. For example, we can say *val studentNames = new array[String](20)*. This will create an array object containing *20* strings that can be accessed by *studentNames(i)*, where *i* is an index variable of the type integer.

Lists can be declared as *List(1, 2, 3)* or as multiple elements concatenated using the symbol '::'. For example, *List(1, 2, 3)* can also be written *as 1::2::3:: Nil*. Two lists *xs* and *ys* are appended using the symbol ':::'. For example, *List(1, 2)* ::: *List(3, 4, 5)* derives *List(1, 2, 3, 4, 5)*.

Scala supports *if-then-else, case statement, while-loop, do-while-loop, iterator foreach-loop, definite iteration for-loop*, and recursive function calls. By passing a function as a parameter to another function, composition can be simulated. Scala supports destructive update in index variables making it possible to develop programs using regular iteration. Scala also supports operator override like Java does.

Scala functions are defined using the reserved word *def* and use standard parameter passing. Type is explicitly declared and verified. Scala uses Java classes, can import Java classes, and can access methods using *<class-name>.<method-name>*, where method is a function declaration. Following is a simple example of a function definition:

```
def factorial(n: Int): Int =
{if (n == 0) 1
 else n * factorial(n - 1)
}
```

The function describes the *factorial* function using Scala syntax. It declares type of the argument as integer and type of the function as integer and calls the function recursively. Note that the syntax is very similar to traditional programming languages.

Scala is a block structured language, and functions can be nested inside another function. Local variables have scope within the blocks where they have been declared. Inside the nested block, the variables that are declared in outer blocks are shadowed. Scala supports the module concept using Java *packages* and can import methods using import clause. Scala imports all the class libraries of Java and any predefined library in Scala by default before executing a program. Scala uses both call-by-value and call-by-name for parameter passing. The syntax for call-by-name is *<identifier>* : '=>' *<Type>*, whereas the syntax for call by value is *<identifier>* : *<Type>*. Scala utilizes Java's exception handling capability.

Figure 9.3 shows some of the abstractions for functional programming using Scala syntax. The syntax mixes up destructive updates of variables along with functional programming. However, list is an immutable object and cannot be destructively updated. The programs in Figure 9.3 show Scala's capability and syntax to build functions recursively, iteratively, using composition, using apply-all function-form simulated using map in Scala, and construction functional-form that applies multiple functions on one data element to generate a sequence. The function *add5* adds 5 to an input parameter *n* and returns the value. Note that the parameter type and function type has been explicitly declared. The function body is a block and is enclosed in curly brackets.

The function *square_add* shows composition *square* · *add5* of two functions: *square* and *add5*. First the function *add5* is applied to generate a number that is 5 greater than the input parameter, and then the generated value is squared. For example, *square_add(5)* is equivalent to *square(5 + 5) = 100.*

The function *power_rec* illustrates the use of if-then-else and recursion in Scala. Note that the predicate is enclosed in parentheses, and there is no mutation. In contrast, the function *power_iter* uses local mutable variables *a* and *b* to compute the value of the function x^n. The value is accumulated in the accumulator *b*, and is finally returned after the termination of

```
val x = 2 + 3 // declare a global variable
println("Hello World") // print "Hello World"
def add5(n: Int): Int = {n + 5} // function to add 5 to a number
def square(x: Double): Double = {x * x} // square using double_float
def int_square(x: Int): Int = {x * x} // integer_square
def square_add(x: Int): Int = int_square(add5(x)) // composition

def power_rec(x: Double, n:Int): Double =
   {if (n == 0) 1 else x * power_rec(x, n-1)} // if-then-else and recursion

def power_iter(x : Int, n: Int): Int = //iterative version of power
         {var a = n; var b = 1;
          while(a > 0 ) {b = x * b; a = a - 1}// destructive update of variables
         }

def sum_list(xs:List[Int]): Int = // Example of recursion on lists
         { if (xs.isEmpty) 0
           else xs.head  +  sum_list(xs.tail)
         }

def add_rows(xs : List[Int], ys:List[Int]):List[Int] =
         { if (xs.isEmpty) Nil
           else xs.head + ys.head::add_rows(xs.tail, ys.tail)
         }

def apply_all(my_func:Int => Int,   xs:List[Int]): List[Int] =
   {xs  map my_func}

def construction(my_funcs:List[Int => Int], n:Int): List[Int] = // construction
   { if (my_funcs.isEmpty) Nil
     else my_funcs.head(n)::construction(my_funcs.tail, n)
   }
// Use of iterators in Scala

def test_iterator(args: Array[String])
         { val names  = {"John", "Nancy", "Meera", "Tom"}
                   while (names.hasnext){ println(names.next}
         }
```

FIGURE 9.3 Abstractions and functional programming in Scala.

the while-loop. The functions *sum_list* and *add_seq* illustrate the use of recursion using lists. The function *sum_list* adds all the integers inside a list using recursive call on the rest of the list. The built-in method *isEmpty* is used to verify an empty-list, the method *head* is used to access the first element of a list, and the method *tail* is used to access the rest of the elements in the list. The function *add_row* has been described earlier while discussing Lisp and Haskell. The only change is the use of the methods head and tail to describe the first element and the rest of the elements in the list.

Scala uses a built-in *map function* that provides the capability of apply-all functional form. However, the parameter is written first, followed by higher order function map, followed by the function name. In this case *apply_all(int_square, List(1, 2, 3))* will generate a list *List(1, 4, 9)*. Another important thing to note is the way the type has been declared for the function. Function's type has been declared as *Int => Int*, which means it takes an input argument of the type integer and generates an output value of the type integer. The last function is the *construction* functional that takes a sequence of functions, which work on the same argument to generate a sequence of output values. For example, *construction(List(add5, int_square), 5)* will generate *List(10, 25)*. The program for *construction* function returns a null list if the list of functions is empty. Otherwise, it calls recursively with the list of the rest of the functions and concatenates the output of applying the first function on the given argument with the rest of the output sequence derived by applying rest of the functions on the argument. The last example illustrates the use of associative list of maps that can be used to map any domain value to the corresponding codomain value. The variable *myindex* is an associative list of two maps "A" → 1 and "B" → 2. A new element "C" → 3 is added, and then **println**(myindex("B")) prints the corresponding range value 2.

9.4.6 Abstractions and Functional Programming in Ruby

Ruby is another multiparadigm language that integrates imperative, object-oriented, and functional programming paradigms. It is also used as a scripting language. However, we will study different aspects of Ruby in different chapters. In this chapter, we will study functional programming aspect and integration of mutable objects and immutable objects in functional programming paradigm.

Unlike Scala, which is statically typed, Ruby is a dynamically typed polymorphic language. It supports different types of entities such as integers, floating point, strings, indexible sequences, sets, hash tables, and class. Indexible sequences can be dynamic arrays or vectors. Every element in Ruby is an object. Ruby supports local variable, global variables, class variables, and instance variables. Array is represented within square brackets with the elements separated by commas. Multidimensional arrays are modeled as a sequence of sequences. Arrays support different types of data objects. For example, one can have an array *multiarr A = [[a, b, c, d], [1, 2, 3, 4], ["Hello", "There", "Ladies &", "Gentlemen"]]*. It creates a new array dynamically as any other object using the constructor *Array.new*. A set can be created dynamically using the constructor *Set.new*.

Ruby has a rich library to manipulate matrix, and the overloaded operator '+' can be used to add two matrices. A library is loaded by the statement *require <library-name>*. Strings are also treated as indexible sequences. There are many operations on strings such as (1) concatenating two strings, (2) treating strings as array of characters, (3) length of a string, (4) reversing a string, and (5) chopping the last character of a string.

In control abstraction, it supports all imperative programming paradigm control abstractions such as block, parallel assignment, if-then-else, unless (opposite semantics compared to if-then-else), case statement, for-loop, while-loop, until-loop (equivalent to repeat until), a loop-construct that needs a conditional exit, multiple syntax for iterators, recursion, explicit λ-expressions, and function calls. Functions are declared using the reserved word *def*. Ruby is a block structured language supporting nested blocks. Every control abstraction is terminated by the reserved word *end*. Ruby also supports multi-threading and exception handling. Ruby's programming style is more like traditional programming languages and is intuitive and simple compared to other functional programming languages. Some of the examples of the functional programming style of Ruby

```
def greet # illustrating function
      puts("Name:"); gets(Name ); puts("Hello " + Name)
end
m = [[1, 2, 3], ['a', 'b', 'c']] # Array has different types of objects
m = "cat"; n = 4; m1, m2 = m2, m1 # parallel assignment

def factorial(n) # illustrating recursion and if-then-else
      if (n == 0) then 1
      else n * factorial(n - 1)
      end
end
def fibonacci(n) # Illustrates the syntax of case statement
      case (n)
            when 0 then 1
            when 1 then 1
            else fibonacci(n - 1) + fibonacci(n - 2)
      end
end
def sum_seq(xs) # illustrating iterators and destructive update
      accumulator = 0
      for n in xs do accumulator = accumulator * n end
      return acc
end
def append(xs, ys) # appends two sequences
      zs = xs + ys
end
def add_seq(xs, ys)
      zs = Array.new # creating a dynamic array
      length1 = xs.length - 1
      for n in 0..length1 # another form of iterator
            zs[n] = xs[n] + ys[n]  # expanding dynamic array
      end
      return zs
end
def add_matrix(m1, m2) # use of while-loop
      m_final = Array.new ;  size = m1.length ; index = 0
      while (index < size)     # while loop
            m_final[index] = add_seq(m1[index], m2[index])
      index += 1
      end
      return m_final
end
```

FIGURE 9.4 Abstractions and programming in Ruby.

are given in Figure 9.4. Figure 9.4 illustrates the syntax used in Ruby. Every control abstraction ends with a reserved word *end*.

The function *greet* illustrates (1) the syntax for defining a function; (2) interaction with the user; and (3) the use of the overloaded operator '+' which is being used here to concatenate two strings. The function displays 'Name:' on the screen, accepts the next string—for example, "Arvind"—concatenates with the string "Hello" using overloaded operator '+,' and displays the string "Hello Arvind." The next statement shows that array elements could be any type of data object. Statement # 3 illustrates the *parallel assignment* to swap the value of variables *m1* and *m2*. It shows that a variable can be associated with any value.

The definition of the function *factorial* illustrates the syntax of if-then-else statement and the recursive style of programming in Ruby. The definition of the function *Fibonacci* illustrates the syntax of case statement. The definition of the function *sum_seq* shows the use of an iterator to sum up a sequence. The definition of the function *sum_seq* illustrates the use of accumulators and iterator. The function *append* illustrates the power of overloaded operator '+' that has been used here to append two sequences. The function *add_seq* illustrates (1) creation of a dynamic array *zs* as an object using a constructor and (2) another style of iterator that iterates on a subrange. The function *add_matrix* illustrates (1) while-loop syntax and (2) destructive update of the index variable. The object-oriented programming style and its integration with functional style are described in Chapter 11. The application of Ruby as a scripting language is described in Chapter 14.

9.5 IMPLEMENTATION MODELS FOR FUNCTIONAL LANGUAGES

AOR eagerly evaluates parameter-expressions and binds the variables to the evaluated value to augment the existing environment. NOR defers the evaluation of parameter-expression until needed. Hence they prefer *lazy evaluation.* In order to avoid multiple evaluation of the same subexpression, they use *call-by-need* that caches the values of the subexpressions after the first evaluation. These values are looked up on subsequent occurrences of the same expression. The execution efficiency of *call-by-need* is better than *call-by-name* and closer *to call-by-value.*

9.5.1 SECD Machine and Eager Evaluation

SECD machine is an abstract machine for implementing λ-expressions. It has four stacks: S—evaluation stack for expressions; E—a stack that holds the corresponding environment: *(id, value)* pairs indexible by the id; C—control string; and D—stack of states of the machine at the time of function calls. SECD machine is a state transition machine. SECD machine has two versions: for eager evaluation and for lazy evaluation.

During β-reduction the (id ↦ value) pair is stored in the environment stack E, and the expression is moved one subexpression at a time to evaluation stack S, which picks up the value of the variables from the environment stack E and evaluates the expression. When a function call is made, then the current state of three stacks—S, E, and C—are dumped on the dump-stack D. After dumping, the evaluation stack is made empty, and the control jumps to the called function.

To explain the concepts, we will use stacks as a sequence and represent stack value after *push(<data>, <stack>)* as *<data>::<stack>*; and *stack value after pop(<stack>)* as *rest(<stack>).* The state transitions that are needed in a SECD machine are as follows:

1. The next element in C is a literal *<literal>*. Then it is put on the top of the stack S. The new state becomes *(<literal> ::S, E, rest(C), D).*

2. The next command is an identifier X. The value of the identifier X is looked up in the indexible environment *E*, and the value is pushed on top of the stack S. The new state becomes *(value-of(X)::S, E, rest(C), D).*

3. The next command is a λ-expression of the form *[<bound-variables>, <body>]*, then the closure *[<bounded-variables>, <body>, E]* is put on top of the evaluation stack S, and the new state becomes *([<bounded-variables>, <body>, E]::S, E, C, D)*.

4. The next state is a closure on top of the evaluation stack. In that case, the following actions are taken: (1) the triple *(S, E, C)* is pushed on the dump-stack D; (2) the bounded variables are associated with the parameter value, and the environment stack is updated with these bindings; and (3) evaluation stack is made empty. The state transition is given by *([bounded-variables, body,* E_1*]::<args>::rest(S), E, C, D)* → *(nil, {bounded_variables* → *<args>}* $\uplus$ E_1*, [<body>], (rest(S), E, C)::D)*, where the symbol '$\uplus$' denotes disjoint union.

5. If the top of the evaluation stack is a kernel function *<kernel> <args>*, then the kernel function is applied on the args, and the result is placed on top of the stack. The transition between the states is given as *(<kernel>::<args>::rest(S), E, C, D)* → *(evaluate(<kernel>(<args>))::rest(S), E, C, D)*.

6. If top of the command string is an apply function of the form *apply (<func>, <args>)*, then it is restructured as *<args> <func>* @, where the symbol @ is a delimiter for the apply function, and indicates that the function should be applied on the arguments. The state transition is *(S, E, apply(<func> <args>)::rest(C), D)* → *(S, E, <args>::<func>:: @::rest(C), D)*.

7. The command string is *nil*, and the dump-stack D is not empty. It means that the called function is over, and the control has to go back to the calling function. The dump-stack D is popped to get back the calling function's environment, evaluation stack, and next part of the control string of the calling function. The result from the called function is concatenated on top of the restored evaluation stack. The overall transition is *(<result>::rest(S), E, C, (*S^{prev}*,* E^{prev}*,* C^{prev}*)::*D^{prev}*)* → *(result::*S^{prev}*,* E^{prev}*,* C^{prev}*,* D^{prev}*)*, where *<closure>::<args>::*S^{prev}*,* E^{prev}*,* C^{prev}*,* D^{prev}*)* was the state before execution of the called function, and the evaluation of the *<closure>::<args>* yields the result on top of the evaluation stack S.

The functional forms if-then-else is handled by transforming the if-then-else functional to the form *<predicate> cond <func1> <func2>* @, skipping *<func2>* if the *<predicate>* evaluates to true, and ignoring *<func1>* if the evaluation of the predicate returns a false on top of the evaluation stack S.

9.5.2 Graph-Reduction Strategies

Graph-based reduction uses directed graphs to model expressions such that shared variables are treated as a single node, and common subexpressions are subgraphs that can be accessed using pointers. Graph-based representation supports call-by-need, as the value of a common subexpression is evaluated once and looked up subsequently. Figure 9.5 illustrates the graph-based modeling of λ–expressions, and the application of the operator

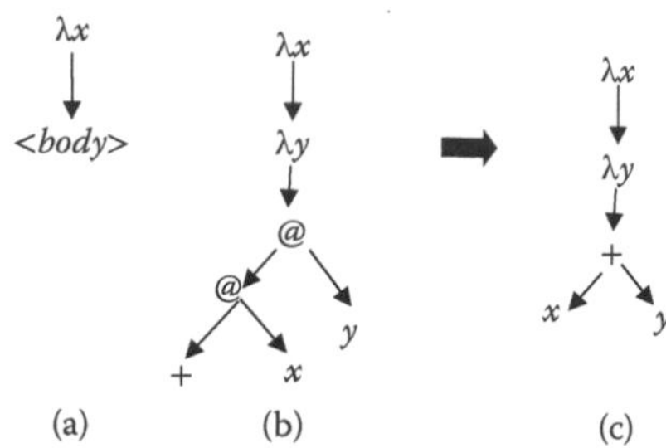

FIGURE 9.5 A direct graph representation of λ expressions. (a) Generic tree for an expression (b) tree for the expression "+ x y" (c) tree after simplification.

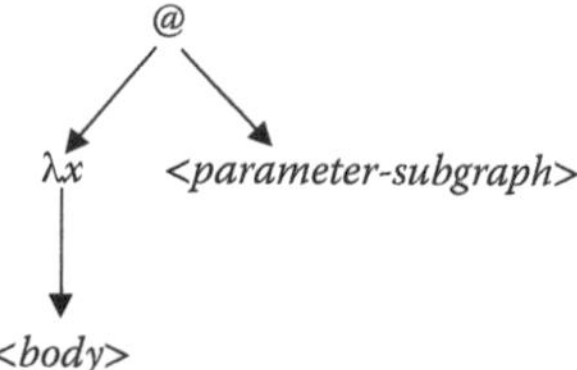

FIGURE 9.6 Representing graph for λ-expression with parameters.

to the arguments. A λ-expression *λx.<body>* is shown in Figure 9.5a. It is a tree representing a vertical directed edge from the declaration to the body of the expression. The nodes in a graph can be a bounded variable, an operator, or an apply-node, denoted by the symbol '@' that applies the argument to the body of the λ-expression using β- and δ-reduction. The left subtree of an apply-node represents the function, and the right subtree represents the argument. The body of λ-expression can contain another nested λ-expression.

The expression *λx. λy. (+ x y)* is given in Figure 9.5b. The simplified form of the graph representation could be to represent as (n + 1) tuple, where n is the number of arguments and the first argument is the function-name. The simplified version of the graph in Figure 9.5b has been shown in Figure 9.5c.

A λ-expression with parameters is represented as a tree as shown in Figure 9.6. The graph-reduction technique for NOR traverses the expression-graph of the form *<subgraph> <exp*$_1$*>, ... <exp*$_N$*>* to the left nodes until the node is not an apply-node denoted by the label @. The four valid possibilities for the graph G are (1) G is a single-node atomic data object, (2) G is another λ-expression, (3) G is a composite tuple with n ≥ 1, or (4) G is a primitive function of arity k. These four conditions of the graph have to be transformed using β- and δ-reductions. β- and δ-reductions are needed in two cases: (1) G is a primitive function, or (2) G is a user defined λ-expression. The β-reduction is done by short circuiting the edges to the body of λ-expressions and altering the edges in the subgraph representing the body of the λ-expression such that nodes corresponding to the substituted variables have an edge to the argument-subgraph. Multiple occurrences of the variables in the body are handled by setting multiple edges to the corresponding argument-subgraph. The edges are realized using pointers. During δ-reduction, the argument-subgraph is reduced. Because the variable nodes point to the subgraph using pointers, the evaluated value is accessible to all the occurrences without additional evaluation. The δ-reduction of the parameter-expression behaves like call-by-need, because the arguments are evaluated once, and multiple occurrences of variables get the evaluated value using pointers. This sharing of the reduced value of the argument-subgraph is the basis of the improved efficiency. However, the shared expression has to be partially *duplicated* if the shared expression is a function that is applied to different arguments.

Example 9.13

Figure 9.7 illustrates the representation and reduction of the λ-expression *λx. (+ x x) 4 * 5 + 6* using a graph. The expression has two occurrence of variable *x*. In NOR, both the occurrences of the variable *x* will be substituted with the expression *4 * 5 + 6* without using δ-reduction first. Thus, β-reduction would derive the reduced expression as *(+ 4 * 5 + 6 4 * 5 + 6)*. According to call-by-need optimization, only one occurrence of *4 * 5 + 6* should be evaluated, and the computed result should be stored for future lookup. Figure 9.7a shows the original graph. The parameter-expression has been converted into tuple form *(+ (* 4 5) 6)* and has been represented as an argument-subgraph, such that the operator has been put as the root node of the tuple representation, as discussed before. The graph-based representation allows this single evaluation of expression, because the expression *4 * 5 + 6* is represented by a single argument-subgraph.

Both occurrences of variable *x* point to the root of the argument-subgraph as shown in Figure 9.7b. Figure 9.7b shows the transformed graph after β-reduction: the edge to the node with *λx* symbol has been bypassed, and two edges that were pointing from the apply-node to variable *x* now point to the argument-subgraph. Figure 9.7c shows the transformed graph after the δ-reduction of the argument-subgraph. Note that the argument-subgraph is evaluated only once, and the use of pointers allows access to the computed value.

9.5.3 Implementing Lazy Evaluation

There are two popular abstract machines and their variations that have been developed to compile directed graphs using graph-based reduction. The abstract machines are G-machine and ABC machine. Graphs in G-machines are either apply-nodes with two arguments or leaf nodes. Graphs in ABC machine have variable arguments. G-machine translates the program to an intermediate level functional languages. A popular variation of G-machine is the *Spineless Tagless G-machine* (STG machine) that has been used to implement Haskell. STG machine uses the pointers as part of the heap-objects that takes to the corresponding code to be executed.

9.5.3.1 ABC Machine

ABC machine translates the NOR reduction of λ-expressions to a sequence of an abstract instruction set that is somewhere in between traditional von Neumann machine for imperative languages we have studied earlier and intermediate level functional language.

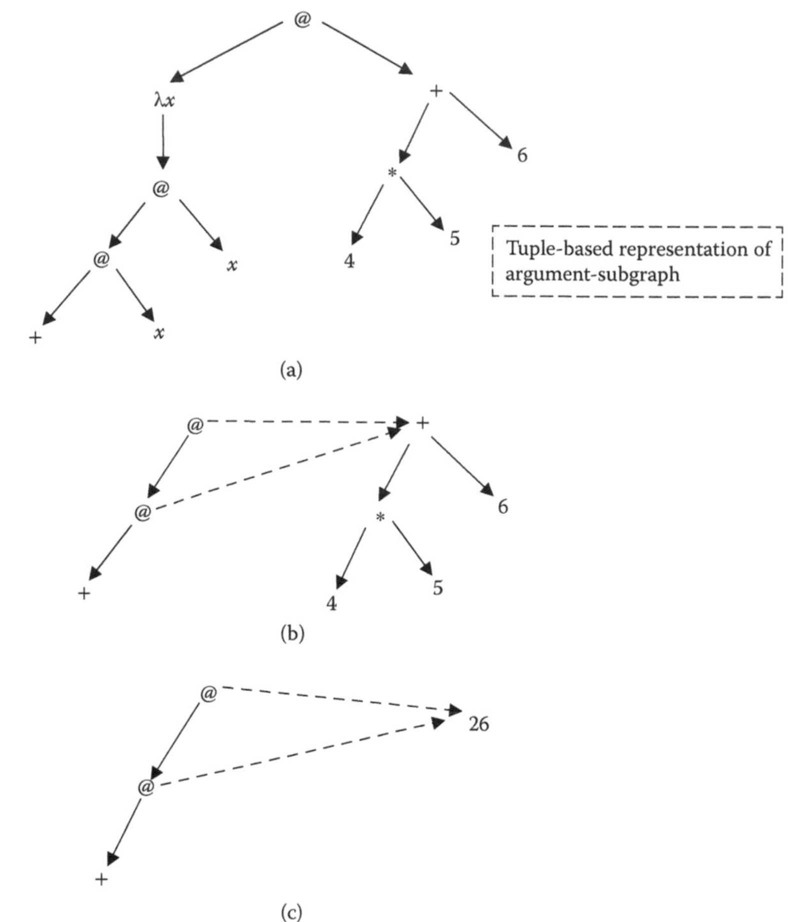

FIGURE 9.7 Graph reduction and argument-subgraph sharing. (a) Graph with expression "+ *x x*" and parameter "+ * 4 5 6"; (b) graph after parameter substitution; (c) graph after simplification.

ABC machine is composed of several types of memory storage. ABC machines have (1) a graph store to store the graph to be rewritten; (2) a program store to store the corresponding instructions; (3) A stack to store reference to the graph nodes; (4) B-stack like an evaluation stack in an SECD machine to handle reduction of basic values; (5) C-stack like a traditional control stack; (6) a descriptor store that translates the coded value to an actual symbol or type information; and (7) an I/O channel to display the results.

The state of an ABC machine is an 8-tuple consisting of the snapshots of various stacks and program store. ABC machine is a state-transition machine, and a microinstruction alters the current state. Microinstructions are classified as (1) get an instruction into a program store; (2) increment and update the program counter; (3) get a node, create a new node, delete a node, and update a node value; (4) extract information stored in a node; (5) redirect an edge to another node; and (6) get the description of a symbol from the descriptor-store. Based upon the micro-instructions, graph in the graph store is reduced. The basic values are evaluated in the B-stack. C-stack is used to perform nested reductions. C-stack also contains return addresses, and program counters can be stored and recovered from this stack.

9.5.3.2 *Strictness Analysis*

As discussed earlier in Section 9.2.1 NOR technique performs β-reduction from the outermost layer toward the innermost layer with demand-driven evaluation; the expression evaluation is deferred to improve the execution efficiency. The deferring of evaluation is called *nonstrictness*, and the strategy to implement *nonstrictness* is called *lazy evaluation.* Lazy evaluation is useful for infinite data structures, as the data structure is extended as needed. Lazy evaluation employs call-by-need to improve the execution efficiency as call-by-need evaluates a subexpression only once and memorizes the subexpression and the corresponding evaluated value to avoid re-evaluation of the subexpression in the future. An abstract technique is to represent the λ-expressions using directed graphs, and compile the directed graph to abstract instructions of underlying variations of G-machine or ABC-machine. The major concern in lazy evaluation is the substitution of complex parameter-expressions with multiple occurrence of bounded variables that are not eagerly evaluated causing memory and execution overhead of additional graph-reduction. Part of this overhead has been reduced by implementing call-by-need using shared argument-subgraphs in graph-based abstract machines.

Lazy evaluation also benefits by *strictness analysis*—a program analysis technique in an abstract domain—that finds out at the compile-time which expressions are potentially safe to be evaluated first to derive the output value before substituting the parameter-expression with bounded variables. An abstract domain is a domain that is used to study the abstract properties of a program using compile-time program analysis. For example, mode and type of values are abstract domains. Instead of deferring the evaluation of such expressions, the function-call evaluates these expressions before passing the parameter. In essence, strictness analysis transforms call-by-need used in lazy evaluation to a call-by-value for the selective arguments improving the execution efficiency.

9.6 INTEGRATION WITH OTHER PROGRAMMING PARADIGMS

Functional programming paradigm has been integrated with imperative programming paradigm in multiple programming languages to support mutable objects that in turn support memory reuse and memorizing the partial computation to be used later. Many languages such as Lisp, Scala, and Ruby have benefitted from this integration. Functional programming paradigm has also been integrated with object-oriented programming paradigm in many languages such as Scala and Ruby that have recently been gaining popularity as scripting languages. The object-oriented aspect of Scala and Ruby is discussed in Chapter 11.

9.6.1 Concurrency in Functional Languages

Functional languages have also been integrated with the concurrent programming paradigm. Concurrency can be exploited in functional programming languages by (1) making parallel binding of the values to the bounded variables, (2) concurrently reducing the arguments, and (3) spawning two separate processes for then- and else-function. Function closure can also be spawned as a separate process. In addition to exploiting concurrency to improve the execution efficient of functional programs, different implementations of functional programming languages also support thread-based programming, where each thread can spawn a separate function.

9.6.1.1 Concurrent Programming in Lisp Family

Different variations of common Lisp interface with a thread-based library such as Java's threads or x86 threads. Multilisp is a variant of Scheme with parallel programming constructs. Multilisp works in a single-shared address space. The idea of Multilisp is to compute in advance and suspend whatever cannot be computed. It has a construct called 'future' that forces eager evaluation of an expression or assignment statement. Owing to the presence of combination of fine- and medium-grain parallelism, Multilisp is suitable for shared memory multiprocessing.

9.6.1.2 Concurrency in Haskell

Haskell supports both implicit parallelism and concurrency. It supports concurrency using a construct called *forkio* that starts a thread that can run concurrently to other threads. It uses a construct called *MVar*. *MVar* is a shared box that is either full or empty. A thread trying to put more in an *MVar* is blocked until *MVar* becomes empty. A thread trying to take from an empty *MVar* blocks until *MVar* is full. *MVar* can be used as (1) a lock, (2) a shared channel between two threads for inter-thread communication, and (3) for asynchronous I/O. Haskell also uses *timeout*, which raises an exception if a thread continues beyond a certain time. The semantics of *timeout* is that an additional thread is started that throws an exception on the first thread after the specified time.

9.6.1.3 Concurrency in Scala

Scala uses both Java's concurrency model that is based upon synchronized methods and a thread-based concurrency library. In addition, Scala has a message-passing library, based upon programming language Actor that can be used to implement concurrent programs. In message-passing paradigm, the send and receive operations are asynchronous: the sender drops message in a system area called mailbox that is read by the receiver without the sender waiting for the receiver to read from the mailbox.

9.6.1.4 Concurrency in Ruby

Ruby has many variations and is still evolving. Different implementations support different forms of concurrency. Standard Ruby interpreter supports multithreading, multiprocessing, mutex locks for synchronization, conditional variables waiting for resources while in a critical section, and pipelining. JRuby—Java-based implementation of Ruby supports libraries for actor-based models and the use of channels.

9.7 SUMMARY

Functional programming paradigm is based upon mathematical function that maps input values to an output value. Functional programming is one of the two major declarative programming paradigms. Pure functional programming does not support destructive update, mutable objects, and global variables.

Mathematical functions are modeled using λ-expressions that have three components: bounded variables, expression-body and input parameters. Input parameter is an expression or another λ-expression. Given an input parameter, the bounded variables are bound left to right, and the resulting expression is simplified. Variables can be bound after simplifying the parameter-expression or before simplifying the parameter-expressions. If the parameter-expressions are evaluated before binding to the corresponding variables, the mechanism is called *eager evaluation*; if the evaluation of expressions is deferred and done only when needed, then it is called *lazy evaluation*.

The process of binding variables to input parameters is called β-reduction, and the simplification of arithmetic expression is called δ-reduction. λ-expressions can be nested, and nested expressions can be solved using two techniques: AOR or NOR technique.

In AOR technique, the innermost rightmost expression is reduced first, and outermost expressions are evaluated at the end; evaluation is done progressively from the inside toward the outside, and the reduction is done eagerly. Eager reduction means that the parameter-expressions are evaluated first using β- and δ-reduction before binding them to the variables within the body of the λ-expression. This scheme has both advantages and disadvantages. The advantage is that parameter-expression is evaluated once despite having multiple occurrences of the same bounded variables in the body of a λ-expression. This reduces the time and space overhead. The disadvantage is that even those parts of parameter-expressions are evaluated that may not be immediately needed causing execution time overhead.

NOR technique progressively evaluates a λ-expression from outermost leftmost layer first to the innermost rightmost layer, and defers the evaluation of parameter-expressions until needed. This kind of evaluation is called demand-based evaluation or lazy evaluation. The problem with NOR is that parameter-expressions may be nontrivial and may get bound to multiple occurrences of the same variable in the body of the λ-expressions. This leads to multiple reductions and increases the number of reduction steps. To reduce the reduction steps, expressions are represented as directed graphs such that functions and arguments are two children of an apply-node. The bounded variables in λ-expressions become a node in the graph, and the parameter-expressions are represented as a shared subgraph. β-reduction removes variable-declaration node, bypasses to the expression-body, and inserts edges from the node representing bounded variables to the corresponding parameter-expressions. Because edges are implemented using pointers and subgraphs are reduced only once, the parameter-expressions are evaluated once. Graph-based reduction of λ-expressions support call-by-need, because the parameter-expression is evaluated only once and used multiple times.

FPS (functional programming system) proposed by John Backus, is another alternate form of expressing functions. The major differences between λ-expressions and FPS are (1) the separation of the parameters from the functions, (2) clear identification of higher-order functional-forms that take functions as parameters to form more complex functions, and (3) clear definition of the set of kernel functions. FPS uses identity function to pull the value of a variable inside the function-definitions and uses constant function to include a constant in the function-definitions. The seven major functional forms are (1) insertion of a functional operator, (2) construction that applies a sequence of function on an input value to generate a sequence of output values, (3) apply-all that applies a function to a sequence of elements to generate a sequence of output values, (4) composition that applies a sequence of function in a specified order on the input values, (5) conditional that applies different functions based upon the outcome of a predicate, (6) iteration that repeatedly applies functions until a predicate is false, and (7) recursion that applies a function recursively.

FPS has become the basis of functional programming languages to define the kernel functions, and higher-order functional-forms. However, all the functional languages use constants and variables inside the definition of functions. Higher-order functions are formed using *map* function (called *apply* in Lisp), which takes a function name as input data along with its arguments and converts it into a function. Using *map* function, functional forms like *construction, apply-all,* and *insertion* can be simulated. Composition is allowed in functional languages, because functions can be called within a function. All the modern functional programming languages support conditional functions, iteration, and recursion. The recursive style of programming and iterators using linked lists is natural to functional programming languages, because sequences are traditionally modeled as linked lists.

The major drawback of the pure functional programming is (1) the lack of storage of the partial computation because of the lack of the support of global variables and destructive updates and (2) excessive use of recursive programming because of the lack of iterative style of programming. Many programming languages such as the Lisp family of languages have supported limited amount of destructive update and global variables. Most of the functional programming languages support while-loop functional and iterators that step through every element of a sequence. Common Lisp has an extensive iteration handling capability and supports definite-loops like *dotimes.*

Lisp is one of the earliest programming languages that mixes functional programming paradigm with the limited imperative programming paradigm. Lisp is a well-developed language with an extensive library, and it has multiple variations and descendants that integrate with other programming paradigms. Lisp supports implicit parametric polymorphism. Although traditionally the major representation of sequence is a linked list, arrays are also used extensively. Lisp also uses an associative property-list that supports (key, value) pairs. Lisp uses applicative order reduction with eager evaluation.

ML and Hope are polymorphic functional programming languages. Declarative parametric polymorphism was first introduced in ML. Both ML and Hope make extensive use of declarative polymorphism, Hope allows for the declaration of type variables and user-defined polymorphic types.

Haskell is a pure functional programming language that uses NOR reduction technique with lazy evaluation. Haskell is a polymorphic language that explicitly declares the signature of a function. Haskell uses a higher-order *map* function and *foldl* functional forms to derive new functions using functions as argument. Haskell also supports guard-based programming in addition to other abstractions.

Ruby and Scala are two modern multiparadigm polymorphic languages that integrate functional programming, mutable objects, and iterative constructs with an object-oriented programming paradigm. Scala is a statically scoped, strongly typed polymorphic language, whereas Ruby is a dynamically typed polymorphic language. Ruby and Scala are being used as scripting languages. Each value in both the languages is treated an object. Ruby extensively makes use of iterators and arrays to process the collection of data elements, Scala is a language built on top of Java, and uses JVM for a compiled version of the program. It interfaces well with Java and uses Java memory model for execution and concurrency in addition to actor-based message passing model. Scala supports higher-order functions, while-loop and conditionals like if-then-else.

Functional programming paradigm has been implemented using multiple different types of abstract machines. There are three machines, and their variations that have been used extensively: SECD machine, G-machine, and ABC machine. SECD machine has been used both for eager evaluation and lazy evaluation; G-machine has been used for graph-based reduction and uses an intermediate macro-based languages; and ABC machine mixes up low-order stack-based instructions with intermediate macro-based instructions for graph reduction.

SECD machine has four types of stacks: evaluation stack (S), environment stack E, command stack C, and dump-stack (D). Evaluation stack is used to evaluate expressions; environment stack is used to store the current environment; command stack is used to store the commands and sometimes to restructure the functions in the form that is more suitable for evaluation stack; and dump-stacks is used to store the state of (S, E, C) stacks before a new function is called. SECD machine has multiple reduction strategies to evaluate expression and call embedded λ-expressions.

ABC machine has eight important data structures: graph store, program counter, program store, A-stack for holding references to graph store, B-stack for holding basic values and evaluation, C-stack for control, program store, symbol-description store, and I/O channel for displaying the information. It has the capability to create, modify, and delete graph nodes for β- and δ-reduction.

The major drawback of lazy evaluation is increased overhead of expression reduction because of delayed evaluation of multiple occurrences of the same expression after β-reduction. The problem is handled partly by using call-by-need, where the value is cached after the first call-by-name evaluation to reduce the reduction overhead. This is easy in graph-based expressions, as multiple occurrences of variables are treated as edges to the same parameter-expression that is evaluated only once. The efficiency can be improved further by using strictness analysis that finds out the arguments that are always needed for the output value, and those are evaluated first. These parameter-expressions are evaluated as call-by-value to reduce the number of expressions being deferred.

Functional programming has been integrated with various programming paradigms. The most notable integrations with (1) imperative programming to exploit memory reuse and storage of previous computations for efficiency purpose, (2) object-oriented paradigm, (3) logic programming paradigm, and (4) concurrent programming paradigm. The major advantage of integrating functional and object-oriented programming is that methods are functions. The discussion of the functional and logic programming paradigms is in Chapter 10, and the discussion of integration with object-oriented paradigm is deferred to Chapter 11.

Concurrency has been exploited to incorporate (1) implicit parallelism in functional programming languages, (2) multiple threads, and (3) the "Actor" based model of message passing. Multilisp—a variation of Scheme—uses a construct *future* to evaluate the arguments to be reduced in advance provided the values for the variables are available. Haskell uses (1) forkio to spawn a thread; (2) uses an abstract construct called *MVar* that can be used for multiple purposes such as use of locks, and channels between two communicating threads; and (3) transactional memory for atomic operations. Scala uses actor-based message passing model for asynchronous communication between threads. Ruby supports extensive level of concurrency including multithreading, multiprocess invocations, mutual exclusion using locks and conditional variables, and pipelining.

9.8 ASSESSMENT

9.8.1 Concepts and Definitions

ABC machine; α-substitution; AOR; apply; apply-all; apply-node; β-reduction; call-by-name; call-by-need; call-by-value; composition; concurrent functional programming; conditional statement; construction functional form; δ-reduction; eager evaluation; forkio; FPP, functional form; future; G-machine; graph reduction; guarded functions; higher-order function; insertion functional form; iterator; kernel functions; λ-calculus; λ-expression; lazy evaluation; map function; MVar; nonstrictness; NOR; recursive function; SECD machine; STG machine; strictness analysis; traits; while-loop.

9.8.2 Problem Solving

1. Solve the following λ-expression using both AOR and NOR technique. Show the β- and δ-reductions clearly for each step of reduction.

   ```
   λx.λy.λz.(λa.λb.a*a + 2*a*b + b*b) x+yy+z)1 2 3
   ```

2. Solve the following λ-expression using both AOR and NOR technique. Show the β- and δ-reductions clearly for each step of reduction.

   ```
   (λx.λy.(λz.λw.w + 2*z - 2 + z*w) x-y x+y)1 2
   ```

3. Solve the following λ-expression using both AOR and NOR technique. Use α-substitution to resolve the name conflicts of variables. Show the β- and δ-reductions clearly for each step of reduction.

   ```
   (λx.λy.(λz.λx.(x*x + 3*x*z) 3*y 4*x)2 3
   ```

4. Use call-by-need to reduce the δ-reduction steps in NOR technique for problems 1 and 2.

5. Write a small function to pick up all the numbers that are smaller than a given number. For example, given a sequence *< 10, <7, 12, 9, 3, 15, 2, 16>>* will give a sequence *<7, 9, 3, 2>* by comparing all the elements in the sequence *<7, 12, 9, 3, 15, 2, 16>* with the number 10 and picking up all the elements smaller than 10.

6. Write a program to multiply two N × N matrices using at least one application in apply-all and then in Haskell using map construct.
7. Write a program to add two polynomials using a Lisp or Scheme program, and present a trace for the following polynomials:

   Polynomial 1: $4\,x^{20} + 5\,x^4 + 6x - 10$

   Polynomial 2: $5x^{18} - 5\,x^4 + 7\,x^2 + 8x + 5$

8. Write and execute a merge-sort program for Scala, Lisp, Hope and Haskell using the maximum amount of functional forms and compare your programs.
9. Write a recursive version of a function and then an iterative version of the function partitioning a sequence into two subsequences such that all the elements in the left part are smaller than the right part. Code the two versions in Lisp. Use dolist in the iterative version.
10. Write an FPS program to add a list of numbers.
11. Show a directed graph for the λ-expression $\lambda x.\ (+\ x\ x)\ 5 + 2$, and show the reduced graph after β- and δ-reduction.

9.8.3 Conceptual Type

12. What are different functional forms in FPS? Define each one of them with a clear example using FPS.
13. What are the major differences between Haskell and Lisp? Explain.
14. Explain the different classes of kernel functions in FPS.
15. Explain AOR and NOR techniques using a simple example.
16. Explain the functionality of SECD machine.
17. Explain the functionality of ABC machine.
18. Compare the control abstractions and data abstractions in Haskell and the Lisp family of languages.
19. Explain how NOR technique contributes to additional reduction steps, and how call-by-need and strictness analysis save the additional reduction steps. Explain by using a simple example.
20. Compare the control and data abstractions related to functional programming in Scala and Ruby.

FURTHER READING

Abelson, Harold, Sussman, Gerald J., and Sussman, Julie. *Structure and Interpretation of Computer Programs*, 2nd edition. Cambridge, MA: MIT Press. 1996. Also available at http://mitpress.mit.edu/sicp/full-text/book/book.html.

Backus, John. "Can programming be liberated from the von Neumann style? A functional style and its algebra of programs, ACM Turing Award lecture." *Communications of the ACM*, 21(8). 1978. 613–618.

Burstall, Rod M., MacQueen, David B., and Sannella, Donald T. "HOPE: An experimental applicative language." In *Conference Record of the 1980 LISP Conference*. 1980. 136–143.

Collingbourne, Huw. *The Book of Ruby*. San Francisco, CA: No Starch Press. 2011.

Daum´e III, Hal. "Yet another Haskell tutorial." 2002. Available at http://www.umiacs.umd.edu/~hal/docs/daume02yaht.pdf

Field, Anthony J. and Harrison, Peter J. *Functional Programming*. Wokingham, UK: Addison Wesley. 1988.

Gilmore, Stephen. "Programming in standard ML '97: A tutorial introduction." Laboratory for Foundations of Computer Science, The University of Edinburgh. Available at http://homepages.inf.ed.ac.uk/stg/NOTES/notes.pdf. September 1997, revised March 2004.

Halstead. Jr., Robert H. "Multilisp: A language for concurrent symbolic computation." *ACM Transactions on Programming Languages and Systems*, 7(4). 1985. 501–538.

Hudak, Paul. "Conception, evolution, and application of functional programming languages." *ACM Computing Surveys*, 21(3). 1989. 359–411.

Hudak, Paul, Hughes, John, Jones, Simon P., and Wadler, Philip. "A history of Haskell." In *HOPL III: Proceedings of the Third ACM SIGPLAN Conference on History of Programming Languages*. 2007. 12-1–12-55.

Hutton, Graham. *Programming in Haskell*. Cambridge University Press. 2010.

Jones, Simon P. "Parallel and concurrent programming in Haskell, a tutorial." Available at http://community.haskell.org/~simonmar/par-tutorial.pdf

Jones, Simon P. and Singh, Satnam. "A tutorial on parallel and concurrent programming in Haskell." In *Proceedings of the 6th International Conference on Advanced Functional Programming, AFP'08*. Springer-Verlag. 2009. 267–305.

Ng, Kam W. and Luk, Chi-Keung. "A survey of languages integrating functional, object-oriented and logic programming." *Microprocessor and Microprogramming*, 41(1). 1995. 5–36.

Odersky, Martin, Spoon, Lex, and Venners, Bill. *Programming in Scala*, 1st edition. Artima Press. (2nd edition), Walnut Creek, CA, USA, 2010.

Plasmeijer, Rinus and Eekelen, Marko van. *Functional Programming and Parallel Graph Rewriting*. International Computer Science Series. Boston, MA: Addison-Wesley. 1993.

Pucella, Riccardo. *Notes on Programming Standard ML of New Jersey*. Department of Computer Science, Cornell University, USA. 2001. Available at http://www.cs.cornell.edu/riccardo/prog-smlnj/notes-011001.pdf

Touretzky, David S. *COMMON LISP: A Gentle Introduction to Symbolic Computation*. Redwood City, CA: Benzamin/Cummings Publishing Company. 1990.

Ullman, Jeffrey D. *Elements of ML Programming*. Upper Saddle River, NJ: Prentice Hall. 1997.

CHAPTER 10

Logic Programming Paradigm

BACKGROUND CONCEPTS

Abstract computation and information exchange (Chapter 4); Abstract concepts in computation (Section 2.4); Abstract implementation (Chapter 5); Concurrent programming (Chapter 8); Data structure concepts (Section 2.3); Discrete structure concepts (Section 2.2); Functional programming paradigm (Chapter 9); von Neumann machine (Section 2.1).

Like functional programming, logic programming is a declarative programming paradigm. However, unlike functional programming, which is based upon theory of mathematical functions and λ-calculus, logic programming is based upon the theory of *predicate calculus*. As discussed in Section 2.2.2 predicate calculus is based upon *propositional calculus* and the *quantification*. Most of the logic programming languages are based upon *first order predicate calculus*. *First order predicate calculus* treats data and functions separately and does not have the capability to treat relations as logical terms, whereas higher-order logic programming can also treat relations as logical terms. *First order predicate calculus* is based upon the integration of *existential* and *universal quantification* with *propositional calculus*. Propositional calculus is based on declaring the axioms that we believe to be true and the use of logical operators, such as logical-AND, logical-OR, implication, and negation. The process of writing the program is to represent rules that can (1) derive new axioms using the given the rules and axioms or (2) solve a query given the rules and axioms.

The logic programming paradigm is quite comprehensible to humans, because it is based upon simple mathematical logic. Logic programming currently has many extensions such as *constraint logic programming, temporal logic programming, higher-order logic programming*, and *inductive logic programming. Constraint logic programming* includes constraints along with other predicates connected through logical operators. In a constraint logic program, both the predicates and constraints have to be satisfied. Constraint logic programming is used for optimization problems and it will be discussed in Section 10.5. *Temporal logic programming* also incorporates the notion of time interval and the ordering of events in the logic programming, and it will be discussed in Section 10.5. Logic can be *deductive* or *inductive. Deductive logic* derives new facts by deducing information from the existing facts. *Inductive logic* uses the rules and background information to derive general rules that govern all the given examples. Logic programming also supports meta-programming capability that treats a program as data, and it can be used to derive the properties of the program using program analysis in an abstract domain.

Logic programming has found applications in many areas, such as artificial intelligence, expert systems, genome comparison, optimization problem, game playing, drug discovery, and complex expert systems. Logic programs have the capability of generating multiple solutions because of nondeterministic nature, mostly because of the commutative nature of logical operators—logical-AND and logical-OR—in expressing a logic program. This capability of connecting several rules through logical-OR gives logic programs the capability to generate multiple solutions for a problem, because each fact can potentially generate a solution.

At the implementation level, logic programs use exhaustive search techniques, mostly depth first search, to derive a solution. Popular logic programming languages such as Prolog use a depth first search tree to implement the logic. Because a depth first search on a specific branch may not yield a solution, a mechanism to try other branches has to be designed. Prolog uses backtracking to explore alternate branches. More about backtracking will be discussed in Section 10.2.2.

10.1 LOGIC PROGRAMMING FUNDAMENTALS

A logic program is a set of procedures. Each procedure has a name and a finite number of arguments. Pattern matching and binding variables are used for parameter passing. In classical logic programming, parameter passing does not allow any evaluation of the actual parameters. Rather, it uses a technique called *unification* to exchange the information back and forth to (and from) the called subprogram and the calling subprogram. More about unification is discussed in Section 10.1.4.

10.1.1 Facts and Rules

Logic program contains two types of information: *facts* and *rules. Facts* are beliefs that the axiom is true, and rules are used to break up a complex query to simpler queries that can be progressively decomposed to yet simpler subqueries until a subquery can be pattern-matched with the given facts. Given an input from outside world, rules can also be used to derive new facts. There can be multiple rules connected through logical-OR to give multiple definitions to solve a complex problem. Every rule has a left-hand side (sometime also called Clausehead) and a right-hand side (sometimes called tail). The left-hand side is separated from the right-hand side using an implication symbol '←' (or ':-' in Prolog), which means that the left-hand side is implied by the right-hand side. Alternately, we can say that the left-hand side can be derived by the right-hand side. *Left-hand side* is a single logical term with probably multiple arguments (possibly 0), and *right-hand side* is a conjunction (logical-AND) of logical terms. The rules are connected to each other through logical-OR. However, two rules do not share their environment. The right-hand side of facts is trivially true, and no further reduction is needed.

Example 10.1

Let us study the example of rules and facts using a simple logic program expressed using predicate-calculus symbols: the symbol '∧' denotes logical-AND, the symbol '∨' denotes logical-OR, the symbol '¬' denotes negation, and the symbol '←' denotes implication. For our convenience, we will use the Prolog convention that variables start with a capital letter, and a constant starts with a lowercase letter. Note that this convention of variables in Prolog is different than classical logic, where variables start with lowercase letters. The denotation of variables is not the same as the notations used in mathematical logic and is specific to Prolog-like language. Let us take the following example:

```
1)((∀X ∀Y sibling(X, Y) ← ∃Z parent(X, Z) ∧ parent(Y, Z)
                               ∧ ¬(X == Y) ) ∨
2)(∀X ∀Y sibling(X, Y) ← ∃Z fraternity(X, Z) ∧ fraternity(Y, Z)
                               ∧ ¬(X == Y))
3)(parent(tom, mary) ∨
4)parent(neena, mary) ∨
5)parent(tom, john))
```

The above program shows facts and rules in a logic program. There are two rules (clause #1 and #2), and three facts (clause #3, #4, and #5). Clause #1 states that for all *X* and *Y*, *X* is a sibling of *Y* if there exists a variable *Z*, such that *Z* is a parent of *X* and *Z* is a parent of *Y* and *X* is not the same as Y. Clause #2 states that for all *X* and for all *Y*, *X* is a sibling of *Y* if there exists a variable *Z* such that *Z* is the fraternity of *X* and *Z* is the fraternity of *Y* and *X* is not the same as *Y*. Both the clauses are connected through logical-OR. Each clause has a left-hand side and a right-hand side. The right-hand side implies the left-hand side. The left-hand sides of clause #1 and #2 have the same name and same number of arguments. Programmatically, they form two clauses of a logical procedure. The three facts also have the same name and same number of arguments, and are connected to each other through logical-OR.

If we look closely, the variables *X* and *Y* on the left side of the rules are universally quantified, and the variable *Z* that occurs exclusively on the right side of a rule is existentially quantified. The predicates on the right side are connected through logical-AND. This is the repeat pattern in every rule. Logic programming language designers have simplified this predicate calculus version to more human-comprehensible version by (1) dropping universal and existential quantifiers, (2) putting a comma instead of logical-AND, (3) putting a period '.' instead of logical-OR to separate the rules and facts, (4) replacing the negation symbol '¬' with 'not,' and (5) replacing the implication symbol '←' with ':-'. Using these substitutions, the program in example 10.1 will look as follows:

```
sibling(X, Y):- parent(X, Z), parent(Y, Z), not(X = Y).
sibling(X, Y) :- fraternity(X, Z), fraternity(Y, Z), not(X = Y).
parent(tom, mary).   parent(neena, mary).   parent(tom, john).
```

The above program is a typical program syntax used in language like Prolog. In the future, we will use this syntax to write a Prolog program.

10.1.2 Forward and Backward Reasoning Systems

These rules and facts in a logic program can be used to solve a problem in two different ways: (1) forward reasoning and (2) backward reasoning. Forward reasoning system takes the known facts and uses the rules to derive new facts that are stored in the database. Backward chaining system takes a query and progressively resolves the query to a combination of simpler queries connected through logical operators, until the subqueries can be solved by low-level kernel functions or by matching the facts.

In a forward chaining system, the process of applying rules to derive facts is done *eagerly*, whether the derived facts are useful to solve the problem at hand or not. That means lots of redundant computations are also done. This becomes an efficiency issue. Despite that forward reasoning systems are good for many types of applications that try to predict all possible outcomes if a particular condition develops. For example, if we are monitoring a nuclear power plant, and the temperature crosses a threshold, we would like to know if this change is causing any disaster in the system. Forward chaining systems are good for such systems. However, as we can see that forward chaining systems are computationally expensive, as they apply the rules to large number of facts. New derived facts become part of the fact database. Example 10.1 illustrates a simple example of forward chaining system. The program derives a new fact *sibling*(*tom*, *neena*) that becomes part of the database.

A backward reasoning system is more focused. It uses depth first search to find a solution and returns a value as soon as the query is satisfied. A backward chaining system starts from the final state to be derived and keeps breaking up the query to smaller queries using pattern matching and parameter passing using rules that have the same name and matching arguments. In the end, if the smaller subqueries can be satisfied by the facts or kernel functions, then the query is satisfied, and the answer is returned. Unlike forward chaining systems that derive all possible outcomes, backward reasoning systems answer a specific question in a focused way.

There are programming languages that use either a forward chaining system or a backward reasoning system to implement their abstract machine. Prolog—a popular logic programming language—uses a backward reasoning system. An example of forward reasoning languages is OPS5.

10.1.3 Data Representation

The primary data representations in logic programming are sequences, tuples, n-ary tree, and facts. However, many implementations of logic programming include in their data representations very rich set of data structures such as dynamic arrays, associative maps, graphs, blackboards, unordered sets, and ordered sets. A list is included within the square brackets, and elements are separated by comma. For example, *[a, b, "Arvind Bansal"]* is a list. An empty list '[]' does not have an element. An n-ary tree is represented as a functor of the form *<functor-name>(Arg1, … ArgN)*. A tuple is represented within left and right parenthesis. For example, *(4, 5, 6)* is a tuple. Most of the programming either uses functors or lists. A list is internally represented as a binary tree such that the left-hand side child is the first element of the list, and the right subtree is the rest of the list. Concatenation

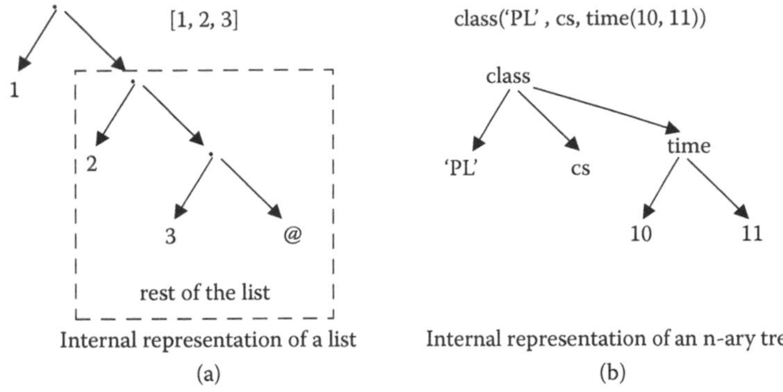

FIGURE 10.1 Internal representation of logical terms.

between the first element and the rest of the list is denoted by a vertical bar '|.' The list *[1, 2, 3]* can be represented in many ways: *[1| [2, 3]]*, *[1, 2 | [3]]*, and so on. Basically it is the same binary tree representation, and all three representations are syntactic sugar. Figure 10.1 illustrates the internal representation of a linked-list and a functor.

A logical variable is an immutable object at the programmer level. However, the implementation engine can unbind the value, as discussed in Section 10.2.3. We will use Prolog convention to represent variables. A variable is represented with the first letter as an uppercase letter followed by zero or more letters, digits, or underscore symbol. A literal that starts with a lowercase letter. The constants starting with uppercase letters have to be enclosed between single quotes. For example, *Xs* is a variable, cs is a literal, 'PL' is a literal, and 1 is a literal. An atom could be a literal, an integer, and a floating point number.

A logical term is defined recursively: a logical term can be an atom, a variable, a list, a tuple, a functor, or a combinations of logical terms. Binding a variable in a logical term is called instantiation of the logical term.

Example 10.2

For example *class('PL,' cs, time(10, 11))* is an instance of the functor class(Course, Department, Time): the instantiation has been derived by binding variables Class, Department, and Time to the values 'PL,' cs, and time(10, 11), respectively. A logic term is called a *ground term* if it contains all constants and is called a nonground term if it contains at least one unbound variable.

10.1.4 Unification—Bidirectional Information Flow

Unification is a means of equating two logical terms by finding out the minimal set of variable bindings, which when applied to both the logical terms would make them identical. For example, let us consider two logical terms: *a(1, X)* and *a(Y, 2)*. If we take a set of variable bindings *{X/2, Y/1}* and apply both the logical terms, then the instance of both logical terms would become *a(1, 2)*. In this case, we say that two logical terms are unifiable with the unifier *{X/2, Y/1}*.

Unification is a very different notion compared to assignment used in imperative languages and functional programming languages. The typical assignment operation evaluates the right-hand side expression and binds the resulting value to the memory location of the variable on the left-hand side. Information flow in assignment operation is one directional: from the right-hand side to the left-hand side. Unification is different from assignment operation as follows:

1. There is no evaluation of an expression in unification.
2. Both the left-hand side and right-hand side have equal status.
3. The information flow is bidirectional in unification.
4. Unification performs both pattern matching of constants and binding of variables to a logical subterm.
5. When a variable is bound to a value, then all the occurrences of variables in both the left-hand side term and right-hand side term get bound to this value.

The difference will be clear if we unify *X* and 3 + 4. In an assignment statement, the memory location corresponding to the variable *X* will be bound to value 7. In unification, the right-hand side is not evaluated and the variable *X* is bound to the term 3 + 4 as it is. If we unify 3 + *X* and *Y* + 6, then *Y* will be bound to value 3, and *X* will be bound to value 6. Unification of (a, *X*, *X*) and (a, 3, *Y*) will yield the set of bindings {*X*/3, *Y*/3}. First *X* gets bound to 3. Because all occurrences of *X* will bind to 3, the third argument of the first logical term also gets bound to value 3. When third arguments of both the logical terms are matched, then the variable *Y* in the second logical term also gets bound to value 3.

Example 10.3

Unification of logical terms b(*X*, *Y*, *[3, a(d, e)]*) and b(*N*, *M*, *[M, N]*) would give the binding set {*X*/a(d, e), *Y*/3, *M*/3, *N*/a(d, e)}. The whole substitution steps are given in Table 10.1.

Aliases are denoted by $\text{Var}_1 \parallel \text{Var}_2$. In this example, the variables *X* and *N* are aliases, and the variables *Y* and *M* are aliases. Substitution of a variable by a subterm will be shown as Var/subterm. In case of aliases, we can substitute logical terms with any of the aliased variables. When one of the variables is instantiated in set of aliased variables, then all other

TABLE 10.1 Unification Process of Two Logical Terms

	Binding	Binding Set	Instantiated Term 1	Instantiated Term 2
1		{}	`b(X, Y, [3, a(d, e)])`	`b(N, M, [M, N])`
2	`X ‖N`	`{X‖N}`	`b(N, Y, [3, a(d, e)])`	`b(N, M, [M, N])`
3	`Y ‖M`	`{X ‖N, Y ‖ M}`	`b(N, M, [3, a(d, e)])`	`b(N, M, [M, N])`
4	`M/3`	`{X ‖N, Y/3, M/3}`	`b(N, 3, [3, a(d, e)])`	`b(N, 3, [3, N]))`
5	`N/a(d, e)`	`{X/a(d, e), Y/3, M/3, N/a(d, e)}`	`b(a(d, e), 3, [3, a(d, e)])`	`b(a(d, e), 3, [3, a(d, e)])`

variables in the set of aliased variables also get instantiated, because aliasing is an equivalence relationship that means aliasing is reflexive, symmetric, and transitive.

We can define the unification process as comparing the corresponding logical subterms position by position, and perform one of the following operations:

1. If both the subterms are uninstantiated variables, then they are aliased to each other, and either variable can be used for substitution in the logical terms. All the occurrences of variables are substituted.
2. If one the logical subterm is an unbound variable, then it is substituted by the other logical subterm, and all the occurrences of the variable are substituted by the corresponding logical subterm. In case the variable is aliased, then all the aliased variables in the set are bound to the same logical subterm.
3. If both the logical subterms are ground terms, then they are matched. The unification fails if they do not match. If the unification fails, then a substitution set is not created.

An algorithm for unification process is given in Figure 10.2.

The logical term T_1 and T_2 are copied into instances I_1 and I_2, which are progressively instantiated further when a new substitution is derived after matching the next position. The Boolean variable *unified* is initialized to *true* and is made *false* when two logical ground subterms do not match. The program exits prematurely out of the loop if the Boolean variable *unified* becomes false. After getting out of the loop, if the Boolean variable *unified* is true, then the binding set *S* and the most general common instance *I* that is the same as I_1 and I_2 is returned in the form of a tuple (S, I). The mathematical operator '⊕' denotes the insertion of the new binding in the binding set. The mathematical operator ⊙ denotes the application of a binding on a logical term to find a new instance.

```
Algorithm unify
Input:  Two logical terms: T1 and T2
Output: Binding set S;

{   I1 = T1; I2 = T2; S = {}; unified = true;
    for each position  p in I1 and I2
    {   if (is_variable(I1(p) ) ∧ is_variable(I2(p)) then
                      S = S ⊕ I1(p) || I2(p);
                  elseif (is_variable(I1(p)) ∧ non-ground(I2(p))) then
                             S = S ⊕ I1(p)/ I2(p);
                  elseif (non-ground(I1(p)) ∧ is_variable(I2(p))) then
                             S = S ⊕ I2(p) / I1(p);
                  elseif (ground(I1(p)) ∧ ground(I2(p)) ∧ ¬ (matches(I1(p), I2(p))))
                             {unified = false; exit}
                  else {Sp = unify(I1(p), I2(p)); S = S ∪ Sp;} % unify non-ground terms
                  I1 = I1 ⊙ S; I2 = I2 ⊙ S;
    }
    if unified   then return  S;
}
```

FIGURE 10.2 An algorithm for unification of two logical terms.

Insertion of new bindings can perform two operations as follows:

1. If the variable is part of an aliased set, then a transitive operation is applied on every variable on the aliased set to bind every aliased variable in the set to the corresponding value.
2. If the variable is not part of an aliased set, then the binding is just included in the binding set.

For example, $\{X||N, Y||M\} \oplus X/3$ is equivalent to $\{X/3, N/3, Y||M\}$. The operator $\odot$ shows the application of the bindings to instantiate the logical terms. For example, if the logical term is $b(X, Y)$ and the set of bindings is $\{X/4, Y/[a, b, c]\}$, then $b(X, Y) \odot \{X/4, Y/[a, b, c]\}$ will instantiate the logical term to $b(4, [a, b, c])$.

10.1.4.1 Assignment in Logic Programs

Logic programs also support assignment statements in addition to unification. An assignment statement is used to evaluate arithmetic expressions and bind the resulting value to the variable in an assign-once manner. Unlike imperative languages, assignment in logic programs is not destructive, because variables are not mutable in standard logic programming. However, there are some implementations of Prolog that allow mixing of mutable variables. The symbols for unification and assignment are quite different than imperative style of programming. Prolog—a popular logic programming language uses the '=' sign for unification and 'is' to denote the assignment. For example, the statement *M1 is M − 1* evaluates the expression *M − 1* and binds the value it to the variable *M1*.

10.1.5 Representing Logic Programs

Logic program consists of *rules* and *facts*. The right-hand side of rules has one or more predicates. The right-hand side of facts is always true and is omitted while writing a program. The rules with the same name and number of arguments are grouped together and are called a *procedure*. Each rule is called a *clause*. Formally we can define a logic program as a set of *procedures*. Each *procedure* is a set of clauses that are related to each other through logical-OR. Each *clause* has a left-hand side that is implied by the conjunction of the right-hand side. The left-hand side of a clause is called clausehead, and the predicates on the right-hand side of a clause are called *subgoals*.

Example 10.4

In the following example, we are using standard Prolog syntax for convenience.

```
factorial(0, 1).
factorial(M, N) :- M > 0, M1 is M - 1, factorial(M1, N1),
                   N is M * N1.
```

```
fibonacci(0, 1).            fibonacci(1, 1).
fibonacci(M, N) :-
     M > N, M1 is M - 1, M2 is M - 2,
     fibonacci(M1, N1), fibonacci(M2, N2),
     N is N1 + N2.
```

The above program has two procedures: *factorial/2* and *fibonacci/2*. The number 2 shows the number of arguments. The procedure *factorial/2* has two arguments: the first an input argument and the second an output argument. It has two clauses: the first clause is a fact showing the base case, and the second clause is a rule showing the recursive definition. Similarly, the procedure *fibonacci/2* has two arguments: the first is an input argument, and the second is an output argument. It has three clauses: the top two facts are base cases, and the last clause is a recursive clause.

The first clause of the procedure *factorial*/2 reads as "factorial(0) is 1." The second clause of the procedure *factorial*/2 reads as (1) *factorial(M)* is defined for $M > 0$, computes the value of the expression $M - 1$ and stores it in the assign-once variable *M1*, recursively compute the factorial of *M1*, and multiplies the value of the variable *M* with the value of the output variable of the recursive call. There are four subgoals in the recursive clause of the factorial.

10.1.6 Properties of Logic Programs

Logic programs are inherently nondeterministic and support parametric polymorphism. The nondeterministic computation is present because of commutativity in logical-OR between the rules and commutativity in logical-AND between the subgoals. Logic programs also support inherent parametric polymorphism. The logic programs use four logical operators extensively. The operators are *logical-OR, logical-AND, implication,* and *negation*. Logical-OR is used between the clauses, logical-AND is used between the subgoals on the right-hand side, implication is used between the right-hand side and the left-hand side, and negation is used with the subgoals to negate the outcome of a predicate.

Example 10.5

Let us take the following logic program to append two linked-lists. The program can be used for any two types of lists. The polymorphic type for the program is List(α) × list(β) → List(α ∪ β). That means the first list could be of any type, and the second list could be another type. The output list contains elements of both the types.

```
append([], Ys, Ys).
append([X | Xs], Ys, [X | Zs]) :- append(Xs, Ys, Zs).
```

In the above example, the predicate append/3 has three arguments: first two arguments are input arguments and the third argument is an output argument. The first clause of the above program reads that appending an empty list to another list gives the second list. The second clause says, if the first list is nonempty, then the first element *X* of the first list is the first element of the output list using unification and rest of the output list is built recursively by appending the rest of the first list with the second list.

The program works in two different modes, which are as follows:

1. In deterministic mode, we can give any two lists and derive the third list. For example, append (*[1, 2]*, *[a, b]*, *Zs*) will give the value of *Zs* as *[1, 2, a, b]*.
2. In a nondeterministic mode, we can give a query append(*Xs*, *Ys*, *[a, b]*), and the program will generate values for the first two arguments. The possible values are [] and *[a, b]*; or *[a]* and *[b]*; or *[a, b]* and [].

10.2 ABSTRACT IMPLEMENTATION MODEL

The implementation of a given logic program is at two levels: (1) mapping the execution of a logic program to an *AND-OR tree* and (2) mapping the AND-OR tree on a low-level abstract machine supported by von Neumann machine. *AND-OR tree* is a logical tree. The properties are as follows:

1. There are two types of logical nodes: *AND-nodes* and *OR-nodes*. *OR-nodes* are true if one of their children is true. *AND-nodes* are true if all their children are true.
2. The levels of AND-nodes and OR-nodes alternate: the children of AND-nodes are OR-nodes, and the children of OR-nodes are AND-nodes.

Figure 10.3 shows a schematics of AND–OR tree. The node at level 0 is an OR-node, the nodes at level 1 are AND-nodes, the nodes at level 2 are OR-nodes, and so on.

10.2.1 Query Reduction

Logic programs solve a problem by asking a query in which the variable that needs to get a value is not bound to any value in the beginning. The complex query is matched with the left-hand side of a rule having the same relation-name and number of arguments, and the set of bindings is added to the environment. In classical logic programs, unification is used for pattern matching. After the unification, the set of derived bindings are applied to the right-hand side subgoals to instantiate subgoals that split the query to a conjunction of simpler subqueries. Each subgoal is solved repeatedly, unless facts are reached or a kernel function is reached. If the unification with a fact succeeds, then the corresponding subgoal returns *true*. If all the subgoals are true, then the left-hand side of the corresponding rule

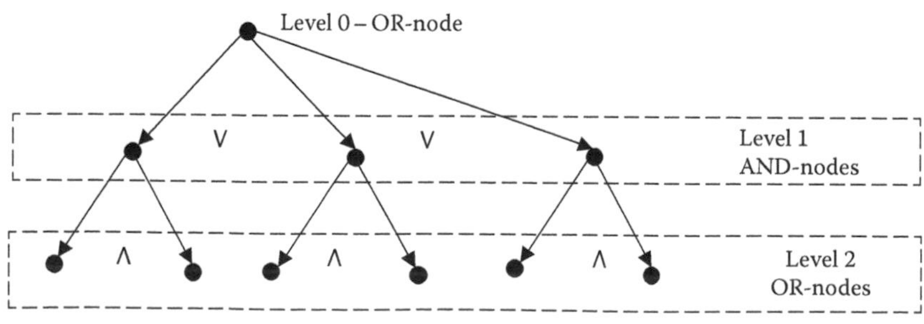

FIGURE 10.3 An AND-OR tree.

becomes *true*. The process is repeated. If the root node is true, then the query succeeds, and the value of the variable is picked from the environment. Note that the environment of the each rule is separate from other rules' environments, and the multiple values may be bound to a variable: one or more values for each successful rule.

Example 10.6

Let us consider the logic program given in Example 10.1:

```
sibling(X, Y) :- parent(X, Z), parent(Y, Z), not(X = Y).
sibling(X, Y)  :- fraternity(X, Z), fraternity(Y, Z), not(X = Y).
parent(tom, mary).    parent(neena, mary).    parent(tom, john).
```

Assume that we want to derive siblings of Tom. We will ask a query: Who is a sibling of Tom? A query is stated as *sibling(tom, M)?*. This query is progressively reduced using rules to the fact level. First we look for rules that have the same name 'sibling' and two arguments. Two rules are identified. One can take either of the rules first because of logical-OR between the two rules. Let us assume that the first rule is selected first. Currently, the environment is empty. The query *sibling(tom, M)* is unified with the left-hand side of the rule, and the set of bindings are put in the environment. So the environment becomes *{X/tom, Y/M}*: the variable *Y* has been substituted with the variable *M*, because *M* is existentially quantified and *Y* is universally quantified. This set of bindings *{X/tom, Y/M}* is applied on the subgoals on the right-hand side of the clause. The three subgoals are instantiated to *parent(tom, Z)*, *parent(M, Z)*, and *not (tom = M)*. At this point, there are two ways to solve the subgoals: (1) solve in the left to-right order, and follow the data dependency caused by producer–consumer relationship because of shared variable *Z* or (2) solve the subgoals *parent(tom, Z)*, and *parent(M, Z)* individually, and then take the intersection of the set of values for variable *Z* returned from both the subgoals. Logic programs use the first approach of left-to-right order and follow the data dependency. When we solve the query *parent(tom, Z)*, it can unify with two facts giving two possible values: *mary* and *john*. Let us assume that we pick up the correct value mary so that the variable *Z* is bound to the value *mary*. The new environment is *{X/tom, Y/M, Z/mary}*. This binding *Z/mary* is applied to all the subgoals on the right to make the remaining subgoals *parent(M, mary), not (tom = M)*. The subgoal *parent(M, mary)* is unified with the corresponding facts. Only two facts unify giving the possible values of *M* as *tom* or *neena*. If we bind the variable *M* to the value *tom*, then the third subgoal *not (tom = tom)* will fail. Let us assume that we again make the correct choice, and bind thc variable *M* to the valuc neena. The new environment is *{X/tom, Y/M, Z/mary, M/neena}*. The binding *M/neena* is applied to the third subgoal that gets instantiated to *not (tom = neena)*, which is true. That all three subgoals are true implies that left-hand side of the rule is true. That means the top-level query is true. The value of *M/neena* is picked up from the environment and returned.

Although in theory logic programs can be executed in nondeterministic order, it is not always possible because of (1) the dependences introduced by producer–consumer relationship between the shared variables and (2) the nontermination of recursive clauses in some cases.

For example, let us take the logic program for factorial/2 in Example 10.4. The variable *M1* is a shared variable, and unless its value is computed, recursive definition of factorial cannot be invoked.

10.2.2 Mapping Query Reduction to AND-OR Tree

Solving a logical query can be mapped on AND-OR tree as follows:

1. Top-level query is mapped on an OR-node or a conjunction of OR-nodes.
2. Clausehead is mapped on an AND-node.
3. The subgoals on the right-hand side of a clause are mapped to the next level OR-node.
4. Unification occurs on the edges connecting OR-node to an AND-node.

Using this mapping, a query-reduction process basically becomes an AND-OR tree. The schematic is show in Figure 10.4.

Figure 10.4 shows the AND-OR created during the reduction of the first clause. The part of the AND-OR tree made by query reduction using the second clause is similar to the reduction of the first clause. The solid arrows show the edges in the AND-OR tree, and the dashed arrows show data dependency caused by producer–consumer relationship between the subgoals because of shared variables. The facts are shown by the symbols *F1, F2*, and *F3*: *F1* denotes the fact *parent(tom, mary)*; *F2* denotes the fact *parent(neena, mary)*; and F3 denotes the fact *parent(tom, john)*. The thick solid arrows show successful unifications, and shaded single arrows show failed unification.

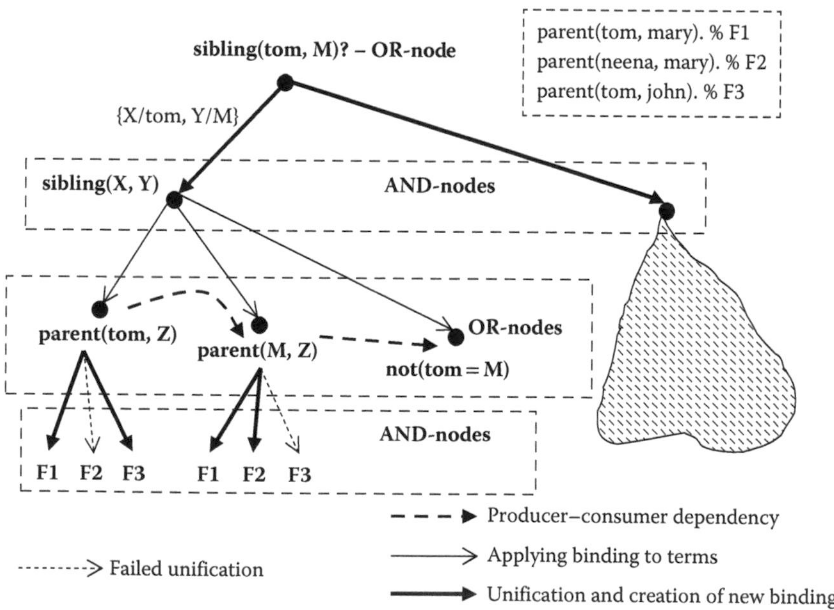

FIGURE 10.4 Mapping query reduction to an AND-OR tree.

For example, the subgoal *parent(tom, Z)* unifies successfully with facts *F1* and *F3*. The unification of *parent(tom, Z)* and *F1* derives a possible value for the variable *Z* as *mary*, and the unification of *parent(tom, Z)* and *F3* derives a possible value of the variable *Z* as *john*. Similarly, the subgoal *parent(M, mary)* successfully unifies with either of the facts *F1* or *F2* giving the possible values of the variable *M* as either *tom* or *neena*, which is passed to the third subgoal.

Till now we have been assuming that the underlying engine that implements an AND–OR tree is capable of picking up the correct bindings that can be applied to the right-hand side subgoals for a successful solution. In reality, finding out a solution is a search space problem, and exhaustive search is used to find out a solution. Depth first search has been used to implement sequential version of logic programs because depth first search moves in a focused way to find out the solution quickly and needs less memory to explore. Variations of depth first search and breadth first search have been used to implement AND–OR trees for concurrent implementation of logic programs. Here, only depth first search-based implementation is discussed, because Prolog uses depth first search. Depth first search means executing the subgoals in left-to-right order and selecting the clauses in the textual order from top to the bottom.

A regular search tree has the same type of nodes. However, an AND–OR tree has two different types of nodes. OR-nodes have option of exploring subtrees rooted at alternate children, whereas AND-nodes provide only substitution to their children. OR-nodes are called *choice-points*, because they have alternatives to explore, and a *trail* of choice points is kept to facilitate the exploration of the alternate paths in case the previous exploration does not yield a solution.

10.2.3 Backtracking

The mechanism to go back to the previous *choice-points* using the trail and to explore alternate possibilities in the search tree is called *backtracking*. To store the trail of choice-points, a *trail-stack* (or *OR-stack*) is needed in addition to a control stack and a heap. The purpose of the trail-stack is to mark all the choice-points in reverse order and unbind all the variables that were bound after or during unification in previous choice-points. Each trail is a pair of the (choice-point marker and a sequence of references to the variables that have been bound). Abstract operations performed by backtracking are as follows:

Step 1: Pop the last choice-point from the trail-stack.

Step 2: Unbind all the variables that were bound between the last choice-point (including the binding caused by unification between OR-node and AND-node at the last choice point) and the failed subgoal.

Step 3: Pick up next unexplored AND-node, and unify the OR-node corresponding to the choice-point and the next AND-node to pick up new set of bindings.

Step 4: If all the alternatives are consumed at the last choice-point, then go back to step 1 and keep repeating until the trail-stack is empty or a solution is found.

Example 10.7

Let us understand backtracking using the following program, and a query *sibling(billing, W)?*

```
sibling(X, Y) :- parent(X, Z), parent(Y, Z), not(X = Y).
parent(billy, linda). /* fact F1 */
parent(billy, bill). /* fact F2 */
parent(john, bill). /* fact F3 */
```

In the above program there are two procedures: *sibling/2* and *parent/2*. The procedure *sibling/2* has one clause, and the procedure *parent/2* has three facts. The three facts will be denoted as *F1, F2*, and *F3* for the convenience of illustration in Figure 10.5; the fact *parent(billy, linda)* is denoted as *F1*; the fact *parent(billy, bill)* is denoted as *F2*; and the fact *parent(john, bill)* is denoted as fact *F3*. Figure 10.5 illustrates an AND–OR tree to solve the query *sibling(billy, W)?*

We assume left-to-right traversal of the AND–OR tree. The dashed arrows show the backtracking to the previous choice-point. The list of choice-points keeps changing dynamically on the stack. Given the query *sibling(billy, W)*, the logical terms *sibling(billy, W)* and *sibling(X, Y)* are unified to derive the set of bindings as *{X/billy, Y/W}*. The right-hand side subgoals become *parent(billy, Z)* ∧ *parent(W, Z)* ∧ *not (billy = W)* after the application of the set of bindings. The subgoals are executed in left-to-right order. The first choice-point is the OR-node *parent(billy, Z)*. The subgoal can unify with the facts *F1* and *F2*. First we unify the subgoal *parent(billy, Z)* with the fact *F1*. The unification binds the variable *Z* to the constant value *linda*. The binding *{Z/ linda}* is applied on subgoals to the right. The remaining subgoals become *parent(W, linda)* ∧ *not (billy = W)*. The subgoal *parent(W, linda)* is the second choice-point. It can unify with the fact *F1*. This unification derives the binding *{W/billy}*. This binding {*W/billy*} is applied on the remaining subgoal *not (billy = W)* to get the instantiation *not (billy = billy)*, which fails. The program backtracks

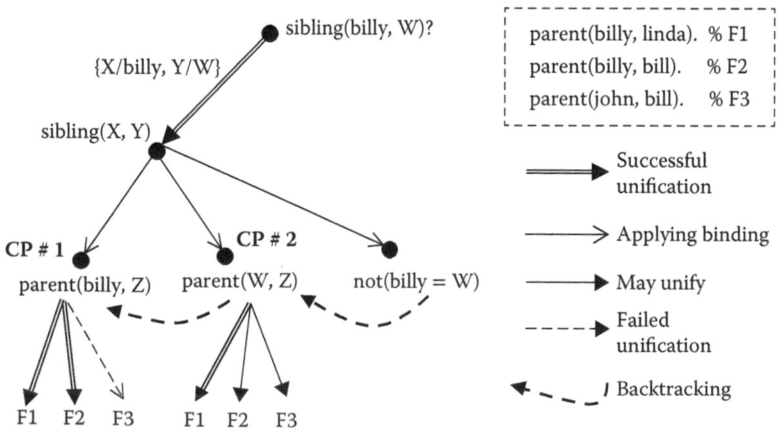

FIGURE 10.5 Illustrating backtracking in AND-OR tree.

by popping the second choice-point and unbinding the value of the variable *W*. Because there is no other alternative for CP #2 (choice-point #2) that would successfully unify, backtracking pops CP #1 from the trail-stack, and unbinds the variable *Z*. Now the next clause *F2* is tried. This returns the value of the variable *Z* as *bill*. The choice-point CP #1 is pushed back again on the trail-stack, and the binding *{Z/bill}* is applied on the subgoals on the right to derive the instantiated subgoals *parent(W, bill)* ∧ *not (W = bill)*. The second subgoal succeeds with the binding *{W/billy}*. The third subgoal fails because *not(billy = billy)* is false, and the program backtracks again unbinding the variable *W*. The new value of the variable *W* is *john*. The third subgoals succeeds with this new binding as *not (billy = john)* is true.

10.2.4 Warren Abstract Machine

Warren abstract machine (WAM) is a low-level abstract machine to compile AND-OR trees on a von Neumann abstract machine. It uses a hash table to jump to specific procedure, if-then-else statements to try out alternative clauses, registers to hold the references to the arguments, and a trail-stack in addition to control stacks and heaps to support backtracking. The major difference between an abstract machine for imperative languages and the WAM is the presence of the trail-stack to handle backtracking.

Local variables bound to constants are stored in the control stack or processor registers, and complex logical terms and other variables involving lists and n-ary structures are stored in the heap. The information about the choice-points is stored in the trail-stack. Variables bound to complex logical terms are references to the corresponding logical structure in the heap. An n-argument nested structure is represented as multiple cells connected through chain of pointers. For example, a structure *f(a, b(M, N), L)* will be represented as $register_1 \rightsquigarrow f(a, X1, L), X1 \rightsquigarrow b(M, N)$, where the symbol '$\rightsquigarrow$' shows the reference. The above representation means that register_1 points to a structure *f(a, X1, L)*, which has three arguments: the first argument is a constant, the second is a reference to another structure *b(M, N)*, and the third is an unbound variable. Each n-argument structure has $n + 2$ cells, where the first two cells contain the name and number of arguments of the cell, and the remaining cells carry reference to the arguments or the basic values. A list is modeled as a two-argument structure: reference to the first element and reference to the rest of the elements.

The code area corresponding to each clause contains a label to make the jump to that clause. Calling a procedure is very similar to the implementation of procedure calls in the imperative languages, as described in Chapter 5. Arguments (or references) are held in registers for execution efficiency. Only those registers that are altered in the called procedure corresponding to a subgoal are stored in the control stack to save the state of the calling procedures; unaltered registers are not saved. The type of instructions supported in WAM are *hashing on the procedure-name, arity and first argument, set_structure, set_value, get_structure, get_value, put_value, conditional jump, arithmetic operations, unify_variable, and unify_value, try_me_else <Label> retry_me_else <Label>, trust_me_else_fail*, and *proceed*. The instructions *get_structure* and *get_values* are used to get the information from the arguments of the calling subgoal, and the instructions *put_structure* and *put_value* are used to set up the argument values of the subgoals on the right-hand side subgoals before invoking them. Given a rule of the form

```
clausehead(Args0)  :- subgoal1(Args1), ... subgoalN(ArgsN),
```

when translated in WAM will have following pattern:

```
Allocate N % for at least N variables
get_arguments of Clausehead in the registers
put_arguments of subgoal1
call subgoal1
...
put arguments of subgoalN
call subgoalN
```

The instructions *try_me_else <label>* is a composite instruction that stores the label *<label>* in the control stack and executes the first clause. If the first clause fails, then it pops the label *<label>* from the stack and jumps to the next clause. The instruction *retry_me_else <label>* is similar to *try_me_else <label>* except it is used for compiling the clauses other than the first clause and the last clause. The instruction *try_me_else_fail* is used for the last clause and states that if the clause does not succeed, then treat the calling subgoal as having failed, which means go back to the previous choice-point by performing a *pop(trail-stack)* operation and unbind the variables between the previous choice-point and the failed subgoal.

10.2.5 Program Analysis

Logic programs do not support memory reuse. Hence, they need to perform smart program analysis to identify the variables that can be reused. Fortunately, tail-recursive procedures use many such variables that can be reused in the following recursive calls. Compilers for logic programs use many program analysis techniques to improve the execution efficiency of the compiled programs. The general technique for program analysis is called *abstract interpretation*, which means the program is interpreted in some *abstract domain* such as type domain (or input/output mode domain) to infer type-related information (or mode related information) about variables. Using these abstract interpretation techniques, the execution efficiency of the programs is improved significantly. Abstract interpretation has also been used to derive producer–consumer relationship for automated parallelization of logic programs.

10.3 PROGRAMMING USING PROLOG

Prolog is a popular backward reasoning based logic programming language that can be implemented on a uniprocessor machine. It is based upon depth first search of the AND–OR tree and uses backtracking to explore the alternatives and simulate nondeterminism. Backtracking can be enforced programmatically by failing after getting a solution to generate multiple solutions. The clauses in Prolog are called *Horn clauses. Horn clauses* have only one n-ary structure ($n \geq 0$) on the left-hand side.

Prolog is based upon first-order predicate calculus. Negation in Prolog is based upon closed world assumption using *negation-by-failure*, which means Prolog assumes that if the solution cannot be derived, it does not exist. This is a serious limitation.

There are four popular implementations of Prolog: *Sicstus Prolog, Bin Prolog, GNU Prolog*, and *SWI Prolog*. There are also implementation such as Visual Prolog that integrate visual and logic programming. All these implementations are descendants of the standard Edinburgh Prolog. However, different implementations have enhanced their capabilities by incorporating various libraries for system-based programming, network-based programming, array-based programming, visual programming, object-oriented programming, use of blackboards and global variables, use of modules and libraries, etc. Although Sicstus Prolog uses blackboards and mutable dynamic arrays for allowing limited imperative style programming, GNU Prolog supports mutable global variables and mutable dynamic arrays for limited imperative style programming.

We have already seen basic Prolog style of programming in the definitions of *sibling/2*, *factorial/2*, and *fibonacci/2*. Prolog programming is quite different from the imperative style of programming for many reasons, which are as follows:

1. Logic programming is a declarative style of programming and does not explicitly support destructive update, memory reuse, and global variables for storing partial results. Owing to this, Prolog has developed more recursive style programming, including a tail–recursive style of programming. Tail-recursive programming is used to simulate indefinite iteration, as shown in Example 10.8. Internally, the index-variable N and other memory locations are reused using smart compile-time program analysis.

2. The rules in Prolog are not mutually exclusive and are not simply if-then-else statements, as if-then-else statements do not allow the execution of alternatives after successful condition is executed. In Prolog, even if a clause succeeds, backtracking can be forced to try out alternate clauses.

3. The imperative style of programming does not support nondeterministic programming and do not support implicitly multiple solutions.

4. Prolog has operators to build predicates as data and then convert them into predicates.

5. Prolog supports meta-programming to reason about the programs.

6. The pointers and pointer-based declarations are not explicit. List is a very high-level construct closer to the abstract definition of sequence.

7. Prolog supports tuples. However, there is no explicit named tuple like 'struct' or 'record'. However, structs and extensible data entities can be easily constructed.

Prolog supports destructive updates through the use of (1) *assert* and *retract* predicates, (2) the use of *blackboards* as in Sicstus Prolog, and (3) global variables as in GNU Prolog. Many variations of Prolog also support *dictionaries* in the form of *associative maps*.

Assert and *retract* predicate can change the rule database and facts database. *Blackboard* is a global key-value pair, where the value can be accessed by giving the key. Blackboard has operations like *bb_get(+Key, -Term), bb_put(+Key, +Term), bb_delete(+Key, ?Term), bb_update(+Key, ?OldTerm, + NewTerm).* Here the symbol '+' means input mode; the symbol '-' means output mode; and the symbol '?' means the mode is not determined and can be either.

Example 10.8

The following program illustrates the use of the blackboard (supported by Sicstus Prolog) to store the result of previous computations to improve execution efficiency. There are two procedures: *factorial_bb/2* and *factorial/2*. The function *factorial/2* computes factorial using the blackboard. The function *factorial_bb/2* has two clauses: the first clause looks up the blackboard for the stored result that has been previously computed. If the value is not stored, then it calls back the factorial in a mutually recursive manner. The use of blackboard provides efficiency, because the computed value is looked up in the blackboard.

```
factorial_bb(N, V)  :- bb_get(fact:N, V), !.
factorial_bb(N, V)  :- factorial(N, V), bb_put(fact:N, V).


factorial(N, M)  :- N > 0, !, N1 is N - 1,
                     factorial_bb(N1, M1), M is N * M1.
factorial(0, 1).
```

10.3.1 Cuts—Programmer-Directed Efficiency

Solving a Prolog query is a depth first search of the AND–OR tree. Many times, we need just one solution. For example, let us take the following procedure for membership test. The underscore is used in Prolog to denote anonymous variables that are not needed in the procedure.

```
member(X,  [X | _]). % Use unification to match arguments
member(X,  [_ | Xs]) :- member(X, Xs).
```

The above procedure states that the given value is a member of the sequence if the given value is the same as the first element of the sequence. Otherwise, the second clause steps through the rest of the sequence using tail-recursion.

If we give a query *member(4, [3, 7, 4, 9, 4, 12])?*, then once the membership test succeeds, there is no further need to go through the wasteful searching of a choice-point for alternate solutions. To improve the execution efficiency, a barrier is put programmatically to restrict the exploration of the choice-points. This programmatic barrier is called *cut* in Prolog. Placement of a cut puts the restriction of exploring the choice-points generated in the subgoals within the procedure in alternate clauses and the subgoals before the occurrence of the *cut* in the same clause. Once the *cut barrier* has been crossed, the backtracking in that procedure to any choice-point before the "cut" within the same procedure will fail the subgoal and take the control to the choice-point to the subgoal that called the procedure with an embedded "cut".

The deterministic version of the procedure *member/2* with "cut" will look like the following:

```
member(X, [X|_]) :- !.
member(X, [_ | Xs]) :- member(X, Xs).
```

In the above program, the program succeeds once after the first match of the value in the list. Subsequent backtracking will return failure.

Cut should be used very carefully, because it prunes the search space. Unsafe use of cut may change the control flow missing the solution, because the solution may lie in the pruned search space. Sometimes the unsafe use of "cut" is called *red cut*, and the safe use of cut is called *green cut*. A "cut" is *safe* if it does not prune the search space that contains the solution. Generally, placing a "cut" in a procedure that is being used deterministically to generate only one solution is safe; any procedure being used non-deterministically to generate more than one value will need its choice-points making the use of "Cuts" unsafe. "Cut" has been used effectively in modeling if-then-else and case statements.

10.3.2 Programming with Sets

Because Prolog is capable of generating solutions from different clauses, it can be used for set-based programming. It supports a predicate *setof/3* that takes three arguments: (1) the variable whose multiple values have to be generated, (2) the top-level predicate that needs to execute to generate multiple solutions, and (3) the variable representing the set of solutions. One can define set operations like union, intersection, and Cartesian product to perform complex operations. Set-based programming is quite a powerful way, as illustrated in Example 10.9.

Sets are generated using backtracking and collecting solution in the factbase, using assert/retract predicates, that remains unaffected by backtracking. Each time a choice-point allows exploration of a new part of the AND–OR tree, it potentially produces a new solution. This solution is collected using assert/retract predicates, and the process is repeated by backtracking. After the backtracking returns failure, all possible branches have been explored, and the set of solutions is collected from the database.

Example 10.9

The following program illustrates the use of set-based programming in Prolog. Given the students with multiple majors, we want to find out the students who are only cs_majors or double majors who are doing both computer science and biology. The procedure *cs_majors/1* uses the predicate *setof/3* to find out the set of cs-scientists, set of mathematicians, and set of biologists; removes the union of the set of biologists and mathematicians from the cs-scientists to remove the possibility of double majors. The procedure *compbio_majors/1* finds out the set of cs-scientists and the set of biologists, and takes their intersection to find out the set of double majors doing both computer science and biology.

```
cs_majors(Students) :-
   setof(C, cs_majors(C), CS_Scientists),
   setof(B, biology_majors(B), Biologists),
   setof(B, math_majors(B), Mathematicians),
   union([Biologists, Mathematicians], Set1),
   subtract(CS_Scientists, Set1, Students).

compbio_majors(Students) :-
   setof(C, computer_science_majors(C), CS_Scientists),
   setof(B, biology_majors(B), Biologists),
   intersection(CS_Scientists, Biologists, Students).

cs_majors('Ahmad').    cs_majors('Kevin').
cs_majors ('Shivani').
biology_majors('Ahmad').     biology_majors('Julie').
math_majors('Tom').    math_majors('Kevin').
```

10.3.3 Nondeterministic Programming

Logic programs support nondeterminacy because of (1) the use of logical-OR between the clauses of the same procedure, (2) logical-AND between the subgoals of the same clause, and (3) bidirectional information flow between goal and clausehead during unification. Arguments can be used in multiple combinations because of the bidirectional informational flow of unification. For example, the *member/2* procedure without the cuts in Section 10.4.1 can be used in two mode combinations: (1) *member*(+, −) mode, where '+' means input and '−' means output and (2) *member*(−, +) mode, where the second argument is given and the first argument value is being generated. Although the mode *member*(+, −) is deterministic, the mode *member*(−, +) is nondeterministic, because it can generate multiple solutions by backtracking and applying different clauses. The query *member(X, [a, b, c])* will generate the first solution 'a,' and if forced to backtrack will successively generate 'b' and 'c.' Similarly, the procedure *append*/3 in Example 10.5 can be used in two different modes: (1) deterministic mode *append(+, +, −)*, where it generates only one solution and (2) nondeterministic mode *append(−, −, +)*, where given the third argument, it can generate multiple possibilities of first and second arguments. A query like *append(X, Y, [1, 2, 3])* will generate four possible combinations of *X* and *Y*: (1) *X* = [], *Y [1, 2, 3]*; (2) *X* = *[1]*, *Y* = *[2, 3]*; (3) *X* = *[1, 2]*, *Y* = *[3]*; and (4) *X* = *[1, 2, 3]*, *Y* = [].

10.3.4 Abstractions and Meta-Programming

Logic programming languages support n-ary structures, linked-lists, and unnamed tuples that can be separated using different operators. Certain implementations of Prolog languages such as Sicstus Prolog support an extensive library for system and shell-based programming, dynamic array-based programming, multithreaded programming library, set-based programming, graph-based programming, and associative maps. List of key-value pairs can be converted into associative maps, and associative maps can be converted back into list of key-value pairs.

The named tuples such as structs or records are not explicitly supported. However, the combination of named n-ary structures and tuples can be used to simulate any complex struct (or record) effectively, as illustrated in Example 10.10.

Example 10.10

```
% mode student_new(+, +, +, +).
student_new(Name, Age, Course, Id) :-
    student_template(Template),
    Template = student(name(Name)-age(Age)-course(Course)-id(Id)),
    assert(Template), !.

% mode student_info(+, +, ?).
student_info(age, Name, Age) :-
    student(name(Name)-age(Age)-_-_).
student_info(course, Name, Course) :-
    student(name(Name)-_-course(Course)-_).
student_info(id, Record, Id) :-
    student(name(Name)-_-_-id(Id).
student_template(student(name(_)-age(_)-course(_)-id(_)).
```

The above program shows how a record can be dynamically created and accessed. The above program can be extended to update dynamically the data fields. The first procedure *student_new/4* is a constructor that gets a template of the form *student(name(_)-age(_)-course(_) – id(_))* already in the database, uses unification to insert the input values given in the arguments of the predicate *student_new/4*, and uses predicate *assert/1* to insert the new structure in the run-time database that can be accessed during run time. The operator '–' separates the fields. The procedure *student_info/3* has three clauses to extract specific fields: *age, course,* and *id*. The first argument of the corresponding clauses specifies the field-name for efficient indexing and hash-table lookup.

An if-then-else statement in Prolog is of the following form:

```
(Predicate_1 → then-subgoals_1
; Predicate_2 → then-subgoals_2
...
; Predicate_N → then-subgoals_N
; otherwise → catch-all subgoals
).
```

The right-hand side can contain another if-then-else statement. Different implementations of Prolog use different syntactic sugar for if-then-else.

Because variables can be bound to nonground terms, the feature can be used to build a list or n-ary structure dynamically by unifying a uninstantiated variable to a structure containing variables that can be extended again by binding the variables to another structure. Let us understand this feature using two examples: Examples 10.11 and 10.12. Example 10.11 builds an output list using recursion. Example 10.12 builds an extensible tree that can be treated like a dictionary.

Example 10.11

```
filter([], []).
filter([X|Xs], Ys) :- integer(X), X > 0, !, Ys = [X|Zs],
                      filter (Xs, Zs).
```

The above program filters the input list in the first argument to pick up the positive integers. The variable *Ys* denotes the output list. The query *filter([a, 4, −3, 5, 9, 0], Ys)?* returns the value of the variable *Ys* as *[4, 5, 9]*. To start with, the variable *Ys* is unbound. After the value 4 is inserted, the variable *Ys* gets instantiated to a nonground term *[4|Zs]*, and the recursive clause builds up *Zs*. When the value 5 is inserted, the variable *Zs* gets instantiated to a nonground term *[5|Zs1]*. The process continues until the input list becomes empty, and the variables expansion is terminated by putting an empty list at the end. So *Ys* = *[4|Zs]*, *Zs* = *[5|Zs1]*, *Zs1* = *[9|Zs2]*, and *Zs2* = *[]*. It is equivalent to *Ys* = *[4, 5, 9]*.

Example 10.12

```
% mode lookup(+,?, ?).
lookup(Key-Value, Dictionary) :-
    var(Dictionary),
    format("Do you want to insert in dictionary?", []),
    read(Answer),
    (Answer == 'y' → Dictionary = (Key-Value, _, _)
    ;otherwise → true
    ).
lookup(Key-Value, Dictionary) :-
   Dictionary = (Key1-Value1, _, _), Key == Key1, Value = Value1, !.
lookup(Key-Value, Dictionary) :-
   Dictionary = (Key1-_, Left, _), Key < Key1, !,
                lookup(Key-Value, Left).
lookup(Key-Value, Dictionary) :-
   Dictionary = (Key1-_, _, Right), Key >= Key1,!,
                lookup(Key-Value, Right).
```

The above program defines a dictionary that utilizes a binary tree of the form (*Key-Value, LeftSubtree, RightSubtree*). The leaves of the dictionary are unbound variables that can be extended by instantiating the variable to a tuple of the form (*Key-Value, LeftSubtree, RightSubtree*). The first clause extends the leaf node by checking if the node is an unbound variable then instantiating the unbound variable by unifying to (*Key-Value, LeftSubtree, RightSubtree*).

The second clause checks if the given *Key* = = *Key1*, then it unifies the given input value associated with the input key with the value associated with the key in the dictionary. The unification process allows it to retrieve the value stored in the dictionary if the input variable *Value* is unbound, and matches it with the value stored in the dictionary if the input variable *Value* is bound. The third clause compares the input key with the dictionary key and traverses to the left subtree if the input key is smaller than the key stored in the current node. The last clause traverses to the right subtree if the given key is greater than or equal to the key in the dictionary.

Prolog uses a combination of "cut" and tail-recursive programming to simulate equivalence of if-then-else statement and repeat-loop. The following code simulates if-then-else statement:

```
If-then-else(Predicate, ThenGoal, ElseGoal) :-
    call(Predicate), !, call(ThenGoal).
If-then-else(Predicate, _, ElseGoal) :- call(ElseGoal).
```

The first clause calls the predicate. If the predicate is true, then the control goes past the cut and executes then-part. If the predicate fails, then it backtracks to the choice-point, picks up the second clause and executes else-part.

Negation of a goal can also be implemented using "cut" as follows:

```
\+(Goal) :- call(Goal), !, fail.
\+(Goal) :- !.
```

The operator '\+' is used in Prolog to denote negation. Note that the use of a combination of "cut" and "fail" forces the control out of the procedure by forcing failure. The above program states that upon successful execution of the goal, failure is forced using a cut, and the calling subgoal fails. If the subgoal before the cut fails, then the second clause is trivially true.

Iteration is managed using tail-recursive programming and smart compiler analysis that identifies the variables that can be reused in the next cycle, as shown in the following example.

Example 10.13

The following example writes "Hello" N times. The first clause reads the number of times to say "Hello" and then calls the procedure *write_hello/2* with two parameters: (1) the number of times to say "Hello" and (2) *index-variable Index* initialized to value

1. The clause *write_hello/2* keeps writing "Hello" until the value of the index-variable *Index* is greater than the upper-limit *Max*. There are two major differences compared to the imperative style of programming: (1) a new variable *NewIndex* is created every time, and (2) tail-recursion is used to loop back. The memory reuse of the memory location corresponding to the variable *Index* is handled at the compiler level. The variable *NewIndex* uses the same memory location occupied by the variable Index. All the occurrences of the index-variables are mapped to the same memory location allowing for memory reuse.

```
print_hello :- read(N), write_hello(N, 1).
write_hello(Max, Index) :-
          (       Index =< Max → write("Hello"), nl,
                  NewIndex is Index + 1,
                  write_hello(Max, NewIndex)
          ;otherwise → true
          ).
```

Prolog also has an explicit *repeat construct* to write iterative programs. The construct for the repeat structure is as follows:

```
repeat,
  <subgoal₁>, <subgoal₂>, …, <subgoalN>, !.
```

The abstraction keeps backtracking upto a *repeat* reserved word, to force the control to move forward. Because backtracking causes all the variables to get unbound, a new alternative is explored. Generally, the last subgoal repeatedly fails to process all the data items. After the last subgoal *<subgoal$_N$>* succeeds, the control goes past the cut and does not come back. Repeat-loop has been used to frequently to read and process the data from files, where the last subgoal checks if the end of file has been reached.

10.3.4.1 Operators for Higher-Order Programming

Logic programming has a mechanism that converts dynamically a predicate built as data and executes it as a subgoal. The operator in Prolog is " =..". For example, *Subgoal =.. [length, [a, b, c], X], call(Subgoal)* will call a subgoal *length([a, b, c], X)*, which will return $X = 3$.

10.3.4.2 Modules

Many versions of Prolog support the notion of modules. The procedures can be exported by specifying the predicate module/2, where first argument is the module name, and the second argument is a list of predicates to be exported. Other modules can import all the exported procedures or selected exported procedures from another module by using predicate *use_module/2* either from a specified library or a user module. The first argument is the module name, and the second argument is the list of predicates. An example from Sicstus Prolog syntax is given below.

```
module(<module>, [<pred1>, …, <predN>]) /* export */
use_module(library(<module>), [<pred1>, …, <predN>]). /* import */
use_module(<module>, [<pred1>, …, <predN>]). /* import */
```

10.3.5 Limitations of Prolog

Prolog is a backtracking-based language based upon unification. Unification process does not evaluate an expression. This causes an extra step of flattening the expressions. The following syntax in a program would be more expressive. However, because unification does not support evaluation of an expression, complex expressions in parameters cannot be evaluated.

```
/* following programs are not allowed in Edinburgh Prolog */
factorial(0, 1).
factorial(N, M) :- factorial(N-1, M1), M is N*M1.

fibonacci(0, 1).                fibonacci(1, 1).
fibonacci(M, N) :- fibonacci(M - 1, N1), Fibonacci(M - 2, N2),
                   N is N1 + N2.
```

Another problem is the *occurs-check* in unification. For example, if we unify X and f(X), it would lead to indefinite loop. The third problem with Prolog is the overhead of storing the choice-points even for the deterministic programs with multiple clauses. A program is deterministic if the clauses are mutually exclusive, and only one clause can succeed at a time. However, unless augmented with complex program analysis techniques, all the choice-points generated for deterministic procedures cannot be removed. The use of *negation as failure* is incomplete, as the lack of information in Prolog database does not mean that the 'negation of the goal is true.' Pure Prolog does not support the concept of *temporality*—the notion of time and events occurring before and after other events.

10.4 EXTENDING LOGIC PROGRAMMING PARADIGM

Classical logic programming paradigm has been extended by (1) incorporating the notion of time and ordering of the events; (2) incorporating constraint-based programming in logic programs; (3) inferring the general rules from given positive and negative examples, which means using *inductive reasoning* instead of traditional deductive reasoning; (4) incorporating *higher order logic* such that predicates themselves can be treated as terms; and (5) integrating the logic programming paradigm with other programming paradigms—more specifically functional programming paradigm, object-oriented programming, visual programming, event-based programming, and concurrent programming.

10.4.1 Temporal Logic Programming

Temporal logic programming includes the concept of truthfulness of a predicate in a time interval: a predicate may be true in one time interval and false in another. There may be temporal relationship between two predicates. For example, if a predicate is true at time interval n, then another predicate may be true at time interval (n + 1). There are many logic programming languages based upon extending Prolog with temporal operators such as *Templog* and *Tokio*. Some of the operations for temporal logic implemented in the programming language *Templog* are as follows:

1. Predicate p is true at the next time instant.
2. Predicate p is always true.
3. Predicate p is never true.
4. Predicate p eventually becomes true.
5. Predicate p precedes predicate q.
6. Predicate q is true until predicate p becomes true.

Using these operators, such programs can be written about scheduling queue, message passing, and serving a request, because all these programs need a notion of chronological order. For example, an element inserted in the queue before another element also gets processed first; an acknowledgment for a message can only be sent after the message is received. A server can only serve the request after the request is made.

10.4.2 Constraint Logic Programming

Constraint-based programming puts extra restrictions to prune the search space in deriving the solution. An example of constraint-based programming is the *Sudoku*—a constraint-based puzzle of a two-dimensional block with a size that is a multiple of 3. It has several constraints: (1) numbers 1..9 should occur in each row; (2) numbers 1..9 should occur in each column; (3) numbers 1..9 should occur in the diagonal; (4) square blocks of 3 × 3 starting from the left-hand side top corner should have numbers 1..9, and (5) in any row, column, diagonal, and 3 × 3 blocks, the numbers should not be duplicated. Another example of constraint-based programming is a *map coloring problem*, where a map has to be colored with a minimum number of colors such that no two adjacent regions have the same color. Another example of constraint-based programming is to place N-queens on the chess board such that no two queens attack each other.

There are many applications of constraint logic programming in the real world such as space optimization problems, time-scheduling problem, planning, resource allocation, message routing to balance the traffic load, load balancing, and profit-maximization problems. A real-world example of a space-optimization problem is to fit all the circuits on a PC board such that total space is limited to box size. An example of a scheduling problem is to finish all the work within a deadline. Another example is to manage all the expenses within a budget.

Constraint logic programming integrates logic programming with an embedded constraint solver. Constraint solver uses a generalized form of unification, because unification performs pattern matching only by using an *equality operator*, whereas constraints also use inequality operators such as '<', '>', '≤', '≥', belonging to a range, etc. In a clause of a constraint logic program, there are two types of subgoals: *constraints* and *logical* subgoals. Constraints are stored in a constraint store. Constraints are solved using a constraint solver during a query reduction. If the constraint solver shows that constraints are incompatible with the values, then the program backtracks to find an alternate solution. Constraint logic programs are solved like regular logic programs except that constraints must be satisfied. For example, we can write a clause of the following form:

```
Clausehead(X, Y) :- constraint(X), predicate1(X, Z),
                    predicate2(Z, Y).
```

Here the constraint *constraint/1* would be stored in the constraint store at run time and solved using constraint solver, and *predicate1/2* and *predicate2/2* will go through the standard unification process. All three will return truth values. So the reduction process is still modeled by an AND–OR tree. WAM has been extended to handle the constraints in addition to regular predicates.

Many constraint programming languages have been developed such as Chip and ECL^{i}PS^{e}. Constraint programming has also been incorporated on top of the popular implementations of Prolog such as GNU Prolog, Sicstus Prolog, and SWI Prolog.

ECL^{i}PS^{e} is a Prolog-like language that supports constraints and modules. ECL^{i}PS^{e} supports different types of constraints using constraint libraries such as *interval constraints, linear constraints, integer constraints, global constraints* and *user-defined constraints. Interval constraints* provide the capability to bind a logical variable to an interval such as -4..20. The syntax to set up interval constraint is *<variable>* ::= *<interval>*. ECL^{i}PS^{e} supports a variety of control abstractions such as if-then-else statements, case statements, iterative loops, and iterators such as foreach-loop. It supports data abstractions like lists, arrays and *shelf*. Like blackboards, *shelf* is a persistent object with multiple slots that preserves the written value even during backtracking unless explicitly deleted by a programmer action. A sample program in ECL^{i}PS^{e} is explained in Example 10.14 and Figure 10.6.

```
% handling interval logic

:- lib(ic). % use interval constraint library
:- use_module(library(ic)). % load interval constraints

year(M) :- M :: 1..12. % assign interval 1 .. 12 to the variable M
winter(4).
fall(9).
summer_months(M) :- % use of interval comparison
  year(M), winter(W), fall(F), M #> W, M  #< F. % returns 5..8
```

FIGURE 10.6 An example of constraint-based programming in ECL^{i}PS^{e}.

Example 10.14

The library *lib(ic)* is a library to handle interval constraints. The library and the corresponding module are loaded. The predicate *year/1* sets up an interval 1..12. The predicates *winter/1* and *fall/1* are predicates that store the starting months of the winter season and fall season. The predicate *summer_months/1* calls three predicates and two interval constraints. The subgoal *year(M)* that returns the interval 1..12, the subgoal *winter(W)* returns value of the variable *W* as 4, and the subgoal *fall(F)* returns the value of the variable *F* as 9.

The constraint *M* #> *W* returns the interval 5..12, which is further refined by the constraint *M* #< *F* to derive the value of the variable *M* as the interval 5..8.

10.4.3 Inductive Logic Programming

Most of the programming system is based on *logical deduction*, which means that given a rule-base, rules are applied to break up queries into simpler forms connected through logical operators. However, deductive logic is not good for learning, because learning is based upon collecting many positive and negative observations and infers a general rule that explains all the observations. This is called *inductive reasoning* and is the basis of *computational learning*. The logic programming paradigm that uses induction instead of deduction is called *inductive logic programming*. *Progol* is an inductive logic programming language that builds rules given examples and dynamically asserts the generalized rules to be used later.

The rules are formed using positive examples, negative examples, and background knowledge that include backgrounds facts and rules in the database. The ground terms are generalized to variables, and if two ground terms share sufficient frequency of the same values at different arguments, then a rule is formed. This rule may be too general and is made more specific using negative examples.

Example 10.15

Let us take an example where there are many positive facts as follows:

New facts:

```
parent(tom, mary). parent(joe, mary). parent(cathy, john).
parent(nina, john).
```

An analysis of these examples will reveal that second arguments of fact1 and fact2 are the same. Similarly, second arguments of fact3 and fact4 are the same. A general rule can be formed using these examples:

```
my_relationship(X, Y) :-
    parent(X, Z), parent(Y, Z), not (X = Y).
```

All four facts satisfy this rule. This would be inductive logic. If there were a negative example that did not satisfy this rule, then it could be added as an exception in the rule.

10.4.4 Higher-Order Logic Programming

We saw in Chapter 9 that function programming supports higher-order functions called *functional-forms* that take functions as arguments and dynamically form a complex function that can be executed. An example of functional-form was *apply* in Lisp and *map* in Haskell. A simple ad hoc form of support for higher-order logic programming is present in Prolog with the operator '=..', which takes a list of data elements and treats the first element as functor-name and the remaining elements as the arguments to convert a list into a predicate that can be executed.

Example 10.16

The following program takes a list of functions that work on a list of data elements, and returns a list of output values such that ith element of the output argument *ResultList* is generated by applying ith function in the argument *FunctionList* on the input argument *DataList*.

```
construction(FunctionList, DataList, ResultList) :-
        (FunctionList == []→ ResultList = []
        ; otherwise →
                FunctionList = [Function | Fs],
                ResultList = [Result | Rs],
                append(DataList, [Result], ArgsList),
                Goal =.. [Function | ArgsList], call(Goal),
                construction(Fs, DataList, Rs)
        ).
```

The program reads as follows: if the first argument *FunctionList* is empty, then return an empty *ResultList*. Otherwise, pick up the next function *Function*, make an *ArgsList* by appending *DataList* with [*Result*], and convert the list [*Function*| *ArgsList*] into a dynamic goal *Goal* using the operator ' =..,' and then call this goal. After that, build the remaining output list *Rs* recursively, using the rest of the list of functions *Fs*.

For example, *construction([length, sum_list], [4, 5, 6, 7], Results)* will return a list *[4, 22]*, where *4* is derived by a dynamically created subgoal *length([4, 5, 6, 7], Result1)*; and the value *22* is derived by a dynamically created subgoal *sum_list([4, 5, 6, 7], Result2)*.

Example 10.17

The following program is equivalent to *apply_all* in functional programming. It takes three arguments: a function; a list of data elements on which function has to be applied; and a uninstantiated variable that will return a list of data elements, such that the ith element of the variable *ResultList* is derived by applying the function on the ith element of the list of data elements *DataList*.

```
apply_all(Function, DataList, ResultList) :-
        DataList == []→ ResultList = []
        ; otherwise →
                DataList = [Data | Ds],
                ResultList = [Result | Rs],
                Goal =.. [Function, Data, Result],
                call(Goal),
                apply_all(Function, Ds, Rs)
        ).
```

The program reads as follows: The procedure *apply_all/3* returns an empty list in the output variable *ResultList* if the input argument *DataList* is empty. Otherwise, it takes the next data element *Data*, applies function *Function* on it to produce a value *Result*, and builds the rest of the list *Rs* by recursively calling *apply_all/3* with the rest of the data list *Ds*.

We can also write Prolog predicates to handle functional-forms such as *composition* and *insertion*. However, this mechanism of incorporating higher-order logic programming is ad hoc, and there are efforts to integrate higher-order logic programming within the framework of logic programming. Two such efforts are *λ-Prolog* and *Escher*. These languages embed functions as arguments to Prolog predicates, and this λ-expression can be called as a regular function. The unification in λ-Prolog also needs to unify λ-terms. There are many efforts to integrate functional and logic programming paradigms, as described in Section 10.5.

10.5 INTEGRATION WITH OTHER PARADIGMS

Another effort to enrich the logic programming paradigm is to integrate other programming paradigms within the framework of logic programming using Horn clauses. The major efforts in integrating other paradigms are (1) integration of the functional programming paradigm, so that arguments can be functions and expressions that can be evaluated before passing the parameter; (2) concurrent execution of logic programs; (3) extending logic programs by incorporating objects in the logic programming paradigm; and (4) interfacing logic programming with visual programming languages. In addition, there are efforts to integrate multiple paradigms together into one language.

10.5.1 Integration with Functional Programming

The motivation behind integrating functional and logic programming paradigms is to exploit the advantages of both functions and relations. Relations have bidirectional information flow, and functions complex expressions as arguments to another function. However, functional programming is based upon evaluation of expression and unidirectional information flow. The major problem is the incorporation of logical variables in expressions, because the presence of uninstantiated logical variables makes the expressions hard to evaluate.

The programming languages that integrate functional and logic programming paradigms use different approaches to integrate bidirectional informational flow present in unification with unidirectional information flow and expression evaluation in assignments. The integration approaches can (1) interface two languages or (2) modify unification to handle λ-expressions. A combination of three techniques, namely, *narrowing, residuation,* and *semantic unification*, are used to integrate unification in the logic programming paradigm and term-reduction in the functional programming paradigm.

Narrowing finds out minimal substitution between two terms using unification so that a term can be reduced using term-reduction techniques. *Narrowing* has been used in integrated functional and logic programming by unifying the subgoal with the clausehead and applying the bindings to reduce the right-hand-side equation. Note that right-hand side of the rule is an equation that can be evaluated.

Example 10.18

This example illustrates the integration of unification of the clausehead and the function term on the right-hand side, followed by the evaluation of expression.

```
sum_list(nil) <= 0.
sum_list(X::Xs) <= X + sum_list(Xs).
```

The above program reads that if the argument is an empty list, then evaluate the right-hand side, and return the result that is trivially 0. The second clause reads that if the input list is nonempty, then add the value of *X* with the result returned by *sum_list(Xs)* and return the result. Note that right-hand side is an expression, and the recursive call *sum_list(Xs)* uses unification with the clausehead to pass the substitution.

An alternate approach to narrowing is to delay the evaluation of expressions containing uninstantiated variables until the logical variables get instantiated. This approach is called *residuation.* Languages like Funlog use residuation. Residuation is sufficient for the functional programming style. However, if the language uses the logic programming style, then the *residuation* technique may fail, and the terms will never be reducible. This failure is caused by the presence of *do not care variables* that are not needed in the goal-reduction process.

Semantic unification finds out the semantically equivalent expressions and performs minimum reduction until the two terms can be unified. For example, semantic unification of $(X, X) = (5, 2 + 3)$ will succeed, giving the set of binding $\{X/5\}$; the expression $2 + 3$ has been evaluated to give 5, and then unification of two terms succeeds.

Some of the languages that integrate functional programming and logic programming are *Curry, Funlog, LogLisp, F*, Qlog, Eqlog,* and *Escher.*

10.5.2 Integration with Object-Oriented Programming

There are multiple languages that integrate the logic programming paradigm and object-oriented paradigm. There are different approaches to integrate logic programming with object-oriented programming, which are as follows:

1. A program in the object-oriented logic programming paradigm is a group of objects that have methods that are like Prolog procedures instead of imperative style procedures and are accessed as *<object-name>.<method>*. The languages that support this approach are *LogiC++*, *KSL*, and *Orient84*.

2. Languages that incorporate objects, classes, and inheritance within the framework of logic programs. They use a preprocessor to translate an object-oriented logic program to a regular logic program. The languages that take this approach are *Intermission, Vulcan, CPU, OOPP,* and *OLPSC*.

3. Many languages have a library to incorporate object-oriented programming. For example, Sicstus Prolog takes this approach.

There have been attempts to integrate logic programming, functional programming, and object-oriented programming. There are many languages such as *G, FLOOPS, UNIFORM, PARADISE*, and *LIFE* that integrate all three programming paradigms.

Oz/Mozart is a multiparadigm language that integrates logic programming, object-oriented programming, concurrent programming, and supports both deterministic and nondeterministic programming.

10.5.3 Concurrent Logic Programming

Logic programming languages naturally support nondeterministic programming because of built-in logical-AND and logical-OR. The clauses can be executed concurrently. Similarly, subgoals can be executed concurrently. The concurrent execution of clauses is called OR-parallelism, because different subtrees rooted at an OR-node are being executed concurrently. The concurrent execution of subgoals of a clause is called AND-parallelism, because subgols are the children of an AND-node. AND-parallelism, has two approaches to handle the shared variables: (1) use streams between producer and consumer occurrence of shared variable to pass the values to the consumers and (2) use intersection of the values generated by concurrent execution of subgoals with shared variables. There have been many attempts to exploit concurrency in logic programs. Most notable are the OR-parallel language *PARLOG* and the *AND-parallel* language *Concurrent Prolog*. Both these languages use guards and commit to the guard after successful execution of guards to restrict backtracking.

Another aspect of incorporating concurrency in logic programming languages is the incorporation of multithreaded programming. Most of the logic programming languages such as Sicstus Prolog and Oz support a library module supporting multiple threads and lock mechanism.

10.6 SUMMARY

Logic programming is a declarative programming paradigm that is based upon first-order predicate calculus. First-order predicate calculus is an integration of propositional calculus, existential quantification and universal quantification.

A logic program is a set of procedures. Each procedure is a set of *Horn clauses* that are related to each other using logical-OR. Each clause has a left-hand side and a right-hand side such that the left-hand side is implied by the right-hand side. The predicates on the right-hand side are called *subgoals*, and the term on the left-hand side is called a *clausehead*. Logic programs are inherently nondeterministic because of commutativity of logical-OR and logical-AND operators and bidirectional information flow in unification. However, the presence of shared variables in a subgoal introduces data dependency because of the producer–consumer relationship. A logic program has rules and facts. Rules have a non-trivial right-hand side, and the right-hand sides of facts are trivially true.

There are two types of reasoning systems in logic programs: *forward reasoning systems* and *backward reasoning system*. A forward reasoning system applies rules on facts to derive new facts, and keeps repeating the process until no new facts can be derived. Backward chaining systems take a query and find out whether it can be derived using the set of rules and facts in the program. Because a forward chaining system applies rules on all the facts, it is inherently slow in nature. However, it is good for monitoring systems where the implication of new information has to be figured out. Backward chaining systems perform focused searches on part of a search space for a solution and are inherently faster than forward chaining systems. There are languages both for forward chaining systems and backward chaining systems. For example, the OPS5 family uses forward reasoning, and Prolog uses backward reasoning.

Classical logic programming is based upon query reduction using *unification*. Unification is a pattern matching process with two-way information flow without any expression evaluation. It matches position-by-position the logical subterms and upon successful matching, returns the set of substitutions. There are four possibilities in unifying two logical subterms. If the corresponding subterms are constants, then they are simply matched. If one of them is a variable, then all the occurrences of that variable are bound to the corresponding logical subterms in the other logical terms. If both of them are unbound variables, then both the variables are made aliases. Aliasing is an equivalence relationship, and binding to any variable in the alias set binds every variable in the set with the same binding. If the unification fails, then the substitution set is not generated.

The execution of logic program is mapped on a logical tree called an AND–OR tree. An AND–OR tree has two types of nodes: AND-nodes and OR-nodes. All the children of AND-nodes have to be true for the AND-node to be true. At least one of the children of an OR-node has to be true for an OR-node to be true. The level of OR-nodes and AND-nodes alternate in an AND–OR tree. Backtracking is used to simulate nondeterminism in logic programs. The process of backtracking takes the control back to the previous *choice-point*, and unbinds all the variables that have been bound between the failed subgoal and the previous choice-point. An alternative rule of fact is selected for unification, and the search on an alternate part of the tree is done to derive a new value for the variables that were unbound.

Prolog is a popular logic programming language that is based upon first order predicate calculus. It supports backtracking and simulates nondeterministic programming using backtracking. Prolog supports many features such as set-based programming, use of "cuts" to prune the search space, and meta-programming. However, because of the lack of mutable variables in the standard definition of languages, it has to resort to alternate mechanisms to store partial computations. There are three techniques to store partial computations that do not undo the bindings with backtracking. The techniques are (1) use of blackboards, as in Sicstus Prolog; (2) a combination of assert and retract predicates in all the implementations; and (3) the use of global variables in GNU Prolog. Blackboard stores key-values pairs and provides operations to access and update the values using a key-match. The predicate assert inserts a fact in a Prolog database, and the predicate retract removes a fact from the Prolog database. Both these techniques can be used to simulate the effect of global variables in logic programming. Prolog also supports higher-order programming using the operator =.. that is used to dynamically build predicates. Using this facility, functional-forms like *apply_all, composition, insertion,* and *construction* can be simulated.

The implementation of logic programs is based upon mapping the AND–OR tree into a low-level abstract machine WAM (Warren Abstract Machine). WAM supports depth first search and many abstract instructions to compile logic programs. It uses hashing on procedure-name and number of arguments to jump to specific procedures and uses a combination of *try_me_else*, *retry_me_else*, and *trust_me_else_fail*, along with getting and storing the complex structures in the heap. The references to the arguments are stored in the processor registers. An extra stack called *trail-stack* is used to handle backtracking. The trail-stack stores the choice-points (OR-nodes with alternate nondeterministic clauses), so that an alternate path can be explored in case of backtracking. Backtracking traverses to the previous choice-point by popping the last choice-point from the trail-stack and unbinding the last bound variable.

The major limitations of the classical logic programming is in (1) handling the nested expressions, (2) passing functions as parameters, (3) lack of evaluation of expressions while passing parameters for calling functions, (4) lack of handling temporality present in real-world events, (5) lack of handling constraints, and (6) the lack of a theoretically clean model to handle higher-order functions. The major advantage of logic programming is to provide relations.

In the last two decades, a lot of effort has been made to extend the classical logic programming paradigm by (1) incorporating constraints, temporality, and higher-order functions and (2) integrating with other programming paradigms.

The integration of functional and logic programming provides the advantages of relations and equations in one language. The approach taken is to interleave unification in logic programming with expression reduction in functional programming. *Narrowing* provides partial minimal unification until the resulting terms can be reduced. Semantic unification provides solving semantically equivalent equations before unification of logical variables. Another scheme is to delay the process of reduction unless the logical variables in the terms are instantiated.

Temporal logic programming incorporates operators in logic programs to reason about ordering the events using temporality. Different predicates can become true at different time intervals and after certain sequences of events. TempLog and Tokio are two temporal logic programming languages.

There has also been effort to integrate object-oriented programming with logic programming and languages that integrate functional and logic programming. The most notable languages are Hope, Oz, and Escher. Hope is a polymorphic language that integrates functional and logic programming. Oz is a mutiparadigm language that integrates logic, functional, object oriented, and concurrency. Escher also integrates functional and logic programming.

Logic programs inherently support concurrency because of the presence of AND-nodes and OR-nodes. The concurrency that is exploited by concurrently executing the subtrees rooted at OR-nodes is called OR-parallelism and is equivalent to executing the different clauses of a procedure concurrently. The concurrency that is exploited by concurrently executing all the children of AND-nodes is called AND-parallelism. Because each subgoal of an AND-node is capable of producing multiple solutions, streams are used to connect the subgoals, or an intersection is used to find out common values generated by two subgoals for a shared variable. Both these approaches have been used in implementing AND-parallelism. Parlog exploits OR-parallelism, and Concurrent Prolog exploits AND-parallelism. Many versions of Prolog and the multiparadigm language Oz also support thread-based programming through the use of a library module.

10.7 ASSESSMENT

10.7.1 Concepts and Definitions

Abstract interpretation; AND-node; AND–OR tree; AND-parallelism; arity; assert; backtracking; bidirectional information flow; blackboard; choice-point; clause, clausehead; closed world assumption; concurrent logic programming; constraints; constraint logic programming; cut; deductive logic; deterministic program; dictionary; goal; goal reduction; green cut; ground term; higher-order logic programming; inductive logic; instantiation; interval constraint; logical variable; inductive logic programming; meta-programming; narrowing; negation as failure; nondeterministic programming; nonground term; occurs check; OR-node; OR-parallelism; program analysis; query; query reduction; red cut; residuation; retract; safe cut; semantic unification; set-based programming; shelf; subgoal; substitution set; temporal logic programming; trail; trail stack; unification; unsafe cut; WAM

10.7.2 Problem Solving

1. Give the tree structure for the following logical terms:

```
a. class(cs, time(10), location('MSB 124')
b. [a, b, 1, 20]
c. "Program" /* string is a list of ASCII characters */
d. instructor(course(cs), [michael, tom]).
```

2. Unify the logical terms f(*X*, *X*, *[4, 5, 6]*, *L*) and f(M, *Y*, *[M, 5, B]*, *B*) one step at a time, showing the intermediate substitution set and instances of logical terms. Give the final substitution set and the common instance.

3. Draw an AND–OR tree for the following program and a query *uncle(tom, U)*. Clearly mark the OR-nodes and AND-nodes and identify the edges and bindings for successful unification.

```
uncle(X, Y) :- parent(X, Z), brother(Z, Y).
brother(X, Y) :- parent(X, Z), parent(Y, Z), male(Y), \+ (X = Y).
parent(tom, neena).   parent(neena, joe).   parent(clark, Joe).
parent(tom, ted).   parent(ted, mary).
```

4. Show the choice points and backtracking for the following program. During backtracking, clearly identify the variables that will be unbound, and trace the forward execution with the new value for the query provide_education(tom, *X*).

```
provide_education(X, Y) :-
    parent(X, Y), values_education(X), values_education(Y),
    has_finance(Y).
parent(tom, mary).   parent(tom, john).   parent(cathy, mary).
values_education(cathy).   values_education(john).
values_education(tom).   has_finance(john).
```

5. Write a recursive declarative style logic program for merge-sort, and execute using GNU Prolog.

6. Write a recursive declarative program to perform quicksort, and execute using GNU Prolog.

7. Write a nondeterministic program to find out a subsequence of a sequence.

8. Simulate a struct professor that has fields name, course, room, office-hours, and access various fields.

9. Read more literature on the Internet about temporal logic programming and explain how you would represent using temporal logic insertion of an element in the queue only after the first element has been taken out.

10. Write and execute the following programs in Prolog.

    a. Adding the list of numbers

    b. Adding two matrices of numbers

    c. Polymorphic program to remove duplications from a list of data elements

    d. Adding two polynomials

10.7.3 Extended Response

11. What do you understand by AND–OR tree? How it is used in the query reduction? Explain using a realistic example.
12. How is backtracking implemented using a trail-stack? Explain using a simple example of logic program.
13. What is unification? Explain using a nontrivial example.
14. How does logic programming support nondeterministic programming? How has it been simulated using backtracking in Prolog? Explain using a simple example.
15. Explain the effect of placing a "cut" in terms of choice-points and trail-stack.
16. What are the mechanisms in Prolog to store partial results, and how do they improve execution efficiency? Explain using an example by writing and executing a program not given in the book.
17. What is the difference between forward and backward reasoning systems? Explain using a simple but nontrivial example.
18. Explain with examples and code how Prolog supports higher-order functional-forms.
19. Explain different techniques used to integrate one-direction information flow in functional programming with two-direction information flow in the unification process in logic programming. Give examples.
20. Read more literature and tutorials on the programming language ECL^{i}PS^{e} from their website (http://www.eclipseclp.org), and discuss the use of constraints using two nontrivial programs.

FURTHER READING

Abadi, Martin and Manna, Zohar. "Temporal logic programming." *Journal of Symbolic Computation*, 8(3). 1989. 277–295.

American National Standard Institute. *Programming Language PROLOG PART I*. ISO/IEC 13211-1. 2012.

Apt, Krzysztof and Wallace, Mark. *Constraint Logic Programming Using Eclipse*. Cambridge, UK: Cambridge University Press. 2007.

Clocksin, William F. and Mellish, Christopher S. *Programming in Prolog* (5th edition). Berlin, Germany: Springer Verlag. 2003.

Diaz, Daniel. "GNU prolog." Available at http://www.gprolog.org/manual/gprolog.html

Gavanelli, Marco and Rossi, Francesca. "Constraint logic programming." *25 years of Logic Programming, LNCS 6125*. Berlin, Germany: Springer-Verlag. 2010. 64–86.

Hanus, Michael. "The integration of functions into logic programming: from theory to practice." *The Journal of Logic Programming*. 19,20, 1994. 1–48.

Haridi, Seif and Franz, Neils. *Tutorial of OZ*. Available at http://www.mozart-oz.org/documentation/tutorial/

Lloyd, John W. *Declarative Programming in Escher*. Technical Report ACRC-95:CS-013. Department of Computer Science, University of Bristol. 1995. 121, Available at http://www.cs.bris.ac.uk/Publications/Papers/1000073.pdf

Marriott, Kim and Stuckey, Peter J. *Programming with Constraints*. Cambridge, MA: MIT Press. 1998.

Muggleton, Stephen. "Inductive logic programming." *New Generation Computing*, 8(4). 1991. 295–318.

Nadathur, Gopalan and Miller, Dale. *Programming with Higher Order Logic Programming*. New York, NY: Cambridge University Press. 2012.

Ng, Kam W. and Luk, Chi-Keung. "A survey of languages integrating functional, object-oriented and logic programming." *Microprocessor and Microprogramming*, 41(1). 1995. 5–36.

Sicstus Prolog Manual. Available at http://www.sics.se/isl/sicstuswww/site/documentation.html

Sterling, Leon and Shapiro, Ehud Y. *The Art of Prolog*, (2nd edition). Cambridge, MA: MIT Press. 1994.

Warren, David H. D. *An Abstract Prolog Instruction Set*. Technical Note 309. SRI International. Menlo Park, CA. 1983.

CHAPTER 11

Object-Oriented Programming Paradigm

BACKGROUND CONCEPTS

Abstract concepts in computation (Section 2.4); Abstractions and information exchange (Chapter 4); Abstract implementation and low level behavior (Chapter 5); Data structure concepts (Section 2.3); Dynamic memory management (Chapter 6); Distributed computing (Section 8.4); Polymorphism (Section 7.5).

Object-oriented programming was born out of need to develop large-scale software. Object-oriented programming has been applied to the development of large databases including multimedia databases and major commercial software. In order to develop large-scale software, one has to build upon software developed earlier to avoid duplication, make it robust, and maintain it. Reusability and maintainability is a major factor for the large-scale software development to (1) reduce the cost of development and evolution and (2) keep the number of bugs relatively low. Major interrelated components of object-oriented programming are *modularity*, *reusability*, and *information-hiding* of the implementation-level details of modules.

Modularity means breaking up large software into smaller modules that can be placed as a class library and loaded in other modules when needed. This supports better reuse and maintainability. Only a limited public part of a module is visible outside. This public part is necessary for an object to communicate with other objects. Hidden parts of the module are needed for the implementation and can change with time to incorporate new technology or redesigned software to improve the execution efficiency. *Reusability* means that proven debugged and robust software is used again to avoid the time and cost overhead of development and maintenance. Information hiding means that the details of implementing various entities declared in one module is invisible to other modules unless exported intentionally by the programmer and intentionally imported in other modules.

Class-based languages extend the notion of modular programming by merging the passive concept of module boundaries with the active notion of instances of classes called *objects*. Like modules, classes are passive templates that group data objects as well as subprograms working on them with a difference that multiple instances of objects can be created at run time, and these objects can communicate to each other by sending messages to invoke the public methods associated with the objects at runtime. Messages are different from program invocation because messages invoke only public methods of other objects; private and protected methods cannot be invoked by objects that are instances of other classes and do not belong to a class-hierarchy. These public methods in turn can invoke private methods within the same object. These multiple objects have individual states at a particular time-instance, and the total computational state of the program is the cumulative sum of the individual states of the objects. The communication between objects changes the overall cumulative state of the computation, because the object receiving the message changes its state to another state, performs more computations, and transmits a new message.

Reusability requirement has necessitated two additional facilities: (1) development of off-the-shelf software developed as class libraries that can be included in other class-definitions and (2) hierarchical structure of class templates, such that previously declared templates can be reused by refining the templates. The hierarchical structure of the class allows descendants to inherit unrestricted data entities and the unrestricted methods declared in the ancestor-classes. The root node is the *root-class*, also known as the *super-class*, and all its children are called *subclasses*. Parent of a class is called the *parent-class*, and the children are called the subclasses. The notion of the parent-class and subclass are relative, which means a node in a directed acyclic graph (DAG) can be both a subclass for its parent-class as well as a parent-class for its children. An object is an active instance of any node in a DAG and is involved in computation, as shown in Figure 11.1.

A subclass (also called a *derived-class*) and a parent-class are linked through an *inheritance* link. *Inheritance* means if a subclass does not contain a data-entity or a method, then the hierarchy will be traced back until the definition of the method or the data-entity is found. In one hand, the use of inheritance promotes reusability by allowing tested off-the-shelf class libraries. On the other hand, inheritance causes dependencies on the parent-class or other ancestor-classes: any change in the library may affect all the programs importing from class-libraries, and many times this affects the program maintenance adversely. A change in the definitions of the member-entities that can be inherited by the subclasses can also adversely affect the behavior of the subclasses. Another problem of inheritance is in the object migration in remote method invocation. A migrating object has to carry all the member-entities it uses,

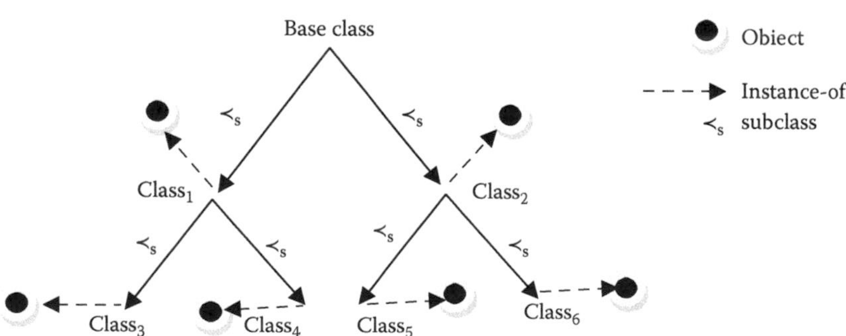

FIGURE 11.1 Tree-based hierarchy of subclass and single inheritance.

including inherited methods and class-variables. If the methods used are defined in ancestor-class, then the ancestor-class definitions also have to be migrated, adding to the data-transfer overhead. Owing to this problem, many languages such as Emerald and Javascript have used *flat definitions of objects* and do not support an hierarchical definition of inheritance.

Most of the modern object-oriented languages support the concept of class, class-inheritance, polymorphism through the use of generic methods and inheritance, encapsulation, and abstract types. The methods embedded in the class-definition could be functions or procedures. The notion of object and class varies from language to language. In some languages like Ruby, every data-element is an object. For example, an integer value 10 is an object in Ruby. Built-in operations can be possibly overridden to provide flexibility and extensibility of the class-definitions.

In object-oriented languages, polymorphism is present by the use of generic methods that use type-variables. It is also present in multiparadigm languages such as Ruby and Scala that support generic functions. Polymorphism is also present because of the use of inheritance, where a subclass can inherit the attributes of a class. Thus, the same operator defined once in a base-class can be used by all the descendant classes using inheritance. Polymorphism can also be present using subtyping in object-oriented languages.

Object-oriented programming has been integrated with the imperative programming paradigm, concurrent computing, and distributed computing for a long time such as in C++, C#, Java, Modula-3, and Emerald. It also has been integrated with the function programming paradigm such as in common Lisp object-based system (CLOS), Ruby, and Scala, and with the logic programming paradigm such as object-oriented variations of Prolog and Oz—a recent language that integrates object-oriented and logic programming paradigms.

11.1 CLASSES AND OBJECTS

The notion of class serves many purposes in object-oriented programming. Like module, it becomes the natural boundaries for objects: variables and other declared entities. Methods cannot be automatically seen outside the class, unless they are made public. Some languages like C++ allow the notion of *friends* to provide limited visibility of member-entities declared in one class to another even if a member-entity is not public. The property of tree structure supports inheritance of the entities declared in ancestor-classes by the descendant classes. Class also acts as a template for run-time creation of objects. Owing to the hierarchical nature of the class, subclasses can extend the template of the parent-class either by (1) redefining (called *override* in OO-jargon) a data-entity, or (2) redefining a subprogram entity (called method in object-oriented programming jargon) including built-in operators. A class can be defined partially in the base-class and progressively extended in subclasses. Such class-definitions are called *partial class.*

A class structure can be nested, which means a class may be defined within another class. Some languages like C++ and C# support this type of nesting. The nested classes are called *inner classes*, and the class in which an inner class is nested is called an *outer class.* An inner class can access only the static members of the outer classes in C++ and C#. However, in Java, inner classes can access all the members of the outer class.

An entity declared in a class, if redefined in subclass, is *shadowed* (hidden) in favor of the new definition in the subclass. Different subclasses may redefine an already defined method having the same name in an ancestor-class. These methods are called *virtual methods.* A method is accessed using the class-hierarchy followed by method name. A class-hierarchy is the path from the root-class to the current class. For example, a class-path *system.lang.thread* defined in Java has a three-level hierarchy of classes: *system*, *lang*, and *thread*; lang is a subclass of system, and thread is a subclass of lang. The methods taken from the subclass thread would be addressed as *system.lang.thread.<method-name>*. Alternately, the library *system.lang.thread* can be included in the software being developed. Assuming that there is no name-conflict, just the *<method-name>* can be used after loading the corresponding class-library.

An *abstract method* is declared only in a class. Its implementation—the body of the method—is missing. The purpose of an abstract method is to provide the correct position of the method in the class-hierarchy, so that it can be correctly inherited. A subclass may implement an abstract method in its own way, including the possibility of not defining it at all. An *abstract class* is a class that contains one or more abstract methods or a subclass that inherits at least one abstract method without providing its implementation.

Like abstract classes, *interfaces* also contain the declarations. However, unlike interfaces, abstract classes can contain fields that are not static and final, and they can contain some implemented methods or partially implemented methods leaving it to subclasses to complete the implementation. If an abstract class contains only abstract method declarations, it should be declared as an interface instead. Interfaces have been used extensively in remote method invocation in distributed object-oriented programming and event-based programming.

The following sections discuss these concepts and implementation of object-oriented languages, and describe distributed model of object-oriented programming.

11.1.1 Object—An Instance of a Class

An *object* is an active instance of a class. Objects are allocated memory in a heap at runtime, and can have a state. The method that creates an object and initializes the instance is called a *constructor* in object-oriented programming jargon. When an object is created, it creates its own data area in the heap and uses the methods declared in the class. An *instance variable* belongs to an instance of a class, whereas a *class variable* is part of the corresponding class, and is common to every instance of the class. Each object has a separate copy of instance variables and executes the methods (including virtual methods, explained in Section 11.2.1) defined in the corresponding class or ancestor-classes based on class-inheritance. A computational state changes when the value of one of the variables accessible from an object is updated.

Two objects that are instances of the same class are different because (1) instance variables occupy different memory locations and (2) two objects may be in different states at a given time. A variable inside an object can be altered if (1) it is a public variable and is altered by an object including other objects, (2) is a protected variable and is altered by an object within the class-hierarchy of the actual class where the variable has been declared, and (3) a private variable that is altered by the object's method. A private-method in an object can be invoked only by a method declared within the same class of the object.

Depending upon whether the method is static or virtual, the access mechanism is different. Depending upon the object-oriented languages, both mutable and immutable objects are supported. In modern multiparadigm languages such as Ruby and Scala, lists and strings are immutable objects, and arrays and vectors are mutable. The public data-entities in an object are called *outer subobjects*, and the protected data-entities in the instance of a class are called *inner subobjects*.

An object is created and initialized using a special initialization method(s). The method allocates the corresponding memory locations in the heap and stack or both, depending upon the language, and initializes the values of the member-variables of the created object. This special method can be called an *initialization method* as in Ruby or a *constructor* in some object-oriented languages such as C++, Java, and C#. An initialization can have zero or more parameters that are used to pass the initial values to the public members of an object. Generally, created objects have a default initialization value. However, parameters can be used to pass the initialization values. For example, one can create a circle object by passing the parameter values of a center and the radius. There can be more than one initialization method, depending upon how many initialization values are passed during the creation of an object.

Example 11.1

Figure 11.2 shows a class-definition for stack in Java-like syntax. The class stack is public, which means every class can access the stack and its public entities. The stack-class has been implemented using an indexible sequence (in this case, an array of objects).

```
public class stack
{   protected Object[  ] mySeq;  // a vector of references
    protected static maxsize = 256;  // maxsize is a class variable
    protected int top = 0;  // top is an instance variable
    public stack(  ) {  this.mySeq = new Object[ ]; }  // constructor
    public boolean isEmptyStack  ( )   { return  (top == 0)}
    public boolean isFullStack  ( )  {  return  (top == maxsize) }
    public void push(Object next _ element)
    {   if  (top == maxsize) throw new OverflowException( );
        mySeq[top++]  = next _ element;

    }

    public Object pop( )
    { if  (top == 0) throw new UnderflowException( );
          Object popped = myseq[--top];
          myseq[top]  = null; // make the reference null
          return(popped);
    }

    public Object top(  ) { return  (mySeq[top - 1]) }
}
```

FIGURE 11.2 An example of class for stack using Java syntax.

The class *stack* has three types of entities: variables, protected data-entities, and public methods. Each abstract operation on stack is written as a public method. Objects are accessed using a reference stored in an array element, and objects themselves are stored in the heap. In order to garbage-collect an object, the reference is assigned the value 'null.' The constructor *public stack ()* creates a new instance of the *stack*. The method *isEmptyStack* returns true if the value of the variable *top* is equal to zero. The method *isFullStack* returns true if the value of the variable *top* is equal to the maximum size of the stack. The method *push* takes an object-reference and inserts it in the stack *mySeq* after incrementing the index *top* by one. The method *pop* decrements the value of the index-variable top, and removes the element indexed by the variable *top.*

11.2 CLASS-HIERARCHY AND INHERITANCE

Class-structure is arranged in the form of a DAG (directed acyclic graph), and *tree* is a special case of DAG that allows clear separation of various subclasses. In a tree-based hierarchical structure, a subclass inherits from only one parent-class and its ancestors as shown in Figure 11.1. Note that there is only one root-class. Subclass relationship is denoted by the symbol '$\prec_s$.' All the nonleaf nodes can be both subclasses of the class corresponding to the parent-node and parent-class, for the classes represented by their children-nodes. The leaf-nodes are the final subclasses and have no children.

11.2.1 Subclasses

The subclass relationship is transitive, which means that class $A \prec_s$ *class B* $\bigwedge$ *class B* $\prec_s$ *class C* implies *class A* $\prec_s$ *class* C. It is also called an *is-A relationship* using artificial-intelligence terminology. A subclass is also called a *derived class.* In Figure 11.1, class$_3$ and class$_4$ are subclasses of class$_1$; class$_5$ and class$_6$ are subclasses of class$_2$; and class$_1$ and class$_2$ are subclasses of the root-class. Alternately, the root-class is a parent-class for the classes class$_1$ and class$_2$; class$_1$ is a parent-class of the classes class$_3$ and class$_4$; and class$_2$ is a parent-class of class$_5$ and class$_6$.

A subclass may declare more entities than its parent-class or any of the ancestor-classes or can override the definition of a method in any of the ancestor-class if it is permitted to do so by the ancestor-class. A method cannot be overridden by a subclass if it is *sealed* inside the class where it has been declared. Unless declared as private, inherited entities are accessible to the descendants for reuse. Derived classes are different from parent-classes because (1) derived class can inherit the member data-entities and methods using an inheritance link. However, inheritance is antisymmetric, and parent-class cannot inherit methods and data-entities of the derived class, (2) derived class can redefine (called *override* in object-oriented jargon) an already defined method or define an abstract method in the parent-class, (3) a parent-class can seal its member-entities to avoid being redefined by the subclass, and (4) a member-entity in a parent-class can be made private to make it invisible in the subclasses.

Example 11.2

We extend the class *stack* given in Example 11.1 to an indexible stack that can act as a stack as well as an indexible sequence, with an additional method *get_data* that picks up any *data_element* from a given valid index between 0 and top. The subclass *indexibleStack* will extend the class stack using the following syntax:

```
class indexibleStack extends stack
{ public Object getData(int index)
      { if (index < 0 || index > = top)
            throw new StackRangeException();
        else return (mySeq[index])
      }
}
```

All other methods already declared in the class *stack* are inherited by the subclass *indexibleStack*. Different languages have different ways of declaring a subclass. For example, Ruby notation will be *indexibleStack* < *stack* to state that the subclass *indexibleStack* is a subclass of the class *stack*.

A subclass (derived class) can inherit all the member-entities (other than the sealed and private entities) of the ancestor-classes. An ancestor-class is a parent-class or an ancestor of a parent-class up to the root-class (or base-class). Given a method or a data-entity in an instance of a class, it is first looked up in the class. It is used if it is found in the class. Otherwise, it uses the inheritance link to go to the next level of parent-class to search for the definition. The process is repeated until the member-entity is identified.

An object, which is an instance of a class, can use an inherited method from one of the ancestor-classes with a data-entity declared in the class, provided the data-type of the data-entity is compatible with the data-type of corresponding data-entity used in the inherited method. Similarly, an object that is an instance of a subclass can apply a method on an inherited data-entity, provided the method defined in the subclass is type-compatible with the data-type of the inherited data-entity. Under certain conditions, the type-compatibility condition can be relaxed using *casting* as described in Section 11.3.3.

11.2.2 Virtual Methods

A method may be (1) defined in a base-class and inherited by a subclass (2) just declared in the base-class and defined in the subclass, (3) defined in the base-class and redefined differently later in different subclasses. If a method is simply declared in the base-class and later used in the subclass, then it is called an *inherited method*. If a method is just declared in the base-class and defined in different subclasses differently based on their requirement, then the method in the base-class is called an *abstract method* or *pure virtual function*. After defining a method in a class, there may be a need to change the method definitions in different subclasses. Thus, a method may have the same name but different definitions in different subclasses. Such redefined method is called a *virtual method*.

An abstract method can be used to define hierarchical position; only derived classes provide implementations for pure virtual functions. A pure virtual function that is not defined in a derived class remains a pure virtual function. A derived class containing only abstract methods is also an *abstract base-class*.

Virtual methods refer to different functions depending upon the corresponding subclass. Object-oriented languages that support class-hierarchy also support virtual methods. Different languages have different syntax to declare virtual methods and to specify whether a virtual method is being overridden in a subclass. For example, virtual methods are annotated with a reserve word "virtual" in C++ and C# to separate them from other *inherited nonvirtual methods*—methods that have unique names in class-hierarchies and are not redefined in subclasses. Java uses the reserved word "@ override" before the method definition to show that the method is being redefined in a subclass. Similarly, C# uses "override" before the method name in the subclass. In the case of extensible libraries and classes that can be derived further, the methods have to be declared as virtual methods.

Example 11.3

Let us consider an example to explain the concept of virtual method. A company has many types of employees such as executives, monthly salaried employees, and daily workers. All three subclasses of employees' salaries and benefits are paid using different equations. Yet they all are employees. By default, they are treated as regular employees who get monthly salary. Let us write a simple intuitive declaration for virtual method using C# syntax.

```
using System;
using System.Collections.Generic;
namespace Salary
{ public class Employee
  {public virtual void salary()
           {Console.WriteLine("Regular monthly salary");}
  }
  public class Executive: Employee
  {public override void salary()
           {Console.WriteLine("Gets executive bonus");}
  }
public class DailyWorker: Employee
{public override void salary()
           {Console.WriteLine("Gets only hourly wages");}
 }
public class OtherEmployee: Employee
{… //no overriding inherits parent-class method
}
}
```

Methods with unique names that are declared in a class and never altered are called *static methods* and can be given a unique identifier in the compiled code. However, the methods that are defined in different subclasses cannot be given a unique identifier at compile time and need to be bound at run time after the object belonging to specific subclass is known. Thus virtual methods need dynamic binding to the code-area.

11.2.3 Multiple-Inheritance

A subclass may also use multiple-inheritance, as there can be more than one parent-class. An example of multiple-inheritance is in the definition of dolphin, which would inherit some properties from land-bound creatures such as breeding and some properties from fish such as swimming.

In a subclass that inherits from more than one parent-class, the inherited entity should have a unique name across the parent-classes and their ancestors to avoid any ambiguity. Occurrence of two entities with the same name causes ambiguity because of the name-conflict and needs to be resolved. However, it is difficult to resolve name-conflicts if the libraries are imported.

Example 11.4

Let us take an example of teaching assistants who are teaching basic-level courses. They act both as instructors and students. They take and teach courses. They inherit the pay-related information from the class *employee*, the course-teaching-related information from the class *instructor*, the tuition-payment-related information from the class *student*, and the course-taking-related information from the class *student*.

Definitions containing multiple-inheritance are modeled as DAGs, because there are classes that have more than one incoming edge due to the presence of multiple parent-classes. Multiple-inheritance can cause ambiguities in the inheritance of methods. For example, let us take the multiple-inheritance structure given in Figure 11.3.

A virtual method *m* declared in the base-class *P* is inherited by its subclasses *Q* and *R*. Virtual methods are accessed using a *Virtual Method Table* (VMT) that is an array of references to the code area of virtual methods in the current class and its ancestor classes. The virtual method table (VMT) used to access virtual methods of *Q* and *R* has reference entries for method *m*. Since the subclass *S* inherits from both the parent-classes *Q* and *R*, there is ambiguity whether to accept virtual method *m* from *Q* or *R*. Similarly, suppose *Q* does not declare a virtual method. Rather, it inherits the definition from P. Because *Q* is using the virtual method of P, again there is a conflict as to whether *P's* virtual method should be used or *R's* virtual method should be used by S.

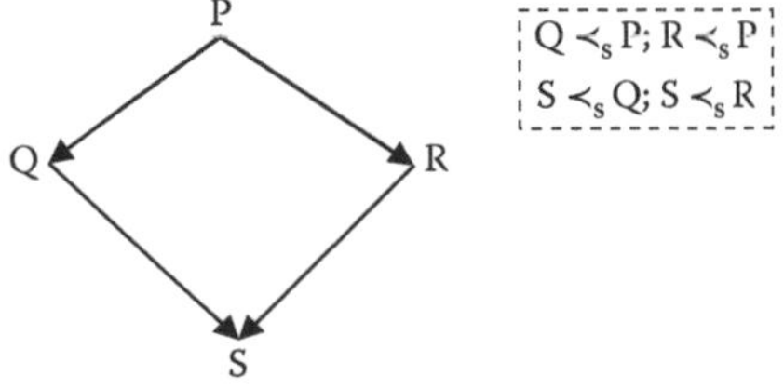

FIGURE 11.3 An illustration of multiple-inheritance complexity.

Depending upon the programming languages, such name-conflicts are treated differently. Although one language may treat such name-conflict as an error, other languages may use some heuristics to resolve the conflict. One technique to resolve the conflict is to explicitly specify which parent-class is being used when a method with a common name in two parent-classes is being used. This explicit specification can be done in two ways (1) by concatenating the parent-class name before the method name as *<parent-class-name>*::*<method-name>* and (2) a reserved word *using* can be used in the subclass-definition followed by *<parent-class-name>*::*<method-name>*. Some languages like Scala do not support multiple-inheritance. Instead, they use *traits* to include common features across different classes. Languages like C# and Java do not support multiple-inheritance of classes. Some languages like CLOS, C#, Python, and Ruby use *mixins*. *Mixin* is like an abstract class-definition that is added to other classes to enhance their functionality. Mixins are implemented in the classes where they are included.

11.3 VISIBILITY AND INFORMATION EXCHANGE

Unlike procedure-based languages where the scope was limited to procedures, class-based languages have variables that are associated with class, method, and objects. Object-oriented languages have four types of variables: global variables, class-variables, instance variables, and local variables. Global variables are available across the classes. Class-variables have limited scope within a class and descendant subclasses and are available to different methods inside the class and the subclasses. There is a single copy of the class variable that is shared across multiple objects of the same class and the descendant subclasses. Instance variables are specific to every object-instance of the class. Every time an object is created, a new instance variable is created. Since instance variables are specific to dynamic objects, they occupy a memory location when an object is created at run time. Local variables are similar to local variables in imperative languages and have limited scope within the methods they are declared in.

In Example 11.1, the variable *top* is an instance variable. The variable *maxsize* is a class-variable, has only one copy within the class *stack*, and is shared among various objects of the class *stack*. There is a need to separate these variables while declaring inside a class. Different programming languages use different reserved words. As shown in Figure 11.1, Java uses the reserved word "static" in front of the variable declaration to mark it as a class variable. Ruby uses '@@' sign in front of a variable to denote that it is a class variable.

11.3.1 Visibility of Member Entities

A class contains three types of entities and three types of visibility of the entities. A class contains variables, data-entities, and methods. A *method* is a procedure or function whose scope is limited within a class unless it is declared public. There are three types of visibilities associated with entities. The visibility classifications are *public, protected*, and *private*. A *public entity* is visible outside the class. Declaring an entity as public has the same effect as exporting the entity. A *protected entity* has different meanings in different languages. In C++ and C#, the declaration "protected" limits the visibility within the subclasses and their descendants using inheritance hierarchy. This is more like creating nonlocal variables in nested procedures in the procedural programming paradigm. The declaration "protected" in Java is a little more liberal, and its scope is restricted within the corresponding package. That means the other classes within the same package can also see a protected entity. A *private entity* has limited scope within the class it is declared. In addition, a method may be *sealed* so that its original definition may not be overridden. For some languages, the default mode of a method is "public", whereas for others default mode is "private". For example, the default mode of C++ and C# is "private": unless specifically declared "public", a class or its entities are not visible to instances of other classes. However, the default mode for the programming languages Ruby and Scala is "public". Some languages such as C# allow access to private methods using access methods *get* and *put*: the method *get* reads the value of a private data-entity, and the method *put* writes into a private data-entity.

A subclass can regulate the visibility of the members declared in the parent-class. This regulation varies from language to language based on their design philosophies. For example, in C++ a subclass can further restrict access only to the protected and public members of the parent-class. A subclass in the object-oriented language Eiffel can also increase the visibility of the members of the parent-class. In C++, protected or public members of a protected parent-class are protected members of the corresponding derived class, and protected and public members of a private class are private members of the derived classes. In C++, a derived class can restore the visibility of the members of the parent-class by inserting a declaration "using" followed by the public or protected portion of the derived class declarations.

11.3.2 Encapsulation

The visibility of the entities is controlled by two features of object-oriented languages: *package boundary* and *class boundary*. We have already discussed the visibility of the variables and methods declared within a class boundary. *Encapsulation* provides means to put multiple classes inside a package boundary, and the encapsulated classes cannot be used until the corresponding library is loaded. In some languages like Java, package is also a boundary for the protected variables declared in a class as described in Section 11.3.1. Packages serve many purposes (1) allow the reuse of the name of the classes and interfaces—names of the entities declared in a package are specific to a package and can be reused in other packages, (2) provide regulated reusability, and (3) regulate the visibility. When a package is imported by a program, only the public entities in the loaded package can be used by other programs. A method in a package can access a public method in another package by providing the hierarchical description of the method that includes *<package-name>*: *<class-hierarchy>*.*<method-name>*. Alternately, by importing the package, short names can be used. However, the use of short names has a possibility of name conflict if two or more packages are imported and they use methods having the same name.

11.3.3 Information Exchange

We have already seen information exchange between the procedures in traditional procedure-based languages. However, object-class-based languages provide an additional layer of class and then package around the classes. The use of class creates a new type of variable called *class variable* that is visible to every object-instance and is used for information exchange between object-instances of different subclasses and the class. Information can also be accessed from individual object-instances using public variables. An instance variable of an object can also be read or updated using *attribute-access methods* or *attribute readers.*

The information-exchange mechanism between the methods typically uses the local variables and parameter-passing mechanisms. As explained in Chapter 4, object-oriented languages use two types of parameter passing: *call-by-value* and *call-by-reference.* Call-by-value is used to pass the information about the simple data-entities stored in the control stack and for copying the reference of an object stored in the heap. Call–by-reference is used to create a reference for complex data-entities stored in the control stack. Parameter passing is also used to exchange the information to a constructor when a new object is created.

11.4 POLYMORPHISM AND TYPE CONVERSION

In Chapter 7, we have seen the power of polymorphism and different types of polymorphism. Most of the modern object-oriented languages also support different types of polymorphism. The use of overloading, coercion, and inclusion is common.

The use of *generic templates* (or just *generics*) in many object-oriented languages or the use of dynamic typing also supports parametric polymorphism. Generic class-definitions use type-variables in the formal parameters. During an object creation, the type of the parameter is passed to specialize the generic object to a specific type. The use of parametric polymorphism makes an operation within a class behave differently, and this binding is done dynamically at runtime.

Inclusion polymorphism is supported through the use of inheritance. A subclass inherits the definitions of the parent-class and can use the member-entities. Subclass allows "override" in addition to inheritance because a virtual method can redefine a method. Different subclasses redefine the virtual methods thus *overloading* the method name.

11.4.1 Parametric Polymorphism and Generic Templates

Generic methods use a generic declaration utilizing *type-variables.* A *type-variable* is instantiated with a specific type at the time of invocation, and the method becomes specific to the type passed as parameter. A general abstract syntax for generic method in C++ using template construct is illustrated in Examples 11.5, 11.6 and Figure 11.5. Similarly Ada, Java, and C# support parametric polymorphism through the use of generic methods. Example 11.6 illustrates the use of generic methods in C#.

Example 11.5

The following code illustrates parametric polymorphism through template declaration. The variable *Typevar* is a generic type-variable that can be bound to any type passed as parameter. The operator '>' is overloaded and can be used to compare integers as well as floating point numbers. For a call maximum<*int*>*(a, b)*, the type-variable *TypeVar* will be bound to "int", and the function will behave as an integer type. For a call maximum *<float> (a, b)* the function will behave as a "float" type.

```
template <class Typevar>
Typevar maximum (Typevar first, Typevar second)
{
     Typevar MaxVal;
     if (first > second) MaxVal = first;
     else MaxVal = second;
     return (MaxVal);
}
```

Example 11.6

The following example shows the use of generic method in C#. The procedure swap has a generic type *T* that can be instantiated at run time to different type parameters, such as *int, float*. For example, a call *swap<int>(ref a, ref b)* will bind the type-variable T to *int*, and a call *swap<float>(ref a, ref b)* will bind the type-variable T to float.

```
static void swap<T>(ref T first, ref T second)
{T temp; temp = first; first = second; second = temp;}
```

Example 11.7

Let us consider the C++ program given in Figure 11.5 in Section 11.5.2. The declaration "Template <class Typevar>" defines a generic class using the type-variable *T* that can be specialized at runtime by creating an object and passing the specific value for the type. For example, if the type-parameter *integer* is passed, then the object will behave as a stack of integers, and if the type-parameter *float* is passed, then the stack will behave as a stack of floating-point numbers. Similarly, the use of type-variables *T* in front of the stack elements is altered at run time, when the type-variable is instantiated to a specific type.

11.4.2 Casting

Liskov's behavioral subtyping principle states that "if a data-entity has a subtype of another data-type, then the object of the original data-type can be substituted with an object of the subtype without loss of behavior-compatibility and without loss of desirable properties of the program." This notion of type compatibility has been used in object-oriented programming to perform type-conversion of data-entities between parent-class and subclass while using an inherited method in a subclass or an inherited data-entity in a subclass. This notion of type conversion is called *casting.*

Casting allows the use of an object of one type in place of another compatible type, among the objects permitted by class-inheritance rules. Casting should make sure that data-types are compatible. If data-types are not compatible, then casting would give an incorrect result. A type is compatible if there is no information loss. For example, an *integer* attribute is compatible with a *long integer* because there is no information loss. However, a *long integer* object cannot be transformed to an *integer* because of the information loss. There are two possibilities of type conversions involving the use of inherited member-entities in an instance of a subclass.

Case 1: An inherited method is used on an object that is a subtype of an object originally used in the inherited method. This case is called *upcasting*.

Case 2: A virtual method in a subclass uses an inherited data-entity such that the corresponding data-entity in the subclass is a subtype of the inherited data-entity. This type of transformation is called *downcasting*.

11.4.2.1 Upcasting

Upcasting allows the promotion of an attribute to the corresponding attribute in the ancestor-class of an object to utilize an inherited method declared in the corresponding ancestor-class of the object. Upcasting is a special case of coercion because it preserves the information during the transformation of the data-type of a data-entity declared in subclass to the corresponding data-entity in the parent-class. Upcasting is safe because a class contains all the attributes of the ancestor-class (parent-class or classes further up in the hierarchy), and a method in the ancestor-class will already have all the needed attributes. After upcasting, the corresponding methods available to the ancestor-class can be used. Upcasting can be done automatically by the compiler, or it could be user-defined. Compile-time checking is not always possible. For example, the use of Object-type in Java requires runtime type checking.

11.4.2.2 Downcasting

Downcasting is not safe because (1) a subclass may add additional attributes including new data-entities and methods in a subclass, and a virtual method may be operating on attributes that are defined only in a subclass and do not exist in the transformed version of the actual object in the parent-class and (2) the parent-class may not be able to cast type to a subtype without information loss. Owing to this inherent safety consideration, downcasting has to be done explicitly through a programmer action.

11.4.3 Subclass versus Subtyping

An interesting question is asked: Does subclass support subtyping? Certainly a subclass supports upcasting and downcasting. An inherited method in the parent-class can be called in an instance of the subclass with a compatible object-entity declared in the subclass during upcasting. There is a common notion of inheritance in both subclass and subtype. As discussed in Chapter 7, a subtype inherits all the operations of the original type. Similarly, subclass inherits all the member-entities including methods from the ancestor-classes including parent-class. The corresponding data-entity in a subclass can be a compatible subtype because during upcasting no information will be lost, and the inherited method could work on the subtypes after upcasting the subtype declared in a subclass to the type declared in the parent-class. Another similarity is the notion of defining additional operators in the subtypes and the notion of defining additional member-entities including methods in the subclass. However, subtypes do not allow redefining the meaning of an operator in the traditional sense, whereas subclass allows the *override* of a method. For example, Ada is a language that supports subtyping. If we say that *probability* is a subtype of *float*, then the subtype *probability* will inherit the addition, subtraction, multiplication, and division operation defined for the type *float*. The meaning of the operators is not altered. Overriding a virtual method in a subclass is a case of overloading—an ad hoc polymorphism. In that sense, subclass is different from subtyping. Another major difference is the regulation of visibility between the parent-class and subclass. A parent-class can have private member-entities: a private member entity is invisible to the subclasses and other descendant classes in the hierarchy. However, the notion of subtyping does not restrict the entities not to be inherited by the subtypes.

11.5 CASE STUDIES

In this section we will study the concepts of object-oriented languages utilizing four popular object-oriented languages: C++, Java, Scala, and Ruby. All these languages support multiple programming paradigms as shown in Appendix I. Although the basic concepts of class-based hierarchical structure is common in all these languages, only C++ supports *multiple-inheritance* in true sense. Other languages support single inheritance along with some features such as *mixin* (used in Ruby) or *traits* (used in Scala) to add common features in multiple classes. Scala and Ruby treat every entity as an object. Even functions are objects in Scala that provide very interesting features that are different from procedural programming style. Similarly, iterative constructs are methods for data-entities.

11.5.1 Abstractions and Programming in C++

C++ is a programming language that has object-oriented programming features built on top of the language C and uses many control and data abstraction features of C. It uses the concepts of modularity; visibility using private, protected, and public entities; and packages. The default visibility in C++ is that the entities are private unless declared otherwise. The visibility-type is added before the method declaration, and there is no restriction on the order of method declarations based on the visibility.

In addition to traditional data-types such as built-in types, arrays, structs, pointers, recursive data-types using pointers, enumeration types, and string types, C++ supports polymorphic types such as generic methods using type-variables, inclusion polymorphism through subclasses and inheritance. The subclasses can extend the parent-class-definitions by (1) redefining the methods having the same name, and treating them as dynamic virtual methods and (2) declaring new methods in the subclasses. C++ supports overloaded operators and coercion.

A simplified version of an abstract syntax for a C++ programs is given in Figure 11.4. The abstract syntax states that a C++ program is zero or more library inclusions followed by class-definitions followed by the main-body of the program.

```
<C++ program> ::= {<library-declaration>}* {<class-definitions>}* <main-body>
<library-declaration> ::= # include <library-name> ;
<class-definition> ::= <class-declaration> {<methods>}*
<class-declaration> ::=
          class <class-identifier>
          '{' {<visibility> <class-members-declarations>';'}*'}' |
          '{' {<visibility>:{<class-members-declarations>';'}* }*'}'
<class-member-declaration> ::=
          <class-variable-decl> | <data-abstractions-decl> |
          <method-decl> | <constructor-decl>
<visibility> ::= public | private | protected
<method-decl> ::= <function-decl> | <procedure-decl>
<class-variable-decl> ::= <type-info> <identifier> {',' <identifier>}*;
<constructor-decl> ::= <identifier>'('{<type-info>}* ')'
<function-decl> ::= <type-info> <identifier>'('{parameter-decl;}*')'
<procedure-decl> ::= void <identifier> '('{parameter-decl;}*')'
<parameter-decl> ::= <type-info> <identifier> {','<identifier>}*';'
<type-info> ::= <concrete-type> | '<'<type-variable>'>'
```

FIGURE 11.4 An abstract syntax for a class-declaration in C++.

A *<class-definition>* is defined as a *<class-declaration>* followed by the bodies of the methods in the class. A *<class-declaration>* includes the reserved word *class* followed by the *<class-identifier>* followed by *<visibility>* of the class members followed by *<class-member-declaration>*. A *<class-members-declaration>* can be class-variables, data-abstractions, and method-declarations in the class. A method-declaration can be either a function or procedure. A method is defined as type information followed by an identifier showing the name of the method followed by parameter information for the methods. The parameters passed have the type information and the identifier-name. The type information can be concrete as well as a type–variable as in the parametric polymorphism, and type of the object can be passed as parameter. Similarly, polymorphic functions also include generic type-variable as their type and the type of formal parameters so that they can be instantiated to the concrete type of the passed parameters during run time.

C++ uses all the control abstractions of C and other traditional imperative programming languages such as if-then--else statements, case-statements, while-loops, do-while-loops, 'break' to exit nested blocks, and user-defined functions.

The command to create objects is included in the body of the main program. A class-object can be *automatic* or *static. Automatic objects* are allocated dynamically in the memory area when the control reaches the statement that creates it and destroyed when the control exits the block. A *static object* is destroyed after the program terminates. A method is accessed using *<object-name>.<method-name>*. The symbol '.' is the member-access operator. Objects are allocated in the heap. However, data structures and variables that are not part of the class-definitions are created like traditional imperative programming languages and use control stack as well as heap. All the local variables in a method are allocated in the control stack, and recursive data-types and dynamic objects are allocated in a heap. A subclass-definition in C++ is declared as follows:

```
class <subclass-identifier>: [<visibility>] <parent-class-identifier>
```

A public or protected method can also be accessed inside a class-definition by giving the full class-pathname starting from the root-class followed by the method name. The nodes in the path are separated by the symbol '::'. For example, a method area described in the subclass square in the hierarchy quadrilateral, rectangle will be given as *quadrilateral:: rectangle:: square::area* (*<parameters>*).

Constructors—C++ jargon for a method to create an object—are declared as a public methods in the class-definition. There is more than one way to declare a constructor. The constructors without parameters are called *default constructors.* A constructor has the following attributes: (1) it has the same name as the class name, (2) it just allocates memory in the heap and has no type, (3) a class can have more than one constructor based on how the parameters are passed, and (4) constructors are automatically executed when an object is created. The general abstract syntax for a constructor is given below:

```
<constructor> ::= <path>::<class> '('{<parameter>}*')'
      {<local-var> = (<init-value> |<parameter-value>)}*
```

An example of a subclass *indexibleStack* is described for C++ in Figure 11.5. All the class declarations are grouped together, and the program-body comes after the class declarations. The subclass *indexiblestack* is defined after the definition of class *stack.* The subclass *indexiblestack* describes a new method, *dataItem*, that takes an index as parameter. The subclass *indexedstack* inherits information including the constructor from the parent-class *stack.*

In C++, a problem can be solved using struct-based programs as well as class-based programs. The advantage of class-based programming over struct-based programming is information-hiding provided by the class-definition. By default, the entities declared in class-based definition are private and invisible to outside world unless declared public. All entities declared within "struct" are always public.

11.5.2 Abstractions and Programming in Java

Java is a strongly typed language and supports multiple paradigms. In this section, we limit our discussion to abstractions related to object-oriented programming. As shown in Figure 11.1, Java supports classes and inheritance. However, Java has limited support for parametric polymorphism through "Object" declaration and subtyping hierarchy.

"Object" declaration creates a reference-type that can be associated with any object in the heap. Objects in heap exist as long as there is a reference from the control stack to the heap. The reference has to be assigned null value to recycle the memory occupied by an object. Java supports both mutable and immutable objects. Strings are immutable objects: an operation such as "concatenation" on a string returns a new string using new memory locations.

Java is a single-inheritance language and does not support multiple-inheritance like C++. Java class starts with a reserved word "class". The subclass uses the reserved word

```
template <class T> class stack {
   protected:
        T [ ] mySeq; // mySeq is an array of references
        static maxsize = 256; // maxsize is a class variable
        int top = 0; // top is an instance variable
   public:
         stack( ) ;
         ~stack( );
         boolean isEmptyStack ( ) ;
         boolean isFullStack ( ) ;
         void push(<T>);
         T pop( );
         T top( );}

template <class T> indexedstack : public stack
{ public T dataItem( int ); }

template <class T> stack <T>::stack( )
{ this.mySeq = new T [ ]; } // constructor

template <class T> boolean stack::isEmptyStack ( ) {return (top == 0); }

template <class T> boolean stack::isFullStack ( ) { return (top == maxsize); }

template <class T> void stack::push(<T> next _ element) {
         if (top == maxsize) throw OverflowException( );
                   mySeq[top++] = next _ element; } // insert the element

template <class T> stack <T> ::pop( )
   T popped;{
              if (top == 0) throw UnderflowException( );
              popped = myseq[--top]; // pull out the top element
              myseq[top] = null; // make the reference null for garbage collection
              return(popped);}

template <class T> stack<T>::top( ) { return (mySeq[top - 1]); }

template <class T> indexStack<T>::dataItem(int Index ) {
              If (index < 0 || index == maxsize) throw RangeException( );
              return (mySeq(index));}
```

FIGURE 11.5 Object-oriented programming in C++ using polymorphism.

"extends" followed by class-name. Methods are declared like a function preceded by the visibility declaration. A Java method can be public, private, or protected. A virtual method uses "@override" before the method definition to declare that virtual method definition is being redefined.

11.5.3 Abstractions and Programming in Scala

Scala is a multiparadigm statically strongly typed polymorphic language that supports functional programming, object-oriented programming, and concurrent programming. It supports both mutable and immutable objects. Lists and strings are immutable objects, whereas arrays are mutable objects. Scala is built on top of Java. So it implicitly uses many of the features of Java. In fact, Scala automatically loads the Java package *java.lang* by default. Like any other object-oriented language, Scala imports the library of already developed classes using "import" command.

Scala is a pure object-oriented language. Everything is an object including numbers and functions. Because numbers are objects, they also have methods. All the arithmetic operators such as +, −, *, and /, are methods. For example, 2 + 3 is equivalent to 2.+ (3). Because functions are also objects, they can be stored as objects in variables and passed as parameters.

A function definition is preceded by the reserved word 'def'. Class-definitions are preceded by the reserve word 'class'. A class can prevent further creation of subclasses by using the reserved word 'final'. Like C++, Java, C#, and other similar object-oriented languages, an instance can refer to itself using 'This'. Scala allows overloaded methods. Like Java, the reserved word "extends" is used to declare that a class is a subclass of another class.

Scala does not support multiple-inheritance. Instead it uses *traits* to include the common attributes in multiple subclasses. Traits can extend classes or other traits. A subclass in Scala can inherit methods both from parent-class as well as traits. Traits can also have parent-class and subclass relationships. Traits are declared using the reserved word 'trait' followed by trait declaration. The body of a trait is executed when an instance using the trait is created. The default parent for a class or trait is *AnyRef*, a direct child of class *Any*.

A constructor in Scala is the whole body of the class-definition. When an instance of the class is created, each field corresponding to a parameter in the parameter list is initialized with the parameter automatically. An explicit constructor is not needed. The constructor in a derived class must invoke one of the constructors in the parent-class. Traits use only primary constructors without any arguments. Traits cannot pass arguments to the parent-class constructor.

Scala's visibility rules are similar to Java with some variations. Unlike C++, the default visibility in Scala is "public." The visibility is regulated using the keywords *protected* and *private*. Visibility declarations appear at the beginning of function, variable, or trait declarations. Scala also uses *private[this]* to make the declaration private to a specific instance within a class.

Method definitions are regular functions and use reserved word 'def' followed by optional argument lists, a colon character ':' and the return type of the method. Methods are treated as abstract if the method-body is missing. Methods can be nested.

The class *stack* and *queue* are built-in classes in Scala. In order to illustrate the concept, we take another simple example of abstract data-type *complex number*.

Example 11.8

Figure 11.6 illustrates the definition of class *complex* to define an abstract data-type complex number. Complex numbers are modeled as a pair of them (*real-part, imaginary part*). The class-definition includes just two fields: *realpart* and *imgpart*. There are three abstract operations: *add, subtract*, and *multiply*. The operation

```
class Complex(first: Double, second: Double)
  {       val realpart = first
          val imgpart = second
          override def toString = realpart + " + " imgpart + "i"

  def add(that: Complex): Complex =
        new Complex(realpart + that.realpart, imgpart + that.imgpart)

  def subtract(that: Complex): Complex =
        new Complex(realpart - that.realpart, imgpart - that.imgpart)

  def multiply(that: Complex): Complex =
        new Complex(
              realpart * that.realpart - imgpart * that.imgpart,
              realpart * that.imgpart + imgpart * that.realpart)
}
```

FIGURE 11.6 An example of object-oriented programming in Scala.

toString has been redefined using the override feature for ease of printing the result. The three abstract operations have been written as three methods as described earlier. Every method creates a new complex number that is automatically printed out on the console using the redefined method *toString*.

After loading this program in Scala interpreter, if we give the following commands:

```
val c1 = new Complex(3, 4) //print out a complex number 3.0 + 4.0 i
val c2 = new Complex(4, 5) //prints out a complex number 4.0 + 5.0 i
c1 add c2 //print out a complex number 7.0 + 9.0 i
c1 subtract c2 //prints out a complex number -1.0 + - 1.0
```

11.5.4 Abstractions and Programming in Ruby

Ruby is a multiparadigm, dynamically typed interpretive language that integrates functional, imperative, object-oriented, and concurrent programming (including threads and locks) paradigms. We have already seen the functional and concurrent features of Ruby. In this section, we will study the object-oriented features of Ruby.

Like Scala, every entity including numbers and files is an object. It would be valid to write an expression 2 + 3 as 2.+(3) since addition is treated as a method to object 2. The class-definition is preceded by the reserved word 'class'. The class-variables have a prefix '@@' to separate them from the instance variables, which are annotated with a prefix '@.' An instance of a class is created by a built-in method *new.* A method *initialize* (if defined) is executed whenever an instance is created, and sets up the initial state of an object.

A subclass relationship is set by the symbol '<.' A subclass is declared as *<class>* '<' *<parent-class>*. All the methods declared in the parent-class or ancestor-classes can be used by a subclass. Ruby does not support multiple-inheritance. However, Ruby can include features using *partial classes.* This controlled acquisition of features is called *mixin.*

A *partial class* need not define all its methods at once. The methods may be defined in separate parts of a program. Thus partial class can be used to enhance the capability of a class. A partial class is declared like a subclass.

Instance variables in Ruby are not shared with other objects even of the same type. Class-variables are shared across the instances of the class. An instance variable of an object can be read using *attribute-readers.* Attribute-readers are methods that return value of an attribute when called, or they can be grouped together using a shortcut called *attr_reader.* It is placed after the class-name. The syntax for the shortcut is *<attribute-reader>*:: **attr_reader** {':'*<instance-variable>*}$^{+}$. Attribute-readers can also be used to write into the instance variables.

Ruby methods are public by default. In order to limit them within the class, they need to be declared *private*, and in order to limit to the subclasses only, methods have to be declared *protected.* There are two ways to declare the visibility: (1) using visibility-reserved words before defining the method and (2) at the end of all definitions using declaration like *<visibility-declaration>*:: (**private** | **protected**) {':'*<method-name>*}$^{+}$. Ruby allows accessing a method (having the same name as in the current class) of the parent-class using the reserved word 'super'.

Example 11.9

Figure 11.7 shows the same example of defining a class *stack* and then defining a subclass *indexedstack* that adds an additional method. The example illustrates the syntax to create and use a class, subclass, methods, initialization, class-variables, and instance-variables. The instance variable *@my_stack* is a reference to a dynamic array. The methods use the built-in Ruby-operations on indexed-sequence such as *length, pop, push*, and *last* to define the methods in the class *stack.*

The method *push* takes a single parameter *newdata* as an element and passes it on to the built-in method. Note the absence of parenthesis in the statement *@my_stack.**push** newdata* which is Ruby style of programming. Every method starts with a reserved word 'def' and is ended by a reserved word 'end.' The class-definition is terminated by the reserved word 'end.'

The statement *IndexedStack* < *Stack* declares the subclass *IndexedStack.* Its initialization method has the reserved word 'super' embedded in it. At the end of the class-definitions there is a group of statements to test the stack operations. The statement *st* = *IndexedStack.**new*** creates a new stack *st* that is a dynamic array. The statement ***puts st.**empty* prints true if the stack is empty. The statement *st.**push**(4)* puts the element 4 on top of the stack. The statement *st.push(5)* puts the element 5 on top of the stack. The statement ***puts*** *st.top* prints the top element of the stack, that is, value 5. The statement *puts st.pop* pops and prints the top element from the stack. The statement *puts st.full* returns true prints true if the stack is full.

Many popular programming languages such as C#, F#, and Python are object-oriented languages. Many other languages such as Lisp, Prolog, Modula-3, Fortran 90 onwards, COBOL, and Oz also support object-oriented programming. The main

```
class Stack // Defines a stack using an indexed sequence
@@maxsize = 256 // Declare a class variable maxsize
      def initialize // Constructor
              @my_stack = Array.new // Create stack as a dynamic array
      end
      def push(newdata) // Push a data element
              @my_stack.push newdata // Ruby has built-in push for sequence
      end
      def pop // Pop an element out of the indexed sequence
              @my_stack.pop // Ruby has built in pop to take out last element
      end
      def count
              @my_stack.length // Ruby has built-in length function for sequence
      end
      def empty
              @my_stack.length == 0 // Check if length is zero
      end
      def full
              @my_stack.length == @@maxsize // Check if length is maximum size
      end
      def top // Top element of the stack
              @my_stack.last // Last element of a sequence is the top element
      end
end
class IndexedStack < Stack
      def initialize
                 super
      end
      def data(index) // Access an element by giving the index of the sequence
             @my_stack[index]
      end
end

st = IndexedStack.new; puts st.empty; st.push(4); st.push(5);
puts st.top; puts st.pop; puts st.top; puts st.full; puts st.data(0)
```

FIGURE 11.7 Object-oriented abstractions and programming in Ruby.

features of object-oriented programming as described earlier are class and subclass definitions, inheritance, visibility within class-hierarchy and outside, capability to override method definition, mechanism to create an instance, and mechanisms to access instance variables and class-variables. Some languages such as Javascript do not support the notion of inheritance and are often called object-based languages.

11.6 IMPLEMENTATION OF OBJECT-ORIENTED LANGUAGES

Object-oriented programming is different from traditional procedure-based programming because of the presence of classes—an additional encapsulation, inheritance, virtual methods, and the dynamic nature of objects—instances of classes that receive the message from other objects. Action to invoke a public method is taken by the receiving object based on message. In turn, the receiving method may also invoke other protected and private methods that are invisible to other objects. Static procedure call in procedural languages is replaced by the dynamic calls of the objects. The dynamic calls are initiated by (1) calls to create new objects, (2) messages sent by other objects, and (3) calls by other methods within the same object. When a dynamic call is made, the control flow jumps to another function or procedure that is decided by the receiver object. This phenomenon of jumping the control flow based on messages and receiver object is called *late binding*. The control stack is a sequence of the frame of the methods that are invoked. Note that the sequence of frames in the stack includes frames belonging to methods from different objects that are instances of different classes.

There are three broad classifications of compilation schemes: (1) separate compilation coupled with dynamic loading/linking, as in Java and .NET, (2) separate compilation with global linking as in C++, and (3) global compilation as in Eiffel. Separate compilation scheme is most general, where each class or group of classes in a package is compiled separately irrespective of the knowledge of other classes and without any anticipation about the inheritance mechanism, such as single inheritance or multiple-inheritance from other classes. An important advantage of separate compilation is localization of error and the effect of modifications. Thus separate compilation is in the spirit of object-oriented software development.

Different implementations are quite different because of difference in philosophy of overhead of time and space management. As we can see that identifying shared entities and grouping them together will save space. However, it would also add an extra level of indirection and thus slow down the execution because of an additional memory-access cost. We will discuss a general schema for the implementation.

Some of the features of object-oriented programming languages are similar to the compilation and implementation of procedural languages and have already been discussed in Chapter 5 in detail. Those features are (1) compilation of nonvirtual functions and procedures, (2) parameter passing, (3) allocation of global variables, (4) normal overloading of operators, and (5) shadowing of class-variables.

The major issues in the implementation of object-oriented programming languages are (1) allocation of the data structures of a dynamically created object in a heap, (2) accessing the data-entities and methods of the parent-class and other ancestor-classes, (3) handling of multiple-inheritance in the languages like C++, (4) handling of virtual methods, and (5) dynamic subtyping. We will discuss the implementation of these features and a general schematic of implementation in the following sections.

11.6.1 Storage Allocation and Deallocation

When compiling an object-oriented program, methods form the code area, and the data-entities in the objects make the data area. Although objects are created dynamically, the code area does not change; only the data area and the pointers to the compiled version of the virtual methods are created dynamically. In addition, the upcasting and downcasting of an object is done dynamically. Because the code part does not change, it is compiled statically at compile-time. The pointers to the nonvirtual methods are also computed at compile-time. However, the compilation of polymorphic types requires dynamic switching to a different code area based on the data-type tags as described in Section 7.8.2.

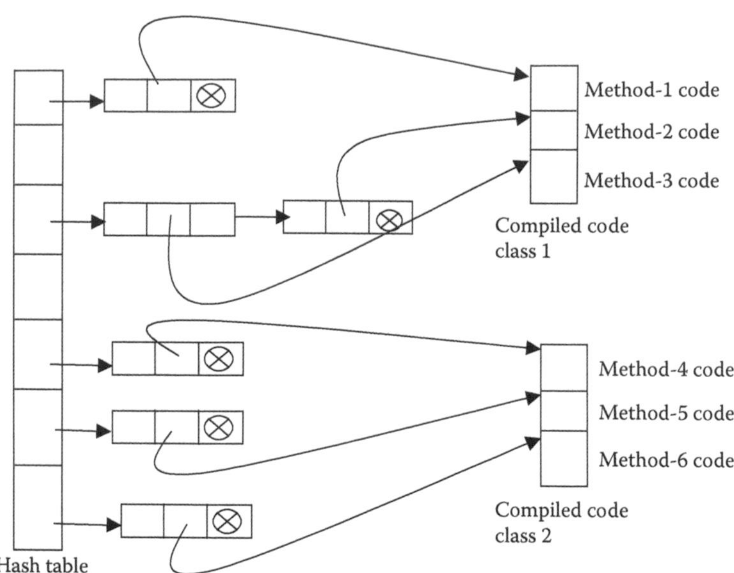

FIGURE 11.8 Schematics of code lookup using hash tables.

11.6.1.1 Compiled Code

The code area of an object-oriented program is the union of all the methods declared in various classes and subclasses and is identified uniquely by the concatenation *<package-name>::<class-hierarchy>::<method-name>*. There are many ways to refer to a specific method. One such technique is to map the full method name to some internalized symbol. The function table is a hash table where the internalized symbol is the key, and the address is stored along with the key in a tuple of the form (*internalized-symbol, start address of the compiled function*). The hash function is used to look up the start address of the compiled method in near-constant time, and the control is passed to the compiled code of the corresponding method. A schematic of the function table and the compiled code is given in Figure 11.8. The assumption is that each class is compiled separately.

There are many hash-function techniques that are used to ensure that the collision among the keys is kept to minimum. Since the name of the methods tagged with class-hierarchy is unique, a *perfect hashing function* can be devised. A *perfect hashing function* maps to the value in key-value pair without any collision. However, here we will assume that collision is possible. The unique symbol corresponding to the method is looked up using a hash-function. Collision after the application of hash-function has been handled using a linked list of tuples (*symbol-name, function-address, link to next tuple*). Each tuple carries the address to one compiled method. When a class library is loaded at run time, the corresponding symbol-name and the address of the compiled codes for methods are inserted in the hash-table.

11.6.1.2 Data Allocation

There are many types of data in an object-oriented language. Data is allocated based on their scope and dynamic nature. For example, a data can be (1) a global variable, (2) a class variable, (3) a local variable in a method, (4) a parameter, (5) attributes of an object,

(6) pointers to dynamic attributes such as virtual methods, (7) pointers to the attribute-table of the ancestor-classes, and (8) pointers to the functions-table in the code area. In addition there may be constants that are only read. We can separate the data into multiple parts: (1) data that can be included as part of the code area because it is read-only, (2) data such as local variables in a method that can be made part of the control stack, (3) data such as class-variables that can be made as part of the attribute-table of a class in the heap, and (4) data-entities in an object that are stored in the heap with the object area.

There are two models of variables depending upon object allocation. In the *value-based model* of variables, dynamic objects having the same life time as the method are allocated in the frame of a stack. In the *reference-based model* of variables, dynamic objects are allocated in a heap, and the stack only stores a reference to the object. Many languages like C++ that have extended an imperative programming model to allocate some data-objects in stack while allocating the other dynamic objects in heap. Generally, those objects that have scope and lifetime within a given method are allocated in the frame of the method and reclaimed from the control stack as soon as the method is over. Many languages such as Smalltalk, Python, Ruby, and Java use the *reference model*, whereas languages like Ada, C++, and Modula-3 also allow for a value-based model. There are advantages and disadvantages in both the allocation schemes. Value-based allocation uses control stack for the allocation, which is efficient for storage allocation as well as memory reclaim as discussed in Chapter 5. Reference-based allocation uses heap that has (1) additional overhead of allocating objects in heap, (2) an overhead of indirect access through the use of pointers, and (3) the overhead of garbage collection. Thus reference-based model has an overhead of execution efficiency. However, the mechanism of dynamic memory management of objects in heap is straightforward.

Object-specific declaration in a class is stored in the heap using a template called *attribute-table*. An *attribute-table* includes (1) information related to data-entities associated with the class-instances, (2) pointer to virtual-method-table (VMT) to access the virtual functions, (3) pointer to the inherited data entities from the ancestor-classes, (4) pointer to the attribute-tables of parent-classes and subclasses for upcasting or downcasting, (5) pointer to shared data-entities such as class-variables, and (6) pointers to inner subobjects. Inner subobjects are protected subobjects in the class-hierarchy of the base-class. The mechanism to access the data-entities is similar to the allocation of data in the heap already discussed in Chapter 6. Internally there is a set of offsets for entities in the object; there is one offset for each declared aggregate entity. The offset is computed at compile time based on order of the attributes, and does not change at run time. Since the offset does not change at run time, some of the information is embedded in the compiled code of the methods. The attribute-table of subclass can be merged with the attribute-table of the parent-class if a subclass does not override any method and does not insert any additional method. Such merging provides space efficiency.

11.6.1.3 Object-Allocation

The creation of the first instance of a class using a constructor in C++ or through the use of reserved word 'new' in other languages creates an attribute-table that includes the *invariant offsets* to various attributes and the memory locations for various attributes

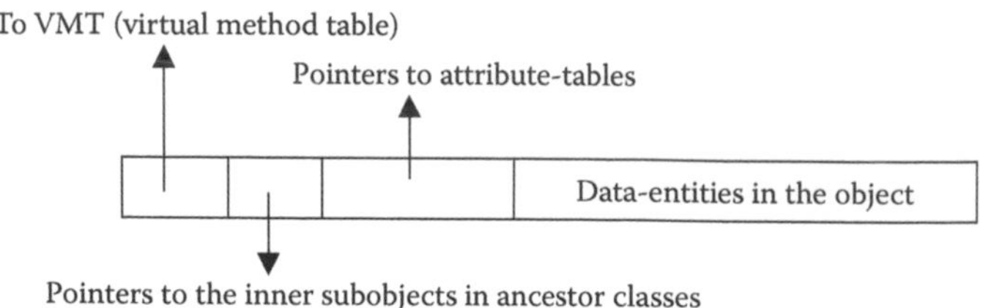

FIGURE 11.9 Schematics of object allocation in heap.

in the heap. If a data-structure does not include generic type or extensible type then it is allocated within the attribute-table. Otherwise, a reference for the generic type or extensible type is stored that points to the object stored in the heap. A processor-register points to the attribute table in the heap.

A generic object-allocation for a single-parent-class is given in Figure 11.9. There are different types of pointers followed by the allocated memory for the data-entities that have been constructed using the information given in the attribute-table of the class. The major classes of pointers are (1) pointer to VMT, (2) pointer to the memory location of the inner subobjects, (3) pointer to the required attribute-tables to retrieve the relative offsets needed for upcasting or downcasting.

A VMT-pointer, as the name suggests, points to a table that contains pointers to the compiled code area of the virtual methods. Pointers to the memory-locations of inner subobjects are used to point to the inherited data-entities, such as class-variables and the inherited methods code area, in the class-hierarchy, starting from the root-class.

11.6.1.4 Deallocation

The objects are deallocated when (1) a frame is released after the execution of a method or (2) the object is released by an action of explicit destructor or delete operation as in C++. Java and C# use a *finalize* method that is the last method called before the release of an object for garbage collection.

11.6.2 Implementing Casting

Both upcasting and downcasting have additional overheads of accessing information because the attribute table of a subclass and parent-class are different from the attribute table of the current class due to: (1) additional methods (including virtual methods) and data-entities in a subclass, and (2) some of the methods and attributes like class-variables are inherited from the parent-class and not declared in the subclass.

This difference in attributes including method declarations results in the variation of the offsets of the equivalent attributes in two attribute-tables and has to be readjusted using the knowledge of the amount of offset-shift. This offset-shift can be handled by (1) calculating the offsets at the compile-time, embedding the offsets adjustment in the compiled code, and adjusting the offset when going across the class-hierarchy or (2) using specific access-functions to access the attributes in the attribute-table. In the offset-method, an extra table of the relative offsets is kept between two classes in the same hierarchy as part of the attribute-table. If the casting is dynamic, then the offset information has to be available at runtime for the proper transformation, and type-checking of the data-entities has to be done at runtime before casting.

11.6.3 Implementing Multiple-Inheritance

Handling multiple-inheritance is a bottleneck in the implementation of the object-oriented languages because of (1) name-conflict of the inherited attributes in the ancestor-classes, and (2) conflict of relative offsets of the same inherited attributes accessible from different paths. Name-conflict is itself sufficient for many object-oriented languages to propose alternate mechanisms such as traits or mixins.

One scheme for class-object layout for an object-allocation in the subclass S shown in Figure 11.3 is given in Figure 11.10. There are as many VMT pointers as the incoming edges to the class S. In Figure 11.3, there are two paths, P → Q → S and P → R → S, that include edges Q → S and R → S. The virtual methods can be redefined along any of these two paths. Hence there are two VMTs: the first one contains the virtual methods in P and Q and S; and the second one contains the virtual methods in the classes P and R and S, in that order. The pointer to VMT is followed by the pointers to the inner subobjects of the corresponding ancestor-classes in the order starting from the base-class P.

The allocation scheme in Figure 11.10 has a duplication of the VMT pointers and the pointer to inner subobjects of P. This duplication can be removed, as illustrated in Figure 11.11, by keeping only one copy of the VMT of P in the first VMT and by altering the first pointer to inner subobjects of P to point to the field that stores the second copy of the pointer to inner subobjects of P.

The first VMT contains all the virtual methods of class P, Q, and S, in that order, whereas the second VMT contains only the virtual methods of class R. Similarly, the first field pointing to inner objects of P now points to the second field of the inner objects of P as shown by the dashed pointer.

11.6.4 Implementing Virtual Entities and Methods

Virtual methods are handled using the integration of two techniques: (1) addition of the class-hierarchy as prefix to the objects and methods, and compiling each of the methods with this unique, long name and (2) performing run-time resolution of the method

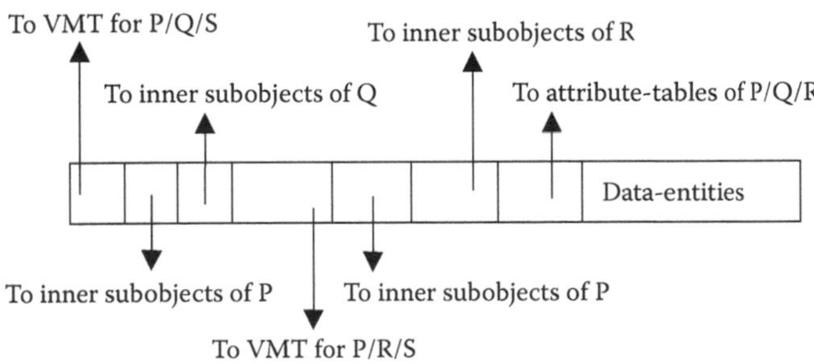

FIGURE 11.10 Object-allocation schematics for multiple-inheritance.

being called. Since objects are created dynamically, the reference to actual method is also generated at run time. The presence of virtual methods creates a need for run-time reference to the compiled code area. If all the objects of a subclass create references in their area in the heap, then there will be excessive number of pointers pointing to the same compiled code area corresponding to the same virtual methods. In order to reduce the wastage of memory, a dynamic shared table is created in a heap for every subclass. This shared table is called VMT. The table is an array of pointers (or array of unique symbolic-names to be used by hash-table lookup) of all the virtual methods declared/accessible in a subclass. This includes the unshadowed (not overridden) virtual methods in the ancestor-classes. The pointers point to the start address of the compiled virtual methods in the compiled code area. The concept is illustrated in Figure 11.12.

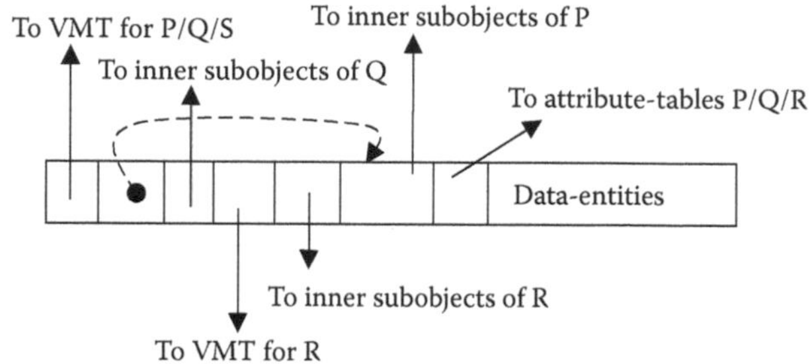

FIGURE 11.11 Optimized object allocation for multiple-inheritance.

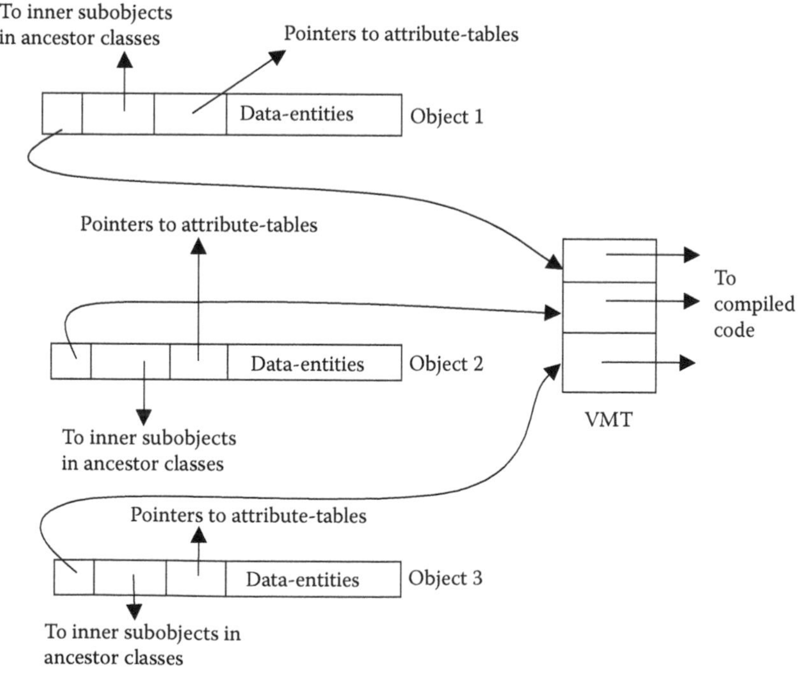

FIGURE 11.12 Schematics of objects pointing to VMT.

Multiple objects that are instances of the same subclass share the same VMT. In a hash-table-based scheme, if the code area is relocated, only the address in the tuple is altered accordingly; VMT remains unaffected.

11.6.5 Overhead Issues and Optimizations

The implementation of object-oriented languages suffers from the overheads caused by the excessive number of pointers to implement inheritance, virtual methods, and offset shifting to access the corresponding data-type during upcasting and downcasting. One way to reduce the overhead of the indirect access is to use caching of the addresses of the corresponding functions. The change in offset during casting can be handled using *attribute-access functions* instead of transforming the offset. An attribute-access function can be optimized by caching the exact address of the data-entity. The VMT of a subclass can be merged with the parent-class if it does not redefine any of the virtual methods in the parent-class. The redundant pointers to the virtual methods declared in ancestor-classes can be removed from the VMT of the current class and the subclasses of the current class if the inherited virtual method is not used during execution of all the object in the subclasses of the current class.

11.6.6 Run-Time Behavior

The run-time behavior of the program involves interaction of the control stack, the objects allocated in the heap, and run-time jumping of the control to various compiled methods. The overall integrated schematic is illustrated in Figure 11.13. The figure shows that there are two types of references from the control stack to the heap: (1) *references to the extensible data-entities in the heap* and (2) *references to object-allocations.* The object-allocation contains two types of pointers to the function table: (1) pointers for VMT along with the offset of the virtual method in the table and (2) pointers for nonvirtual methods.

The main program creates a new object in the heap using the corresponding attribute-table, and sends a message to execute a method in the corresponding class. If it is the first object created in the class, then the corresponding attribute-table is also loaded in the heap, and the corresponding VMT is created. The object uses the address of the nonvirtual methods or the pair (*address of the VMT, offset within VMT*) to pass the control flow to the corresponding method. The frame of the new method that consists of the local variables and parameters is placed on top of the control stack. Any variable bound to a dynamic data-entity is a reference to the corresponding heap allocation. A variable bound to basic data-type is allocated on the control stack for efficiency, as the memory location can be easily reclaimed after the method has completed its execution. After the termination of a method, the corresponding frame is recovered from the control stack, and the corresponding reachable objects in the heap are also released for garbage collection. If a method calls another method, then the frame of the called method is put on top of the frame of the calling method, like execution in the imperative paradigm discussed in Chapter 5. Parameter passing between the methods and the main program to the first method is handled conventionally as discussed in Chapter 5.

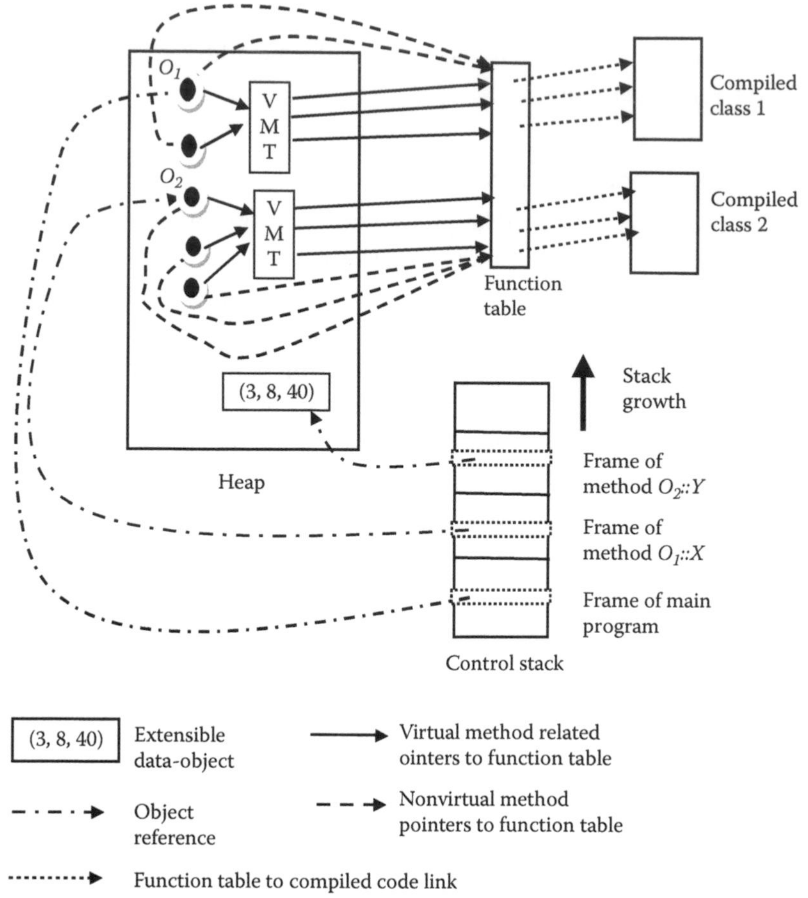

FIGURE 11.13 Schematic of run-time behavior.

Example 11.10

Figure 11.13 shows two classes, each having a VMT. There are two instances of $Class_1$ and three instances of $Class_2$. The object O_1 is an instance of $Class_1$, and object O_2 is an instance of $Class_2$. The main program creates an object O_1. At this time, the frame of main is on top of the stack, and one of the locations in the main-frame is pointing to the object O_1.

The main program now invokes a public method $O_1::X$ in the object O_1. In response, the frame of the method $O_1::X$ is put on top of the control stack, and the control is passed through pointers in object-allocation to the code area of the method *X*. The method $O_1::X$ calls a public method $O_2::Y$. The frame of the method $O_2::Y$ is allocated on top of the frame, and the control passes to the compiled code of $O_2::Y$ through the pointers stored in the object O_2 and the function-table. After the completion of method $O_2::Y$, the memory locations corresponding to the frame of the method $O_2::Y$ are recovered from the control stack as described in Chapter 5, and the control is passed back to the next instruction in the method

O_1::X. All the reachable objects in the frame of O_2::Y that have a local scope are released for garbage collection. A similar process is repeated after the method O_1::X terminates.

The object-oriented languages continuously create dynamic objects, and most of the objects are dynamically and automatically disposed when they are not needed. There is a heavy need for garbage collection in heap to keep the space available: 30–40% overhead in Java and 20–30% overhead in C++. However, for large-scale software development, object-oriented programming has established its usefulness because of modularity and reusability. One of the goals we have is to reduce the overhead of dynamic memory management in object-oriented languages. There have been hardware and software attempts to make dynamic memory management efficient.

11.7 DISTRIBUTED OBJECT-ORIENTED MODELS

Programming languages like Emerald and Java use a network of processors to execute object-based programs. Emerald is a strongly typed programming language suitable for distributed object-based programming. Unlike other object-oriented programming languages, Emerald does not have a class structure and does not support inheritance but supports fine-grain distributed computing. Java uses Java virtual machine (JVM) to execute and interface with multiple processors. Each processor has the corresponding JVM engine and or uses JIT compiler to execute Java programs on local machines for efficient execution. In this section, we will study both types of object representation and the distributed method invocation and information-exchange mechanisms.

11.7.1 Distributed Objects in Emerald

Emerald objects are like modules that can export the operations. Emerald objects are created using a reserved word 'object.' Operations are equivalent to methods in typical object-oriented programming languages. Emerald objects are independent of their location. This independence allows a programmer to develop a program similar to the program development on a uniprocessor machine. Emerald objects consist of private data, private operations, and public operations. Only exported objects can be accessed. A constructor is used to create an object.

Objects can be embedded inside a control abstraction. For example, if a constructor is placed inside a loop, then multiple objects can be created. Emerald objects are flat and do not support any hierarchical structure. There is no notion of inheritance: there are no parent-classes or the notion of virtual methods or overriding or inherited methods.

Emerald separates the type of information as *abstract type* and *implementation type. Abstract types* are a group of *signatures* where a *signature* is defined as a name and type of the arguments in an operation. The difference between abstract type and implementation type is that abstract type can be implemented in different ways. Emerald types can be passed as objects, and thus a type can be constructed or refined and can be used to implement polymorphism.

Emerald uses the notion of *conformity* instead of inheritance. *Conformity* is related to the notion of inclusion. The property of conformity states that an abstract type T_1 conforms to another abstract type T_2 if object of type T_1 can substitute for an object of type T_2. An object of type T_1 confirms to an object of type T_2 if the following three properties are satisfied: (1) T_1 has all the operations of T_2, (2) for every operation in T_2, the corresponding operations in T_1 have same number of arguments, (3) the types of arguments and results of T_1 conforms to the types of arguments and results in T_2, respectively.

The difference between conformity and inheritance is that conformity allows inclusion and thus the substitution. However, the notion of hierarchy is missing: conformity, unlike inheritance, does not support code sharing. The major advantage of not sharing the code and the lack of the hierarchical nature of class is that code mobility of ancestor-classes is not needed when Emerald objects move from one processor to another.

Emerald supports concurrency: multiple objects can be working at the same time. An active object contains a process that is started after an object is activated. A control-thread can spawn multiple objects that may be working on separate threads independently providing fine-grain concurrency.

11.7.2 Distributed Objects in Java

With the availability of network of processors, there has been a clear need for the development of distributed object models for the web-based languages like Java. Java is supported by JVM, which makes Java portable on almost every operating system and architecture. Although Java programs may be running on a heterogenous network of architectures and operating systems, the whole system works as if it is working in a homogenous environment because of the portability of JVM. The methods in Java can be executed on remote processors without any problem of format conversion because of uniform representations in JVM.

11.7.3 Remote Method Invocation

Remote method invocation (RMI) uses remote procedure call with a difference that it is built on top of JVM, transmits an object over the communication channel, and uses built-in Java operators for type matching of the remote objects and casting. RMI layer is built on top of the JVM, and the application layer is built on top of the RMI layer. The overall system architecture for RMI is: Client → RMI layers → Server. The JVM layer takes care of *classloader, security management,* and *distributed garbage collection.* The RMI system is distributed into multiple layers: (1) stub/skelton layer, (2) remote reference layer, (3) transport layer, and (4) reference-counting distributed garbage collection layer. Distributed reference-count garbage collection is used for garbage collection. *Stub* is a reference to the remote object. A remote object can be persistent. A call to method on a *stub object* sends a request to the corresponding *remote object* after setting up the communication channel. The *server-side object* executes the method and then sends the result back to the originating JVM.

The parameters are passed either using call-by-reference or call-by-value in the RMI system. If the parameter being passed is a remote object, then its reference is passed. Otherwise, a copy of the object is passed to the remote method. *Object serialization* marshals and unmarshals the parameter values and results returned from the remote method; the object is encoded into a stream of bytes from one of the JVMs during "marshalling" and back to a copy of the object during "unmarshalling." Because an object often contains references to other objects, the process of serialization traverses the graph of objects during serialization, and maintains the graph structure at the other end. Serialization iterates through the class-hierarchy structure from the root-class to the actual class of the object and writes all the class information and nontransient and nonstatic public, protected, and private fields including the fields holding references to other objects into the stream. At the other end, reading the stream involves building the class-hierarchy after reading the class information, creating an instance of the object, restoring the contents of each field of the object, and matching the type of the fields with the created fields. After an object is created at the remote processor for the first time, a *handle* (collection of pointers to the object) is created for the instance. Next time while referring to the object, the handle is used to minimize the overhead of the object serialization. Because objects are serialized, the true Java type of an object can be transmitted to the remote processor.

To invoke a method on a remote object, a client obtains a reference to the object using a name-server provided by the RMI system. The class-definition *java.rmi.Naming* provides *uniform resource locator* (URL)-based methods to look up, bind, store, rebind, unbind, and list the name-object pairings on a particular host and port.

11.7.4 RMI-Based Programming

In this section we will discuss simple format and scheme for RMI-based programming using Java. Clients of remote objects interact with remote interfaces, and local arguments are passed by copy rather than by reference. A remote object is passed by reference. Clients have to handle exceptions that occur during a remote method invocation. The Java library to handle remote interfaces is called *Java.rmi.remote*, and the client side classes extend *java.rmi.remote* by additional remote methods. The remote methods have two components: "method declaration" and "capturing remote exception". The abstract syntax for remote method interface is as follows:

```
<remote-method-interface>::=
            public interface <interface-name> extends
              <interface-name>, java.rmi.Remote
            '{'    {public (void|<function-type>) <identifier>
                       '('<parameter-list> ')'
                              throws java.rmi.RemoteException ';'}+
            '}'
```

In the above declaration, the first interface-name is like a class-definition, and the second is like a parent-class-definition that is being extended. There are two additional remote interfaces: *java.rmi.Remote* and *java.rmi.RemoteException. RemoteException* class is a super class of remote exception that includes communication failure, protocol errors, failure during serialization process, and failure of the method at the remote end.

The class that implements the remote method imports the following two libraries: *java.rmi.RemoteException* and *java.rmi.server.UnicastRemoteObject*. The abstract syntax for the implementation of remote class is as follows:

```
<remote-class> ::= public class <identifier> extends
                   Unicast RemoteObject implements
                   <interface-name>'{'{<remote-method>}+ '}'
<remote-method> ::=
         (public | protected | private) (void |<function-type>)
           <identifier>'('<parameter-list>')'
         throws RemoteException,{<Exception-name>}*
           '{'<method-body>'}'
```

The abstract syntax states that a remote implementation class implements the given interface, and it describes one or more methods. The method definition includes the exception *RemoteException* in addition to other exceptions. The super class *UnicastRemoteObject* provides support for point-to-point active object references (invocations, parameters, and results).

11.8 SUMMARY

The major goal of object-oriented programming is the development of large-scale software. Large-scale software construction requires modularity, reusability, robustness, and ease of software evolution and maintainability. Object-oriented programming achieves this goal by encapsulating functions along with data entities within an object; generic template-based programming called class, notion of packages(modules), regulating visibility of the functions, and procedures in a class-definition using private, protected, and public declarations, and inheritance. The class libraries can be imported for software reuse. Although there is some debate about the use of inheritance in reusability because of the propagation of effect of changes in the methods modified in the ancestor-classes as is evident in the development of inheritance-less languages like Emerald, inheritance has the advantages of using the previously declared class library. To avoid name-conflict while using different classes in a program, full-name prefixed by the class-hierarchy is also used to describe methods. An object is created and initialized using constructors that are like special methods in a class. Constructors can be initiated explicitly or invoked when a new object is created using the reserved word 'New'. An object can be explicitly destroyed by a programmer action as in C++ or automatically destroyed. Treating every entity as an object has resulted in an interesting style of programming in multiparadigm languages, such as Ruby and Scala as basic arithmetic operations and control abstractions can be treated as methods.

The object-oriented programming paradigm supports the notion of polymorphism using the notion of generic-templates to support parametric polymorphism, notion of subclass to support inclusion polymorphism, and through the use of overloaded operators that can be overridden using a new method definition in a subclass.

Multiple-inheritance means that a subclass has more than one parent-class. Handling more than one parent-class causes problems of name-conflict. Name-conflict occurs because both the parent-classes may have the same name methods. Name-conflict can be removed by qualifying the inherited method by the proper-class name. However, most of the languages other than C++ do not support multiple-inheritances; they use single inheritance or some restricted version of multiple-inheritance. Languages like Java and Scala use single inheritance with a capability to enhance the capability of a class.

The implementation of object-oriented programming languages has additional complexity because of run-time memory allocation of objects and their interaction with the functions and jumping of the control from one method of an object to a method in another object. The compiled code can be generated separately one class at a time, or globally, which involves a library of classes and includes their hierarchy. The compiled code is accessed using a function-table that can use the pointers or hash-functions along with symbolic references to access the start-address of the compiled-method. The attribute-table of a class stores the type information and offsets needed for memory allocation of an object in the heap. An object-allocation has multiple pointers to access the virtual methods of the current class up to the base-class, inner subobjects that can be inherited from the base-class up to the parent-class, pointers to remove duplicate entries because of multiple-inheritance, and memory locations for the declared data-entities in the corresponding class.

The run-time behavior of the object-oriented programming languages involves object-allocations in the heap as they hold the pointers to the method being executed. The frames for the methods being executed are placed in the control stack. The basic data-entities, references to complex objects in the heap, and pointers to return back to the calling method are allocated in the control stack. The objects local to a method are deallocated from the heap after the method has finished execution and later reclaimed by the garbage-collection process.

The distributed object-oriented programming is based on migration of objects to the remote processor and executing. Emerald uses a flat object structure like a module that includes data-entities and operations on data-entities. The objects easily migrate to the remote processor because of their flat structure. To support the inclusion of objects, Emerald uses abstract-types. Emerald executes objects transparently by embedding objects inside the control abstractions and using thread-based programming. An object is created when an object is referenced for the first time.

Java uses remote method invocation analogous to the remote procedure call studied in Chapter 8 in terms of the use of stubs, skeletons, and connection through the ports. However, it is built on top of the JVM that provides homogenous treatment of the implementation across different operating systems and architectures. Remote method invocation transmits an object and the corresponding class structure by traversing the class-hierarchy and serializing the object including the class structure. Once an object is transferred to the remote processor, it is rebuilt by a deserialization process, and a handle is created for the reuse of the object in the future. RMI uses both references and copying of the object for parameter passing.

11.9 ASSESSMENT

11.9.1 Concepts and Definitions

Abstract class; abstract method; abstract type, ancestor-class; attribute-table; base-class; casting; class, class-hierarchy; class-library; class-variable; code migration; conformity; constructor; data migration; derived class; deserialization; distributed objects; downcasting; encapsulation; function table; generic method; global compilation; implementation type; information hiding; inheritance; inner subobject; instance, instance-variable; interface; maintainability; marshaling; message passing; method; mixin; module, modularity, multiple-inheritance; object; operation; override; package; packing; parent-class; partial class; perfect hash function; private, protected, public, pure virtual function; receiver; remote method invocation; reusability; root-class, separate compilation; serialization; signature; skelton layer; stub; subclass, subtype; super; super class; trait; transport layer; unmarshalling; unpacking; upcasting; virtual method, Virtual Method Table (VMT); visibility.

11.9.2 Problem-Solving

1. Write an object-oriented class-definition in C++ that can perform the following operations on strings where strings are modeled as an indexible sequence of characters.
    a. Comparing two strings for equality
    b. Comparing two strings to identify the string that precedes the other using alphaneumeric comparison based on ASCII value
    c. Finding out if the two strings are palindromes
    d. Finding out the position where two strings differ
2. Write a Ruby class-definition for handling priority queues. It should support the following operations:
    a. Get next element based on the priority
    b. Insert an element based on the priority
    c. Insert at the end of the queue
    d. Count the number of elements in the queue
    e. Return a Boolean value if the queue is empty
    f. Return a Boolean value if the queue is full
3. For the class-definition *indexedStack* defined in Figure 11.3, give a possible object-allocation scheme.

4. Write a simple class-definition exhibiting upcasting. Rewrite the class-definition using access methods to access the attributes of classes.

5. Write a simple class-definition exhibiting downcasting. Rewrite the class-definition using access methods to access the attributes of the classes.

6. Self-study the remote method invocation of Java, and write a simple Java program that finds out the maximum value in a sequence of numbers on a remote processor and returns the result.

7. Write a simple Emerald program to perform a bubble-sort of a sequence.

11.9.3 Extended Response

8. What do you understand by upcasting and downcasting? Explain using simple examples.

9. What are the overheads of implementing upcasting and downcasting in object-oriented programming languages? Explain.

10. What are the different restrictions on visibility of declared entities? Explain each one of them using a simple example.

11. What do you understand by inheritance? How does it affect reusability? Answer both pros and cons of using inheritance in an object-oriented language.

12. Explain the concept of virtual methods and their advantages in object-oriented programming languages. How are the virtual methods different from abstract methods and inherited methods? Explain.

13. What do you understand by VMT? What is the need for VMT? Comment on the time and space efficiency of using VMTs.

14. Compare and contrast multiple-inheritance in C++ and "single-inheritance + traits" in Scala in terms of modeling real-world phenomenon, limitations, and implementation overheads.

15. What is the difference between conformity in Emerald and inheritance in other typical object-oriented languages? Explain.

16. Read about the advantages and disadvantages of Java Object type using the Internet, and discuss using your own examples.

17. How is the object-oriented implementation model different from the procedural model of implementation discussed in Chapter 5? Explain.

18. What are the problems in implementing multiple-inheritance? Explain using a simple example in C++, represent the class-definition as a DAG, and then give the corresponding object-allocation scheme.

19. Compare and contrast object definitions in Emerald with object definitions in Java.
20. What are the inner subobjects in a hierarchical-class? Explain using a simple example that has at least two levels of subclasses, with virtual and nonvirtual methods in each of the classes.
21. Explain the advantages of JVM in handling remote method invocation.
22. Explain the mechanism of remote method invocation including remote exception handling.

FURTHER READING

Bacon, David F., Fink, Stephen J., and Grove, David. "Space- and time-efficient implementation of the Java object model." In *16th European Conference on Object-Oriented Programming*, Springer-Verlag, Berlin, Hydelberg. *LNCS 2374*. 2002. 1–21.

Bloch, Joshua. *Effective Java*. 2nd edition. Stoughton, MA: Addison-Wesley. 2008.

Corson, Tim and McGregor, John D. "Understanding object-oriented: a unifying paradigm." *Communications of the ACM*, (33)9. 1990. 40–60.

Ducournau, Roland. "Implementing statically typed object-oriented programming languages." *ACM Computing Surveys*, (43)3. 2011. 18:1–18:48.

Gil, Joseph Y., Pugh, William, Weddell, Grant E., and Zibin, Yoav. "Two-dimensional bi-directional object layout." *ACM Transactions on Programming Languages and Systems*, 30(5). August 2008. 28:1–28:30.

Gil, Joseph Y. and Sweeney, Peter F. "Space-and time-efficient memory layout for multiple inheritances." In *Proceedings of the 14th Annual Conference on Object-Oriented Programming Systems, Languages, and Applications, OOPSLA '99*. ACM SIGPLAN Notices 34(10). 1999. 256–275.

Gosling, James, Joy, Bill, Steele, Guy, and Bracha, Gilad. *The JAVA Language Specification*. 3rd edition. Addison-Wesley. 2005.

Lippman, Stanley B. *Inside the C++ Object Model*. Redwood City, CA: Addison Wesley. 1996.

Liskov, Barbara and Guttag, John. *Program Development in Java: Abstraction, Specification, and Object-Oriented Design*. Upper Saddle River, NJ: Addison-Wesley. 2000.

Liskov, Barbara and Wing, Jeannette. "A behavioral notion of subtyping." *ACM Transactions on Programming Languages and Systems (TOPLAS)*, 16(6). 1994. 1811–1841.

Maasen, Jason, Nieuwpoort, Rob V., Veldema, Ronald, Bal, Henri, KIelman, Thilo, Jacobs, Ceriel, and Hofman, Rutger. "Efficient Java RMI for parallel programming." *ACM Transactions on Programming Languages and Systems*, 23(6). 2001. 747–775.

Meyer, Bertrand. *Object-Oriented Software Construction*. 2nd edition. Upper Saddle River, NJ: Prentice Hall. 1997. 1370.

Nierstrasz, Oscar. "A survey of object-oriented concepts." In *Object-oriented Concepts, Databases and Applications*, ed. W. Kim and F. Lochovsky. New York, NY: ACM Press and Addison-Wesley. 1989. 3–21.

Odersky, Martin, Spoon, Lex, and Venners, Bill. *Programming in Scala: A Comprehensive Step-by-Step Guide*. 2nd edition. Mountain View, CA: Artima Incorporation. 2011.

Philippsen, Michael and Zenger, Matthias. "JavaParty: Transparent Remote Objects in Java," *Concurrency Practice and Experience*, 9. 1997. 1225–1242.

Raj, Rajendra, Tempero, Ewan, Levy, Henry M., Black, Andrew, Hutchinson, Norman C., and Jul, Eric. "Emerald: A general-purpose programming Language." *Software: Practice and Experience*, 21(1). 1991. 91–118.

Stroustrup, Bjarne. *The Design and Evolution of C++*. Reading, MA: Addison-Wesley. 1994.

Wegner, Peter. "Concepts and paradigms of object-oriented programming." *ACM SIGPLAN OOPS Messenger*, 1(1). 1987. 7–87.

CHAPTER 12

Web and Multimedia Programming Paradigms

BACKGROUND CONCEPTS

Abstract computations (Section 2.4); Abstract implementation of programming languages (Chapter 5); Abstractions in programming languages (Chapter 4); Concurrent programming (Chapter 8); Distributed computing (Section 8.4); Distributed objects (Section 11.6); Operating system concepts (Section 2.5); von Neumann machines (Section 2.1).

Since the early 1990s, with the ubiquity of Internet connectivity, the need for *code* and *data mobility* has taken off due to the need of resource sharing. Ubiquity means the availability of resources on the Internet irrespective of their physical locations and the locations of clients. The presence of wireless computing has further facilitated ubiquity. Many application areas have been positively impacted by the Internet connectivity also called *Web*. *Web* is a network of information nodes such that each node is capable of (1) performing computation, (2) storing and retrieving information from an information node, (3) requesting data or code to be transmitted from a remote node, (4) transmitting data or code to a remote node, and (5) setting up connection with remote nodes for code and data transfer. The remote node location is called the *uniform resource locator* (URL) in web-programming jargon.

Web-based programming intertwines *code mobility* and *data migration* including multimedia transmission and rendering at the client-end. Multimedia information includes a combination of text, images, sound, audio clips, video clips, and audiovisuals that integrate and synchronize both audio streams and video streams to give a minimally perceptually distorted visualization of real-time events as they occurred. Multimedia information is important for better comprehension of real-world events, because we perceive the world by using multiple sensors and modes.

Different applications have different needs based on the size of code, size of data, security needed for code or data, and privileges provided to the client or the server for the use of code or data. For example, a banking transaction related code and data cannot be transmitted to remote processors without high security. However, a software code for calculating a monthly mortgage by a consumer can be transmitted to the client-end.

Application areas that have benefitted by the presence of the *World Wide Web* (WWW) include banking; telephony; transportation; stock market investments; sale of consumer products; the entertainment industry such as game-playing and demand-based movies; information archives and exchanges; the education industry; and collaborative design and modeling. Web-based applications have also improved the speeds with which we do business and the quality of service (QoS). Textual information has been replaced by real multimedia, which involves audiovisuals, movies, and clips. Books are being replaced by e-books, and availability of e-books is becoming *pervasive*—available at any place at any time. Information retrieval and processing has become demand driven. There is automated software that is used to collect and index information from various computers, and the indexing is used to perform search by search engines like Bing, Google, and Yahoo.

If we look closely, all these applications require an integration of *computation*, *resource sharing*, *visualization*, and *security* during information transmission and sharing. The computation has to be distributed over the Internet on different information nodes of the WWW to reduce overload on servers. Either code or data or both can be transferred from one web-node to another. If the data is transferred, it is called *data mobility* (or *data migration*), and if code is transferred, it is called *code mobility* (or *code migration*). In recent years, multiple web scripting and multimedia languages have been developed.

This chapter shows models of computations involving code and data mobility, the new language constructs to facilitate web-based and multimedia programming, and various virtual machines and languages that support this integrated web-based programming.

12.1 CODE AND DATA MOBILITY

There are four models of mobile computing based on how data and code migrate over the web. The models are (1) *client-server*, (2) *remote evaluation*, (3) *code-on-demand*, and (4) *migrating agents*. Depending upon the supported models, different languages support different constructs and security mechanisms.

Client-server model assumes that one of the web-nodes needs information from other web-node. The requesting node is called the *client*, and the provider node is called the *server*. A request is transmitted from the client to a server. The server performs computation, retrieves information, and sends the result to the client. There is no mobility of the code. In the *remote evaluation model*, the data is transmitted from a web-node N^C to another web-node N^R that has the resources and code to process the data transmitted by N^C. This paradigm supports data mobility. In the *code-on-demand model*, the client node N^C has the resources to execute mobile code and the needed data. The node N^C requests mobile code from the remote node N^R to process the data. After receiving the requested code, the mobile code is executed using local resources and data.

In the *migrating agent model*, the code may be sent from one node to different web-nodes. The mobile code is self-sufficient and uses the local resources of the remote nodes to process the data present at the remote node. The code has the capability to migrate from one node to another node and may have an embedded plan and itinerary to visit different nodes based on certain criterion. During that visit, the migratory code may process data resident on the remote nodes and transmit the result back to the node that initially originated the migratory code.

Most of the web-based information retrieval uses client-server model. The programs have two parts: *client-end* and *server-end*. Client-end and server-end programs communicate to each other using *web-forms* and HTTP protocol—a web-based communication protocol built on top of TCP (Transmission Control Protocol). TCP is a low-level communication protocol to transfer data over the Internet. Languages like Java support applets—a mobile code that migrates and executes at the client node.

12.1.1 Issues in Mobile Computing

The major issues in mobile computing are (1) handling heterogeneity of environments, (2) handling overhead of execution and transmission caused by code and data mobility, and (3) handling security problems caused by code and data mobility. Different computers have different architectures with different internal data formats and have different operating system. Such a network is called *heterogenous*, where data representation, computer hardware, and operating systems are different.

Mobility of code and data causes inefficiency because of packing and unpacking of the code and data, as we discussed in the remote procedure call section in Chapter 8 and the remote method invocation in Chapter 11. In addition, because web-based programming also involves rendering the data to the user in a perceptible multimedia form, it slows down the overall execution significantly. In a client-server model, a server can be heavily loaded, because there are too many clients making requests.

The last major issue is problems with security because of (1) infection of mobile code by malicious code during transit to the client, (2) attack on the processors by malicious code disguised as mobile code, (3) attack on the client-end processor because of infection of the return data from an infected server, and (4) leak of the sensitive information to some malware during transmission. The following sections discuss approaches to handle problems of code and data migration to remote computers.

12.1.1.1 Handling Heterogeneity

Heterogeneity problems are caused by (1) difference in computer architecture; (2) difference in hardware configurations, such as available memory and performance capability of the machine where code is being executed; (3) difference in operating systems and their versions; (4) difference in the available compilers to compile the mobile code; and (5) difference in the available system libraries and language-specific libraries needed for program execution. Because of this heterogeneity, a mobile code may not execute on a remote node. In order to remove the problem of heterogeneity, there have been two approaches: (1) interoperability to translate the format of the source node to the format of the destination node and (2) virtual machine implementation on every architecture and operating system to enforce homogeneity of the environment.

Interoperability is concerned about transforming code and data from one architecture and operating system to other architectures and operating systems. This can be done by the source node before transmitting the code and data to the target node or by the target node after receiving the code and data. Either way, one of the machines (source machine or the target machine) has to be aware of the organization of the other machine to perform the transformation. To handle this problem of transformation, a *common language interface* has been used. The *common language interface* translates an outgoing data to a *common metadata representation,* and incoming metadata is translated back to the destination machine's native representation.

The second approach implements a *common virtual machine* on every web-node to provide homogeneity, and the high-level language is translated to this low-level common instruction set of the common virtual machine. In essence, the use of a common instruction set used in a common virtual machine transforms the network of heterogenous computers to a network of homogenous computers.

The problem of availability of resources on the remote computer can be handled if the machine transmitting the code is aware of the resource requirement and the capabilities of the remote machine and can match the capability before requesting for code migration. The problem can also be caused because of changing load conditions or the network conditions. If the load condition in a remote machine is quite high, then code migration to the remote node should be avoided. The compatibility problem between compilers and libraries is serious and can cause run-time failure. Compatibility across compilers or interpreters and the built-in libraries has to be maintained when the virtual machines evolve.

12.1.1.2 Handling Execution and Migration Efficiency

Efficiency has been handled at different levels (1) using a just-in-time (JIT) compiler to transform into efficient native machine code; (2) using cache to reduce data transfer overhead; (3) using stream-based transmission of multimedia objects, so that rendering of video and audiovisuals at the client-end can be interleaved with transmission; (4) run-time adjusting the resolution based on the communication channel capability and traffic congestion; and (5) keeping multiple copies of the frequently used code and data at various nodes called *mirror-sites.*

12.1.1.3 Handling Safety

With the availability of mobile code, the chance of malicious code affecting the host machine, the chance of a malicious host corrupting the mobile code, and the chance of a malicious host making a mobile code a carrier of hostile code that can affect other hosts has increased tremendously. For example, mobile malicious code can delete private files, system files, and classloader to crash a machine. Even worse, it may corrupt the boot area, so that machine cannot boot again. Many times mobile malicious code is transmitted as data files and can become active once in the host machine.

A mobile code can be corrupted during transmission or become malicious in an intermediate node during transmission from the originating node to the destination node. Privileged mobile data can be read by snooping software or the intermediate nodes during transmission from the originating node to the destination node. Worse still, a malicious website may disguise itself as a genuine website, stealing all the privileged information. This phenomenon is called *phishing.*

Safety is handled at multiple levels: (1) at the client-end, (2) during transmission, (3) at the server-end, and (4) verifying the identity of the client and the mobile code. The first level of safety is that any active mobile code is not given full privilege to access a computer's address space and file system. Rather it is *sandboxed*—placed in a protected area, where the mobile code has very limited access to the computer's file structure, has no access to the system area, and may not share the address space with software on the host machine. In addition, the transmitted data is encrypted at the source-computer and decrypted at the destination computer. Each packet carries the identity of the originating web-node and intermediate web-nodes to ensure that data did not originate or pass through known malicious sites. Section 12.2.4 deals with security in more detail.

12.2 WEB-BASED PROGRAMMING

We need programming languages and extensions to existing programming languages that are capable of transferring code and data between information nodes over the web and process the resident data or the data retrieved from the remote node using the resident code on the local node. We also need capabilities to encode, visualize, embed, and render complex multimedia data such as images, sound, audiovisuals, text, and their integration over the web. In the past 20 years, because of the invention of WWW, new visualization languages as well as the new languages and extensions of existing languages for web-based programming have been developed.

The first web-based language was HTML (Hyper Text Markup Language), which could embed the resource locator (hyperlink) of the multimedia object including formatted text such as tables and text with various styles, images, sound, video and display at the client-end. The client would send a request by clicking an embedded hyperlink, and the request would be sent to the corresponding server. The data would be transmitted from the server to the client, and the browser located at the client-end would display the information to a user. A schematic of the client-server model using HTTP (Hyper Text Transmission Protocol) is given in Figure 12.1. The HTML browser includes many decoders to display various media objects embedded in the document. A decoder is a software that interprets the format of multimedia files for rendering at the client-end. In addition to the browser, the client-end also carries a cache that archives the data stored in the recently visited sites so that a revisit to that site can avoid the overhead of data transfer; data is pulled out from the local cache instead of the server during revisits.

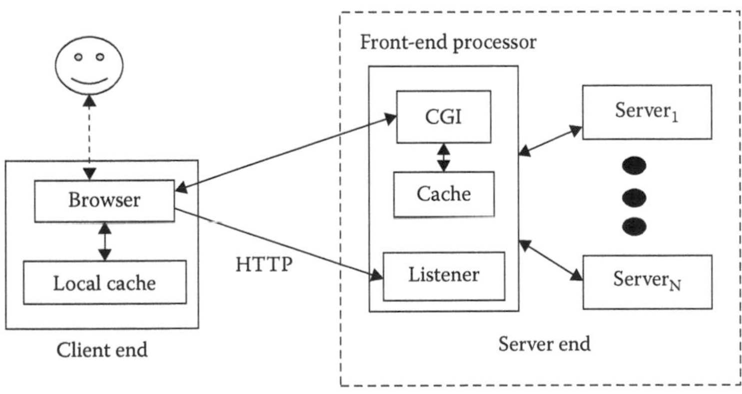

FIGURE 12.1 A schematics of client-server model of web browsing.

At the server-end, there may be more than one server involved in retrieving and processing the data from the database. The servers pass the retrieved information to a common gateway interface (CGI) that reformats the retrieved data into the format acceptable to the browser and transmits the result back. The server keeps a cache of retrieved data, so that the next request for the same data can be retransmitted with little retrieval overhead. Many times, on the Internet, more than one intermediate node is involved in data transmission between the client and server. In that case, intermediate nodes may also cache the data to lower the data transfer overhead.

12.2.1 HTML

The language HTML has three major components: head, body, and hyperlinks to other media objects. The head part includes the title of the documents, transmission protocol name, format-related information, author-related information, and keywords used by the search engines. The transfer-protocol could be (1) http—for web-based browsing of data; (2) ftp—used in file transfer without visualization; (3) file—denotes a file on local computer; (4) telnet—used for remote login; (5) tel—used for telephone dialing; and (6) modem—for modem-based connection. The most common use is http, ftp, and file for information exchange. Usually, the resource is a media file, such as text-file, or image-file, audio-file, or audio-visual file.

The body part contains the annotated document. The hyperlinks are embedded inside the documents to take to another document or media object that may be on a local computer or a remote computer. The locator is of the form *<transfer-protocol>://<node-name>:[<port-name>]/<path-name>/<resource-name>*. A file resource is written in the form *<file-name>.<media-format-type>*. Some of the common media formats are .jpg (image format), graphical information format (.gif) (image format), .wav (sound format), .mpeg (audio-visual clip format), *.avi* (Microsoft audio-visual format), .pdf (printable document), and .mov (movie format).

HTML text formatting is done using a group of tags that annotate the text to (1) change the text style such as color, boldness, size, or heading; (2) make and format a table; or (3) embed and align media objects inside a table or within the text. The general format of the tagged text is '<' *<tag-name>* '>' *<tagged-text>* '</' *<tag-name>* '>'. The tags can have multiple attributes embedded into them. For example, a table can have attributes such as *border, size of the cells,* or *padding from the top and the bottom*. The abstract syntax for attributes is *<attribute-list>* = {*<attribute-name>* ' = ' *<value>*}*.

Example 12.1

```
<HTML>
 <Head>
  <META HTTP-EQUIV = "Content-Type" CONTENT = "text/html;
    charset = iso-8859-1">
  <META NAME = "Author" CONTENT = "Arvind Bansal">
  <META NAME = "Keywords" CONTENT = "programming languages, text
    book">
  <TITLE> Programming Language E-book</TITLE>
```

```
</Head>
<Body BGCOLOR = "#F2FFFF">
      <H1 ALIGN = CENTER > Introduction to Programming
        Languages </H1>
      <TABLE border = 0 cellspacing = 0 cellpadding = 2
        bottommargin = 0 topmargin ="0"bgcolor="white">
              <TR>
                <TD VALIGN = TOP ALIGN = LEFT>
                  <img SRC = "book.jpg">
                <TD VALIGN = TOP Align = Right>
                    <A HREF = "http://www.crcpress.com">
                      CRC Press </A>
              </TR>
      </Table>
</Body>
</HTML>
```

The above code illustrates the overall organization of an HTML document. The HTML document is divided into two parts: (1) the document-head, which contains the metadata such as content-type, character-set, author, keywords, and the title of the document and (2) the document-body, which contains the annotated document. For example, the tag <H1> tells the type of heading; the tag <Table> describes the format and content inside a table; the tag <TR> describes a row; and the tag <TD> describes a data element in the data cell. There are multiple attributes associated with different tags. The hyperlink tag **<A HREF** = "http://www.crcpress.com> CRC Press**</A>** puts a hyperlink under the text *CRC Press*. Most of the tags end with an end marker. In case of <**TD**>, the end tag is optional, because it can be terminated by another <TD>, <**TR**> or </**TR**>.

HTML works well for visualizing formatted documents. However, it has the following limitations: (1) tag and attribute values are mixed with the text making rearrangement and style change difficult; (2) a tag cannot be defined by a user; (3) HTML does not support any computation and text processing; (4) HTML has limited capability of any action based on events; and (5) HTML lacks the capability of dynamically changing the text style based on some condition. That means that e-commerce organizations have to make lot of effort to reformat the website.

The above limitations have been solved by the development of new languages that (1) facilitate dynamic transformation of tag-attributes, (2) allow user-defined tags, (3) integrate HTML with web scripting languages having computation capability, and (4) transmit the code to the client-end and use the local resources at the client-end to process the data. The language XML extends HTML by allowing user-defined tags. XHTML integrates HTML browser capability and XML's (eXtended Markup Language) user-defined tag.

In recent years, many web programming languages have been developed to integrate web-based information retrieval, visualization, computing, and event-based programming. Some of these languages are Javascript, PhP, Obliq, Microsoft ASP, C#, AJAX (Asynchronous Javascript AND XML), SMIL (Synchronous Multimedia Integration Language), Telescript, TCL, and Java.

12.2.2 XML as Middleware Interface Language

There are two major advantages of XML-based representation: (1) an arbitrary directed acyclic graph (DAG) (includes tree) structure can be represented into flat textual format, using tags and (2) user-defined tags can represent any complex object if the object can be decomposed as DAG. This representation is easily serialized and recovered at the remote processor. XML flattens nested database tables to facilitate transmission over the Internet. Similarly, XML has been used to annotate and transmit complex 3D objects and 3D animation over the Internet using languages like X3D.

XML description has two parts: (1) document-type definition (DTD) and (2) a document. DTD describes the constraints on the structure and values of the documents. DTD is described like extended BNF, and does not follow XML tag structure. DTD description can be associated with an identifier, and the reference to the description is provided by the identifier in place of the description. DTD can be stored either along with the document or as a separate file with suffix '.dtd'. An XML document is a valid document if it meets the restrictions imposed by DTD. An XML document has user-defined attributes associated with tag words.

Example 12.2

The example shows the structure of an XML document. The first line states that it is an XML document. The second line shows how the comments are encoded. The third line shows the declaration and the file-name *book.dtd*, where the declaration is stored.

```
<?xml version = 1.0?>
<!-- comment - DTD description starts here -->
<! DOCTYPE book SYSTEM "book.dtd">
<!-- comment - document description starts here -->
<book>
    <author> John Doe </author>
    <publisher> XYZ-publisher </publisher>
    <year> 1980 </year>
</book>
```

XML-based languages such as "Mathematics Markup Language" (MathML) and "Chemical Markup Language" (ChemML) utilize user-defined tags and their interpreter to describe complex equations and chemical formulae for transmission over the Internet and rendering at the remote node.

Example 12.3

The following example shows a 2 × 2 matrix expressed in MathML. The MathML interprets these tags and invokes the corresponding rendering software routines to display the matrix. For example, <mo> denotes a math operator, <mtr> denotes a matrix row, <mtd> denotes a matrix data, and the tag <mrow> denotes one-row visualization. The first row of this matrix contains two data elements—10 and 20—and the second row contain the elements 40 and 60.

```
<mrow>
  <mo>(</mo>
  <mtable>
        <mtr> <mtd> <mn> 10 </mn> <mn> 20 </mn> </mtd> </mtr>
        <mtr> <mtd> <mn> 40 </mn> <mn> 60 </mn> </mtd></mtr>
  </mtable>
  <mo>) </mo>
</mrow>
```

12.2.2.1 DTD—Document Type Definition

DTD consists of *element structure definition*, *attribute-list structure definition*, and *entity definitions*. The type information describes the possible structure of an element or attribute list. For example, the document-type declaration for the XML description in Example 12.2 is as follows:

```
'<!' ELEMENT book (author | publisher | year)* '>'
```

The above declaration states that the element book can have one or more occurrences of three subelements: *author*, *publisher*, or *year*. A simplified syntax grammar for DTD is given in Figure 12.2.

```
<dtd-definition> ::= <element-dtd> | <attrList-dtd>
<element-dtd> ::= '<!' Element <element-name> (<element-type> |
                                  EMPTY | ANY | '(' #PCDATA')' ) '>'
<attrList-dtd> ::= '<!' ATTLIST <element-name> { <attrName-Type-Default}+
                   '>'
<element-type> ::= '(' <type> {( ',' | '|' ) <type>]* ')' | <identifier>
<element-name> ::= <identifier>
<attrName-Type-Default> ::= <attrName> ' '<attrType>' ' <default>
<attrName> ::= <identifier>
<defaultValue> ::= '"' <value> '"' | #REQUIRED | #IMPLIED | #FIXED <value>
<attrType> ::= CDATA | {<entity> ' ' }* | <enumeration> | <identifier> |
                        { <idRef> ' '}+ | {<NMToken> ' '}+ | NOTATION
<entity> ::= <general-entity> | <parameter-entity> | <embedded-media>
<general-entity> ::= '<!' ENTITY <entity-name> <visibility> <definition> '>'
<parameter-entity> ::= '<!' '%' ENTITY <entity-name> <visibility>
                          <definition> '>'
<embedded-media> ::= '<!' ENTITY <entity-name> <visibility> <definition>
                            NDATA <format> '>'
<idRef> ::= <identifier>
<format> ::= .jpg | .gif …
<visbility> ::= PUBLIC <FPI> | SYSTEM
```

FIGURE 12.2 A simplified abstract syntax grammar for DTD in XML.

Elements can pick up a combination of tag value defined within the document, 'EMPTY,' or a user-supplied value denoted by #PCDATA, or could be declared of any type denoted by the reserved word 'ANY'. An attribute list is qualified by the element name followed by a list of triples (*attribute-name, attribute-type, and default-value*) separated by white spaces. An attribute type could be a character data (CDATA) or a defined entity, or a reference to an already defined entity or an enumeration type, name token, or notation. Notation is used to define user-defined notations using a declaration like '<!' Notation *<notation-name>* SYSTEM *<external-identifier>* '**>**'. Attributes can be given default values. Default values can be specified as required (#REQUIRED), can be a constant declared using the reserved word **'FIXED'**, or could be implied **(#IMPLIED)**. The reserved word '#REQUIRED' specifies that the default value could be any value but must be specified. The reserved word '#IMPLIED' states that the value is optional. The purpose of declaring entities is reuse of the entity definition elsewhere in the XML program. An entity could be a general entity, a parameter, or an embedded media object.

A general entity has a name and a definition. An entity could either be public or private. An example of an entity is <! ENTITY publisher "CRC Press">. Now the entity-name publisher can be used elsewhere using a reference '& publisher.' A parameter-entity binds a name to the whole definition statement so that the entity-name can be used in place of the definition. Parameter-entities use an additional '%' symbol that precedes the entity-name to separate it from general entities. For example, '<!' Entity %mybook "<! *Element book Any*>" '>' will declare an element with a parameter-entity that can be used anywhere using the name '%mybook.' A media-entity can be embedded in an XML document using a declaration of the general-entity, visibility of the object, and the format of the media object.

12.2.2.2 Style Sheets

Style sheets have been used to separate the attribute values of the tags from the HTML and XML files. Although the tag in the original text remains the same, changing the style in the style sheets alters the attribute values of the tags. The style sheet carries multiple options for the attribute-value pairs of the tags. Different attribute-value pairs can be picked based on some conditions such as time of the day, season, month, country where the browser is being displayed, the rendering device such as laptop or mobile phone, the bandwidth of the communication link. Changing the style of browsed text needs two files: (1) the original XML document and (2) a style sheet description. A program that transforms XML file to another XML file based on a different style sheet file is called "eXtensible Style Language Transformation" (XSLT).

The style sheet description is kept either internally inside the head of the XML document, or in a separate external file and can be altered easily, because it is relatively small: it does not contain the document text. The attributes of different tags are specified in the style sheet files. In the styling sheet file, different templates associated with tag words are given a name, and in the XML file, these template-names are used to apply the attribute-values.

Cascading style sheets have a suffix '.css' and extensible styling sheets have the suffix. 'xsl'. The XSL templates are described as **.template-name** '{' {<*attribute*>:<*value*>}* '}'.

The XML documentation on which XSLT performs the transformation is enclosed between the tag '<' xsl:template-match <*node-name*> '>' and '<' xsl: template-match/'>'. Transformation is applied on the template using the command '<' xsl: apply-templates '>' after the text document. XSLT also has *if-then-else* and *for-each loop* operation to check the attribute types and values and loop through multiple matching tag names to substitute the attribute values in each of them.

12.2.2.3 Modeling 2D and 3D Objects in XML

A two-dimensional region can be decomposed into multiple homogenous regions. A two-dimensional object can be abstracted as a pair (polygon, attribute-value pairs). Polygons are represented as a sequence of vertices such that there is an edge between the adjacent vertices and the first and the last vertices. Thus a complex region is modeled as follows:

```
<complex-region>
            <region name = "Ohio" points = 3 color = "red">
                   <elements> x-value y-value </elements>
                   <elements> x-value y-value </elements>
                   <elements> x-value y-value </elements>
            </region>
            ...
            <region name = "Oregon" points = 4 color = "green">
                    <elements> x-value y-value </elements>
                    <elements> x-value y-value </elements>
                    <elements> x-value y-value </elements>
            </region>
</complex-region>
```

Alternately, a two-dimensional image can be modeled as a group of images such that each image is identified by its center of gravity (cog) and its image-URL.

```
<complex-region>
            <region name = "Ohio">
                     <cog> x-value y-value </cog>
                     <image> url </image>
            </region>
            ...
            <region name = "Oregon">
                     <cog> x-value y-value </cog>
                     <image> url </image>
            </region>
</complex-region>
```

These techniques can be extended to model 3D objects. A realistic 3D objects will have many attributes such as *shape*, *texture*, *luminosity*, and *connectedness to other subparts*. This information can be represented as a list of attribute-value pairs in each node. Homogenous objects like cones, spheres, and cylinders are modeled easily by knowing the properties that characterize the objects. For example, a sphere can be characterized by center of gravity, number of dimensions, and radius. A cube can be characterized by coordinates of the leftmost point, number of dimensions, and length of the side.

We can model a complex 3D object using a graph such that each vertex in the graph is associated with a component 3D shape, and edge is associated with the points joining two 3D shapes. For example, a table is a two-level tree with a root node associated with the table top, and four of the leaf nodes as legs of the table. The four leaf nodes can be associated with a cylinder or some 3D image.

The idea of modeling 3D objects programmatically using regular basic objects such as cones, cylinders, prisms, and spheres came from "Virtual Reality Markup Language" (VRML). It was integrated with 3D representation capability of XML to generate language like X3D. X3D has the capability to transmit 3D objects over the web for remote visualization.

12.2.2.4 Embedding Computation in XML

There are two ways to incorporate control abstractions: (1) interface with code written in a web-based programming language such as Javascript, PhP, Microsoft ASP, or C# using a special tag and (2) annotate the control abstractions as tags, and features of the control abstraction as attributes of the tags. Both the approaches have been used to extend the capabilities of XML. Section 12.2.3 shows the integration of code written in web-based languages and XML. In this section, we will study how control abstractions can be added using user-defined tags.

"Voice eXtensible Markup Language" (VoiceXML) is used for the voice interaction over the telephone and uses the second approach. The declaration of variables, blocks, if-then-else statements, foreach-loop statements, data, and expressions are used as XML tags. For example, a variable can be declared as the following:

```
'<' var name = <identifier> expression = <url> '>'
```

This statement binds a variable-identifier to an URL, and this variable-identifier can be used later to refer to the URL. Similarly, if-then-else statement has the following structure:

```
'<' if cond = <conditional-expression> '<' audio src = <url> '/>'
           <else/> '<' audio src = <url>/>
'<'/if '>'
```

The above structure states that if the conditional-expression evaluates true, then select the first URL. Otherwise, select the second URL for audio.

12.2.2.5 Embedding Communication in XML

"Simple Object Access Protocol" (SOAP) is a XML-based protocol to exchange information between two Internet-based applications. It solves the compatibility problem between the heterogenous processors by using a flat XML format. Soap messages use an XML format without DTD description. The tags have a prefix of soap namespace: the header tag is called <soap:Header>, and the body is called <soap:Body>. The whole message is embedded between the tags <soap:Envelope> and </soap:Envelope>. The <soap:Body> block has the information that includes the URL name where the message is being sent and the message within the user-defined tags. The <soap:Body> block also has an optional embedded <soap:Fault> block to handle the faults. There are four subelements in the <soap:Fault>: (1) <faultcode>—code to detect the faults; (2) <faultstring>—an explanation of the fault; (3) <faultactor>—what caused the fault; and (4) <detail>—application-specific description. The overall template for the soap message is given below.

```
<?xml version = "1.0"?>
<soap:Envelope
       xmlns:soap = "http://www.w3.org/2001/12/soap-envelope"
       soap:encodingStyle = "http://www.w3.org/2001/12/
        soap-encoding">
       <soap:Header>
       ...
       </soap:Header>
       <soap:Body>
       ...
           <soap:Fault>
                   ...
           </soap:Fault>
       </soap:Body>
</soap:Envelope>
```

12.2.3 Web Scripting

Till now we have discussed the advantages of representing the structure of the information and objects in XML. However, XML alone does not have the capability to compute and interact with the user or process complex data abstractions, such as collection of URLs, each holding different media objects or processing the strings and data. Web scripting languages provide that capability. Java provides additional capability of mobile code through *applets*. This section discusses the features of web scripting languages that facilitate web programming, followed by discussion on the integration of XML and web scripting through XML extensions.

Web scripting languages have the capabilities: (1) to create a web document on the fly, (2) to modify the web document based on user interaction or an event, (3) to interact with the user to fill up a form, (4) to provide password security, (5) to provide data and control abstraction capabilities to process the data, (6) to interface with databases such as SQL or Microsoft Access, (7) to change the style of the text, and (8) to create and access abstract graphics objects such as radio buttons, box, or canvas, for user-friendly interactions. If a web programming language is used to render a scene involving multimedia objects, then many web languages provide (1) concurrent constructs to render multiple streams simultaneously and (2) time-delays between streams to simulate causality between the events.

12.2.3.1 Integrating XML and Web Scripting

Languages like PhP, Javascript, and Java have been used for web programming. All these languages support the notion of events, objects, and capability to interface with web documents to provide web programming. They support history, navigation, form, and windows as objects for ease of dynamic access and manipulation. Various fields of these objects can be accessed using the format *<object-name>.<field-name>*. The scripting language codes are embedded within XML using a tag '<' **script** src = *<file-name>* '>' … </script>. The function name of the scripting language can be called from elsewhere by specifying the name and the parameters.

12.2.3.2 XAML—eXtensible Application Markup Language

XAML is a Microsoft version of XML to support images, objects, and animation that embeds images and objects inside XML. XAML integrates Microsoft's graphics toolkit and XML by embedding special tags that invoke the methods to draw the graphics objects with various parameters. The objects could use the built-in library or customized objects defined by the XAML files. All name declaration use a prefix 'xmlns:x = ' to specify that it is an XAML tag.

```
<page>
  xmlns = http://schemas.microsoft.com/winfx/2006/xaml/
    presentation
  xmlns:x = "http://schemas.microsoft.com/winfx/2006/xaml"
...
</page>
```

For example, we can make a button using the following XAML statement:

```
<button x:name = "btnExit" Height = "50" Width = "50"
 Background = "white" Foreground = "Red" Content = "Exit"
 click = "btnExitApp"/>
```

The name of the button is "btnExit." Its size is 50 pixels × 50 pixels, and its color is red. The content displayed on top of the button is the string "Exit," and the event name method associated with it is called "btnExitApp." Clicking on the button will send a signal to the event handler (described in Chapter 13) that will pass the control to the method "btnExitApp." By default all methods associated with XAML events are public. XAML supports nesting of the definitions.

12.2.3.3 Javascript

Javascript is an object-based dynamically typed web programming language. It does not support class inheritance. However, it treats the forms, navigation pane, and history as objects and can use the construct *<object-name>.<field>* to go to any embedded field. It has the capability to embed any URL for rendering multimedia objects. Javascript is supported by XML: a Javascript function can be embedded in the client-side XML code and be invoked in response to an event, such as mouse-motions over a media object or a region in the window pane, filling up a form, and clicking on a hyperlink. Javascript programs are used for (1) client-end user interactions, (2) simple authentication of the user at the client-end, (3) form checking at the client-end, (4) providing simple code such as displaying a clock or a calendar on the webpage, (5) changing the display style of HTML code, and (6) interfacing with the CGI. A language-specific description of Javascript is given in Section 12.6.4.

12.2.3.4 AJAX—Asynchronous Javascript and XML

Ajax is a programming tool that integrates Javascript with XML (or XHTML) for dynamic display and user interaction over the web. Although XML provides the capability of data-interchange and information retrieval, Javascript provides the computation capability and user interaction capability at the client-end. Thus there are two layers of data transformation: (1) user-interface to Javascript and (2) Javascript to CGI through HTTP request. The AJAX engine has two functions: (1) transforming Javascript call to HTTP request and (2) transforming XML data coming from the server-end to XHTML + CSS data. The Javascript layer can perform validation and error checking on the input data before sending the HTTP request to the server. This validation on the local processor reduces the overhead on the server.

12.2.4 Applets

An applet is a mobile code that migrates to the client-end on demand and is executed on the host machine either as a compiled code using JIT compiler or interpreted as instructions of the virtual machine to remove the problem of heterogeneity. Applet is supported by *Java* and many agent-based languages like *TCL* and *Odyssey*. An applet migrates to the client-end when the webpage containing the applet is loaded. Applets are embedded in HTML and XML files using an <applet> tag. As discussed before, applet goes through code verification and is *sandboxed* for security against malicious code, as discussed in Section 12.2.5.

Applet has five methods: *init(), start(); paint(), stop(), and destroy(). Init method* is executed when an applet is invoked. *Start method* is executed when the user returns to the page containing an applet or after the applet is initialized using *init method.* Start method is run under a new thread that is destroyed by the *stop method. Stop method* is activated when the user moves to another page that does not contain the applet. *Destroy method* is executed when the browser shuts down. *Paint method* is used to repaint media objects during an applet's execution.

Applets have multiple attributes such as codebase, code, object, archive, alt, and name, and media layout primitives such as width, height, align, vspace, and hspace. Applets are either pulled from an URI or are part of an XML file. Java virtual machine (JVM) run-time environment is needed to visualize applet in Java. Applet uses the optional tag '<' param name = <*identifier*> value = <*value*> '>' embedded inside the HTML or XML document to pass the parameter values. The syntax for the applet tag is as follows:

```
<Applet-tag> ::= '<'  Applet
                      [codebase = <URI>] code = <Applet-file>
                        [alt = <Alternate-text>]
                      [name = <AppletInstance>] width =
                        <pixel-value> Height = <pixel-value>
                      [align = <alignment-type>] [vspace =
                        <pixel-value>]
                      [hspace = <pixel-value>] '>'
                      {'<' param name = <Attribute-name>
                        value = <Attribute-value> '>'}*
                      '</' Applet '>'
```

The attribute *codebase* describes the URI where the code is located. The attribute *code* describes the file-name that contains the compiled subclass of the applet. The attribute *alt* describes the alternate text to be displayed if the applet cannot be executed. The attribute *name* describes the name of the applet instance so that other applets can access this applet using that name. The attributes *width* and *height* describe the size of the applet display area. The attribute *align* describes the alignment of the display in the viewing area. Alignment could be top, bottom, left, right, middle, and so on. The attributes *vspace* and *hspace* describe the spacing around the applet.

Java applets need to import two class libraries—*java.applet.Applet* and *java.awt.Graphics*—to execute. Applets are treated as a subclass of *java.appplet.Applet* class in Java. The applet base class has many methods that are inherited by the applet subclass such as (1) URL of the file containing the applet and the applet class definition and (2) fetching and rendering the media objects. Applet supports event-based programming using the container classes *java.awt.event.MouseListener* and *java.awt.event.MouseEvent.* There are multiple events such as *clicking of the mouse*, *mouse over a media object*, or *mouse leaving a region.* Images are retrieved from a URI by method *getImage* and are rendered in a graphics window using the method *drawImage.* A text string is displayed by using a method *drawstring.*

Example 12.4

The code below shows embedding of an applet in XML and the corresponding code for the applet *gps.Java.* The applet code has four major methods: *start, stop, run,* and *paint.* The applet *gps* class is a subclass of applet class and starts a thread called *gpsthread.* The thread *gpsThread* wakes up every 200 milliseconds, picks up the new coordinates, and repaints the graphics window. The method paint picks up the relevant coordinates (omitted here), and method *run* repaints the value in the XML designated window. The parameters declared in XML code are pulled in the applet by using a method *getparameter(<Parameter-name>)*.

Browser end embedding

```
<applet code = gps height = 200 width = 200>
<param name = font value = "Arial">
<param name = size value = "36">
<param name = style value = "bold">
</applet>
```

The corresponding code for gps.Java will be as follows:

Applet-code

```
import java.applet.Applet;
import java.awt.*;
...
public class gps extends Applet implements runnable {
protected Thread gpsThread = null;
...
public void start() {
if (gpsThread = = null) {
         gpsThread = new Thread(this);
         gpsThread.start();
}
}
public void stop() {gpsThread = null;}
public void run() {
        while (Thread.currentThread() == gpsThread) {
        repaint();
        try {Thread.currentThread().sleep(200);} catch
          (InterruptedException e) {}
        }
}
public void paint(Graphics g) {
            string myFont = getParameter("font");
            string myStyle = getParameter("style");
            int mySize = Integer.parseInt(getParameter("size"));
            Font f = new Font(myFont, myStyle, mySize);
        ...
        }
}
```

12.2.5 Security in Web Programming

This section deals with handling security against malicious mobile-code in web programming. To handle security threats caused by mobile code, four techniques are used: (1) ensure minimal trust threshold in files and use of digitally signed applets, (2) safe interpreters, (3) fault isolation, and (4) code verification.

A server should meet a minimal trust threshold such that it is not seen as frequently sending malicious code. No server or intermediate nodes used during transmission of an applet can be guaranteed to be devoid of malicious software. If the severity and the frequency of malicious software transmitted from a server crosses a threshold, then the mobile-code from that server is avoided. An applet is considered safe if it is digitally signed by the trusted producers, and the public cryptographic key can ensure that they originated from the same originating source.

In the safe-interpreter, the applet is put in a protected area and is interpreted using a master interpreter instead of executing a compiled code. Communication between the applets is regulated to make sure they do not collude to perform a malicious action. Aliases are used outside the protected area to communicate with the applets. To regulate the excessive use of resources by the applets, each applet is allocated some electronic cash upon entry that is deducted after every use; upon excessive use, the remaining electronic cash is forfeited.

Code verification ensures that the JVM bytecode matches with the Java language specifications and does not perform unsafe operations. Unsafe operations are (1) operations causing stack overflow or underflow; (2) namespace violation; (3) forging of pointers, security managers, and classloaders; (4) local disk accesses; (5) opening its own windows; and (6) communication with processors other than the host processor or the originating processor. The classloader separates the namespace of the local trusted bytecode and the downloaded applet. Java loads a security manager to perform code verification when an applet is loaded.

Fault-isolation identifies the untrusted and unsafe code and makes sure that it does not access any area permitted only for trusted code. Rather it accesses the local allocated address space called *fault-domain*. This process is called *sandboxing*. *Sandboxing* is achieved by (1) inserting a conditional statement to raise exception in case an address space is accessed outside the fault-domain and (2) separating the addressing mechanism that fault-domain has different higher order bits than the remaining memory space. However, the second scheme makes the code architecture dependent, as different architectures have different word-sizes.

12.3 VIRTUAL MACHINES AND RUN-TIME INTERFACE

In order to provide a homogenous environment across heterogenous nodes two approaches are used: (1) development of virtual machines and translating web programs to the common instruction set of the virtual machines or (2) development of a common low-level interface and all the high-level code is translated to low-level intermediate code that can be later compiled to individual machines. There are two virtual machines that are used commonly for Internet-based computing: (1) JVM to execute Java programs and (2) Microsoft's stack-based virtual machine CLR for .NET framework used to execute programs written in C#, Visual C++, Visual Basic, and F#. "Common Intermediate Language" (CIL), also called Microsoft intermediate language (MSIL) is an object-oriented intermediate language like the JVM instruction set.

The execution of instructions in virtual machines is slow because of the use of interpreters and needs to be compiled to the native machine code for faster execution.

Run-time compilation of a common instruction set is called *JIT compilation*. Both JVM and CLR use JIT compilation on frequently executed segments of the interpreted instructions to speed up the execution. The languages like C#, Visual C++, and Visual Basic are first converted into CIL, and then use JIT-compiler at native machines to improve the execution of frequently executed code segments. CLR provides additional service such as garbage collection and memory management.

12.3.1 Java Virtual Machine

JVM is a zero-address machine that uses a stack-based implementation. The instructions are classified as *load instructions, store instructions, instructions for arithmetic and logical operations, setting up a reference to an object in the heap, stack management operations, object management operations, method invocation operations, jumping to another instruction including subroutines, returning from a subroutine*, and *instruction to support widening (coercion) and narrowing (casting)*. In narrowing, lower order bits are discarded. There are multiple stack instructions such as *push, pop, duplicate*, and *swap*. The instruction swap interchanges top two elements. The instruction set also supports multiple conditional branches and unconditional jump to implement higher-level control abstractions such as if-then-else and iterative loops.

The instruction set also supports the creation of new objects and arrays and returns the reference to the objects. To invoke methods, there are five instructions: *invokevirtual, invokeinterface, invokestatic, invokedynamic*, and *invokespecial*. The methods *invokevirtual, invokestatic,* and *invokeinterface* invoke a virtual, static, and interface method, respectively. *Invokespecial* is used to invoke methods that need special handling such as *constructor* or private methods where visibility bit has to be set.

A Java instruction has a one-byte opcode followed by the operands. The instruction length is one byte to three bytes long depending upon the number and type of the operands. This code is called *bytecode*. Java instructions carry the type information with them, and there is no type check at JVM level. Because only one byte is used for opcode, all Java types cannot be encoded in opcodes, and many types have been mapped to instructions handling other types. For example, when loading Boolean type, short integer type, and character type all are mapped to the integer instruction *iload*.

There are three types of references: *class type, array type*, and *interface type*. Data areas are created when the corresponding thread starts, and are destroyed when the thread exits. JVM utilizes a heap that is shared among all the Java threads. A reference to an object points to a pair of references in the heap: (1) the first reference points to the table of references pointing to the corresponding methods in the object and (2) the second reference points to the memory block allocated for the object. A method is polymorphic if the formal parameter and return values are of the type *Object []*.

The method area is common to all the Java threads, and the method area is started when JVM starts. Every class has a symbol table called *runtime constant pool*. Each thread has its own stack to support native methods—methods written in languages other than Java. A new frame is allocated every time a method is invoked. Each frame has its own local variable allocation, a reference to the runtime constant pool and its own *operand stack* to perform arithmetic and logic operations for zero address instructions and to prepare the parameters to call other methods. After a successful execution of a subroutine a value is returned, the control stack is restored back to the computational state of the calling method.

An exception handler specifies (1) the offset of the code within a method where the exception-handler is active and (2) the reference to the code for the exception handler. When an exception occurs, the offset is matched with the current execution location. The control is passed to the code for the exception handler upon match. If no exception handler is found, then the current frame is removed, the control is passed back to the calling method, and the exception is thrown again in the context of the calling method.

JVM supports *synchronization* in synchronized methods using a single construct called *monitor*. When invoking a synchronized method a flag *ACC_Synhronized* is tested and set. The flag is checked by other synchronized methods before executing the method.

In Chapter 11, we briefly mentioned *attribute-tables* that carry the information related to class attributes. An attribute-table in JVM carries the following information: (1) a magic number—a globally unique identifier—specifying the class file format; (2) a version of the class file format; (3) information related to runtime constant-pool (symbol table related to the class), (4) access flags for private and protected; (5) references to the symbol table of the current class and the parent class; (6) count and references to the interfaces in the constant pool; and (7) count and references to the various fields, methods, and other attributes in the class.

The compiled code in JVM has an abstract syntax *<instruction>* ::= '*<index-value>* *<opcode>* [*<Argument1>*] [*Argument2*]'. *Index value* is used for jump and branch instructions. The compiler translates methods in each class into bytecode instructions, as illustrated in Example 12.5, and places the bytecodes for all the methods in a separate file.

Example 12.5

Let us take a simple program and compile it. The comments show the effect of each compiled instructions, and the explanation is built into the comments.

Source code

```
x = 4; y = 5;
If (x > y) y = y - x;
```

Compile code

Index Instruction **Comment**

```
1 bipush 4        % push constant 4 on top of the stack
2 istore_1        % store top of the stack in local variable 1
3 bipush 5        % push constant 5 on top of the stack
4 istore_2        % store top of the stack in local variable 2
5 iload_2         % load local variable 2 on top of the stack.
6 iload_1         % load local variable 1 on top of the stack
7 cmple 10        % compare if top(Stack)< second(Stack) then
                      % go to instruction 10
8 swap            % swap the top two values on the stack
9 isub            % subtract top of the stack from the next value
10 istore_2       % store top of the stack in local variable 2
```

12.3.2 Just-in-Time Compilation

JIT compilation optimizes the execution of the JVM instructions and the CIL to the native machine code at run time for efficient execution. JIT compiled code is slower than static compilation because of the lack of extensive optimization and run-time compilation overhead. Compiled code is cached to avoid the overhead of run-time recompilation.

One of the major optimizations in JIT compilation is to break away from the linear execution of stack-based machines to DAG (Directed Acyclic Graph)-based representation of expressions to reduce the execution overhead. Using graph-based representation, the order of JVM instruction is transformed to improve (1) register-based operations using native code, (2) reorder and regroup multiple interpreted instructions to map to one native code instruction, and (3) remove redundant computations. Some of the redundant computations in zero address machines are (1) storing (or loading) the intermediate results many times to (from) local variables that can be left temporarily in the processor registers to reduce memory access overhead and (2) array bound checks for every array access that can be avoided if the first array access is within bounds and we know the extent of array access. The generation of the equivalent machine instructions is delayed until a generated value has to be consumed. The code-segment between the producer and consumer of values is reordered and regrouped to generate a smaller number of native machine instructions. Intermediate results are stored in unused registers instead of storing back in local variables.

Code compiled using JIT should interact with JVM in many ways: (1) method calls in JVM may indirectly call a native method and (2) JVM should be able to locate stack frames for the compiled methods and understand the format within those frames. JVM uses two stacks for each thread: (1) Java interpreter stack for interpreted methods and (2) stacks for the local evaluation within a method. After JIT-compilation, part of the interpreter stack is not needed because of compilation of the code-segment.

One of the problem of JIT compiler is that only frequently executed parts of code-segments are compiled to avoid the overhead of run-time compilation, and there is continuous transition between the format changes in handling the compiled format and interpreted stack format, especially for parameter passing: compiled code frequently uses processor registers for parameter passing, whereas Java-interpreted code uses Java stack for passing the arguments. The performance results show that JIT compilation improves the performance from two to ten times of the execution time of the interpreted code.

Example 12.6

Let us compile the code in Example 12.5 using JIT compilation. We assume that the host machine has a three-address instruction set. The values of local variables are first used in instruction #7. Thus the instructions #1 to instruction #6 are rearranged to avoid the use of local variables. One possible rearranged instruction sequence can be as follows:

```
1.  bipush 4
2.  istore_1
3.  iload_1
```

```
4.  bipush 5
5.  istore_2
6.  iload_2
```

Now we can replace first three instructions by one native machine instruction for a three-address machine as move #4, R1. Note that storing and loading overhead has been removed by storing the constant 4 in the register R1. Similarly, instructions 4–6 are replaced by another native machine instruction move #5, R2. The instruction 7 can be replaced by *branchle R2, R1, 5.* The new JIT-compiled code will be as follows:

JIT compiled code

```
1.  move #4, R1 % move constant 4 to register R1
2.  move #5, R2 % move constant 5 to register R2
3.  branchLE R2, R1, 5 % jump to index 5 if R2 = < R1
4.  subtract R2, R1, R2 % R2 = R2 - R1
5.  store R2, Y % store the new value of R2 in variable Y
```

JIT compilers also do limited dependency analysis to reshuffle the code sequence to improve performance and remove redundant codes. However, extensive dependency analysis is not possible because of the run-time overhead of JIT compilation.

12.4 COMPONENTS OF MULTIMEDIA SYSTEMS

A multimedia system is a combination of multiple media to facilitate realistic human perception. Humans understand an event by integrating the sensor-acquired perception of multiple media: text, still images, vocal sound, music, video-clips, audiovisuals, and gestures. Without a meaningful integration, perceptual distortion causes confusion instead of comprehension.

Multimedia systems can come in many forms: (1) multimedia over the local machine, where all the media objects are residing on one or more local databases; (2) multimedia where the remote media objects are being pushed by the server to the client; (3) demand-based multimedia, where the remote media objects are being pulled by the client from the server on demand; (4) multimedia objects are picked up from various URIs from remote servers; (5) real-time multimedia objects are transmitted over the communication channel; (6) multimedia objects are being synthesized using a multimedia modeling language and transmitted; (7) multimedia objects are extrapolated at the client-end from the previous transmissions using cache and an animation library; and (8) there is a combination of synthesized and real-time multimedia objects. Most of the time multimedia systems combine two or more of these eight schemes to provide a realistic perception. The key is to preserve the QoS (Quality of Service) while maintaining efficient transmission and rendering to avoid perceptual distortion.

In recent years, multimedia systems have become important because of the pervasiveness of the Internet that has made possible: web-based movies; animated clips; instructions; e-commerce; on-demand news; real-time news; multimedia stories for child education; and augmented reality in surgery, transportation, battlefield situation awareness and; Internet-based collaboration to name a few. There is a growing interest in developing new languages and extending existing languages augmented with constructs to embed, visualize, and understand multimedia.

12.4.1 Representation and Transmission

The basic goal of multimedia programming is to use multiple synchronized media to describe a story. A story is made of multiple causally related activities in different scenes where multiple objects interact. Activities in the same scene have similar backgrounds. Multimedia programs allow creation of 3D objects, creation of scenes, and animation of 3D objects in a scene. Motion involved in activities is created either by object transformation programmatically or by rendering images of activities that have already occurred.

Rendering images of activities over the web is done using multimedia streams. Each stream is further divided into a sequence of smaller time-stamped units, such that the objects occurring at the same time-stamp can be rendered at the client-end simultaneously. The basic unit in a video stream is called a *frame*, which is a snapshot of a real-life activity. A video stream is a sequence of frames. For the perception of motion, 30 frames are rendered per second. Sound has to be segmented accordingly such that each segment corresponds to a video-frame with the same time-stamp. To encode video-frames, each video-frame is divided into multiple n × n pixel two-dimensional regions called *mosaics*. The properties of pixels in each mosaic are averaged to achieve better compression. With some loss of accuracy, the multimedia information can be compressed and transmitted over the Internet. There are various formats for multimedia representations: .jpg (JPEG) and .gif are used for still images and drawings; .mvi and .wmv are used for movies; and .mpg (MPEG) is used for video clips. The meta-information for a video-frame is *palette of colors used in the frame, frame-rate, aspect-ratio, mosaic-size, image-resolution, vertical size of image, number of bytes in the image*, etc.

An audio signal is sampled for amplitude. The sampling rate depends upon the frequency of the sound and the desired quality of rendered sound at the client-end. A higher sampling rate increases the quality of sound while increasing the overhead of data-size to be stored, processed, and transmitted. The sampled amplitude is digitized into a byte. If the samples' rate is twice the base frequency of the wave, then the sound can be constructed at the client-end. For vocal sound, one needs around 4000 samples per second, whereas for hi-fidelity instrumental music, one needs 40,000 samples per second. After digitization, a sound wave becomes a stream of bytes that encode the sound-segment. Some of the popular formats for sound are .wav and .avi. The meta-information for an audio signal is *sampling rate number of bytes per sample, number of channels, protection bit, size of the segment*, etc.

12.4.1.1 Coding and Transmitting Complex Objects

Images are compressed using various techniques: (1) identifying only those colors that are used in the image and creating a map between smaller actual color-index space and true-color space, (2) grouping the pixels into a mosaic, averaging the mosaic features, and transmitting only mosaic features instead of pixels, (3) modeling complex objects as a graph of simpler images, and (4) transmitting only modified mosaics incrementally in the following frames.

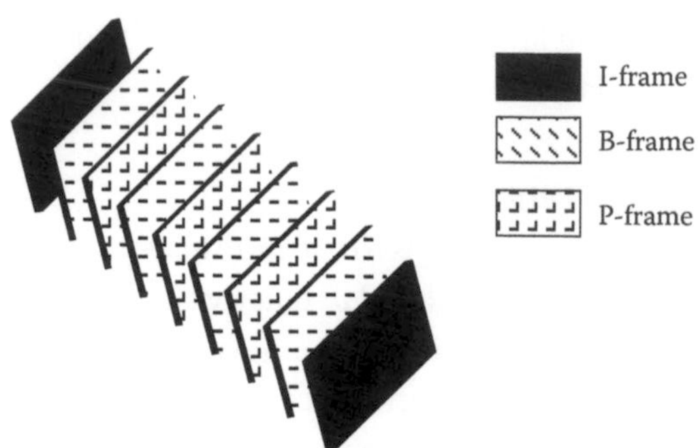

FIGURE 12.3 Frame encoding in MPEG using motion vectors.

Moving subobjects are transmitted regularly, while still parts of images are transmitted once. Between two adjacent frames, only the motion vector of altered mosaics is transmitted. The motion-vector is a pair of the form (difference of x-coordinates, difference of y-coordinates). Because the number of mosaics that do not change the position is quite large, the overhead of transmitting motion is significantly reduced by the use of motion-vectors. MPEG uses three types of frames, with varying overheads of transmission. The frames are *I-frames* (inter-frame), *P-frame* (predicted-frame), and *B-frame* (bidirectional-predicted-frame). I-frame carries all the spatial information of every mosaic in the frame. P-frame carries only those mosaics that have moved along with the motion vector with respect to the last transmitted I-frame or P-frame. B-frame carries the motion vectors with respect to the last I-frame (or P-frame) and/or future I-frame (or P-frame). The advantage of B-frame is in (1) providing fault tolerance in case the previous frame is lost and (2) reducing the size of the transmitted data compared to I-frame and P-frame, because B-frame contains the motion vector even for objects that appear in future frames. Figure 12.3 shows a transmission scheme involving I-frames, P-frames, and B-frames. Note that more B-frames are sent because of their smaller size. I-frames have the largest size and are sent infrequently only for reference.

12.4.1.2 Segmenting Complex Multimedia Objects

MPEG 4 and MPEG 7—the current formats to transmit audiovisuals—decompose a complex object as a graph of simpler objects such that each node is a smaller part of the overall images. The decomposition of complex objects is done based on the attributes and the usage of the simpler objects. In a complex object, some parts are moving, some are high-resolution stills, and some need low-resolution rendering only. A server sends the XML-coded graph of a complex object along with the images of subobjects and the corresponding metadata. The client-end rendering software uses this graph information and the image of the subcomponents to rebuild the object at the client-end.

Example 12.7

Let us take an example of a news announcer. The announcer's most focused part is his or her face. The background and hair are almost invariant and do not change much. Thus the image of the announcer can be split into three parts: *background*, *hair*, and *rest of the face*. The background is sent once as a low-resolution image, hair is sent infrequently when the head tilts. Changing mosaics in the face are sent in every frame, because expression and lip position changes frequently. Actually, the face itself can be split further in three parts: lips, eyes, and rest of the face. Lips and eyes can be transmitted regularly with very high-resolution mosaics of smaller sizes, and the face can be transmitted relatively infrequently with lower-resolution mosaics.

12.4.1.3 Animation and Transmission

Animation includes object image and its motion. The information about motion can be sent in multiple ways: (1) by sending the mosaics that have been altered from the previous images along with their displacement-vector or (2) by sending a parametric equation with respect to time in the form of $motion = f(t)$ instead of motion vectors. The information about changes from one frame to another can be sent using a text-stream where the text is encoded using XML. At the client-end, new frames can be constructed using the last reference frame. This scheme saves the transmitted data and transmission time. However, it pays some overhead in run-time analysis of the images to derive the information about the attributes of displaced mosaics.

Each moving subobject in a scene is a combination of two streams: one video stream and one audio stream that are synchronized using time-stamps. These streams may be sent separately using different communication channels, or the same communication channel alternately one frame-segment at a time.

12.4.2 Perceptual Distortion

A real-world action is decomposed into a combination of multiple, separate media activities. Each media activity is modeled as a stream. A stream can be a video stream, an audio stream, or XML documents for text transmission. The synchronization among these streams is important to perceive the real-world activity. Otherwise, different streams representing part of the same activity occurring at the same time in real life will be rendered at different times causing perceptual distortion.

Example 12.8

Let us consider the case of lecture delivery over the Internet. Delivery of lecture involves (1) slide presentation and (2) hand movement, gestures, lip movement, facial expressions, and sound of the speaker. The sound has to be segmented so that it corresponds to the right slide, right movement, and right gesture. Sound should synchronize with lip movement and facial expressions. Slide presentation should correspond closely to hand movement and gestures that point toward different parts of the slides. There are three different streams: video stream for slide presentation, video stream for facial expressions and lip movement, and audio-stream. If we want to optimize, then slides can be split in text and images and embedded in an XML document. The synchronization of these three streams is very important. Otherwise, a perceptual disconnect will occur making comprehension very difficult.

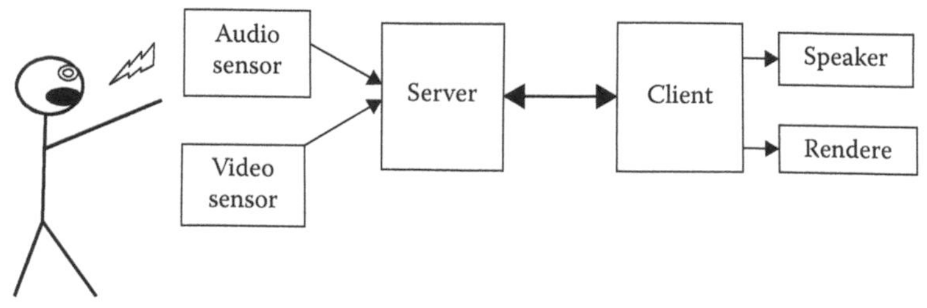

FIGURE 12.4 Sources of perceptual distortion in multimedia delivery.

Figure 12.4 describes four major steps: multimedia capture, archival, transmission, and rendering. During capture and rendering, the information is not digital and involves different sensors. During transmission, different packets go at different speeds because of (1) traffic congestion at a specific interval of time and (2) different sizes of media objects: small media objects need less than one packet, whereas bigger media objects may need more than one packet. There is a natural time mismatch between the synchronized actual events and the corresponding multimedia events rendered at the destination.

The distortion can be contained by time-stamping the media-segments and displaying them within a reasonable interval. The reasonable interval is different for different media streams. For example, if we are listening to a speech, then the maximum allowed time difference for rendering the synchronized video-frame and the corresponding sound segment is ±30 milliseconds. Higher delay will cause a sound-segment to be rendered with the previous or the following video-frame causing perceptional distortion. In multimedia rendering, when there are multiple media streams, mutual restriction between streams has to be maintained to avoid perceptional distortion.

12.4.3 Synchronization in Multimedia

Multiple media streams needs to be synchronized to maintain (1) causality of events, (2) frame level synchronization between multiple streams involved in the same event, and (3) the order between the frame-segments within the same stream. The synchronization between the frame-segments within the same stream is called *instrastream synchronization* and is achieved by numbering the frames in chronological order. The synchronization between the different streams belonging to the same object and event is called *interstream synchronization* and is achieved by time-stamping the frames-segments. Both intrastream and interstream synchronization are handled by the MPEG encoder.

12.4.3.1 Synchronization in Multimedia Rendering

To render media frames in order at the client-end, the frames should be received and organized in advance. When the frames start reaching the client-end, the following scenario can happen: (1) media frames arrived in an unordered fashion and (2) some of the frames are missing and need to be retransmitted. Thus, enough time should be allowed to provide for retransmission of the frames from the servers. The received media frames are stored in a buffer and sorted to put them in order.

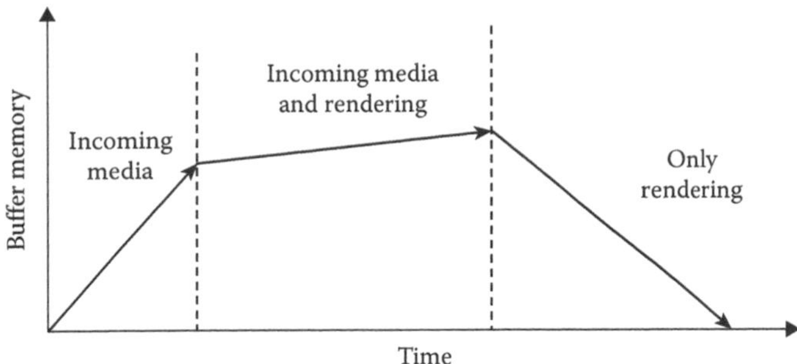

FIGURE 12.5 Buffer requirements during rendering.

There is a time lapse between receiving the frames and rendering. During the rendering, the frames are removed from the buffer, rendered, and discarded. As shown in Figure 12.5, there are three phases in the buffer. In phase I, frames keep accumulating in the client-side buffer before rendering starts. In Phase II, frame reception and rendering occurs concurrently: new frames are accumulated, and rendered frames are being discarded. In the last phase, all the transmitted frames have been received, but the frames are still being rendered. In this phase, the buffer gradually becomes empty.

Memory-size requirement in the client-side buffer also changes during multimedia on-demand when a user presses "fast-forward" or sees the video in slow motion, or temporarily pauses a video display. Fast-forwarding increases the rendering rate, and buffer gets empty faster. Thus the server has to push the data faster. Slow-motion video removes the frames slowly from the buffer. Thus frames in the buffer keep accumulating, and the server has to be requested to slow down the frame transfer rate. Similarly, the buffer keeps accumulating during pause, and client has to request the server to pause transmission and then request fast transfer when the pause is deactivated.

12.4.3.2 Synchronization at Program Level

Synchronization is done at multiple layers as the streams go through multiple layers of transformation before being transmitted over the communication channel. Multimedia synchronization at the programming level (application layer) involves incorporating concurrent rendering of stream for (1) interstream synchronization and (2) synchronization to maintain causality between the events. The synchronization at multimedia programming level is achieved by (1) concurrently rendering multiple streams for overlapping events, (2) aligning the stream-start and stream-ends of concurrent streams, (3) providing relative time-delays between the stream-starts or stream-ends of the two overlapping concurrent streams describing two different events, and (4) providing relative time-delays between the end of a one stream and start of the next to render the streams sequentially. Figure 12.6 shows seven classifications of synchronizing two media streams representing separate events.

The figures are self-explanatory. All seven alignments are expressed using a combination of (1) constructs that spawn multimedia streams concurrently, (2) constructs that impose sequential rendering of multimedia streams, and (3) constructs that introduce delays in starting the second stream from the start (or end) of the first stream. The programming

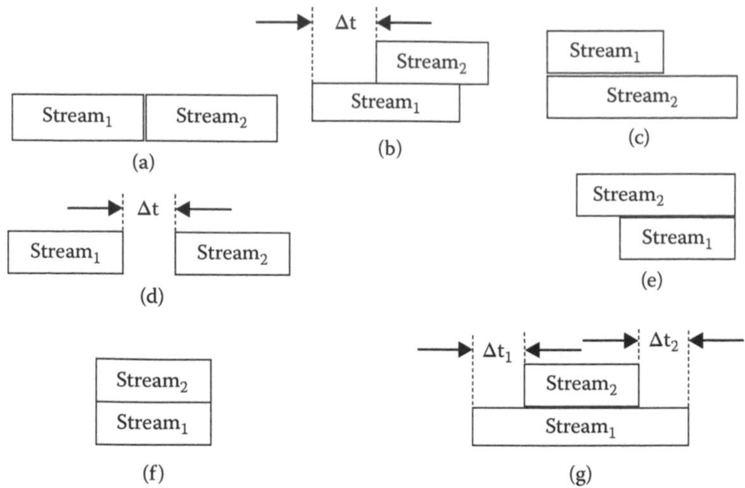

FIGURE 12.6 Multiple streams interval-alignments: (a) nonoverlapping, (b) overlapping, (c) aligned at starts, (d) delayed and sequential, (e) aligned at ends, (f) fully aligned at both ends, and (g) time difference at both ends.

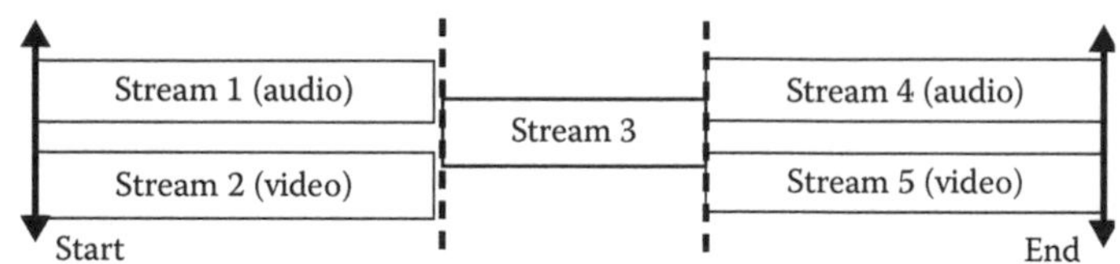

FIGURE 12.7 Multimedia stream synchronization.

constructs of multimedia languages Alice and SMIL support three techniques: (1) associating a name with a media stream, (2) specification of delay-time, and (3) concurrent spawning of media stream to provide event-level synchronization.

Example 12.9

Let us take a scenario where a news clip is being rendered. Before rendering news, an advertisement has to be rendered. After rendering the advertisement, music is played for 2 seconds, and then the news clip is rendered. Both news and advertisement have two concurrent streams: *video stream* and *audio stream*. The overall rendering is illustrated in Figure 12.7. Stream 1 and Stream 2 correspond to the scenario in Figure 12.6f. The rendering of stream 3 corresponds to scenario in Figure 12.6a. Streams 4 and 5 are rendered concurrently and are aligned with respect to stream 3, corresponding to the scenario in Figure 12.6a.

12.5 MULTIMEDIA PROGRAMMING CONSTRUCTS

Objects are constructed programmatically by specifying the graph-relationships between the constituent subobjects and their transformations. Animation of 2D images such as background is supported by incrementally changing the mosaics that change, and the animation of 3D objects is achieved by transformation operations such as *translation*, *rotation*, *shearing*, and their *combinations*. The subobjects can be in a local database or can be downloaded from the web by providing the URLs. Thus a multimedia language should (1) support a library of basic 3D objects; (2) provide capability to combine basic objects to form a complex object; and (3) provide capability to download objects by specifying URLs if it is a web-based multimedia language like X3D.

A scene is created by placing multiple objects in a relative position to each other. The interaction between objects is modeled by providing causality of actions. Many animated actions are repetitive in nature and need iterative constructs in the program. For example, a walking operation is repetitive in nature and can be broken into a sequence of repeat operations, as shown below, using a pseudo code.

```
loop
{lift left leg;  move left leg forward;   put the left leg  down;

lift right leg;  move right leg forward;  put the right leg  down;
}
```

Each of the six actions corresponds to some transformation of the original 3D objects. If two actions are being done concurrently, then they should be supported using concurrent constructs.

The motion of objects is modeled either by (1) program transformation as in Alice, VRML, and X3D or (2) rendering multimedia streams with constructs for concurrency and delay as in Synchronized Multimedia Integration Language (SMIL). Concurrency is implemented by constructs like *parbegin-parend* pair.

Example 12.10

The streams in Figure 12.6 can be rendered using the following parbegin-parend construct. The program uses SMIL syntax.

```
<seq>
   <par> <-- following actions begin concurrently -->
           <video xml:id = stream1 src = video-stream1.mpg/>
           <video xml:id = stream2 src = video-stream2.mpg/>
   </par>
   <-- sequentially execute after stream1 and stream 2 -->
   <video xml:id = stream3 src = video-stream3.mpg/>
   <par>
           <video xml:id = stream4 src = video-stream4.mpg>
           <video xml:id = stream4 src = video-stream5.mpg>
   </par>
</seq>
```

An action can also include a repetitive rendering of a media stream. For example, background music is 2 seconds long. However, while a video is being shown the background, music may be continuously played for more than two seconds using an iterative loop.

12.5.1 Synchronization Constructs

Rendering a multimedia stream can be restricted to time-interval or start with a delay from the start, or from the start of a specific stream, or from the end of a specific stream. This type of limiting the time or providing delay is necessary for the synchronization of media streams.

Example 12.11

Let us take an example of starting a background music that repeats itself four times. There is an event button, "SlideStart" that activates an event upon clicking. The text source starts from the beginning. The music starts from the beginning and ends after the event button "SlideStart" is clicked. The music file has a smaller duration and repeats itself almost four times before it stops. The music also ends after the event button "SlideStart" is activated. After the music ends, the video starts. The lecture starts two seconds after the slide starts. The example involves concurrency, event-based activation, and synchronization of the multiple streams. The whole activity runs for at most 20 seconds. The code is given below in SMIL syntax.

```
<par dur = "20s">
<img xml:id = "slideStart" src = "redbutton.jpg"/>
<text src = "slide1.html"/>
<audio xml:id = "music" src = "music.au" repeatcount = 4
       end = "slideStart.activateEvent"/>
<video xml:id = "video1" src = "video1.mpg",
  begin = "music. end" duration = "10s">
<audio xml:id = "lecture1" src = "audio1.au"
  begin = "video1. begin + 2s" >
</par>
```

12.6 CASE STUDY

We have already discussed XML and extensions of XML such as XAML and AJAX briefly. In this section, we will discuss three multimedia languages: (1) Alice that works on synthesized 3D objects using transformation operations, (2) SMIL that is an XML-based multimedia language for web-based stream rendering and its synchronization, (3) Javascript, and (4) C# constructs that are related to visual display and multimedia programming for .NET programming.

12.6.1 Abstractions and Programming in Alice

Alice is an object-oriented event-based 3D multimedia language used to teach multimedia programming as a first course in programming using animation of synthesized 3D objects and synthesized scenes. It has a library for 3D classes, and an instance of these classes becomes a 3D media object that can be placed in a user-constructed scene. It supports concurrency and sequential execution of methods. Multiple media objects can act concurrently. An action is a method within a media object. The language uses user-transparent transformation techniques to model animation of 3D objects.

The language supports (1) variables, functions, strings, lists, and arrays in data abstractions; (2) nested if-then-else constructs, definite iteration: *loop <m> time,* infinite-loops, while-loops, temporal-loops, function calls, Boolean functions, recursive functions, message-passing capability to user, and ask functions to interact with the user; and (3) capability to create new class after designing an object with customized methods. The elements can be selected randomly from an array. The loop construct can be for a specific duration (including an indefinite duration) or specific number of times. The duration can be defined by an event or action. For example, one can write a condition '*while a key is pressed*' followed by an action. The object will walk until the key is released. Actions can be performed concurrently using a menu-command *Do together.* Similarly, the command '*Do in order*' executes the methods corresponding to actions in a sequential order. An action can be applied on all the objects in the scene or a group of objects. A new method can be defined, or old methods can be altered. The new class can be saved for future use.

For multimedia animation, it supports (1) the ability to insert and add an object in the scene; (2) creating different background effects related to lighting conditions such as fog and brightness; (3) setting visibility of an object; (4) grouping objects to move together, or associating an object with other moving object; (5) moving around an object (including invisible objects); (6) capturing animated snapshots as *poses* and later using those poses; (7) changing the camera position to get a different view of the scene; and (8) performing transformation of one or more 3D objects such as translation and rotation.

To develop an Alice program, a programmer takes following steps: (1) defines a story in natural language; (2) identifies the major scenes in a story because a story is modeled as happenings in a sequence of major scenes; (3) describes a scene background and objects; (4) adds initial objects in the scene and locate their proper position in the scene; (5) writes pseudo code for various actions using a library of actions; (6) adds program actions on the objects or group of objects and (7) changes the parameter values for the scene and object actions. The actions could be about movement of objects and object-parts independently or with respect to other objects. Scene attributes are lighting affect, camera view, skin texture, background sound or music, etc.

An object has six basic motions: *go left, go right, move back, move forward, go up,* and *go down.* That covers positive and negative directions from the current position in all three axes. A motion is defined with respect to the *center of the object* or the *orientation of the object.* The center of the object is defined as its center of gravity, point of contact with the earth, or point of contact with a hinge. Basic motions are grouped to derive complex motions such as *translation, rolling, rotation about its own axis,* or *revolution around another object* while keeping a user-defined distance.

Alice has a graphical interface for scene layout and program development. The graphical interface allows adding new objects in the scene, defining and altering properties and actions of an object, and defining variables and control abstractions involving objects independently or a group of objects. A layout of the graphics editor is given in Figure 12.8.

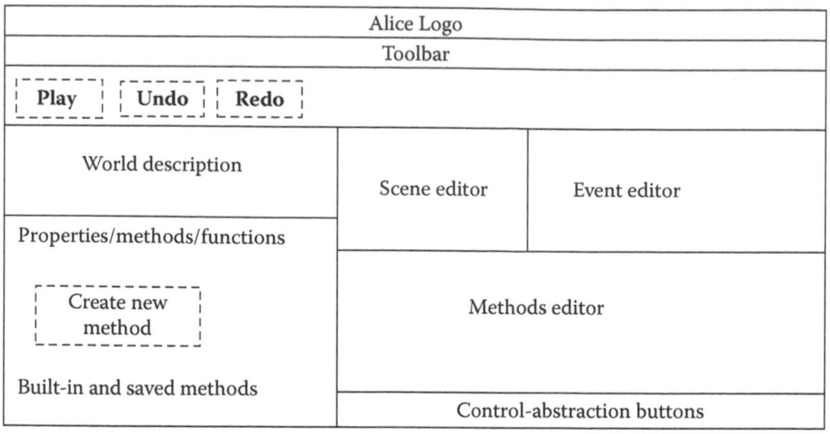

Alice Logo
Toolbar
Play | Undo | Redo
World description
Scene editor
Event editor
Properties/methods/functions
Create new method
Methods editor
Built-in and saved methods
Control-abstraction buttons

FIGURE 12.8 Schematic of graphics editor for program development.

The tool bar contains buttons like File, Edit, Tools, and Help. The button Play runs a created method. The world description contains attributes like *camera-position*, *light-condition*, *ground*, *class directory*, and the *objects* in the scene. The window below the world description window describes available methods in a class and provides a button that facilitates creation of a user-defined method. The window-space 'scene editor' allows addition of new objects and transformation of objects—such as rotation, orientation, and translation—to place the object at a particular location in a specific orientation. The event editor describes the events after the action starts. The methods editor allows the definition of a new method or the modification of an existing method. The control flow abstraction tabs are a group of buttons that facilitate writing the template of the constructs. The control flow abstractions are *'Do in order'*, *'Do together'*, *if-then-else*, *loop*, *'forall in order',' 'forall together'*, *wait*, *and print*. Each control abstraction has multiple attributes that are shown in the window when the corresponding instruction button is clicked.

12.6.2 Abstractions and Programming in SMIL

SMIL is an XML-based web multimedia language that can be used for demand-driven interactive multimedia events to be rendered at the client-end. SMIL is used for on-demand newscasts, advertisement, multimedia interview, multimedia instructions and lectures, and e-books over the Internet.

SMIL has capability to (1) alter the attributes of a region repeatedly over a period of time, (2) provide synchronization of multimedia streams, (3) render multiple streams concurrently, sequentially or mutually exclusively, (4) delay the rendering of stream, (5) invoke methods based on an event, (6) repeatedly render a stream a fixed number of times or for a particular duration, and (7) bind a stream with an identifier that can be used later to refer to the stream. Like XML, SMIL stores meta-information in head-part and SMIL code in body-part. Sample code for SMIL is given in Examples 12.10 and 12.11. SMIL code can be interleaved with XML code using a container 'smil:' as shown in the example below:

```
<ol smil:timecontainer = "seq" smil:repeatDur = "infinite">
<li smil:timeAction = "display" smil:dur = "10s"> Programming
  Languages </li>
<li smil:timeAction = "display" smil:dur = "5s"> Publisher:
  CRC Press </li>
</ol>
```

Each stream can be bound with an identifier, and this identifier can be used later in other stream definitions to align the begin and end of a stream with respect to other streams. For example, a stream can begin with a delay after the begin and end of another stream using the attribute value **'BEGIN'** = *<stream-id>* + *<delay>*. The delay can be added as well as subtracted. To animate an attribute, the syntax used is **'<' animate attributeName** = *<attribute>* **from** = *<from-value>* **to** = *<to-value>* **begin** = *<begin-value>* **dur** = *<dur-value>* **'>.'** The command changes the value of the attribute from *<from-value>* to *<to-value>* starting from the time given by *<begin-value>* for a duration given by the *<dur-value>*.

In addition to the *<par>* … *</par>* and *<seq>* … *</seq>* combination described above, SMIL also supports the mutually exclusive concurrent rendering of a media stream using a tag-pair *<excl>* … *<excl/>*. Any activity embedded within *<excl>* has similar semantics to *<par>*, with an additional restriction that only one embedded media-stream is rendered at a time. If any other stream is rendered, the currently executing stream is suspended.

SMIL supports low-level synchronization of video and audio segments using an attribute *syncbehavior*. The attribute *syncbehavior* can take four alternative modes: *locked, canslip, default*, and *independent*. Under *locked mode*, all the embedded elements inside the *<par>* tag are forced to wait for other media-streams if any of the media-segment is delayed in communication. Under *canSlip mode*, the associated element can slip with respect to other streams under the same **<par>** tag. Under *independent mode*, the associated element is independent of other media streams. Under the *default mode*, the behavior of the associated element is determined by the value of the *syncBehaviorDefault* attribute.

SMIL also has a powerful space layout editor and a graphical editor to specify and visualize the stream order and the intervals. The space layout is declared in the SMIL head and is associated with an identifier so that it can be referenced in the body area to display different multimedia in different regions within a space layout. An abstract syntax for space layout is as follows:

```
<layout> ::= '<' layout ' >'
              '<' root-layout width = <value>
                  height = <value> id = <identifier> '/>'
              {'<' region id = <identifier>  [left = <value>]
                ' ' [top = <value>]
                     width = <value> height = <value>
                      [z-index = <value>] '/> '
              }*
          '</' layout '>'
<value> ::= '"' <integer> '"' | '"' <integer> '%"'
```

The abstract syntax states that there is a root-layout followed by zero or more embedded regions. Each region is a separate window where a different multimedia stream is rendered. The region is specified by the reference-coordinate of the leftmost top-most point, followed by the height and width of the rectangular region. The root-layout and the regions are bound to an identifier, so that they can be referenced in the media stream for rendering purpose. The reference in the media stream is '**region-id =** *<identifier>* . The z-index provides depth of the display. The stream with the highest z-value shows on the screen, in case there is an overlap of the regions.

12.6.3 Abstractions and Web Programming in Javascript

Javascript is an object-based web programming language with the capability to treat graphics window as an object. An object has a hierarchy and within a window object multiple elements are embedded such as navigation, document, history, frame, java packages, location, and various fields within the rendering window. The document element contains many subelements such as forms; anchors; media objects such as image, video, and audio; and applets. Forms include multiple elements such as password, button, textarea, reset, submit, and checkbox. Any subelement can be accessed using the hierarchy *<root-node>*. *<element-node>*.*<subelement-node>*. Sibling nodes can be accessed using the parent-nodes.

Javascript supports many data and control abstractions, as summarized in Appendices II and III. Briefly, Javascript supports multiple basic data types such as Boolean, integer, character, float, objects, arrays, and strings. Arrays are objects too. Strings are treated as immutable objects. Javascript supports multiple control abstractions, such as if-then-else, case statement, for-loop, while-loop, functions, and exception handling. Through the use of applets and documents as objects, it supports web-based mobile computing. The major use of Javascript has been to provide a client-end interface for web-computing. It has been used for small scripts for verifying user information and for formatting data for display. It passes parameter using call-by-value for copying basic data types and copying the reference of the complex objects. It uses call-by-reference to pass the reference of an object.

Javascript also supports event-based programming based on mouse actions and actions on other objects through built-in methods such as onClick(), onBlur(), onFocus(), onLoad(), onMouseMove(), onMouseOver(), onreset(), onLoad(), checkbox.click(), checkbox.blur(), and checkbox.onfocus(). Using these event-methods and windows-objects, any field inside the form such as font, buttons, and checkbox can be verified and modified dynamically based on some condition such as time of day.

Javascript can be embedded in any XML and HTML documents using the *<script>* tag either as a file or as an embedded code. Javascript can also interface with Java programs and use Java packages. The general format for embedding Javascript code in XML is as follows:

```
<script Language = "Javascript">
    //Javascript code
</script >
```

A Javascript file can be included using the following format:

```
<script src = "<filename>.js"/>
```

12.6.4 Abstractions and Web Programming in C#

C# is a high-level object-oriented language preferred for .NET programming, because of its extensive support for graphics objects such as images and 3D objects, and their capability to be embedded in XML document through XAML. C# has extensive support for the code development in Visual Studio software that enhances programmers' productivity. Like Javascript, C# also supports client-end computation, verification, and integration capability with XML. In addition, it supports inheritance and class library, and multiple threads for concurrent programming. It also supports a windows framework as an object hierarchy, as described in the preceding Section 12.6.3 related to Javascript. It is a complete event-based programming language, and, like Javascript, supports many event-based methods for mouse actions and for various form-based objects, such as check-box, buttons, or scroll-bar. It also supports user-defined events. The event actions can be embedded in an XML document similar to Javascript embedding, and the corresponding event-method is activated when that part of the XML file is executed.

C# supports almost a complete set of various data abstractions and control abstractions, as summarized in Appendices II and III. Briefly, C# supports all basic data types such as integer; Boolean; floating point; strings; class inheritance; objects; and class polymorphism including generics. In control abstraction, it supports if-then-else statement, while-loop, for-loop, iterators, functions, threads, mutual exclusion, and exception handling. It has very rich class libraries to handle media objects. It uses call-by-reference to pass the reference of arrays and call-by-value to copy the reference of objects in heap and to pass the values of the basic types.

12.7 SUMMARY

With the increasing ubiquity of the Internet, web and multimedia programming have become popular because of the need to share the resources, computer power, processed data and better visualization. A typical web programming paradigm integrates traditional programming with Internet-based resource and data sharing. Code and data mobility is a common theme in Internet-based mobile computing models. Mobile code is used to avoid the overhead of transmission of very large data.

There are four models that utilize code or data mobility: *client-server*, *remote execution*, *code-on-demand*, and *agent migration*. In a client-server model, a client makes a service request to a server. The server performs the service and sends the results back to the client. The remote execution model reduces the load on a processor by executing the code on another processor. In the code-on-demand model, the server sends the code to the client-end. The code is executed on the client-end using the client's resources and data. Applet-based programming in web-scripting languages uses this approach. This approach significantly reduces the data transmission overhead. The last approach is agent migration, where a mobile code traverses from one node to another node, sending information back to the originating node, based on privileges and authentication.

Mobile computing has many issues such as handling heterogeneity of a network, efficiency issues of transmitting and interpreting the code, and security issues. Heterogeneity problems have been handled by translating the programs in different languages to a common low-level instruction set that is executed using a virtual machine or using JIT compilation. Two approaches are there: (1) JVM used in Java and (2) CLR used in .Net platform of Microsoft. The issue of execution-efficiency has been handled by JIT compilation. The safety issue is quite complex, because both mobile code and host processor can be malicious. Web-based programming uses either (1) a code-on-demand model such as applet and web script languages and (2) the client-server model. In either case, the requesting node (client) has to protect itself against malicious code or potential malicious code embedded inside data. In order to protect against malicious code, multiple steps are taken such as (1) ensuring that the mobile code originated from the actual authenticated node; (2) restricting the mobile code to a limited access depending upon the privileges; (3) accepting only the signed code; (4) verifying that the code is safe and does not try to modify classloader or the security manager; and (5) isolating the potentially unsafe-code using *sandboxing*. Sandboxing is a technique that restricts the access to the address space of the safe code either by checking the addresses accessed by potentially unsafe-code or by restricting the use of higher order bits in address mechanism.

HTML was the first web-based visualization language that used tags to format the data. The tags corresponded to built-in library routines in the browser that are invoked to render the formatted data in a user-friendly way. The browser is connected to CGI at the server-end. CGI formats the retrieved data from the server to HTML format that is displayed by the browser. However, HTML is limited by (1) the lack of support for user-defined tags and (2) fixed format of attributes that cannot be easily updated. Styling sheet language such as CSS and XSLT removed the limitation of fixed format of documents by keeping the attributes associated with HTML tags in a separate file. The CSS file containing tag-attributes is easily modifiable by the user and can be used to transform the original HTML and XML file using XSLT.

XML removed the restriction of rigid HTML tags by allowing user-defined tags. The use of user-defined tags allowed a complex hierarchical structure to be represented as a flat structure. XML became a language of choice for Internet-based information interchange. Relational and nested databases are easily interfaced with XML format to transmit the information to remote nodes. Objects in 3D can be modeled in XML using a combination of structure-graph encoded in the XML document and the URI of the subobjects. However, XML lacked the capability of computation-based interactions and event-based invocation. XML programs are rather large, which can cause errors if the structure is not correctly defined. DTD—document type definition—is used to describe the structure associated with tags. DTD includes (1) the definition of an element that specifies the embedded sub-elements, (2) attribute lists and their possible values, (3) entities and their definitions, and (4) parameter definitions. A structure is valid if it follows the DTD. Parts of XML that interfaces with web programming languages and comments is not controlled by DTD.

Many web-scripting languages like PhP, Javascript, C#, and ASP were developed to interface with XML. XML was augmented to provide computation in two ways: (1) by embedding the control abstractions within XML using XML tags and (2) by embedding the web script using the *<script ...> </script>* tag that called a function written into a scripting language. There are many languages such as Javascript, PhP, ASP, and C# that use this approach to embed computational capability in an XML file. The XML forms and history are treated as objects, and web scripts can be called as methods of the objects. This integration allows event-based programming, and methods related to mouse-movement and mouse-click can be called through XML.

The implementation of web-based languages requires compilation to an intermediate platform independent code to handle the heterogeneity problem. Java programs are compiled to JVM instruction set, and languages like C#, and Visual C++ are compiled to CIL. Both these codes can be either interpreted or compiled at run time to native code using JIT compiler. JVM is an object-oriented abstract instruction set for the interpretation of Java programs. JVM is a zero address machine that makes code portable across all kinds of architectures. The disadvantage is that most of the computers are two-or three-address machines. Instructions with more addresses are computationally more efficient. JVM supports instructions to load, store, compute, swap, conditional jump to an instruction, unconditional jump to an instruction, jump to a subroutine, set up references to heap-objects, return values, and call special methods such as initialization and destruction of objects. The major overhead of JVM is loading and storing the values from the stack to local variables and heap. JIT compiler rearranges the code at run time, identifies the segment of codes that are frequently executed, and transforms them to the machine instructions of the native machines. JIT-compiler has an overhead of dynamic compilation. To reduce this overhead, only frequently executed and computationally intensive parts of the JVM code are JIT-compiled.

Multimedia rendering and web-based programming are closely interleaved. One of the aspects of web-based programming languages is to retrieve multimedia content over the web and render it at the client-end. Multimedia consists of a combination of multiple media objects such as text sound, image, or video. A real-world scene consists of multiple interacting objects, and the action of each object can be modeled as a synchronized group of streams of different media. These media may be stored at one or more URLs.

A real-world scene is captured using sensors that are time-stamped for synchronization, archived, and then transmitted over the Internet. The smallest unit of video rendering is called a *frame*. A frame is a snapshot of the video, and 30 frames per second are needed to give a perception of continuous motion. The packets are sent in a different order than the rendering order to compensate for the extra time taken to transmit larger images. Owing to traffic congestion and transmission order and different packets taking different routes, the packets are reassembled at the client-end and are sorted on time-stamp order before rendering. Lost packets are requested again or reconstructed by extrapolating adjacent frames. The last step requires electronic devices to render different media. Owing to the use of various electronic devices involved in the recording, transmission, and rendering, multimedia production is never completely synchronized. If the rendering is not within a threshold limit, then it causes perceptual distortion.

Different media are archived and transmitted over the Internet using different formats. Image formats require lot of compression. Formats also carry meta-information such as palette of actual colors used in the frame, color-coding scheme, and the mapping between the index of color used and the index of true colors. Mpeg video is sent using three types of frames: *I-frame*, *P-frame*, and *B-frame*. I-frame contains complete spatial information, while P-frame and B-frame contain only motion vector of displaced mosaics. Sound format contains the rate of sampling, channels, resolution, and bytes that show the amplitude of the sampled waveforms.

Synchronization in multimedia is at multiple levels. There are three major synchronization levels: (1) synchronization because of causality of events; (2) interstream synchronization—synchronization between multiple media streams related to activity in the same object at the same time; and (3) intrastream synchronization—synchronization of frames within the same media stream. The synchronization because of causality of events is application layer synchronization and expressed by the programmers in multimedia programming languages. The synchronization is expressed by relating the start and end of the streams using (1) delay mechanism and (2) aligning the start or the end of the two streams using concurrency or sequential execution. A media stream can be repeated multiple times using repeat-counts.

Multimedia programming languages synthesize complex 3D graphical objects programmatically using: (1) transformations on basic 3D objects such as cylinders, cones, cubes, their combinations, and a library of already developed 3D objects or (2) use media streams with synchronization to display actions in a synchronized manner.

Alice is an educational multimedia programming language that synthesizes multimedia objects using a library of 3D objects, and it uses control abstractions along with action attributes to model user-defined animation and interaction with other objects. Control abstractions supported by Alice are (1) loop for certain duration; (2) loop for a certain number of times; (3) if-then-else statements; (4) performing multiple actions concurrently on one object such as bird fluttering both its wings together; and (5) performing actions sequentially. New methods can be built, and library methods can be modified to create new class definitions that are saved for future use. Alice uses a graphical interface for program development.

SMIL is another XML-based programming language that uses multimedia streams for demand-based multimedia rendering. XML embeds SMIL commands for SMIL-based multimedia rendering. SMIL supports concurrent rendering of multimedia streams, sequential rendering of multimedia streams, mutually exclusive rendering of multimedia streams, looping a stream multiple times by duration and by count, event-based rendering, and provides for synchronization between streams by using a delay mechanism. In addition it allows the binding of an identifier with every media stream to be utilized for references during synchronization of media streams.

Javascript and C# are object-oriented event-based multimedia programming languages that treat forms and images as objects to provide animation and render multimedia at the same time. Both languages, especially C#, have extensive event-based programming features such as the use of buttons, mouse actions, signals, and clicking on a specified region. C# also has concurrent thread-based programming capability and supports a very rich library for event-based programming.

12.8 ASSESSMENT

12.8.1 Concepts and Definitions

3D object modeling; Agent migration model; Ajax; Alice; applet; B-frame; bytecode; cascading styling sheet; client; client-server model; code migration; code mobility; code-on-demand model; code verification; CGI; common language interface; data migration; data mobility; Document Type Definition (DTD); event; fault-isolation; heterogeneity; homogenous; HTML; I-frame; interoperability; intrastream synchronization; inter-stream synchronization; Java security; JVM; logical clock; malicious code; malicious host; mobility; mosaic; motion vector; MPEG; multimedia; multimedia programming; multimedia rendering; multimedia stream; multimedia synchronization; .NET programming, object-graph; P-frame; remote evaluation; safe-interpreter; safety; sandboxing; scene; server; SMIL; Soap; stream interval alignment; style sheet; web programming; web scripting; web synchronization; World Wide Web; X3D; XAML; XHTML; XML; XSL; XSLT; VRML.

12.8.2 Problem Solving

1. Write an HTML program to publish your resume on the web using a styling sheet and explain the use of various tags.
2. Write an XML program to represent a table that has four legs and a flat top with a sphere on top of the table at a specific location. Use identifiers for the leg definition for the reuse.
3. Write a DTD for an XML file that checks for 4×4 tables of integer values. The table represents grades taken by four students, each taking at most four courses in a semester. Use your own tags.
4. Write an Alice program that makes a person throw a ball up in the air.
5. Write a SMIL program to render an interview of a person by a news announcer. The announcer first introduces the person for 10 seconds. After the introduction, the announcer asks a question, followed by the person's answer. The announcer takes less than 10 seconds to ask a question, and the person is given almost 60 seconds to answer. The process is repeated for 20 minutes.
6. Write a C# program to draw a rectangle and a square, and then color the square red.
7. Embed a script: Javascript or PhP or C# in an XML program that draws a 3×3 matrix and calls a function-based upon the event *mouseover* to color the table box, and then leave the color permanently colored if the mouse is clicked.

12.8.3 Extended Response

8. What is the need of virtual machines in web-based programming? Explain.
9. What are the security issues with code migration? Explain.

10. Explain the issues with code and data migration in mobile computing. How are they handled.

11. What is the need for JIT compilers? Describe the salient features of JIT compilers.

12. Explain the techniques using examples to integrate computing with XML-based representations.

13. Explain various techniques needed to render causality and synchronization in SMIL.

14. Explain the different classes of instructions in JVM.

15. Explain applets and how are they invoked from XML using an example.

16. What is CGI, and how it is related to web browsing and Ajax? Explain using a figure.

17. Discuss the features in SMIL that can be improved to allow for the declaration of synthetic objects such as in Alice, and to allow for their animation.

18. Read the Internet about sandboxing, and write how potentially unsafe codes are contained using sandboxing.

19. Read the Internet about Java security model, and discuss.

20. Read the Internet about PhP constructs, and compare them with Javascript constructs.

FURTHER READING

Bulterman, Dick, Jansen, Jack, Cesar, Pablo, Bulterman, Dick, Jansen, Jack, Cesar, Pablo, Mullender, Sjoerd, Hyche, Eric, DeMeglio, Marisa, Quint, Julien, Kawamura, Hiroshi, Weck, Daniel, Pañeda, Xabiel G., Melendi, David, Cruz-Lara, Samuel, Hanclik, Marcin, Zucker, Daniel F., and Michel, Thierry. *Synchronized Multimedia Integration Language (SMIL 3.0), W3C Recommendation.* December 2008. Available at http://www.w3.org/TR/2008/REC-SMIL3-20081201/

Conway, Matthew J. *Alice: Easy-to-Learn 3D Scripting for Novices*, Ph D. dissertation. Department of Computer Science. Charlottesville, VA: University of Virginia, 1997. Available at http://www.alice.org/publications/ConwayDissertation.PDF

Cramer, Timothy, Friedman, Richard, Miller, Terrence, Seberger, David, Wilson, Robert and Wocczko, Mario. "Compiling Java just in time." *IEEE Micro*. 17(3). May/June 1997. 36–43.

Flanagan, David. *Javascript: The Definitive Guide* (6th edition). Sebastopol, CA: O'Reilly & Associates Inc. 2011.

Fuggetta, Alfonso, Picco, Gian P., and Vigna, Giovanni. "Understanding code mobility." *IEEE Transactions of Software Engineering*, (24)5. 1998. 1–21.

Gong, Li. "Java security: Present and near future." *IEEE Micro*, 17(3). 1997. 14–19.

Guercio, A., Simoes, B., and Bansal, Arvind K. "Towards large scale voice activated dynamic and interactive Internet based animation and modeling." In *Proceedings of the IASTED International Conference on Software Engineering and Applications*. 2004. 749–754.

Hitzler, Pascal, Krotzsch, Markus, and Rudolph, Sebastian. *Foundations of Semantic Web Technologies*. BocaRaton, FL: Chapman and Hall/CRC Press. 2009.

Kagnas, Kari and Roning, Juha. "Using code mobility to create ubiquitous and active augmented reality in mobile computing." In *Proceeding of the 5th Annual ACM/IEEE International Conference on Mobile Computing and Networking*, Seattle, WA. 1999. 48–58.

Krall, Andreas. "Efficient JavaVM just-in-time compilation." In *Proceedings of the International Conference on Parallel Architectures and Compilation Techniques*, Paris, France. 1998. 205–213.

Lindholm, Tim, Yellin, Frank, Bracha, Gillad, and Buckley, Alex. *The Java™ Virtual Machine Specification, Java SE.* 7th edition. Oracle America, Redwood City, CA. Available at http://docs.oracle.com/javase/specs/jvms/se7/jvms7.pdf.

MacDonald, Matthew. *Pro WPF 4.5 in C# 2012: Windows Presentation Foundation in .NET 4.5.* New York, NY: Apress. 2012.

Moore, Jonathan T. "Mobile code security techniques," Technical Report MS-CIS-98-28. Computer and Information Science Department, University of Pennsylvania. Pennsylvania, USA. 1998.

Roman, Gruia-Catalin, Picco, Gian P., and Murphy, Amy L. "Software engineering and mobility: A roadmap." *The Future of Software Engineering.* Ed. A. Finkelstein. New York, NY: ACM Press. 2000. 241–258.

Sikos, Leslie. *Web Standards: Mastering HTML5, CSS3, and XML.* New York, NY: Apress. 2011.

Simoes, B. and Bansal, Arvind K. "Interactive 3D dynamic object based movies." In *Proceedings of the Fifth International Conference on Internet Computing.* 2004. Vol. II. 708–714.

Suganuma, Toshio, yasue, Toshiaki, Kawahito, Motohiro, Komatsu, Hidekai, and Nkatani, Toshio. "Design and evaluation of dynamic optimizations for a Java just-in-time compiler." *ACM Transactions of Programming Languages and Systems,* 27(4). July 2005. 732–785.

Tanenbaum, Andrew and Steen, Marten van. *Distributed Systems: Principles and Paradigms.* 2nd edition. Upper Saddle River, NJ: Prentice Hall. 2007.

Thorn, Tommy. "Programming languages for mobile computing." *ACM Computing Surveys,* (29)3. 1997. 213–239.

Troelson, Andrew. *Pro C# 5.0 and the .NET 4.5 Platform* (6th edition). New York, NY: Apress. 2012.

Wang, Paul S. *Dynamic Web Programming and HTML5.* Boca Raton, FL: Chapman and Hall/CRC Press. 2012.

CHAPTER 13

Other Programming Paradigms

BACKGROUND CONCEPTS

Abstract model of computation (Section 2.4); Concurrent programming (Chapter 8); Logic programming (Chapter 10); Object-oriented programming (Chapter 11); Code and data mobility (Section 12.1).

This chapter shows programming paradigms that are becoming increasingly popular with the increased ubiquitous use of the networked computers in our day-to-day activity. Computers are no more limited to problem-solving mode; computers are becoming more interactive and have a lot more sensors to sense and to react. The mobile codes are becoming intelligent and purposeful. These mobile codes perform useful activities at the client-end and can migrate from one node to another in a network searching for the appropriate information. With the availability of inexpensive processors, modern-day supercomputers are actually a complex cluster of processors, where a grand challenge problem can be solved by decomposing a task into simpler subtasks and mapping the subtasks on different parts of the cluster. In earlier times, the number of multiple processors was quite limited, and the languages developed were limited to parallelizing the loops or distributing the data on multiple processors. Most of the parallelization was limited to exploiting data parallelism, such as in Parallel Fortran and Parallel C for SIMD machines. With the advent of thread-based programming and message passing based on asynchronous communications, new language constructs for concurrent programming have been developed. With the multiprocessor-core machines becoming available in PCs and the number of processors in supercomputers becoming very large, there is a need to develop languages that facilitate the development of large software for ubiquitous human interaction with minimum effort and be suitable for easy management and evolution as the technology changes.

Programs that support continuous interaction with the real world need to recognize events and react to them. Large-scale software needs modular object-oriented programming. Programming languages are integrating multiple programming paradigms. Agent-based computing is an execution of mobile (possibly intelligent), autonomous, reactive objects that perform computation in a heterogenous address space and return the result to the originating node. X10 and Chapel are two evolving multiparadigm languages for massive parallel supercomputers and integrate imperative programming, object-oriented programming in addition to concurrent programming exploiting both data parallelism and task parallelism. In this chapter, we will study *event based programming*, *agent based programming*, *high productivity programming for massive parallel computers*, and *synchronous programming*. Synchronous languages have been used to model activities that require a signal and causality between various actions in a synchronized manner. There are many applications of synchronous languages such as large-scale integrated logical circuit design.

13.1 EVENT-BASED PROGRAMMING

Till now we have been studying programming styles where the control flow is decided by the programmer either using a single thread or multiple threads of control. In a procedural input-driven model, the interaction of a program is directed by a programmer in input-driven mode; unless a procedure requests to interact and accept the input data, there is no interaction with the outside world.

In reality, a program that interacts with the real world ubiquitously has to continuously keep registering the events that occur in the real world, match with the desired events, and activate corresponding methods (or functions) in response to the desired event. The events in the real world are asynchronous and nondeterministic and can occur in unpredictable order. A program should be able to capture the events and react to process the events. It is a program's choice whether to take an action or ignore the event based on the state of the system. However, the initiator of the computation is the signal from an event-source and not the control thread.

Compared to procedural programming that provides tight coupling between objects, event-based systems provide loose coupling to various objects; the control is decentralized and event based. An object can be activated only if an event occurs; no specific thread controls the events. An event may start a cascade of events, and these events can be nondeterministic depending upon the state of the system.

An event is characterized by a set of identifying features and may involve time-intervals and a region where action occurs. For example, raining or snowing on a specific day and time is an event; an accident is an event; a meeting of two or more people is an event; and a tsunami is an event. Multiple buttons on a graphical dashboard are a source of events, and pressing any button is an event. In terms of computer interaction with the real world and the user, events can be (1) computer-related graphical interaction such as mouse-based interaction, electronic brush related actions, or touch-screen related actions; (2) events related to completion of an I/O activity; (3) Internet-related action such as loading or exiting a website; (4) discrete event simulations; (5) pattern recognition in images, sound video and gesture; and (6) sensor-based applications such as smart homes or modern automobiles that react to the surrounding objects or conditions. As computers become even more ubiquitous, multimedia and gestures will play a major role in interacting with computers, and actions will be taken in response to pattern matching.

An event is characterized by multiple attributes: attributes could be Boolean, or they could be multivalued, or they could be real-valued. Attributes can be color, time, and location (x, y, and z coordinates). Given N attributes, N-dimensional space is defined. An event is a relevant region or a point in this N-dimensional space.

13.1.1 Event Model

There are five components in event-based programming: (1) event-source, (2) event-signal, (3) event-listener and dispatcher, (4) event-handler, and (5) event-object. Given an event, the event has to recognized, the significance of the event has to be established, and depending upon whether the event is a desired event, an action can be taken. An event-source generates an event-signal. For example, when a mouse is clicked on a hyperlink or a check-box, the mouse is the event-source. An event-listener keeps sensing the various emissions and keeps dispatching the characteristics of the captured signal as an argument to back-end software that activates a user-defined event-handling program to invoke an action based on the state of the system. An event-handler will access the characteristics of the event-object as an object-field. Handling an event becomes complex if the events needs to share a common resource or consume information from a common resource. That forces the events to be ordered. Events can be prioritized based on chronology or privileges given to them based on the nature of the events.

The event-handlers are implemented using anonymous objects and private inner classes. Anonymous objects are not associated with variables, and inner classes are embedded inside a class. The user does not have direct access to private inner classes. Event handling includes the notion of an *event adapter*. An event-adapter is an interface with empty methods corresponding to events. An event-adapter connects the event-listener to the event-handler. These methods are overridden by the user-defined event-handlers. Event-adapters are very useful when there are similar methods coming out of the same event-source. For example, mouse is an event-source, and it has multiple similar events such as mouse-clicked, mouse-moved, mouse-exited, etc. After a mouse event is fired, mouse handler embedded inside the mouse-adapter class uses inheritance to use mouse-related methods, and uses override to define the corresponding user-defined method.

A real-world system can be modeled using a finite state machine, where the system can be in different states. The state of the system is altered based on an input signal or occurrence of an event. There are different events associated with each state. Each state is modeled as an object, and the event-handler takes care of events that perform the state transition. The new state and the action taken by the event-handler depend on the state and the event. For example, if a mouse is not within a region, then clicking the mouse will not have any effect. However, if it is hovering over a hyperlink or an object, then clicking the mouse will have an action of loading the web page or rendering the object. A schematic of the event-based model is given in Figure 13.1.

Figure 13.1 states that the listener keeps listening to the event-source. Once the generated signal is captured, event-listener dispatches the event to the proper event-handler through the event-adapter. Depending upon the current system state, an event-handler

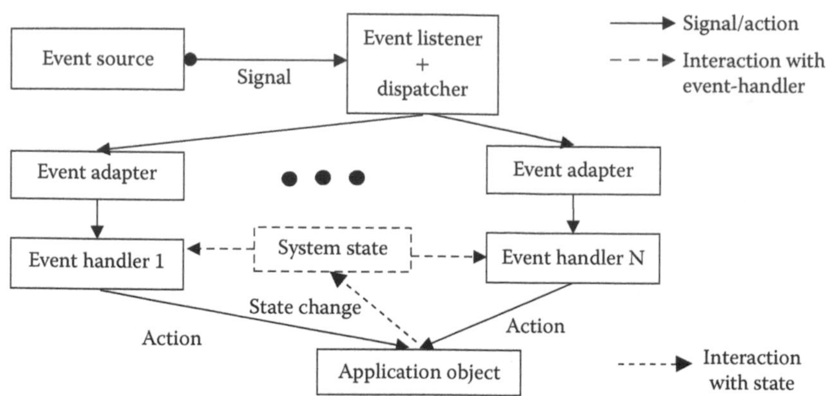

FIGURE 13.1 Schematic of event-based model.

takes an appropriate action on the application object, and the system moves to a new state. Depending upon the new state, more actions can be taken even without any external signal. In graphical user interface (GUI) applications, there may be more potential events. However, an event is discarded unless there is a corresponding event-handler.

The dispatcher sends the arguments to the event-handler. The individual fields in the event-specific argument can be accessed as *<event-argument>.<field-name>*. The event-handler may choose to use or not use an argument.

13.1.2 Developing an Event-Based Program

Many languages including Java, C#, Scala, and Javascript support event-based programming. These languages when interfaced with XML and HTML also support triggering of events from web documents. In this section, we will study developing an event-based program through a mouse interaction program written in C#.

An event-based programming needs (1) the definition of the events and (2) linking of the events to the event-handler that is invoked when an event is captured. There are two parameters in an event-handler: *sender* and *event-arguments*. A form is a collection of such objects. An object is created in the form on which the event-handler acts based on the event. An event can be associated with multiple event-handlers, and multiple event-handlers may be associated with an application object. Event-handler captures the event arguments and executes the method. Once the main program is run, it starts the event-handlers and event-listener.

In visual studio, a canvas (form) is painted with multiple objects each linked to the corresponding event-handlers. Visual studio has a graphics editor to (1) place the objects in a form, (2) create event-handler template, and (3) to automatically generate the main program that invokes the listener and the event-handler. Visual studio automatically translates the visual placement of the objects (on the form) into the textual commands. There are three parts of the event-based code created by visual studio: (1) main program that starts the event-handler and listener, (2) textual creation of the form containing the objects based on the graphical placements of the objects inside the forms, and (3) class definition containing methods for each event-handler.

Example 13.1

Let us understand event-based programming using a simple example given in Figures 13.2 through 13.5. The example illustrates many features of event-based programming such as (1) creation of application objects, (2) dynamic updates of object attributes, (3) linking of events with application objects, (4) state transitions, (5) update of other objects' attribute in response to an event generated by a different source, (6) events generated merely because of state transition, and (7) use of reset button.

The text boxes and the reset button are painted using visual studio form-painter. The textbox has many attributes such as *size, font, background color, foreground color,* and *visibility.* The background color of two boxes was initially set to '*cyan*' using the method *reset* that was invoked by the method *form_load* which is invoked when the form is loaded for the first time. The *RESET* button also has similar attributes. The *RESET* button displays the text "RESET" in bold. The textboxes are named *b1* and *b2*, and the *RESET* button is named "RESET" for use in the event-handlers. There are many mouse and text related events that can be associated with textboxes *b1* and *b2.* Some of the events are *mouseClicked, keyPress, keyUp, keyDown,* and *textChanged.* The overall painted form is illustrated in Figure 13.2.

The textbox *b1* is bound to two events: *mouseClicked* and *textChanged.* The corresponding event-handlers in the program are *b1_mouseClicked(***object** *sender, mouseEventArgs e)* and *b1_textChanged(***object** *sender, mouseEventArgs e).* The argument sender is a reference type of the event-source, and the parameter 'e' is an argument that carries the mouse-movement-related properties such the as x- and y-coordinates of the current position of the mouse.

Figures 13.3 through 13.5 together illustrate the code for the event-based program: (1) Figure 13.3 shows the code for various event-handlers, (2) Figure 13.4 show the main program, and (3) Figure 13.5 shows an object model of designed form. These codes are generated semi-automatically by the code generator in visual studio. Some of the redundant code

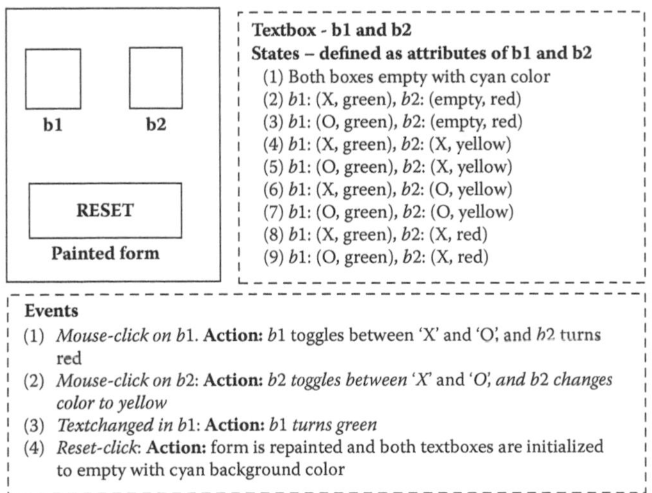

FIGURE 13.2 Illustration of form and event.

```
using System;
using System.Drawing; // needed for color attribute and drawings
using System.Windows.Forms; // needed for forms

namespace CSharpEvents
{public partial class Form1 : Form
  {int turn1 = 0;       int turn2 = 0;
     public Form1( )
     {InitializeComponent( );}

   private void b1_MouseClicked(object sender, MouseEventArgs e)
     {if (turn1 % 2 == 0 )
          {b1.Text = "X"; b2.BackColor = Color.Red;}
      else {b1.Text = "O"; b2.BackColor = Color.Red;}
      turn1++;
     }

   private void b1_TextChanged(object sender, EventArgs e)
     {b1.BackColor = Color.Green;}

   private void b2_Mouse_Clicked(object sender, MouseEventArgs e)
    {if (turn2 % 2 == 0)
      {b2.Text = "X"; b2.BackColor = Color.Yellow; turn2++;}
     else{b2.Text = "O"; b2.BackColor = Color.Yellow; turn2++;}
    }

   private void RESET_Clicked(object sender, EventArgs e)
       {reset( );}

   private void Form1_Paint(object sender, PaintEventArgs e)
     {}

   private void Form1_Load(object sender, EventArgs e)
                  {reset( );}

   void reset()
     {turn1 = 0; turn2 = 0;
      b1.Text = " "; b1.BackColor = Color.Cyan;
      b2.Text = " "; b2.BackColor = Color.Cyan;
      invalidate();
    }
   }
}
```

FIGURE 13.3 Illustration of event-handler definitions.

```
using System;
using System.Windows.Forms;

namespace CSharpEvents
{static class Program
 {[STAThread]
  static void Main( )
    {Application.Run(new Form1( ));}
  }
}
```

FIGURE 13.4 Main program running a thread for new form-object.

has been cleared up in this example, and some of the code for setting up attributes of textboxes and the reset button, and the form painting has been omitted for the sake of clarity. The main program starts a new thread that creates the new form, as shown in Figure 13.4. The object *Form1* is created and initialized by the method *initializeComponent* that sets up the attributes in the object model of the form, and paints the form on the window. Code for painting and setting up the attributes of the textboxes and reset button has been omitted.

```
namespace cSharpEvents
{partial class Form1
  {#region Windows Form Designer generated code
    private void InitializeComponent( )
    {this.b1 = new System.Windows.Forms.TextBox();
      this.b2 = new System.Windows.Forms.TextBox();
      this.RESET = new System.Windows.Forms.Button( );
      this.SuspendLayout( );

        // Automatic setup of b1 and b2 and reset attributes
        ...
        // Automatic setup of form attributes and method to paint
        ...
     } #endregion
    private System.Windows.Forms.TextBox b1;
    private System.Windows.Forms.TextBox b2;
    private System.Windows.Forms.Button RESET;
  }
}
```

FIGURE 13.5 Schematic for object-model of designed form.

The arguments are event-specific. The textbox *b2* is bound to *b2_mouseClicked(object sender, mouseEventArgs e)*. The event-handler for the *reset* button is *reset_Clicked()*. Figure 13.2 describes all the possible states of the form, the events, and the corresponding actions. Briefly, when mouse is clicked on textbox *b1*, it toggles the text between 'X' and 'O', and changes the color of the textbox *b2* to 'red' as shown in the code of the event-handler *b1_MouseClicked*. When the text changes, the state has changed too, and an event *textChaged* is generated for the textbox *b1*. The event-handler *b1_TextChanged* changes the background color of the textbox *b1* to green using the statement *b1.BackColor* = ***Color.****Green*. The event-handler *b2.MouseClicked* toggles the text between 'X' and 'O', and changes the background color of the textbox to yellow. The event-handler *RESET_Clicked* calls the public method *reset* that changes the background colors of textboxes to cyan and clears up the text.

The code in Figure 13.5 shows the creation of objects textbox *b1*, textbox *b2, and* reset-button *RESET*, and declares them private. After the form is created and painted on the screen, the specific event-handlers respond to the dispatch sent by the event-listener. The textboxes *b1, b2,* and *RESET-button* are treated as the fields of the current form, and they are initialized using the corresponding methods.

13.2 AGENT-BASED PROGRAMMING

With the ubiquity of the World Wide Web, code and data have become mobile. Applets are executed on remote computers using the resources of the remote computers. However, applets in Java are not autonomous and do not have the capability to sense the environment around them and react to the environment based on some plan. The concept of applet has been extended to agents for agent-based programming. An agent is an *autonomous, reactive, adaptive,* and *migratory* object. *Autonomy* means that agent is self-contained and does not need the remote system utilities to finish its essential task. *Reactivity* means that an agent can sense the environment around it and change its action based on its response inferred using the sensor information and its knowledge base. An agent can be *adaptive* that means it can evolve its behavior based on the environment. An agent can *migrate* from one information node to another node based on the goals of an agent. For example, an agent that is looking for information about basketball history will be migrating to the websites based on document analysis.

The role of an agent is to act as a representative of humans in the network of computers by migrating from node to node and performing some task using resources of the remote computer. An agent may act individually or in a group of communicating agents based on the overall system design. The task can be as simple as gathering domain-specific data from various web-nodes, or routing packets efficiently on web-node, or displaying some advertisement to the customers. The agent-based programming is not limited to web-based computing. An intelligent multiagent system (MAS), loaded on a loose network of processors, can be used to automatically derive a vehicle, autopilot a airplane, provide intelligence to smart homes, or control an agile robot.

A MAS is a collection of agents that work together to complete a task. Multigent systems can be either *cooperating* or *competing.* Cooperating agents are based on the division of work and information exchange to complete a task. Competing agents are based on notion of *profit maximization* in a limited resource environment. During *profit maximization*, multiple agents take part, and they may *cooperate* or *compete* depending upon whether they are meeting their goal of profit maximization.

Intelligent agents have their own knowledge library of plans and belief systems and form their own goals. Beliefs are like facts to a particular agent. However, they are not universally verified truth. Beliefs of one agent may be different from the beliefs of another agent. Based on their beliefs and logical rules, they can derive new beliefs that may not be universally consistent.

One of the popular models of an intelligent agent system is the *Belief, Desire*, and *Intention* (BDI) system. In a BDI system, each agent has their *beliefs* and *desires.* Based on the situation and resources they may commit to a *desire,* the *desire* becomes the *intention.* Beliefs are updated as the agents interact with the world. Intelligent agent-based systems have also been modeled using biological system to model artificial life forms. Distributed intelligent agent systems have been developed based on the model of cell-based interaction. Figure 13.6 shows a schematic of an intelligent agent.

The knowledge base includes plans, beliefs and itinerary. There are methods that collect the environment information in an event-driven manner and pass on the information to

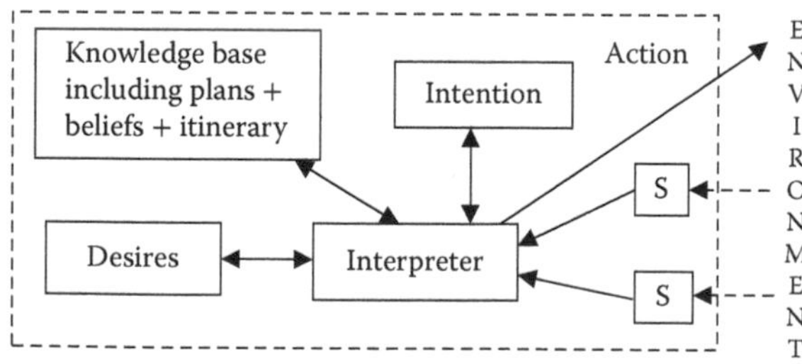

FIGURE 13.6 Schematic of an intelligent agent.

the interpreter. The interpreter derives new inference based on the knowledge base and plans. It may commit to a course of action if a plan matches significantly with the environmental context. In that case, *desires* may become *intentions,* and the *intentions* may be carried out. Some of the action taken by an agent can be (1) sending a message to the originating web-node, (2) sending message to another agent, (3) deactivating itself, (4) archiving new beliefs, and (5) cloning itself for the purpose of migration.

A programming language for agent-based systems has specialized constructs to create an agent, modify a plan, clone an agent for migration to another node, pack and deactivate itself from the current node, use the resources of the host computer to execute its plan, interact with the environment and update its knowledge base, and adopt itself to modify the plan and goal. An agent-based language should support object-oriented programming for agent modeling as well as declarative programming for knowledge management. AgentSpeak is such a language, which is built on top of the logic programming paradigm to implement the BDI model of reasoning.

There are mainly two types of models: *Java-based agent systems* and *BDI model-based systems.* Many BDI-based agent systems are also implemented in Java. The rationale for implementing agent-based languages in Java is their capability to handle heterogeneity of the Internet using JVM. There are other models based on the integration of the object-oriented paradigm and logic programming paradigms that do not utilize the BDI model. There are many agent-based languages that have been developed in the last decade. Some of them are distributed multiple agent reasoning systems (dMARS), AgentSpeak(L), AgentTalk, agent programming language (3APL), Jade, Jack, JDM, Aglet (Agent + applet), Voyager, CLAIM, KABUL, DALI, and ReSpecT, Telescript, and Agent Tcl (Tool Command Language).

13.2.1 Components of an Agent-Based System

An agent has a capability to perform actions autonomously. Since it migrates from one node to another node, it has to have some *itinerary.* During migration, an agent clones its current state, freezes the clone, encrypts the frozen state, and transmits to another node. Upon receiving the frozen clone of an agent, the remote node acknowledges the receipt. Upon successful transmission, the transmitting agent suspends itself or deactivates itself, as the case may be. An agent may also be activated by a request from the originating node to the remote host on which the agent migrates. The remote node buffers the agent until it is ready to be activated, checks its identity with the originating node, and then activates the agent.

An agent system has following components: (1) agent manager, (2) remote node manager, (3) security manager, (4) dispatch manager, (5) persistence manager, (6) event manager, (7) queue manager, (8) global directory service, (9) service bridge, (10) interoperability tools, and (11) agent libraries.

An *agent manager* is responsible for the creation of an agent, destruction of an agent, providing interface between the agent and the host machine, and saving the state of an agent before transfer so that the agent can be recovered later. A *remote node manager* monitors the agents' travels through the Internet, authenticates an agent in response to the request by the remote node, and provides interface with other agent-related components such as agent manager, security manager, queue manager, and persistence manager.

A *security manager* is used to identify the users using the agent, authenticating incoming agents, making sure that the agent does not have access to privileged file system or collaborates with other agents without permission, and authorizes the use of the dynamically loaded Java classes. The *dispatch manager* is responsible for transmitting an agent to the remote host. The dispatch manager uses *agent transfer protocol* (ATP), which provides a machine independent protocol for agent transfer between nodes. An agent is uniquely identified by (*address of the originating machine, agent-id*). To dispatch an agent, the dispatch manager in machine *A* sends a request to the remote machine *B*. After machine *B* approves, the agent is dispatched with its unique identity to machine B and waits for a receive-acknowledgment for a specific time. After machine *B* receives the agent, it sends the status information about the agent.

The *persistence manager* saves the state of the agent periodically using *checkpointing*, so that an agent can be recovered in case of an agent crashes or system crashes. *Checkpointing* is a technique to store the frozen state of an executing program so that it can be recovered later. An *event manager* is used to authenticate and distribute events. A *queue manager* maintains (1) the queue for movement and transition of agents from one node to another and (2) movement of messages between two or more agents to ensure that the message has been received by the intended agents. A *global directory service* is needed to make sure that all the managers and hosts involved can access the itinerary, position, and capabilities of an agent. Different heterogenous machines have different data representation formats. The purpose of the *service bridge* is to provide interoperability and interface between the heterogenous computers and the directory service, agent manager, persistence manager, and security manager. The interoperability tools transform the data formats between the sending machines to the destination machine. Java-based machines use Java-based API called J-AAPI (Java Agents API), which is built on top of JVM and is free from interoperability issues. The agent libraries are carried by an agent to enable its autonomous behavior.

13.2.1.1 Handling Plans and Beliefs

Each intelligent agent interacts with host machines and the world using its own knowledge base. Because new knowledge is derived based on a limited sample size, the derived information is a belief based on limited features extracted from the information. However, this derived inference becomes part of the knowledge base of the intelligent agent and is used for further inference.

A library of plans is stored as part of the knowledge base and directs the course of action an agent has to take. A plan has following components: (1) triggering event, (2) preconditions of plan that should be satisfied similar to guards in guarded programming languages, (3) the actions to be performed if the precondition is satisfied, and (4) a set of actions if the plan fails. A particular plan is activated only if the information collected by the agent matches with the precondition of the plan. This is done by a triggering event. Once an event is recognized, the corresponding plan takes an action, which could be (1) to commit to the plan, (2) send a message to neighboring nodes, and (3) solve a goal. After the commitment to a plan, the action part of the plan solves goals followed by some output data to be sent to the originating node or another cooperating agent.

13.2.2 Agent Security

The security model for Java-based agents uses JVM-based security. All the safety issues raised in applet security in Section 12.2.5 also apply to agent-based security. Actually agents need more security, because agents migrate from one node to another node, transmit the results back to the originating node, and collaborate with other agents. The problem of security is not only because of malicious code but also due to corruption of agents caused by malicious host. There are four major safety concerns: (1) an agent becoming malicious during transition, (2) malicious code affecting the host machine, (3) a malicious agent affecting other agents, and (4) a malicious host affecting an agent.

Encryption and retransmission of a clean copy of an agent from the source node to destination node is used to handle corruption during transition. Protection against malicious code affecting a host machine was discussed in Section 12.2.5. The host machine gives each agent resource access according to its policy. Different types of agents coming from different originating nodes may have different trust levels and priority levels. Based on the backgrounds of agents, different policies are enforced on an agent. Based on these policies an agent can take the following privileged operations, to different extents: (1) obtain file information, (2) read and write in local files, (3) delete files, (4) connect to the originating node, (5) load libraries, (6) open their own windows, (7) clone itself for migration, (8) deactivate itself, (9) dispatch a cloned agent, (10) retrieve a list of agents involved in the same project or working locally on the same machine and the same project, and (11) verify the environment for the suitability of its intentions.

Protection against malicious agents is done by ensuring that an agent can communicate only with authenticated local agents residing on the host machine. If an agent is allowed to communicate with an outside agent, then the outside agent should be authenticated by the host where it is lying, and the remote host itself should have creditable authentication.

The last problem of malicious host affecting an agent is a real threat, because a host that executes a code has complete access to the agent's code and can easily integrate a malicious code-segment without the knowledge of an agent. There are multiple approaches to partially solve this problem as follows: (1) treating agent code as a blackbox and (2) agent code checking periodically its code for (1) alteration of the code size; (2) addition of new classes or methods; (3) using a proof code check where an agent produces the same result on test data; and (4) deactivating the transmitted clone-agent in case of suspicion of a malicious host. It is difficult to treat an agent as a blackbox completely because the remote host has to ascertain that the code itself is not malicious.

13.2.3 Fault Tolerance in Multi-Agent Systems

One of the problems with MASs is that agents may crash because of (1) exceptions, (2) process failure, (3) processor failure, or (4) communication problem. In such cases, the overall MAS can become unstable, as other agents dependent on the crashed agent cannot proceed. Some of the problem can be handled by *checkpointing* the agents. However, checkpointing cannot be done frequently because of significant overhead of archiving the data. That means all the information after the last checkpoint is lost upon an agent crash; information loaded from the last checkpoint may not be consistent with other agents and may have missing beliefs and goals. To alleviate this problem, two approaches are used: (1) keep the information after the last checkpoint in the knowledge base of other agents without mixing with other knowledge bases and (2) keep a shadow agent for every running agent.

In the first scheme, the knowledge base updates are transmitted to other agents piggy-backed along with messages and are saved in a repository distributed among multiple neighboring agents. Upon failure of an agent, a new agent is spawned, and the knowledge base is built using the last checkpoint and incremental knowledge after the last checkpoint kept in distributed repositories with neighbors. This scheme has communication overhead of transmitting knowledge to neighboring agents.

In the second scheme, a *shadow agent* is kept preferably on another processor to protect against the processor failure. The message is transmitted to both the active agent and the shadow agent. The shadow agent updates the knowledge base and remains in *hot state* except it does not send any message outside and does not interact with the environment until the active agent crashes. After the active agent crashes, the shadow agent become an active agent and informs all other neighboring agents to communicate directly. A new shadow agent is created that has the same knowledge base as the currently active agent. This scheme has significant overhead of duplication, as a processor entertains two agents doing the same task: *one active* agent and *one shadow agent.* For large scale agent-based systems, this overhead is quite significant.

13.3 HIGH PRODUCTIVITY MASSIVE PARALLEL PROGRAMMING

As the number of processors in modern-day computers is increasing, the inadequacies of current uniprocessor-based languages for massive parallel processors are becoming obvious. When the multiprocessors were limited to a few hundred processors, limited exploitation of task and data parallelism was complex but manageable. The existing uniprocessor languages were extended to incorporate concurrency constructs such as *multiple threads*, *mutual exclusion* using monitors and locks, and *asynchronous message passing* schemes. Handling synchronizations using locks were quite low-level programming techniques that made the development of large-scale concurrent programs difficult. Programmers are used to a single thread of control flow and found it very difficult to correctly program using multiple threads and low level constructs using locks; programmers make errors that are difficult to bug especially in massive parallel computers running large software.

Currently, message passing interface (MPI) library is a standard middleware for message passing between the processors with different address space. Message passing models are suitable for distributed address space. However, they have significant overhead of message passing. In contrast, in recent years, languages based on *partitioned global address space* (PGAS) model have been proposed that support local *directly addressable memory space* as well as *distributed global address space* that can be asynchronously accessed. Many languages based on PGAS are evolving, as described in Sections 13.3.1 and 13.3.2.

Another aspect for large-scale software development is *modularity* and *reuse.* If we have to develop large-scale software on massive parallel computers, then we have to use object-oriented programming for modularity and reusability, support asynchronous computing, and support constructs for transparent high-level mapping of a task to a part of a massive parallel computer. Two new languages—Chapel being developed by Cray Computer and X10 being developed by IBM—support this philosophy. The following subsections discuss PGAS, constructs to support massive parallel computation and object-oriented programming in these two languages.

13.3.1 Partitioned Global Address Space

Different massive parallel architectures support different types of address spaces such as *shared memory* and *distributed memory.* Distributed memory-based systems use MPI library to communicate between distributed address spaces and have communication overhead. Shared memory address space uses the shared memory locations to communicate between processors and has the overhead of synchronization to handle shared memory locations. To overcome these limitations for massive parallel machines using clusters, the PGAS memory model has been proposed. *Global address space* is the union of all the address spaces addressable by the processors in the massive parallel computer. PGAS provides the programmer-defined capability to partition the overall global address space to solve a task. In PGAS, multiple processors keep working on different partitions. PGAS divides the global address space into multiple partitions such that each partition can be accessed locally by different pools of *activities* (threads). However, they can also access remote locations asynchronously when needed with some additional communication overhead.

In PGAS, there are four types of storage: (1) *local-stack* that is private to each activity, (2) local heap that is shared by a group of activities working together in the same partition, (3) shared global partition that can be accessed asynchronously by other threads remotely using global pointers, and (4) immutable objects that can be copied to any partition. Languages use two types of pointers: (1) *local pointers* that access local partitions and (2) *global pointers* that can access other partitions too.

Figure 13.7 gives a schematic of PGAS. Note that the PGAS is divided into multiple partitions. Each partition has resident pool of threads. Each thread has its own stack and local

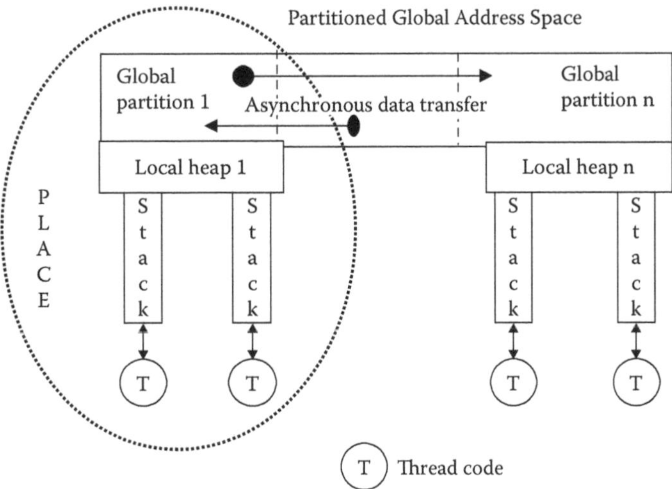

FIGURE 13.7 Schematic of PGAS memory model.

heap and the partitioned global heap. In addition, it can access data from remote partitions if needed using *async* operation, which spawns an activity to access the memory locations in other partitions. However, accessing different partitions is computationally less efficient than accessing a local partition. The pool of threads (*activities* in X10 terminology) and its data in the associated portion are together called a *place*.

There are many evolving languages that use the PGAS model. Some of them are UPC (Unified Parallel C), CAF (Coarray Fortran), Titanium—a Java-based language, X10 (an object-oriented parallel language being developed by IBM) and Chapel (an object-oriented parallel language being developed by Cray Inc.). These languages are evolving based on the application demand.

13.3.2 Constructs for High-Productivity Computing

In this section, we discuss various concurrent and object-oriented abstractions used in X10 and Chapel—two evolving languages that integrate the PGAS model, object-oriented programming, and concurrency for high-productivity massive parallel computing. Both these languages are being developed for large-scale software development on cluster-based massive parallel computers. These languages support both data parallelism and task parallelism in addition to object-oriented programming. Data parallelism is realized by distributing an array across different global partitions of PGAS and spawning multiple threads working concurrently on different partitions simultaneously. Task parallelism is present, because each partition may be involved in different activities.

13.3.2.1 Abstractions and Concurrent Constructs in X10

X10 uses the notion of *activities* instead of threads. A *place* can support multiple activities. A *place* is a group of multiple activities that share a local heap and one of the partitions in the global address space, as shown in Figure 13.7. An activity has its own local stack and shares a local heap with other activities local to a *place*. An *activity* once spawned remains in the same place and does not migrate to other *places*. An *activity* can communicate to other *activities* in other partitioned space using *async operation*. X10 also supports immutable *value classes* that can be copied freely between different places.

X10 supports basic data types that use local storage. It also supports distributed arrays that are allocated in partitioned global address space to exploit data and task parallelism. An array can be logically divided into *multiple regions*, where a *region* is a set of indices. Set operations like *intersection*, *union* and *Cartesian product* are permitted on regions to generate new regions. Strings are allowed as immutable objects.

X10 supports asynchronous execution of a statement at a user-specified place using the construct **async(**<*place*>**)** <*statement*>. The operation *async* has a local scope, which means it cannot access remote partitions during the execution of statement S; a new async operation should be spawned to access the remote locations. Asynchronous activities that return a value to the invoking activities are called *future*. X10 allow the result returned by an invoked async operation to be used by other activities by storing the result of an expression in a designated memory location. Other activities that want to consume the result stored in the location are blocked until the value is available. This sharing is done using an operation *<shared-variable>* ::= **future** (*<place>*) *<expression>* that spawns another activity asynchronously in a place P, and the result is available in the shared variable.

X10 supports the construct **finish**(*<statements>*) that states an activity waits until all other activities are over. Concurrent coordination of activities is regulated by the instruction **atomic** *<statements>*; low-level use of locks is not available. The effect is to provide mutual exclusion from other activities. X10 also supports conditional mutual exclusion using the instruction **when** (*<condition>*) *<statements>*. The statements *<statements>* are executed only when the conditions are satisfied. X10 supports the concept of *clocks* to support completion of more than one activity. If multiple activities are registered with a clock, then the clock does not advance until all the asynchronous activities are completed. Clocks are used to provide *barriers* until all the related activities are completed.

X10 uses the notion of objects and single-inheritance class. An object once created remains locally in the same *place* of creation. Methods support polymorphism. Methods can be *public, private* or *protected.* Methods could be *static functions* or *instance methods*—parameterized code associated with an instance of the class. Methods in a subclass can override the definition of methods in the parent class like other inheritance-based object-oriented programming languages do.

13.3.2.2 Abstractions and Concurrent Constructs in Chapel

Like X10, Chapel is an object-oriented high-productivity, PGAS-based language that supports task and data parallelism in a cluster-based massive parallel computer. Chapel uses a *global view* of an array instead of fragmented view of data taken by single program multiple data (SPMD) model that exploits data parallelism. In the SPMD model, multiple threads are spawned simultaneously that keep working on different portioned datasets of a distributed array. In the *global view model*, a program starts with a single thread, and new threads are spawned dynamically with the parallel language constructs written by the programmer. Data is distributed using the programmer's explicit actions instead of automatic distribution, because high-performance problems need different data distribution for different classes of problems. Chapel's philosophy is that an automated scheme will not be able to analyze the code to identify the appropriate data-distribution scheme. Chapel supports basic data types such as integers, floats, complex, Boolean, and strings. Chapel also supports tuples as well as sequence for ordered collection of data. Chapel also supports union types.

Chapel supports a variety of standard control flow statements such as forall-loop, for-loop, if-then-else statement, and case-statement. Functions can be overloaded. Chapel has a multithread execution model to exploit parallelism and provides high-level abstractions for exploiting data parallelism and task parallelism. Like *region* in X10, Chapel has the capability to define a set of array indices. The set of data elements corresponding to these set of indices are called *domains.* Domain elements can be accessed and mutated in a data-parallel manner. Chapel also supports data abstractions like sets, graphs, and associative arrays using the concept of domain. A domain can have an *infinite set* such as a set of integers or a set of strings.

Example 13.2

The example illustrates the use of infinite domains to map a hash table. *Students* is a variable that is of the type string domain. Course is an array of strings indexed by the elements in the *Students* domain. An element 'Ted' is added to the domain that maps to the value "Programming Languages."

```
var students: domain(string);
var course: [students] string;
students += "Ted";
course("Ted") = "Programming Languages";
```

Subdomains inherit an index set of the parent domain. Chapel allows two types of *subdomains*: *sparse subdomain* and *simple subdomain*. The notion of domain combined with *global view* can be used for data-parallel programming. For example, an assignment statement A = B assigns the value of every element in the domain B to the corresponding elements (elements matching the index value) in domain A. Parallel iteration is supported using *forall-loop* that can iterate over domains, arrays, and other expressions. Chapel supports task parallelism using *cobegin-coend* statements. The parallel subtasks are coordinated with each other using synchronization variables that support *full/empty semantics*: read is done when synchronized variable is in *full state* and write is done when the synchronization variable is in *empty state*. Chapel also supports atomic section like X10 does, instead of low-level lock mechanism.

Chapel supports the concept of *locale*, where a *locale* is a specific node in the cluster of processing nodes. By declaring an array of *locales*, the programmer can specify how much a part of the machine a program will take to execute. An iteration executes in the locale that owns the corresponding index of a domain.

Chapel supports two types of classes: (1) *traditional classes*, as in typical object-oriented programming and (2) *value classes* that act as records but can invoke methods. Chapel does not insist on an object-oriented programming style. Instead programmers can write block-structured programs. Chapel also supports generic methods and functions to support polymorphism. Classes and methods may contain type variables to support generic programming.

13.4 SYNCHRONOUS LANGUAGES

In early 1990s, a new class of language was developed to simulate the clock-based logical circuitry. In a clock-based logical circuit, an action is taken synchronously after every clock signal. The clock-tick is logical rather than physical. However, it can be associated with physical ticks. There is no delay in reaction to the clock-tick. Based on a pattern of clock-ticks, a signal can be emitted, and computation can take place as a reaction to a logical combination of conditions set by emitted signal. Because the computation is based on synchronous clock-tick, the actions are deterministic, and there is no race condition, provided the action is taken instantaneously at the next clock-tick, after the input signal become available.

The major concept in synchronous language is that time advances in lockstep and actions are taken immediately without delay, exploiting deterministic concurrency. Emission of a signal sets another Boolean condition as true. A construct may wait for a combination of one or more input signals that are connected through logical operators. If all the signals are available, the required condition is true. At the next logical clock, another construct waiting for the combination of conditions is executed. An action could simply be an emission of a signal, abortion of an activity, or temporary suspension of some activity. The program reacts instantaneously, so there is no reaction time. The concept is general enough to be used for any system that uses logical clock-tick for synchronization. One such example is multimedia languages, where inter-media synchronization uses the notion of logical clocks. Intermedia synchronization using logical-clock and synchronous programming has been proposed in an XML-based multimedia language TANDEM. Some of the examples of synchronous languages are Estrel, Lustre, Signal, and Quartz. Synchronous languages have been applied in the areas of VLSI design, signal-processing applications, embedded system designs, avionics, life monitoring systems, and automatic control applications. The following sections shows the synchronous constructs of Estrel.

13.4.1 Synchronous Constructs in Estrel

Estrel is an imperative synchronous concurrent language that uses synchronous steps to perform parallel tasks. There are two important components: *sensors* and *signals*. A *signal* is broadcast to all the processes and is received instantaneously by all the processes. Two types of information are broadcast: *values* and *signals*. A value can also be transferred as part of a signal. A signal may contain some value with it. For a signal S, the value is represented by the notation '?S.' The same signal may be emitted by various processes simultaneously. An *event* is a combination of received signals for which the program is waiting. A sequence of events is called a *history*.

The basic programming unit is a module that has a *declaration part* and the *code part*. Estrel is not a comprehensive language and supports only basic types like integers, Boolean, arithmetic operators, and logical operators. Expressions are built using constants, signals, sensor values, and function calls. Estrel supports basic control abstractions, such as assignment statement, if-then-else statement, case-statement, loop-statement, function calls, and parallel tasks in addition to special statements: (1) signal emission, (2) trap handling, (3) exit from a trap, (4) checking for a signal before executing a statement, (5) watchdog statement up to a given time-limit, and (6) await for a given signal combination. The abstract syntax using EBNF for the some of the special statements, is as follows:

```
<signal-emission> :: emit <signal-name>(<expression>)
<signal-statement> :: present <signal> then <statement>
                                       else <statement>
<exit-trap> :: exit <trap>
<watchdog> :: do <statement> watching <signal-occurrence>
<await-signal> :: await <signal-occurrence>
<loop-each> :: loop <statement> each [<count>]
                    <signal-occurrence>
```

```
<every-loop> :: every <signal-occurrence> do <statement>
                end
<multiple-wait> :: await {case <signal-occurrence> do
                                <statement>}* end
<signal-occurrence> :: [immediate] [<count>] <signal>
```

The first rule states that a signal-emission emits a signal that can be a valued signal, and the value is given by evaluating the corresponding expression *<expression>*. The second rule states that if a signal is present, then execute then-part, and otherwise execute the else-part. The third rule states that the trap can be exited using an exit statement. The fourth rule states that a specific statement is executed before exiting. However, if the signal occurs given by *<signal-occurrence>* then exit without executing the statement. The fifth rule states to suspend the execution until a signal occurs. The sixth rule states to execute the body of the loop every time the corresponding signal occurs. The seventh rule is similar, and states to execute the body of a loop statement every time a signal occurs; the eighth rule states that different statements can be executed in response to different signals using a case-statement. The last rule says that signal occurrence may be followed by (1) a reserve word *immediate* that means immediately entertain the signal or (2) an integer count that means wait until the signal has occurred as many times as given by the integer count before triggering an action.

Example 13.3

In this example, we show a simple example of synchronous language that emits a signal every one second:

```
every 30 second do
        emit showPulseRate
end
```

The variables *second* and *showPulseRate* are signals: the signal *second* is being received. After counting 30 such signals, the signal *showPulseRate* is being outputted.

13.5 SUMMARY

In this chapter we discussed four additional programming paradigms. These paradigms are emerging: *event-based programming, agents-based programming, high-productivity massive parallel computing*, and *synchronous programming*. Event-based programming and agent-based programming have established applications and have proved their worth in commercial applications; synchronous programming has found a niche in hardware circuitry simulation and has recently been applied for synchronous multimedia programming. High-productivity massive parallel computing is evolving and if proven, would make the task of large-scale software development much easier on a plethora of massive parallel architectures involving a large cluster of processors. All these technologies are needed in the next decade, as massive parallel computers will be on our desktops, and without programming language support, large-scale software development will be impossible. Event-based programming is already being used in interactive programming involving interaction with sensors and graphical objects connected to the computers. Similarly, MAS-based programming is being used in many tasks including web-based programming for intelligent control and assisting decision-making processes.

Event-based programming is based upon reactive responses to the events in the real world rather than input to the procedural threads. An event-based system has following components: (1) an event-listener and dispatcher that is invoked by the main program, (2) a form constructor and painter that automatically generates the object-model code for each object in the painted form, (3) an abstract interface called event-adapter that links the event to the event-handler written by the programmer, (4) an event-handler written by the programmer for every event that is being registered and needed by the program, and (5) a main program that starts a thread to invoke event-listeners and event-handlers. The main program invokes an initialization method to initialize the form where objects are located and creates the corresponding objects associated with the events. An object may be associated with multiple events. After some characteristics of an object are altered, the state of the system changes. An event is caused either by an external signal or because of the state change. The event-related attributes and information about the event-source are transferred as parameters to the corresponding event-handler.

An event-handler is a regular method that takes an action based on the event-trigger. An event-object such as textbox, button, or scroll-bar is like any other object, and its attributes such as text, background color, foreground color, and text-font are accessed as *<object-name>.<attribute-name>*. An event-handler can alter the attributes of any object in the form using this model of *<object-name>.<attribute-name>*. Event-based programming has found applications in many areas, including monitoring, graphical user interfaces, game-playing, smart homes, sensor-based instrumentation, and web programming.

Agent-based programming takes the concept of code and data migration to another level. In applets, there is limited amount of migration: code migrates to the client-end and gets executed. Agent-based programming extends the concept of code mobility by allowing the mobile-code to migrate to any subset of web-nodes (or nodes in a computer network), get executed, and return the results back to the originating node. In addition, agents are autonomous—knowing when and where to go and having desires and intentions; reactive—respond to an event by performing some computation and taking some action, and can be intelligent. A MAS is a network of agents that cooperate or compete to achieve a goal. An agent contains an itinerary to go to various nodes in the network and has a unique identification number that includes the information about the originating node. An agent can clone itself and freeze its state for migration to another node, The clone carries the code and data area such as heap and stack, and its own frozen state to another node, where it is activated after proper authentication by the receiving node. In case of loss during transmission, a cloned agent is retransmitted. There are many types of managers in an agent-based system to make sure that an agent is created, activated, dispatched, received, and monitored appropriately. Some of the managers are agent manager, dispatch manager, event manager, queue manager, and security manager.

The security and fault tolerance of an agent-based system is quite critical. In addition to security issues involved in applets, an agent can also be corrupted by a malicious host and migrate to new node corrupting the following hosts. There is little protection against avoiding corruption from a malicious host except to avoid known malicious hosts. In addition, an agent can periodically perform some tests to make sure that it has not been corrupted.

The fault tolerance of an agent-based system is important, because crash of an agent may destabilize the whole MAS. There are two approaches to provide fault tolerance: (1) keep the altered knowledge after the last checkpoint distributed in a knowledge base of agents in other nodes or (2) keep a hot shadow agent. In the first scheme, a new agent is created, and the knowledge base is updated in a two-step process: (1) load the checkpointed knowledge into the newly created knowledge base and (2) request for distributed knowledge after the checkpoint from the neighbors. The second scheme keeps an extra agent in a ready state that has all the knowledge as the active agent. However, it becomes active only after the active agent crashes.

There are many agent-based languages that have been developed in the last decade. Languages supporting intelligent agents are based on a popular model called *BDI* (*Belief, Desire and Intention*). Languages supporting intelligent agents integrate logic programming and object-oriented programming paradigms: the logic programming paradigm provides the declarative knowledge base and the inference engine and object-oriented programming provides the capability of agent encapsulation and facilitates agent migration. Many of these languages are built on top of Java, because Java supports object-oriented programming, multiple threads, code migration, virtual machine to provide homogeneity in heterogenous environment, and sufficient amount of security to objects. Some of the historically popular languages and agent-based systems are AgentSpeak(L), AgentTalk, Aglet, APL, dMARS, Jade, Jack, CLAIM, KABUL, and Odyssey.

Most of the languages in the past were designed for uniprocessor machines. Concurrent programming was added to them when multiprocessors became a reality. However, concurrency constructs, data parallelism, and task parallelism developed for multiprocessors with few processors do not scale well because of heterogeneity of address spaces, and the problem of transparently mapping the computations on the address space for efficient computation without significant overhead of data movement. There has been limited effort to make universal high-level constructs for efficient massive parallel computation with the least overhead of data and code migration. Another problem is the lack of support for large-scale software development paradigm on massive parallel computers.

Recently, many new languages are evolving that integrate concurrent programming paradigms and object-oriented programming paradigm for large-scale high-productivity software development on massive parallel clusters of processors. These languages are successors of High Performance Fortran and UPC (Unified Parallel C). Chapel, Titanium, and X10 are based on the notion of PGAS, where each partition can be assigned a thread pool, and an array could be distributed across the partition to exploit both data parallelism as well as task parallelism. Many different data and concurrency abstractions have been developed in these languages.

X10 is a strongly typed language that has the concept of a set of data-elements in an array; place—a combination of activities (threads) and data; clocks—a special value class that advances after registered activities become inactive; async—asynchronously starting an activity at a particular place; unconditional and conditional atomic actions—freezing all other activities if the current atomic activity is executing; finish—activities waiting for other activities to finish if they finish earlier; force—blocking an activity while evaluating an expression requested by it; and future. In addition, X10 also offers some amount of load-balancing using work-stealing by a group of processors from other processors.

Chapel is another high-productivity language for large-scale software development that supports both object-oriented programming and multiple threads based on parallelism on top of the PGAS memory model. It relies on higher level constructs such as atomic action instead low-level locks for synchronization. For concurrent execution, it uses cobegin-coend construct with synchronization variables. Chapel also uses a concept of domain that is similar to region in X10 and allows a subset of a distributed array to be selected for data-parallel computation. Using the notion of domain (possibly infinite), Chapel also supports the notion of sets and associative arrays. Chapel also supports the declaration locale and collection of locales to map the execution of a program to a region of the cluster. Chapel allows iterators over the domain and arrays. Chapel uses two types of definition of classes: *traditional class* and *value class.* Value classes act like records with the capability to invoke methods. Chapel does not enforce an object-oriented format of programming, and programmers can write programs using block structured programming.

Synchronous languages were developed to simulate the large-scale logical circuit and are based on synchronized computation after every logical clock-tick. After a signal becomes available, the next action is taken at the next clock-tick, and the computation is assumed to be instantaneous. The languages have been used to model deterministic systems including clock-based logical circuits.

13.6 ASSESSMENT

13.6.1 Concepts and Definitions

Activities; adaptation; agents; agent-certification; agent-communication; agent manager; agent migration; agent mobility; agent security; agent state; agent transfer protocol; Aglet; async; autonomous; BDI system; belief; blackboard; Chapel; checkpoint; clock; competition; cooperation; data partition; domain; Estrel; events; event-adapter; event-handler; event-listener; event-manager; event-source; future; fault-tolerance; global partition; history; intention; interoperability; J-AAPI, locale; malicious agent; malicious host; MAS; massive parallel; partition; persistence manager; PGAS; place; plan; profit maximization; queue manager; place; reactive; region; remote administration manager; scalability; security manager; shadow agent; signal; service bridge; subdomain; synchronous; trap; X10.

13.6.2 Problem Solving

1. Extend the C# program in Example 13.1 for the game of Tic-Tac-Toe using Visual Studio. A tic-tac-toe program has nine text areas and a reset button, and it toggles between 'O' and 'X' when a mouse is clicked. You may have to keep a 3×3 matrix to test the logic who won and which blocks are occupied, so that mouse-clicks on those boxes can be ignored after they get a value.

2. Write an event-based interface in C# that converts the temperature given in centigrade to Fahrenheit.
3. Write a C# program that highlights a textbox when the mouse moves over the textbox.
4. Write a Javascript program that highlights a state when the mouse moves over the state in a country's map, and takes you to the state website when you click anywhere in the state-area.

13.6.3 Extended Response

5. Explain the difference between the procedural input model of programming and the event-driven model of programming. Give an example from the real world, where one would be preferred over the other.
6. Explain the event-based model and its component using a figure connecting the components.
7. Explain the role of event-adapters in an event-driven model.
8. Explain the difference between applets and agents. What safety issues are added in agent-based models and why? Explain.
9. Explain the different security issues in agent-based systems.
10. Explain different schemes for providing fault tolerance in an agent-based system.
11. Explain partitioned global address space and compare it with shared address memory space and distributed memory space.
12. Explain the salient constructs in X10, and argue how these constructs are useful for large-scale software development.
13. Explain the difference between synchronous languages and other procedural languages.

FURTHER READING

Bansal, Arvind K. "Incorporating fault tolerance in distributed agent based systems by simulating biocomputing model of stress pathways." *SPIE Defense and Security Symposium*, 6201. 2006. 62010801–62010810.

Bansal, Arvind K., Rammohanarao, Koganti, and Rao, Anjana. "A distributed storage scheme for replicated beliefs to facilitate recovery in distributed system of cooperating agents." In *Proceedings of the Fourth International AAAI Workshop on Agent Theory, Architecture, and Languages, LNAI 1365*. Springer-Verlag. 1998. 77–92.

Benveniste, Albert, Caspi, Paul, Edwards, Stephen A., Halbvachs, Nicolas, Guernic, Paul Le, and Simone, Robert De. "The synchronous languages 12 years later," In *Proceedings of the IEEE*, 91(1). 2003. 69–83.

Berry, Gerard and Gonthier, Georges. "The synchronous programming language ESTEREL: Design, semantics, implementation." *Science of Computer Programming*, 19(2). 1992. 1–51.

Bordini, Rafael H., Braubach, Lars, Dastani, Mehdi, Seghrouchni, Amal El Fallah, Gomez-Sanz, Jorge J., Leite, João, O'Hare, Gregory, Pokahr, Alexander, and Ricci, Alessandro. "A survey of programming languages and platforms for multi-agent systems." *Informatica*, 30. 2006. 33–44.

Chamberlain, Bradford L., Callahan, David, and Zima, Hans P. "Parallel programmability and chapel language." *International Journal of High Performance Computing Applications*, 21(3). 2007. 291–312.

Charles, Philippe, Grothoff, Christian, Saraswat, Vijay, Donawa, Christopher, Kielstra, Allan, Ebcioglu, Kemal, Praun, Christoph von, and Sarkar, Vivek. "X10: An object-oriented approach to non-uniform cluster computing." In *Proceedings of the 20th Annual ACM SIGPLAN Conference on Object-Oriented Programming, Systems, Languages, and Applications, OOPSLA '05*. 2005. 519–538.

d'Inverno, Mark, Kinny, David, Luck, Michael, and Wooldridge, Michael. "A formal specification of dMARS." In Singh, Rao and Wooldridge (eds.). *Proceedings of the Fourth International Workshop on Agent Theories, Architectures, and Languages: Lecture Notes in AI, 1365*. Berlin, Germany: Springer Verlag. 1998. 155–176.

Doyle, Barbara. *C# Programming*. 2nd edition. Course Technology—Cengage Learning. 2008.

Halbwachs, Nicolas. *Synchronous Programming of Reactive Systems*. Kluwer Academic Publishers. Revised 1993. Available at http://www-verimag.imag.fr/~halbwach/newbook.pdf

Hansen, Stuart and Fossum, Timothy V. *Event Based Programming*. May 2010, Available at http://ginger.cs.uwp .edu/staff/hansen/EventsWWW/Text/Events.pdf

Jo, Chang-Hyun and Arnold, Allen J. "An agent-based programming language: APL." Proceedings of the *ACM Symposium on Applied Computing*, Madrid, Spain, 2002. 27–31.

Lange, Danny B. and Oshima, Mitsuru. *Programming and Deploying Java Mobile Agents with Aglets*. Boston, MA: Addision-Wesley. 1998.

Luck, Michael and d'Inverno, Mark. *Understanding Agent Systems, Springer Series on Agent Technology*. 2nd edition. Berlin, Germany: Springer-Verlag. 2004.

Lusk, Ewing and Yelick, Katherine. "Languages for high-productivity computing: The DARPA HPCS project." *Parallel Processing Letters*, 17(1). 2007. 89–102.

Oppliger, Ina R. "Security issues related to mobile code and agent-based systems." *Computer Communications*, 22. 1999. 1165–1170.

Rao, Anand S. "AgentSpeak(L): BDI agents speak out in a logical computable language." In W. Van de Velde and J. W. Perram (eds.), *Proceedings of the Seventh European Workshop on Modeling Autonomous Agents in a Multi-Agent World, LNAI Volume 1038*. Spring Verlag. 1996. 42–55.

Shoham, Yoav and Brown, Kevin L. *Multi-Agent Systems: Algorithmic, Game-Theoretic, and Logical Foundations*. New York, NY: Cambridge University Press. 2009.

Starovic, Gradimir, Cahill, Vinny, and Tangney, Brendan. "An event based object model for distributed programming." In *Proceedings International Conference on Object-Oriented Information Systems*, Springer-Verlag, Berlin, Germany. 1995. 72–86.

Synder, Lawrence."The design and development of ZPL." In *Proceedings of the Third ACM SIGPLAN Conference on History of Programming Languages, HOPL III*, San Diego, CA. 2007. 8-1–8-37.

Wong, David, Paciorek, Nick, and Moore, Denarius. "Java-based mobile agents." *Communications of the ACM*, 42(3). 1999. 92–102.

CHAPTER 14

Scripting Languages

BACKGROUND CONCEPTS

Abstractions and information exchange (Chapter 4); Abstractions in functional programming (Chapter 9); Logic programming (Chapter 10); Object-oriented programming (Chapter 11); Operating system (Section 2.5); Web and multimedia programming (Chapter 12).

As operating systems grew, system tasks have become complex and need multiple files and system utilities to complete the tasks. The tasks need a sequence of system commands. These system commands work as an assembly line such that a command executes and generates data that is picked by the following commands. Many commands may also be executed concurrently or in a distributed manner using multiple threads. Many times, commands have to handle a directory of files. Directories are collections of files. Processing complex tasks requires variables to hold the file-names, directories, and file paths, and iteratively process them. The files could be of different types such as text files, image files, data files, and sound files. To process one class of files in a directory, there is a need to selectively filter the files before taking an action. Many times, while executing a compilation command, a new file has to be created automatically that has the same name but a different suffix. For example, compiling a file *myfile.c* will generate compiled code in a file *myfile.o*. That means that there is a need for text processing. Text processing is also needed to parse various switches associated with line-commands. Based on the switches, different system utilities may be called. Again there is a need for conditional statements and text processing. A clear need for a class of high-level languages has emerged that could (1) execute the operating system line-commands, (2) invoke multiple system utilities, (3) integrate the execution files developed in different languages, (4) perform pattern matching on text embedded inside a file to take an action based on the presence or absence of specific text patterns, (5) work on a large collection of files and directories, and (6) work over the web or remotely on various resources such as remote-files and URLs. This class of languages is called scripting languages.

Scripting languages have their history in *Awk* (Aho, Weinberger, and Kernighan) and *Sed (Stream editor)*. *Sed* is an early UNIX command processing language to process data files by looking for a pattern and substituting another pattern. It does not have variables to store any information and has only branch statements. *Awk* is a scripting language, designed in 1970s by Aho, Weinberger, and Kernighan to invoke batch-processing jobs involving UNIX pipes and shell-based programming for UNIX-like operating systems. Later it was adopted for other operating systems. Awk commands include setting up a variable, invoking a system utility, calling functions, text processing, and performing simple computations for searching text patterns. In the current form, it supports many data abstractions such as associative arrays and control abstractions like if-then-else, while-loop, for-loop, and function calls. In Awk, an action is taken based on some pattern match. The main contribution of Awk was to parse the one line-command into words and take an action based on pattern match. Later scripting language like Perl introduced more control and data abstractions; however, regular languages provided library support and built-in support to interface with system utilities and text processing to provide scripting capabilities. In the current form, the gap between the capabilities of the languages for script programming and regular languages has reduced significantly.

An interesting question is, What is the difference between scripting languages and regular programming languages? Can a regular programming language with string-processing capabilities be extended to become a scripting language in addition to being a regular programming language? The answer is 'Yes.' If a language can read a command as a string; parse it to separate the switches and commands; and then, depending upon the command can separate between system utilities or regular programming statements, then it can become a scripting language. However, it should also have the capabilities to (1) call system utilities; (2) perform complex text processing operations; (3) handle complex pattern matching capabilities; and (4) preferably support first class objects—the capability to build system commands as text and then convert them into system utilities and execute. Figure 14.1 illustrates four major capabilities a scripting language should possess. Another capability that can be added for modern scripting languages is the capability to spawn and handle multiple threads and to interface with web programming for web scripting. PHP and ASP (Active Server Pages—Microsoft-supported language), and Javascript are popular scripting languages that support web scripting as well as other major features.

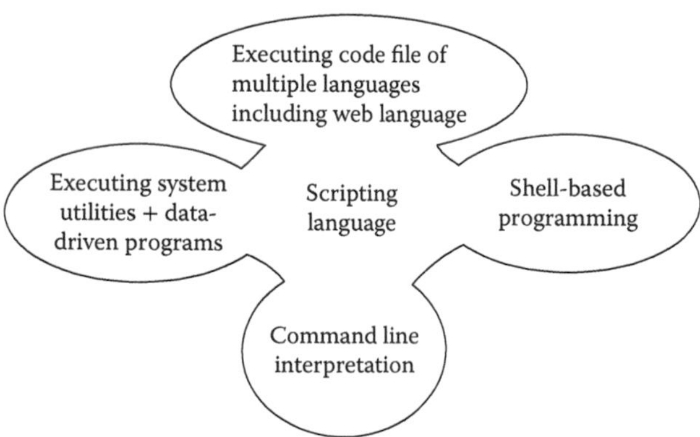

FIGURE 14.1 Major capabilities of scripting languages.

Most of the regular languages are stronger in data abstractions and control abstractions. However, the basic requirement of the scripting languages is to handle collections of files and directories, the use of iterative constructs and conditional statements to selectively perform an operation on these system resources is crucial. The earlier scripting languages were simpler and were generally built on top of the languages that supported system programming. For example, Perl was built on top of C—a system programming language that was used for the development of the UNIX operating system. Later many general purpose control and data abstractions were added to Perl.

As demonstrated in this chapter, augmented with a text processing library, many popular implementations of Prolog are used as a scripting language. The author has developed an extensive text processing library for Prolog and used Prolog for complex data-driven programming and scripting to analyze microbial genomes. The multiparadigm languages Ruby and Python are full-fledged languages. However, because of their extensive built-in support for system utilities, shell-based programming, text processing, and pattern matching, they are extensively used as scripting languages.

14.1 COMPONENTS OF SCRIPTING LANGUAGES

Scripting languages are *glue languages* that integrate the execution of (1) system utilities including compilers, (2) command line interpretation, (3) shell-based programming, (4) execution of codes written in multiple languages including web-based languages, and (5) data-driven execution by invoking a system utility or a user-defined program based on a data pattern in a data file. Figure 14.1 illustrates various components of a scripting language.

14.1.1 Shell-Based Programming

Many times we want to keep different activities of a program in different environments to avoid the mix-up of environment variables and data. That is supported in a UNIX-based operating systems using a concept known as *shell*. A *shell* is a closed environment that has processes and environment variables. A shell is generally self-contained and does not share its environment with other shells. A shell can spawn other shells to perform some computation. However, a new shell does not update the environment of the parent shell unless specifically requested. After the termination of the called shell, the control comes back to the original directory of the invoking shell.

Shell-based programming uses data abstractions, control abstractions, operating system resources, and invokes system utilities in sequential, concurrent, or pipelined order to perform a complex task. Shell-based programming is also known as command language, and it has the properties of both programming language and user interface to the process-related facilities of the underlying operating system. A pipelined execution of two commands C_1 and C_2 is denoted as "$C_1|C_2$," and it means that C_2 does not wait for C_1 to terminate; C_1 generates a stream of data, and C_2 starts executing after some data is available. A command also has directionality denoted by the symbol '>' to show that the output of the execution of command should be archived in a resource following the symbol '>.' For example, a command *dir* > *a* will print out the name and attributes of the files in the current directory into a file-named "a." If the file "a" does not exist, then it is created. Otherwise, it is rewritten.

Every operating system has its own shell-based programming system. There are different types of shells with some variations in capabilities such as C shell, Bourne shell, Korn shell, and Bourne-again shell (Bash). Windows operating systems use Windows Explorer or DOS-like command windows to execute shells. There are two types of suffixes used in Windows operating systems: *.com (command file)* and *.bat (batch file)*. Windows invokes a shell *com.exe* to execute a system utility. Lately, Windows has started developing "Windows Powershell"—a new scripting language with variables and control abstractions.

14.1.2 Data-Driven Programming

An example of data-driven program would be a genome information file that contains different types of information such as (1) a sequence of amino-acids in a protein, (2) information about the pathways where protein is involved, (3) protein nomenclature and functionality, and (4) sequence of nucleotides of a genome. Information is tagged with a specific marker, and different data processing programs are invoked based on the identification of different tags. Such data-driven computing also involves extensive text processing and pattern-matching capabilities such that specific pattern of text may be identified.

14.1.3 Command Scripts

There are multiple system utilities in every operating system. If you are using a Windows system and type help in a command window, then you will see bunch of supported operating system commands that are used to run complex system tasks in ".bat" files.

Example 14.1

Let us write a simple script file *test.bat* for Windows operating system.

```
echo OFF
date /T
time /T
echo "Hello Arvind"
dir /O
echo ON
```

The first command turns off the echo on the screen. However, because the echo is on by default when this statement is executed, the first command '*echo OFF*' will be shown on the computer screen. The next command is *date* followed by a switch "/T." The switch tells the system to display just the date. The next command *time* also has the switch "/T" just to show the time. The next command *echo* displays the message "Hello Arvind" on the screen. The next command *dir* shows all the files in the current directory. The switch "/O" forces the files to be displayed in sorted order, and the last statement turns the echo on again so that everything typed on the keyboard is displayed again on the screen.

A command file like *test.bat* described above is executed in the directory where it is located or by giving the pathname followed by filename. Here we can simply type "test" to run the script file, because Windows automatically puts the suffix ".bat" after test and runs the script program in *test.bat.*

A command can either be given as a command line using a keyboard or a text line in a script file. A scripting language should read the commands as a string, parse the string, identify the commands, and execute the commands by calling the corresponding system utilities. Many times system utilities need to repeat the same operation on a collection of files. In such cases, the script file needs to use the variables and iterative loops. A variable can be associated with data-objects such as filenames, directories, directory paths, accumulators, logical conditions, device names, and index values. Most of the scripting languages are dynamically typed and interpreted for the ease of programming.

14.1.4 Text and String Processing

Text and string processing is very important for scripting languages. Given a command, the string has to be parsed. A command line can have many redundant white spaces such as blanks, tabs, and end-of-line spaces. Many times these spaces have to be ignored, and sometimes they become the proper delimiters. A command can have many white spaces preceding a command or a switch or other arguments that have to be removed. In data-driven programming, a pattern may have a prefix and postfix. Thus, the text-processing capability should be able to splice the text to remove a given prefix and suffix.

A string is a sequence of characters, whereas a command is a list of words with white spaces removed. Scripting languages should support functions to convert a sequence of characters into a list of words. Many times the system utilities process a particular type of file such as an image file. To extract all files of the same type, wildcard characters are needed such that "*.jpg" means any file with a suffix ".jpg" in the file-names. Sometimes filenames specific to certain activities may be identified by a substring embedded within file names. For example, "BookCover.jpg" and "BookCover.gif" are two files that have a substring "BookCover." There may also be cases where multiple files have a different substring, whereas the prefix and postfix may be the same.

14.1.4.1 Pattern Matching and Data-Driven Computation

In data-driven computation, the text in data files may have to be skipped until a relevant substring is identified. All these text-based operations must be supported by a scripting language. This also includes quickly matching unique patterns in large files. The text pattern could be matching a group of characters that may be separated by nonmatching characters. Many scripting languages use different short cut patterns called *regular expressions* for matching the substrings in a file. Efficient pattern matching and data-driven computing is becoming important as the amount of data being gathered automatically is increasing; large amounts of petabyte data is being collected in many projects such as weather analysis, galactic data analysis, and the Human Genome Project. It is difficult to handle such large amounts of data.

14.2 ABSTRACTIONS IN SCRIPTING LANGUAGES

A scripting language needs to support data abstractions such as (1) basic data types like integers and Booleans, (2) strings, (3) dynamic arrays or vectors, and possibly (4) recursive data types such as linked lists. It should also support variables or the capability to simulate environment variables using hash tables, associative arrays, and blackboards: the variables become the domain element and the value becomes the range element.

14.2.1 Control and Data Abstractions

The language should support iterative-loops as scripting languages perform repetitive operations on collections of files and directories. The supported iterative loops include *for-loop* and *while-loop*. Directories are tree structures, any mechanism to traverse the tree such as recursive traversal can benefit scripting languages. However, not all scripting languages support recursion. *If-then-else* statements, *case-statements*, *function calls*, and *mechanism to convert data to function* are also supported by scripting languages. Examples of the control abstractions are illustrated in Examples 14.3 and 14.4.

Most of the glue languages such as Perl support minimal data abstractions such as arrays, scalar variables, and strings. Full-fledged languages like Python, Prolog, and Ruby support other data-structures such as lists. They support extensive operations on strings in addition to multiple pattern-matching operations.

14.2.2 Shell Variables

A variable reuses the associated value at different places in a program, and the value can be updated in destructive programming. Most of the scripting languages support destructive update of variables. The values associated with variables in scripting languages include file-names, path names, message strings, device names, and iteration count. Many times variables are also used for substitution of a string at different places. The value of a variable can be accessed using a prefix symbol specific to a shell being used followed by the variable name. For example, the '$' prefix is used in C-shell.

Example 14.2

The program given below shows a Windows script file (.bat suffix) that illustrates the use of variables and concatenation. First the echo is turned off. In the second line, the variable *mydate* is set of the output of the command "Date/T". In the third line, the variable *user* is set to the string "Arvind." The fourth line displays the value of the variable *mydate*. The command "Date/T" stored as a value to the variable *mydate* is executed and the current date is displayed on the screen. The fifth line illustrates the concatenation of two strings: "Hello" and the value of the variable user. The output "Hello Arvind" is printed on the screen. The last line permanently turns on the echo mode.

```
echo OFF
set mydate = Date /T
set user = Arvind
%mydate%
echo Hello%user%
echo ON
```

The output of the above script is as follows:

```
Sat 10/20/2012
Hello Arvind
```

14.2.3 Type Conversions

Scripting languages such as Javascript and Perl use a very loose definition of type conversion. Instead of showing error when two operands are mixed, they find a compatible type, especially when one of the operands is a string, and convert the other operand into a string. For example, a Javascript expression "abc" + 4 will yield "abc4": the integer 4 is converted into string "4," and the operator '+' is treated as a concatenation operation. However, the expression 3 + 4 will return 7. Similarly, in Perl, the symbol '.' is equivalent to concatenating two strings. For example, "CS". (5 + 6) will add 5 and 6, convert the result 11 into a string, and then concatenate two strings to derive a new string, "CS11."

14.2.4 Regular Expressions and Pattern Matching

Pattern match is used to search for the files having some patterns in the file-names, or in their attributes, or containing a specific text that matches some textual pattern. A regular expression is a collection of alternative patterns such that the match of any pattern implies a successful match. A pattern can be (1) an exact substring occurring anywhere is a text, (2) multiple substrings separated by zero or more substrings that do not match, and (3) a common suffix or prefix. For example, a command 'dir *.pdf' will return all the files in a directory that have the suffix '.pdf.' Similarly, the command 'dir R?se.pdf' will give all the files with '.pdf' suffix having a pattern R_se, where the second character could be any character.

A pattern may match (1) exactly n (n > 0) times in a text to be relevant, (2) at least n times but not more than m times, (3) 0 or more times, (4) 1 or more times, or (5) 0 or 1 time. Matching 0 time means that a text stream does not contain the pattern at all. These different matching criteria are represented differently in regular expressions.

In Perl, membership can be denoted using a square bracket; range using the hyphen sign; and negation by the upward arrow sign '^.' For example, [12a] means if the character matches one of the three characters: 1, 2, or a. The representation [0–9] means characters may match any digit from 0 to 9. Because there are special symbols like '['or '-' that are used in representing the set of characters to be matched, they are preceded by backslash if they are part of the set of matching characters. For example [a–z\–] denotes a character set from a to z and including the hyphen sign '-'. The character '.' matches any character. Some of the characters are used to denote frequently used regular expressions in Perl as given in Table 14.1.

14.2.5 Programming Example

This subsection deals with shell-based programming using the Windows operating system. A similar approach can be taken with Linux or UNIX. This section illustrates the use of various control abstractions and the use of variables using a simple Perl program implemented in Windows operating system. Section 14.3.1 discusses various abstractions in Perl and how they have been used in script-based programming.

TABLE 14.1 Character Meanings in Regular Expressions in Perl

Character	Meaning	Backslash Character	Meaning
{n, m}	At least n and at most m times	\d	[0..9]
{n,}	At least n times	\D	Nondigit
{n}	Exactly n times	\f	Form feed
*	0 or more times	\n	New line
+	One or more times	\r	Carriage return
?	0 or 1 time	\s	White space character
^	Matches prefix	\S	A nonwhite space character
$	Matches suffix	\t	Tab
\b	Matches at word boundary	\w	Word character [0-9a-z A-z]
\B	Matches nonword boundary	\W	Nonword character

Example 14.3

Let us understand the integration of programming abstractions and system command using a simple example written in the language Perl. The example in Figure 14.2 illustrates the integration of control abstractions, array, file-system, system utilities, and shell programming.

The program interactively deletes one file at a time from a given directory 'C:\Users\Arvind\Desktop\testdir.' Note that the directory could be read interactively. If a user types '1' in response to the program interaction, the corresponding file is deleted. Otherwise, it is retained.

There are three scalar variables: *$dir*, *$curdir*, and *$file*. There is an array variable *@ fileList*. The variable *$dir* is bound to the directory from where the files have to be interactively deleted. The directory associated with variable *$dir* is opened using the built-in library function *opendir/2* that creates a stream *DIR*. All the elements in the directory are read using another

```
my $dir = 'C:\Users\Arvind\Desktop\testdir';
opendir(DIR,$dir) || die "Can't open $dir : $!\n";
my @files = readdir(DIR);
close(DIR);

$curdir = chdir(); # get the current directory
chdir $dir; # change the directory to my directory

foreach my $file(@files) # iterate through each file
{   print "Do you want to delete ", $file, " :";
    $answer = <STDIN>; # get the user response
    if ($answer == 1) {
            system "del $file" ; # delete the file
            print "\n > deleted ", $file, "\n"; # inform the user
            }
}
chdir $curdir; # get back the original directory
print "done \n" # inform the user
```

FIGURE 14.2 Illustration of capabilities of scripting language Perl.

built-in function *readdir/1*. Note that the library functions *opendir/2* and *readdir/1* are part of the built-in language library, and are not operating system utilities like *del* that delete a file by invoking a system shell. After all the files are read into the array variable *@fileList*, the directory-stream *DIR* is closed using a library command *close/1*. The path of the current directory is stored in the variable *$curdir* using a statement *chdir()* that returns the path of the current directory. An iterator takes one file at a time in the variable *$file*, and processes it. Each time the user is asked for permission to delete the file, and the file is deleted using a shell command system ("del $file") if the user types in '1'. Otherwise, the file is retained. Note that the print statement can print multiple strings separated by a comma in the same line, and uses '\n' to go to the next line.

14.3 CASE STUDY

This section shows five major scripting languages and their characteristics that make them suitable for scripting. The languages Perl and Ruby started as scripting languages, and added more abstractions as they grew to make them richer. However, Prolog and Python are regular programming languages that have extensive interface with system utilities, string and text processing capability, and file and directory manipulation capability to provide support for script programming. PHP is a language that grew out of Perl to support web-based programming, and provides all the features of Perl. In addition it supports interface to XML, AJAX, and supports the development of server side scripting. There are other languages like *Lua* and *Clojure* that have been developed as scripting languages. All these languages are quite rich and evolving. It is impossible to discuss all the programming features of these languages. We will restrict our discussion to summarizing the features that enables script-based programming, and discuss one running example that has been implemented in Python, Prolog, and Ruby. We have already seen an example of Perl-based script programming in Figure 14.2.

14.3.1 Abstractions and Programming in Perl

Perl started as a project to develop control abstractions and textual pattern-matching capability for high-level system programming scripts. It has extensive pattern-matching support, and has been used in data-driven programming and system programming to handle large-scale data, such as human genome and microbial genome analysis in addition to writing high-level programs to develop system routines.

14.3.1.1 Data Abstractions in Perl

Perl supports literals including integers and floating point numbers, scalar variables, array variables, list-like arrays, associative arrays, and strings. Strings could be both single quoted or double quoted. Both Single and double quote strings are sequences of characters. However, the meaning of backslash in single quoted string and double quoted string is different: in a single quoted string backslash is not a special character but just a backslash. In a double quoted string, backslash is used as prefix for special characters: a single quote is represented by a backslash followed by a single quote, and a backslash in a string is represented by a backslash followed by a backslash. A backslash followed by special characters is used for next line, return, tab, bell, backspace, etc. For example, \n denotes next line, and \a denotes bell, and so on. The language supports arithmetic operators; logical operators; bitwise logical operators; and string operators such as concatenation denoted by '.,' less than denoted by '<,' not equal denoted by '! =,' less than or equal to denoted by ' = <' etc. The operator '+' is converted into concatenation operator if one of the arguments is a string and the other is a number. A scalar variable has a prefix '$' and an array variable has a prefix '@' as illustrated in Figure 14.2. An associative array variable has a prefix '%' followed by an identifier. Each element of an associative array is a scalar variable.

14.3.1.2 Control Abstractions in Perl

Perl also supports multiple assignment statements like $x = $y = "Perl String, which means to assign the string "Perl String" to both the variables $x and $y. An operator of the form $x = $x *<operator>* *<operand>* can also be written as $x *<operator>* = *<operand>*. For example, $x = $x * 4 is the same as $x * = 4. Similarly, $x = $mystring. = "jpeg" is the same as assigning $mystring.jpeg to both the variables $mystring and $x. Perl also supports *autoincrement* and *autodecrement* operations. For example, ++$x has the same meaning as $x = $x + 1, and - $x has the same meaning as $x = $x – 1.

The array representation uses subrange mixed with discrete values. For example, $a = 4; $b = 7; $c = "cs;" $d = ($a..$b, $c, 5) is equivalent to $d = (4, 5, 6, 7, "cs", 5). Perl supports a function call *chop()* that removes the last character from a string, and is used mainly to remove the trailing white space such as blank, linefeed, and tab, from a string. Reading from the standard input is denoted by *<STDIN>*, and writing into a standard output is done using the statement *print*, as illustrated in Figure 14.2.

Perl supports insertion of elements at the end of an array using *push operation*, unshift operation to insert at the beginning of an array, and using assignment operation. For example, *@myarray* = (10, *@myarray*) will put 10 as the first element of the array, followed by the rest of the elements. Similarly, *@myarray* = (*@myarray*, 20) will insert the integer 20 as the last element of an array. Assignment operations can also be used to take element out of an array and to swap elements. For example, (*$first*, *@rest*) = (*1, 2, 3*) will assign the value 1 to variable $first, and assign the array (*2, 3*) to the array variable *@rest*. The operation (*$x, $y*) = (*$y, $x*) will swap the value of the variables *$x* and *$y*. If a scalar variable is assigned to an array, then it picks up the first element of an array. For example, *$x* = *@y* = (*a, b, c*) will assign the value a to the scalar variable $x. *Pop operation* takes an element out of the right side of an array, and *shift operation* takes an element from the left side of an array. In addition, Perl supports other built-in functions such as *reverse* and *sort*.

Perl supports while-loop, for-loop, if-then-else statement, until-loop, and foreach statement. An until-loop is similar to while-loop, except it checks for the negation of a condition to be true. The abstract syntax for the control abstractions are given below

```
<for-loop> ::= for (<initial-expr>; <final-expr>;
                    <increment-expr>) {block}
<iterate-statement> ::= foreach <scalar-variable>
                            (<array-variable>) {<block>}
<until-loop> ::= until <condition> {<block>}
<while-loop> ::= while <condition> {<block>}
```

The use of regular expressions in Perl is defined in Subsection 14.2.4, and an example of Perl script is described in Figure 14.2.

14.3.2 Abstractions in PHP

PHP is a general-purpose scripting language built on top of Perl that, in the current version PHP 5, is used for writing web-server side script and stand-alone graphics programs. PHP code can be freely mixed with XML and HTML code using a PHP tag of the form ‘<’ **?php** *<PHP-program>* ‘**?**>’. PHP program can also be embedded in an XML or HTML program using traditional script tag as ‘<’ **script language** = "php" ‘>’ *<PHP-program>* ‘</’ **script** ‘>’.

PHP, like other traditionally script languages, is dynamically typed: variable-types need not be declared. As in Perl, PHP variables have a prefix of $. PHP supports basic data types such as integers, floating point, Boolean, and nil. PHP supports many data-structures such as dynamic arrays, hash tables, and lists. PHP also supports strings and string operations.

PHP 5 borrows lots of object-oriented programming style and use of reserve words from C++. The objects are reference types, and are referenced through handles. Handles are abstractions that contain the pointers, sizes, and other information to access the objects. However, they do not allow direct manipulation of pointers. PHP supports private and protected member variables and methods. It supports constructors and destructors like C++ does. PHP supports abstract classes and interfaces. Dynamic arrays are treated as objects, and an iterator can iterate through the elements of a dynamic array. Parameter passing is call-by-value to copy the reference of the objects.

In terms of control abstractions, PHP supports if-then-else statements, case-statements, while-loops, function calls, and for-loops. PHP supports web-based interaction through the forms displayed at the client-end. PHP uses $_GET and $_POST functions to interact with the user on the web. For example, a statement like $_GET(name) will get the value for the name field. The same is true for $_POST. However, $_POST supports secure transmission. PHP supports all arithmetic operators, logical operators, and comparison operators. In addition, it supports array operators such as union, equality, identity, and their negations. Union joins two arrays; equality returns true if key-value pairs of two arrays match; identity also checks for the same order in addition to key-value pair equality. PHP supports extensive string operations including encryption, because the information has to be transmitted over the web. The class of string operations is (1) traditional string operations such concatenation, splitting, removing white characters, and length; (2) input-output; (3) encryption and decryption; and (4) parsing and conversion to other types.

PHP supports extensive functions related to dynamic array. An array is used as a hash-table, as an ordered sequence from which elements can be taken out like a stack, elements can be inserted at any location, specific elements and duplicates can be removed, array can be sorted, and so on.

PHP supports extensive file-management; directory management; built-in file-transfer to download and upload file from remote computers; XML interface utilities; and compression utilities. PHP also can work with AJAX because of its capability to interface with

HTML and XML. File management functions can be categorized as reading, writing, reading attributes of a file, modifying attributes of a file, deleting files, creating a unique temporary file, etc. Directory management functions can be characterized as changing directory, reading filenames from a directory, making a new directory, and identifying the path of the current directory.

14.3.3 Abstractions and Programming in Python

Python is a regular programming language with a rich set of data and control abstractions. It supports imperative programming, functional programming, and object-oriented programming. In addition it has elaborate string-processing and string-matching capability, capability to handle regular expressions, interface to operating system commands, interface to socket-based programming for distributed computing and computing over the Internet, and XML for web programming.

14.3.3.1 Data Abstractions in Python

Python supports basic data types such as integer, floating points, characters, and Boolean values. It supports arrays, hash-tables, lists, tuples, and ranges. Range is an ordered set that can be written in the form of lower bounds, upper bounds, and increments. The abstract syntax for range is *<range-definition>* ::= **range**‘(‘ *<lower-bound>*, *<upper-bound>*, [*<increment-expression>*] ‘)’. For example, range(3, 6) is equivalent to an ordered set (3, 4, 5, 6), and range(3, 13, 3) is equivalent to the ordered set (3, 6, 9, 12, 13). Python also supports sets. The abstract syntax for sets is *<set-definition>* ::= ‘{‘ {*<member-element>*}* ‘}’. For example, an abstract syntax for dictionaries (hash-table) is *<dictionary>* ::= ‘{‘ {*<key>*: *<value>*}* ‘}’. An abstract syntax for lists is *<list>* :: = ‘[‘{*<member-element>*}* ‘]’.

14.3.3.2 Control Abstractions in Python

Python supports almost all major control abstractions including an elaborate scheme of iterators; multiple built-in list operations such as append, inserting an element, deleting an element, finding out the first index of an element, stack-based operations by treating list as a stack, queue-based operations by treating a list as queue, sorting a list, and reversing a list; set-based operations such as *union, intersection,* and *difference*; dictionary-based operations such as *inserting a key: value pair, deleting an element given by the key* and, *accessing the value associated with a key.* The iterator can iterate on a set, a list, or a range. It has a powerful comparison operation that compares two sequences element by element and returns false upon first mismatch. For example, (10, 12, 14) < (10, 11, 15) is *false* because 12 > 11; and (a, b) < (a, b, c) is *true*, because the first set has only two elements that are the same as the first two elements of the right-hand side set, and the third element is missing from the first set; the right-hand side set has an additional element.

14.3.3.3 Classes and Objects in Python

A class in Python is a set of simple statements or function definitions. Each identifier on the left-hand side of a statement becomes an attribute of the class and can be accessed as an attribute of an instance of the class. The abstract syntax rules for a class definition in Python is as follows:

```
<Class-definition> ::= class <identifier>[ '('<identifier>')' ] ':'
                       [<InitMethod>] ('pass' | {<Statement>}*)
<InitMethod> ::= def _ init _ '('self{<','Parameter>}*')':
                 {<Statement>}*
```

The first identifier in the class definition is the class name and the second identifier within the optional parenthesis is the name of the parent class. Parent class name is used for subclass definitions to inherit methods. The class name is separated by statements or method definitions by a colon. Python has an optional "__init__" method that is executed as soon as an instance of the class is created. Note the presence of double underscore '__' before and after "init". There are many built-in attributes in class definition such as __doc__ that are used for specific purposes to access attributes within a class.

Example 14.4

The definition given below can be tried using interactive Python. Note that syntax in Python is quite unforgiving. You should not forget (1) the symbol ':' after the class name or the function name and (2) double underscore '__' before and after init.

```
class complex_number:
>>> def _init_(self, realValue, imaginaryValue):
...         self.real = realValue
...         self.imaginary = imaginaryValue
...
>>>
```

After defining the class, an instance is created. The initialization method *_init_* is automatically executed, and a pair with two fields is created as show below:

```
c = complex_number(3.0, 4.0)
```

Here the field c.real has value 3.0 and the field c.imaginary has value 4.0. A class with extensible field can also be created by using a null body in the class as follows:

```
>>> class student:
...        pass
>>>
```

Now we can create an object Mary, and then add on the fields to it as given below:

```
Mary = student();
Mary.major = "CS"
Mary.status = 'junior'
```

14.3.3.4 System and Internet-Based Utilities in Python

Python supports extensively built-in system utilities. Some classes of the system utilities are (1) directory-related utilities such as renaming a directory, deleting a directory, changing the path to a directory, changing protection level of a directory, etc.; (2) system call to invoke any operating system utility; (3) file-related utilities such as renaming a file, deleting a file, changing ownership, and protection level of a file; (4) process-related utilities such as spawning a process, process waiting for other processes, and process sleeping for a specified time; (5) various environment-related utilities; (6) extensive string processing capabilities such as concatenate, substring search, removing trailing and leading spaces, splitting strings at a specific substring or a character; and (7) pattern matching using regular expressions. In addition, Python supports *pickling. Pickling* linearizes a data-structure into a string form that can be recovered by *unpickling.* The *pickling* operation can be used for marshaling to transfer the information over the Internet. Python also supports Internet-based operations such as sending mail and opening a URL for reading the information, and interfaces to render multimedia streams.

Example 14.5

Figure 14.3 illustrates script-based programming using Python by translating the Perl program in Figure 14.2. The program illustrates many concepts and syntax such as (1) definition of functions; (2) text processing; (3) interface with system utilities; (4) use of *os* (operating system) module and *io* (input-output) module; (5) use of control abstractions such as while-loop, iterator, if-then-else, and function call using parameter passing; and (6) use of data abstractions such as lists, strings, and strings as array of characters.

Module os provides three utilities: *getcwd(), chdir/1,* and *system/1.* The utility os.*getcwd()* returns the current working directory, the utility *os.chdir/1* changes the current working directory, and the utility *os.system/1* invokes a shell to execute operating system commands. Module *io* is responsible for built-in utilities for input-output in Python and provides three utilities here: *open/2*, *readline()*, and *close()*. The utility *io.open()* opens a file for reading, the utility *close()* closes the given stream, and the utility *readline()* returns the next line in the currently open file.

Every program file in Python is a module. There are two programs in this module: *readfile/1* and *delete/0.* The functional procedure *readfile(tempfile)* reads the names of all the files in a directory, and returns a list of filenames in the given directory. The procedure *delete()* deletes the needed files after interacting with the user. The global variable *mydir* is used by both subprograms. The variable *tempfile* is local to *delete()*, and has been passed as parameter to the subprogram *readfile/1.*

The local variable *fileover* is initialized to *false*, and is set to *true* after the end-of-file is reached. The variable *filenames* is initialized to an empty list, and keeps gathering one file-name at a time in every iteration-cycle until the end-of-file is reached. The variable *filehandle* is a handle to the stream corresponding to the file, and is used for reading from the stream and closing the stream. Using while-loop, each file-name is

```
import os # import module for operating system utilities
import io # import module for reading and writing in files
global mydir # declare variable mydir as global
mydir = "C:\\Users\\Arvind\\Desktop\\testdir" # assign value

def readfile(tempfile): # reads all filenames from tempfile
        fileover = False # used to get out of while-loop
        filenames = [] # initialize the list of file names
        filehandle = io.open(tempfile) # open a temporary file
        while (fileover == False): # repeat until end of file
                filename = filehandle.readline() # read one line
                if (filename == ""): fileover = True # end_of_file
                else: filenames.append(filename) # insert the filename

        else:
                filehandle.close() # close the file stream
                return filenames # return the list of filenames

def delete(): # deletes the files interactively
        tempfile = "C:\\Users\\Arvind\\Desktop\\temp001"
        curdir = os.getcwd() # get the current directory in curdir
        command = "dir /B " + mydir + " > " + tempfile # make a command
        os.system(command) # collect all the filenames in a file
        os.chdir(mydir) # change directory to mydir
        files = readfile(tempfile) # assign the returned values
        for file in files: # iterate over the list of files
         answer = input("Do you want to delete " + file + " :") # get input
        if (answer[0] == "y"): # get the first letter of the string
              deletecommand = "del " + file # concatenate del with file
              os.system(deletecommand) # delete the corresponding file
                    print("\n deleted file: " + file + "\n") # inform
        os.chdir(curdir) # change back to the old directory
        print("done \n") # inform user that task is done
```

FIGURE 14.3 Illustration of script programming using Python.

read iteratively, because the variable *tempfile* contains only the filenames separated by linefeed character '\n.' After the end-of-file marker is reached, the utility *readline()* returns an empty string, and the Boolean variable *fileover* is set to *true* to take the control out of the while-loop.

The subprogram *delete()* has six local variables: *tempfile*, *curdir*, *file*, *files*, *answer* and *deletecommand*. The names are intuitive. The variable *tempfile* stores the path of the *tempfile* that is created by the system command *os.system("dir /B mydir > tempfile")*. The variable *curdir* stores the current directory. The variable *files* is the list of file-names returned from the subprogram *readfile*. The variable *file* is a scalar variable that picks up one file-name at a time from the list of file-names in variable *files*. The variable *answer* collects the response from the user and stores a string that is an indexible sequence of characters. The subscripted variable *answer[0]* denotes the first character in the user-response. The variable *deletecommand* is bound to a command-string to delete the required file.

The program is straightforward. First the current path is stored in the variable *curdir*, and the command-string is bound to the variable command. Note that command-string is built like a string using multiple substrings, and the operator '+' is used for concatenating the strings. The os utility *os.system/1* takes the command-string and invokes the system command '*dir /B > tempfile*'. The command puts all the visible user-file-names in *tempfile* in the shortest version, without associated attributes. The file *tempfile* is passed to the subprogram *readfile* as a parameter. After returning from the subprogram *readfile/1*, the iterator iterates over the list of files, and each time the file-names bound to variable *file* is displayed to the user, and user response is collected in the variable answer. If the first character of the response is "y" then the file is deleted using the command *deletecommand* that is constructed as a string as explained previously. After the list of file-names *files* is empty, the control comes out of the iterator and the command *os.chdir(curdir)* changes the directory back to the original directory.

14.3.4 Script Programming in Prolog

Prolog is a logic programming language. Different variations of Prolog have different built-in capabilities and syntax to handle system utilities. Sicstus Prolog, Bin Prolog, and GNU Prolog support system library for invoking shell for system routines; library for interfacing with multithread-based programming; iteration capability; and socket-based programming. Text and string processing have been developed using *name/2* predicate, *atom_chars/2* predicate and list-processing utilities.

Different Prolog implementations support different libraries. Sicstus Prolog—one popular commercial version of Prolog developed at the Swedish Institute of Computer Science (SICS), supports extensive libraries for tree-based operations; graph-based operations; list-based operations; set-based operations; system utilities; process creation utilities that spawn a process to execute operating system utilities; file and directory creation and manipulation predicates such as file-delete, make-directory, absolute-file-name, remove a directory, changing ownership, and protection of a files; heap (heap as used in data-structures) operations; association list (hash-table)–based operations; dynamic array operations; object-based operations; socket-based operations; thread-based operation; dynamic generation of XML tags and documents for dynamic XML; interface to databases; and interface to web-scripting languages like TCL. Prolog also supports predicates that can build the predicates as a list of atoms, and then convert them into a predicate to be executed.

Other popular Prolog implementations are GNU and SWI Prolog. GNU and SWI Prolog also support extensive library systems. GNU Prolog has extensive built-in predicates for system programming such as file and directory manipulation, invoking a system call to invoke a shell that will execute underlying operating system utility. In addition, GNU Prolog supports destructive and backtrackable assignment of global variables. A variable can be assigned to any value, and a global array of objects can also be created. SWI Prolog has interface with Internet programming and XML. All three popular versions are continuously evolving, and integrated with built-in database programming capability, and capability to dynamically build predicate.

Prolog's script programming capability has been used in large-scale microbial genome comparison by the author, which involves integration of system utilities, data-driven programming, pattern matching, and string-processing capabilities.

Example 14.6

In the following program (see Figure 14.4), we use GNU Prolog to translate the Perl Program given in Figure 14.2 to selectively delete files in a given directory. GNU Prolog has many essential built-in system predicates to avoid the shell-based invocation of operating system utilities. There are four built-in predicates: *directory_files/2, working_directory/1, change_directory/1,* and *delete_file/1.* The predicate *directory_files/2* creates a list of file-names of all the files in a directory without attributes; the predicate *working_directory/1* returns the current working directory; the predicate *change_directory/1* changes the current working directory to the desired directory, and the directory *delete_file/1* deletes the given file.

The predicate *var/2* picks up the desired directory to be processed: the variable *Dir* gets bound to the desired directory. The predicate *directory_files/2* returns a list of all the files in the variable *FileList.* The predicate *working_directory/1* returns the current working directory in the variable *Curdir.* The predicate *change_directory/1* alters the current working directory to the directory held in the variable *Dir.* The predicate *delete_files/1* interactively deletes the files held by the variable *Filenames* after inquiring the user. After the necessary files are deleted, the directory is changed again to the original directory using the predicate *change_directory/1.*

The procedure *delete_files/1* deletes tail-recursively one file at a time in each cycle. In each cycle, the predicate *delete_one_file/1* is called with the next file. After the

```
:- var(mydir, 'C:/Users/Arvind/Desktop/testdir').

delete:-
        var(mydir, Dir), /* the required directory is in the variable Dir */
        directory_files(Dir, FileList),  /* list of files are in the variable FileList */
        working_directory(CurDir),       /* current directory in the variable CurDir */
        change_directory(Dir),           /* go to the required directory */
        delete_files(FileList),          /* delete interactively the required files */
        change_directory(Current).       /* go back to the original directory */

delete_files(FileList) :-
        (FileList == []→ true            /* exit the tail-recursion if FileList is empty */
        ; FileList = [MyFile|Fs],        /* get the first file-name in the variable File */
        delete_one_file(MyFile),         /* interactively delete that file */
        delete_files(Fs)                 /* invoke tail recursion for rest of the files /*
        ).

delete_one_file(FileName) :-
        format("Do you want to delete: ~a ~n", [FileName]), /* interact */
        read_atom(Answer),                                  /* read the response */
        (Answer == y →
                delete_file(FileName),   /* delete if the response is affirmative */
                format("deleted file ~a ~n", [FileName])   /* inform about deleted file */
        ; true                                             /* otherwise exit */
        ).
```

FIGURE 14.4 An illustration of script writing in GNU Prolog syntax.

predicate *delete_one_file*/1 is over, *delete_files/1* is called tail-recursively with rest of the files. The process is repeated until the list-of-files *FileNames* becomes empty.

The procedure *delete_one_file/1* inquires the user using *format/2* predicate for confirmation to delete a file. Upon an affirmative answer the file is deleted using a built-in system predicate *delete_file/1*; upon negative answer, no action is taken. No action is shown with right-hand side as trivially 'true.'

14.3.5 Script Programming in Ruby

As we have seen, Ruby is a full-fledged multiple paradigm programming language that integrates functional and object-oriented programming. In addition, every data in Ruby is an object. Ruby supports: (1) system calls using the built-in function *system/1*; (2) string processing such as *sub/2* to substitute a substring with another substring, *chop()* to remove the last character, conversion of an object into string using a built-in function *string/1*; (3) signals using a function *trap/2*; and (4) regular expression. In terms of data abstractions, Ruby supports arrays, hash-table of key-value pairs, global variables, strings, and ranges. Ruby has an extensive support of range such as finding out minimum and maximum value, concatenating the ranges, and iterating through the elements in a range using foreach-construct. In addition, we have seen that Ruby is a full-fledged object-oriented language supporting derived class and inheritance.

Ruby has full support for various control abstractions such as if-then-else statements, case-statements, while-loop, for-loop, until-loop, for-each iterators, function calls, and parameter passing. Ruby also supports invocation of threads (fibers), mutual exclusion, socket-based programming for remote procedure calls, and integration to web programming languages. Using its string processing capability, Ruby can be used to dynamically generate and modify XML programs.

In terms of regular expressions and text processing, Ruby supports: (1) matching of any subexpression any number of times; (2) skipping white spaces; (3) finding beginning and end of a string; (4) matching words and characters; (5) converting upper case into lower case and vice-versa; (6) concatenating strings; and (7) identifying special characters in input line such as end-of-line tabs, and end-of-file.

Ruby has extensive built-in library for handling files and directory. Ruby can: (1) read files in a directory and return as an array; (2) change name of a file or a directory; (3) check ownership and change reading and writing privileges of a directory; (4) create and delete a directory; (5) check the path of a file or a directory; (6) navigate through the directories; (7) create temporary files in a directory; and (8) inquire about the existence and status of files in a directory.

Owing to these properties, Ruby satisfies all the requirements of being a scripting language. The example given in 14.2 has been translated to Ruby syntax in Example 14.7.

Example 14.7

The program in Figure 14.5 is a Ruby version of the script program given in Figure 14.2. You can immediately see the integration of object-oriented programming, functional programming, and powerful system utilities in Ruby. Each data-element is

```
def deletefile
        mydir = 'C:\Users\Arvind\Desktop\testdir'
        filenames = Array.new                 # create a new array filenames
        dirstream = Dir.open(mydir)           # open the directory
        dirstream.each {|x| filenames.push(x)} # Read all file-names
        dirstream.close # close the directory stream
        curdir = Dir.pwd # store the current directory name in curdir
        Dir.chdir(mydir) # change the directory to the desired directory
        filenames.each {|myfile|              # for each file do
                puts "Do you want to delete " + myfile + ":" # ask user
                answer = gets.lstrip[0]       # get the first letter
                if (answer == "y") # if the answer is yes
                        File.delete(myfile)  # delete the file
                        puts "\n > deleted "+ myfile + "\n" # inform user
                end
                }
        Dir.chdir(curdir)
        puts "done"
end
```

FIGURE 14.5 An illustration of script writing in Ruby.

an object, and there are many built-in system predicates as part of Ruby. Ruby also uses dynamic arrays to simulate many data-structures like stacks and queues, and sequences on which iteration can be performed. We will discuss the integration of imperative programming, object-oriented programming and functional programming in Ruby.

The program's logic is quite familiar by now. The program uses five built-in modules: *Array, Dir, File, Kernel* and *String.* The function *new* and methods *each* and *push* belong to the *Array* module; the utilities *open, close, chdir* and *pwd* belong to the *Dir* (directory) module; the utility *delete* belongs to the *File* module; the utilities *gets and puts* belong to the Kernel module; and the operator '+' and the method *lstrip* belong to the *String* module. Note that the functions written in a module are written as *<module-name>.<function-name>* as in *Array.new, Dir.open, filenames.push*; and built-in methods for objects are written as *<object-name>.<method-name>*, as in *d.each, d.close,* and *filenames.each.* The symbol '.' is also used for the composition of the functions. For example *gets.lstrip* is a composition of two functions: the function *gets* receives the input string, and the function *lstrip* removes the white spaces from the left end of the string.

The variable *mydir* is assigned the path of my directory. The variable *dirstream* is associated with the stream after opening the required directory. The variable *filenames* is a dynamic array that is assigned the names of all the files in the required directory. The variable *myfile* is associated with the name of a file in the iterator. The variable answer holds the user response.

First a new dynamic array *filenames* is created. All the files in the directory referred in the variable *mydir* are read into the array *filenames* using an iterator, the file-names are pushed in the dynamic array, and the stream *dirstream* is closed. Another iterator takes one file at a time from the dynamic array filenames, and utilizes a string concatenation operator to ask a query from the user. The answer by the user is collected by the built-in method *gets*, the method *lstrip* removes the preceding white spaces, and the index 0 gets the first character of the string. If the character is "y", then the file in question is deleted using *File.delete* command. The program exits after the iterator has processed all the files. Note that each control structure is terminated using the reserve-word 'end'.

14.3.6 Other Scripting Languages

Many other languages such as *Clojure*, *Lua*, and *Scala* also support scripting. We studied the example of five programming languages: Perl, PHP, Prolog, Python, and Ruby. The basic criteria for scripting is the interface with operating system utilities; built-in support for those utilities whose effect is not persistent after the system shell terminates; string and text processing; pattern matching for text patterns using regular expressions, and a rich set of control and data abstractions. As languages get richer by supporting multiple libraries, the difference between scripting languages and regular languages will become less obvious. Interoperability between languages also reduces the need for the development of specific languages.

In future, the complex tasks will also include multithread programming, distributed computing, mobile computing, multimedia programming, web-based computing, and event-based programming. Any language that can be extended to incorporate these programming paradigms, either through the use of libraries or through the use/design philosophies, will qualify for the extended definition of scripting languages.

14.4 SUMMARY

The origin of scripting languages is in solving complex software tasks that involve gluing of one or more aspects of *system programming*; *system utilities*; *shell based programming*; *data driven programming*; *text and string processing*, *pattern matching*; and *control and data abstractions of regular programming languages*. Earlier scripting languages such as Perl started as building up control and data abstractions on top of batch script files to provide more flexibility. Recently, regular languages can interface with system libraries and text processing libraries, and can be used as scripting languages. At the same time, traditional scripting languages have kept improving. Perl that started with limited incorporation of control and data abstraction on top of batch files has added object-oriented programming in addition to a rich set of data and control abstractions reducing the gap between scripting languages and regular programming languages.

Shells usually keep their environment different from each other, unless the results are imported by a shell from other shell's environment. A shell can invoke another shell. However, there is no communication between the calling and the invoked shell. However, results can be imported or written into a persistent object such as file or database to share the information between two shells. Operating system utilities can be executed using a 'execute' call (or system call) in a scripting language that invokes a new shell; the current thread is usually suspended, whereas the invoked shell is executing. There are many types of shells such as Bourne shell, Korn shell, C-shell, tcsh—a variant of C-shell, and Bourne Again Shell (Bash).

Perl is a rich scripting language and supports all the features of a scripting language including object-oriented programming for modular script development. The language supports basic data types; dynamic arrays, where the elements can be accessed by giving index (array feature), and elements can be accessed, retrieved, and inserted from either end (list-like feature); associative arrays; and objects. The language supports standard control abstractions available in regular programming languages such as if-then-else statement, for-loop, while-loop, repeat-until, and iterators. Perl allows shell-based execution of the system utilities by using the statement 'exec', and supports many system utilities as built-in library functions such as opening a file, directory, reading a file, and directory. Perl has an elaborate regular expression and text processing library. Integration of these features makes Perl a powerful scripting language.

PHP is a web scripting language with capabilities of regular programming language. It is built on top of Perl, and the latest revision PHP 5 supports object-oriented programming. PHP supports multiple built-in system handling utilities such as file management utilities, directory management utilities, networking utilities, data base interface, and XML interface.

Python is a full-fledged language that supports imperative and object-oriented programming, regular expressions, string processing, interfaces to operating system commands, Internet socket-based programming, web-based programming, and XML using various modules. In terms of data abstractions, it supports basic data types, arrays, lists, sets, ranges, maps (hash-table). In terms of control abstraction, it supports multiple assignment, if-then-else, case-statement, for-loop, while-loop, iteration on range, lists and sets, thread-based programming, and socket-based programming for remote procedure calls. Thus Python qualifies as a powerful scripting language for complex system involving operating system, networks, and web-based programming.

Prolog is a declarative language with full large-scale programming capability. Many Prolog variations such as GNU Prolog and Sicstus Prolog have capabilities to invoke system commands using the system predicate *system/1*, and support many built-in system level commands. Prolog supports string processing library that has been extended to process an extensive text processing library. By integrating the capabilities of the system library, text processing library, and shell programming to invoke system commands in the underlying operating system, Prolog can be used for script-based programming.

Ruby is a multiparadigm programming language supporting full-fledged integration of object-oriented programming, functional, and imperative programming paradigm. It is easy to write and has extensive built-in capability to process system utilities, strings, pattern matching using regular expressions, and shell-based programming using a built-in function system/1. Thus Ruby has been used extensively for script programming.

There are many other languages such as Clojure, Lua, and Scala that are used for scripting. The purpose of the scripting languages is to be able to transparently execute a complex task by gluing various subtasks. It can be achieved either by integrating multiple paradigms, developing libraries, providing interface with operating system utilities, and the development of interoperability with languages having scripting capabilities. The future of scripting will involve the integration of concurrency, distributed computing, web programming, operating system utilities, networking utilities, remote control of processes, and event-based programming. The major advantages of high-level scripting languages will be in better software maintenance and faster software development time. One of the problems with interpreted scripting languages is the execution efficiency. Thus there is a need to develop interfaces with other programming languages so that executable code in other compiled languages can be called by the scripting languages for better efficiency.

14.5 ASSESSMENT

14.5.1 Concepts and Definitions

Awk; Bash; Bourne shell; C-shell; command language; glue language; Korn shell; pattern matching; Perl; PHP; pickling; Prolog; Python; regular-expressions; Ruby; script file; Sed; shell; shell programming; socket; string processing; system call; system utilities; text processing; unpickling; web-scripting.

14.5.2 Problem Solving

1. Read a Perl manual, and write a program to delete all the files that have a given pattern in their file-names.
2. Read a Ruby manual, and write a program to remove the occurrences of the words given in a data file.
3. Read GNU Prolog manual, write a program to find out directories that have files before a specified date, and print out the list of directories.
4. Read Python manual, and write a Python program that displays the list of all the text files in a directory that contain a pattern of text in their content. The pattern can be defined by the Table 14.1 or any other equivalent table you want to define.
5. Write a Python program to compare two files, and show the position of first mismatch given as a pair (line number, column number). Try it out on a text file that has approximately 20 lines.
6. Write a Ruby program that compresses all the files that are bigger than a specified size in a directory with an option to compress them.

14.5.3 Extended Response

7. Compare the features in Ruby and GNU Prolog for writing system scripts.
8. Compare the regular expression features of Ruby and Perl.
9. Describe various features of a script programming language, and give a real world example of the features in solving complex task.
10. Read the manuals and compare various built-in system utilities, directory utilities, and file-utilities in Perl, Python, GNU Prolog, PHP and Ruby.
11. Read Python manual, and discuss how the language can be used for Web-based programming involving XML files.

FURTHER READING

Bansal, Arvind K. and Bork, Peer. "Applying logic programming to derive novel functional information in microbial genomes." In *Proceedings of the First International Workshop on Practical Aspects of Declarative Languages*, San Antonio, TX. *LNAI 1551*. New York, NY: Springer-Verlag. 1999. 274–289.

Bolsky, Morris I. and Korn, David G. *The New Kornshell Command and Programming Language*. Englewood Cliffs, NJ: Prentice Hall. 1995.

Bourne, Stephen R. "Unix time-sharing system: The Unix shell." *The Bell Systems Technical Journal*, 57(6). 1978. Available at http://www.alcatel-lucent.com/bstj/vol57-1978/articles/bstj57-6-1971.pdf

Diaz, Daniel. *GNU Prolog: A Native Prolog Compiler with Constraint Solving over Finite Domains*. Version 1.4.1. Revised 2012. Available at available on http://gprolog.univ-paris1.fr/manual/gprolog.pdf

Fogus, Michael and Houser, Chris. *The Joy of Clojure*. Manning Publications Company. Greenwich, CT, USA. 2011.

Ierusalimschy, Roberto, de Figueiredo, Luiz H., and Celes, Waldemar. *Lua Reference Manual*. Available at http://www.lua.org/manual/5.1/

Johnson, Mark J. *A Concise Introduction to Programming in Python*. Boca Raton, FL: Chapman and Hall/CRC Press. 2011.

Korn, David G. "ksh: An extensible high level language." In *Proceedings of the USENIX Very High Language Symposium*, Santa Fe, New Mexico. Berkeley, CA: USENIX Association. 1994. 129–146.

Lewis, Mark C. *Introduction to the Art of Programming Using Scala*. Boca Raton, FL: Chapman and Hall/CRC Press. 2012.

Lutz, Mark. *Learning Python*. 4th edition. Sebastopol, CA: O'Reilly & Associates. 2009.

Odersky, Martin, Spoon, Lex, and Venners, Bill. *Programming in Scala*. Mountain View, CA: Artima Press. 2008.

Thomas, Dave, Fowler, Chad, and Hunts, Andy. *Programming Ruby 1.9: The Pragmatic Programmer's Guide*. 3rd edition. Frisco, TX: Pragmatic Bookshelf. 2009.

Wall, Larry, Christiansen, Tom, and Orwant, Jon. *Programming Perl*. 3rd edition. Sebastopol, CA: O'Reilly & Associates Inc. 2000.

CHAPTER 15

Conclusion and Future of Programming Languages

THIS IS THE TIME to sum up the course and summarize what programming languages are about. We can put all the evolution in programming languages and their implementation in perspective as the application and architecture grew with time. We would also like to visualize the direction of the development of new programming languages.

15.1 EVOLUTION OF PROGRAMMING PARADIGMS AND LANGUAGES

Programming is about automating a process. A process is defined liberally as any complex task in daily life. Programs are made of basic operations, repeated operations, and making choices between actions based on evaluation of a condition. Actions are often grouped together and associated with an identifier, so that group of actions can be invoked from multiple locations. The instructions and data have been classified by their common properties. These classifications are called abstractions. We saw various data abstractions and control abstractions. Data-abstractions are single entities, composite entities, aggregations such as collection, aggregation that can be extended dynamically, and entities that can be created dynamically. Control abstractions are classified as assignment statement, sequence of statements, blocks of statements with some common functionality, and functions—blocks of statement bound to an identifier that can exchange information through the use of parameters.

There are multiple aspects of associating values with an identifier; and their interaction with memory location. How the identifier is mapped to values separates an imperative programming from declarative programming. If an identifier is mapped to a memory location, and memory location is mapped to a value, then it becomes an imperative programming paradigm, where the value of a variable is dynamically destroyed and updated. The destructive update causes loss of history, and it cannot be recovered if the future sequence of statements does not yield a solution. However, it allows reusability of memory locations, which is a huge advantage from the implementation viewpoint because of the smaller size of address space required by large software. However, assign-once property of declarative programming reduces side-effects and allows backtracking to recover the past and search alternate search space for solutions. The disadvantage of declarative programming without mutable data-entities is the excessive use of address space because of the memory explosion for repeated actions. Although part of it can be handled using smart compilers that identify memory locations that can be reused, a recent trend is to allow mutable variables and data-structures along with immutable data-structures like lists and strings.

We saw that languages and programming paradigms kept evolving with the evolution of architecture and better understanding of advantages and disadvantages of other paradigms. After debating the advantages of one paradigm over the other paradigms, programmers and language scientists have embraced good features of other languages and paradigms by (1) developing meta-interfaces that allowed code written in one language to be executed in another language, (2) developing virtual machines such that all languages could be translated to common virtual machine, (3) borrowing abstractions in one paradigm and language constructs in one language to other paradigms and languages, and (4) performing deeper integration of paradigms to develop multiparadigm languages. The problem with the deeper integration is the lack of (1) appropriate high-level constructs that will facilitate deeper-level integration and (2) the lack of efficient compilers. This problem of lack of efficient compilers persisted in declarative languages for quite some time, and later the problem of efficient compilation was also there for web programming and multimedia languages. In recent years, the same problem is there with high-productivity languages for massive parallel computing.

One of the controversial issues in programming languages has been the use of static typing. Although the use of typing makes it cumbersome for programmers to track different type of variables in their minds, especially for large-scale software development, there are many advantages of types that help in making program execution easily maintainable, robust, and efficient. The basis of types has been the sets and set-based operations as sets can easily provide classification based on membership of data elements and the well-defined operations associated with sets. Earlier the notion of types was quite fixed; the functions were tightly coupled to concrete types of data elements. Later, this restriction was relaxed by the introduction of type variables in generic functions. By instantiating the type variables at run time to concrete types using parameter passing, the same generic functions can be used for different types of data-entities. Implementation of polymorphic languages has additional overhead of translation. However, the advantages have been well accepted.

As the software complexity grew, the number of identifiers—variables, constants, function names, etc—and the number of statements in software also grew from a few thousand to a few millions of lines. Libraries of code have to be managed and reused for better software maintenance. This gave rise to the notion of modularity and regulating visibility of the identifiers. Visibility of declared entities were controlled using many techniques: (1) providing the scope rule with block, function, and procedures; (2) using module boundaries; (3) using the object boundaries; and (4) using the class boundaries. The idea is to avoid name conflict during the development of large programs. Modularity was also needed to restrict the software library only to the part needed to solve a problem.

Many distributed computing languages such as Emerald use flat structure of objects for the ease of object-migration, and to avoid mishaps caused by altered inherited methods stored in the library. However, inheritance also became a powerful means for the use of off-the-shelf libraries and was invariably used in object-oriented programming languages. By grouping data with the methods working on them, information hiding and modular programming became easier, because the scope of the functions was also limited to data-entities within a class definition. This feature of limiting the scope and visibility of both code and data also provided the capability of objects to be mobile. The notion of inheritance has some bottleneck in mobility, because all the class definitions up to the base class have to be transferred to the remote node. However, class library and the packages have made the software development incremental and facilitated software reuse.

The development of architecture and multiple processors being placed either in hardly coupled or loosely coupled configurations started the development of language constructs that will support multiple subtasks concurrently. There were two ways to exploit parallelism: distribute the data on multiple processors or spawn multiple concurrent threads of activities.

Lot of sequential programs had already been developed, and it is very difficult to develop large scale thread-based concurrent programs because of the issues of deadlocks and racing conditions due to potentially incorrect placement of locks on shared resources. A major effort was started to exploit automatic parallelization of programs to handle the above-mentioned two problems. Most noteworthy was the effort to develop parallelizing compilers for Fortran programs. Automatic parallelization efforts led to exploring various types of dependencies because of the data-flow and control-flow. The dependencies cause the program statements to execute in a sequential order. If the dependency is because of data-flow, then it is called data-dependency, and if it is because of control-flow, it is called control-dependency. A dependency graph shows the partial order of execution of the statements that would be executed with sequentiality embedded because of dependencies. The sequentiality caused by control-dependency can be partially removed by unrolling the loops or by concurrently executing all the possibilities of conditional statements in anticipation, and picking up one that matches with the successful conditional execution. The basic idea in data-dependency is to maintain the sequential consistency. Sequential consistency means that the results returned by the concurrent execution should be the same as returned by sequential machines. One of the problems of automatic parallelization is that the packing (serialization) and unpacking (deserialization) cost of transmitting data-entities between processor. This overhead needs to be minimized. Otherwise, gain achieved by concurrent execution is overshadowed by the overhead of packing-unpacking cost. This asks for exploitation of coarse-grain concurrency rather than fine-grain concurrency.

Concurrency-based constructs are based on (1) spawning multiple threads that may interact with each other, (2) multiple noninteracting activities in a task, or (3) asynchronous actions and messages. When multiple threads share variables, sequentiality is introduced. It can be handled using low-level locks or atomic actions. However, the use of locks is a low-level construct, causes sequentiality overhead if not done optimally, and can cause deadlocks if not done properly. If the subtasks are not interacting, they can be handled using cobegin–coend pairs. Multimedia programming languages such as SMIL use this construct to render the media streams concurrently using <par> - </par> construct.

Declarative programming paradigms—functional programming and logic programming paradigm—evolved because of the need of removing the intertwining of control with logic in the programs by pushing control in the language's implementation engine. The idea was to make the program just logic + abstraction, which would improve comprehension and reduce the code size, thus improving the software maintenance. However, declarative programming paradigm developers got caught into maintaining the purity of the notion of variables to be immutable assign-once, and they used dynamic allocation of objects in the heap. Assign-once property caused inefficiency in the execution of program by increasing the overhead of memory allocation, reducing the reusability of memory locations, and replacing more easily comprehended iterative constructs by recursive programming. This change in programming style made the earlier versions of declarative languages quite different and inefficient from traditional programming style apart, and their use remained limited to niche domains such as artificial intelligence. However, many important constructs and concepts came out of declarative programming style that were later embraced by new-generation languages, especially multiparadigm languages. Some of these concepts are (1) use of iterators, where how the objects were accessed from the aggregate data-entity is invisible to the programmer; (2) the notion of first-class objects, where functions could be dynamically created as data and passed as parameters; (3) development of efficient dynamic memory management techniques; (4) polymorphism that was implicit in many declarative languages, and was declared explicitly in ML for the first time. The programming language ML and its contemporary language Hope illustrated the power of polymorphic declaration. Later, dynamic memory-management techniques and polymorphism were applied in the multiparadigm languages and object-oriented languages successfully.

In the 1990s, as the Internet started growing and became ubiquitous, code and data-mobility became a reality and became a necessity for resource sharing over the web. More and more multimedia was transmitted over the Internet, and with the development of massive storage technology, large-scale multimedia archival and rendering on a computer became a reality. This gave rise to the growth of the web and multimedia-based technology where code and data including multimedia could migrate. At the same time, desktop and laptop technology also saw tremendous growth. To enhance user-friendliness, graphical interfaces were developed. With the growth of graphical interface, the visual programming paradigm and event-based programming paradigm were developed. Visual programming paradigm uses icons to represent a complex object. Visual drag-and-drop technique along with mouse motion and mouse-clicking has been used to access utilities easily compared to the use of commands. The use of signal-based programming has given rise to a new programming style called event-based programming.

In an event-based programming, there is an event-source that keeps generating signals. The application program creates an event-listener that dispatches events to an event-handler. The event-handler, written by a programmer, performs an action based on the state of the system and the event-signal. Many times, just the change in state is sufficient to initiate an action.

The state of the system may change in response to an action taken by the event-handler. Event-based programming is preferred over procedural input-based programming for interaction with the real world because every event is registered, and action is taken as a direct response to an asynchronous event. Many languages support event-based programming. Java, Javascript, and C# are some popular languages for the development of event-based programming.

HTML is the first popular web-programming language. It has the capability of visualization, formatting data using fixed tags, client-server interaction through the use of forms, and hyperlinks to jump to other websites. XML extended HTML by incorporating user-defined tags that allowed flat representation of complex structure. For example, complex images and database can be written using flat structure of XML. Thus, XML became an interface language to transfer data between different heterogenous applications. XML has also been used to translate high-level domain specific web-based languages such as MathML and ChemML, and for transmitting movies and video clips. By embedding the event-triggers in XML and calling script written in event-based languages, web-programming and event-based programming have been integrated. By allowing scripting language programs to be called by XML, XML enjoys computational capability.

With the ubiquity of the Internet, code and data-mobility resulted into migration of objects from one node to another node. This concept was further extended to integrating event-sensing reactive autonomous objects that could migrate from node to node. These objects are called agents, and a new style of agent-based programming has been developed around the concept of data and code mobility. Many agent-based programming languages have been developed on top of Java language. By incorporating knowledge bases and intelligent reasoning, intelligent agents can use their own knowledge-base to infer the situation and take an autonomous action. Many intelligent agent languages have been developed around the BDI (belief, desire and intention) model. Beliefs and desires are also part of knowledge bases, and each agent's knowledge base has a library of plans that guide the agent to react in case of a sensed value and the context. Many intelligent agent-based languages integrate logic programming paradigm for the knowledge bases and object-oriented programming paradigm for object mobility.

After living with different paradigms separately for such a long time, language scientists and programmers have started realizing that a single paradigm is not sufficient for large-scale programming. Multiple-programming paradigms are being integrated into modern-day languages either as add-on libraries or by developing a hybrid paradigm such as integration of functional programming and object-oriented programming in Ruby and Scala or integration of narrowing and unification in the languages integrating functional and logic programming.

As the packing of multiple processors within a computer increased, it became clear that languages developed for uniprocessors are unsuitable for large-scale cluster of computers because of many reasons: (1) the languages used low-level constructs for exploiting concurrency, (2) there was significant overhead of the code and data migration in distributed memory models, (3) there was no support for reusability and software maintainability. This led to the new memory models and new abstractions for high-productivity massive parallel languages that could easily map on massive parallel computers based on cluster-based configuration.

Recently many languages have been proposed for high-productivity massive parallel computing. These languages support different constructs that specify a region of cluster where computation takes place, distributing a collection of data item on different regions of cluster for better data-parallelism, and dividing global memory space into multiple partitions to exploit both data-parallelism and task parallelism. A new memory model PGAS (partitioned global address space) has been used by these languages that supports local computations as well as the capability to asynchronously spawned remote computation. However, whether the mapping of the data-structures should be user-defined or automatically done by the compiler is still a controversial issue. Most of the programmers are not experts in understanding what would be the best way to map a solution, and they do not understand the intricate relationship between data representation and its effect on efficient program execution. This controversy may continue for quite some time. Some of the new languages such as X10 and Chapel are efforts in this direction. Both these languages integrate object-oriented programming for large-scale reusable software development and support constructs that facilitate mapping arrays on PGAS to exploit parallelism. The evolution will keep growing in this direction as more supercomputers become part of global networks. Consequently, web programming and event-based programming will also be integrated with such languages.

With this pace of technological development, we will have current-day supercomputers on our desktops somewhere in the next decade. Without the development of high-productivity languages that are easy to program and maintain, large-scale software development will not be possible. The major issues in the development of high productivity languages are (1) mapping programs automatically to different regions of cluster of computers for efficient execution of programs, (2) developing user-friendly techniques for program development so that programmers do not have a conception-disconnect with the use of new constructs, and (3) reducing the communication overhead of code and data transfer between various processors and memory in any architecture used for supercomputers. At present there are different topologies of interconnection of processors in supercomputers, and there is no standardization. Besides, different classes of problems are better suited for certain class of architectures. Thus, the idea of coming up with a universal high-productivity language for massive parallel computers is quite a challenging task.

15.2 EVOLUTION OF IMPLEMENTATION MODELS AND COMPILERS

Now we turn our attention to the evolution of the implementation models and the development of efficient compilers that supported the efficient execution of these languages. Without such development, many important classes of languages will die. After all, our aim is to execute the software efficiently, as the purpose of any software development is to efficiently solve a real world problem.

With the change in the language paradigms, compiler and implementation technology also evolved. This evolution facilitated the development of future languages as language designers realized the possibility of efficient implementation of various adopted constructs from other programming paradigms. The growth in hardware and architecture has also played a major role in the efficient implementation of compilers and execution of low-level code. The compiler technology and the implementation engines of today would not have been conceivable with hardware available two decades ago.

The first implementation was static allocation of memory, which has the advantages of direct memory access. However, it lacked memory reuse and dynamic allocation of memory. This restriction limited the development of the programming constructs that supported recursive programming and the use of recursive data-structures and dynamic objects. Stack-based allocation quickly followed, which allowed recursive procedure and memory reuse: the whole frame of a procedure could be reclaimed after the called procedure terminated. For the recursive data-structure and dynamic objects, there was still no concept of dynamic memory management. However, many languages used operating system–allocated memory in a heap that was not recycled. C was such a language that was very popular in the late 1970s. Around the same time, dynamic memory management and the garbage collection idea was developed for functional programming languages such as Lisp. The concept of dynamic memory management revolutionized (1) the development of data-abstractions that could be extended at run time; (2) the concept of dynamic allocation and reclaim of memory allocation of objects in a heap; and (3) the implementation of polymorphism, because objects could be created and kept in the heap. As garbage collection techniques improved, the overhead of dynamic memory management reduced. The implementation model of object-oriented programming used extensively the concept of heap and extended the concept of object for runtime access of the code area using the references from the objects. Many languages—such as C++, Java, and C#; functional programming languages; logic programming languages; dynamically typed languages, including multiple web programming languages; and multiple paradigm languages, such as Ruby and Scala—heavily use the concept of heap and dynamic memory management. Logic and functional programming languages used more than one stack for the implementation of the concept. Prolog uses a trail-stack to backtrack in case the exploration of search fails in a branch. The development of low-level abstract machines for functional and logic programming languages improved their execution efficiency significantly to make them mainstream languages instead of fancy prototype languages during 1980s.

The development of the distributed computing paradigm started the concept of serialization and deserialization and interplay of operating systems to connect to remote ports for communication with remote processes. With the development of technology to transparently connect to remote ports, a whole new development of web-based languages became possible: hyperlink-based visualization became possible. Event-based programming started the concept of event-handler instead of sequential thread of execution in procedural languages. Web-based programming showed the importance of virtual machines to handle the problem of heterogeneity. The use of just-in-time compiler improved the execution efficiency lost because of zero address virtual machines.

Owing to this improvement in implementation models, we have many truly multiparadigm languages that integrate multiple paradigms such as functional and object-oriented languages, the concept of mutable and immutable objects, and the integration of unification (concept first proposed for logic programming) with narrowing (concept first proposed with functional programming). Polymorphism and object-oriented programming have become accepted features in most of the modern-day languages and are being extended to new multiparadigm languages for massive parallel supercomputers. C# integrates event-based programming, object-oriented programming, multithread-based programming, imperative programming, and web-based programming.

15.3 CONSTRUCT DESIGN AND COMPREHENSION

One of the important concepts in language development is to design concepts that are comprehensible and easy to use. Another issue is to communicate the meaning of these constructs to the programmers for formal design of compilers and to explain the meaning to the programmers. In the procedure-oriented model for a uniprocessor machine, the control flow was straightforward, except for the unconditional jump statements. To explain procedural languages constructs, three formal notions of semantics—operational, axiomatic, and denotational semantics—were developed. Although these semantics are reasonably well defined for imperative uniprocessor programming languages based on state transitions, they were not good for (1) explaining concurrency constructs, (2) constructs in object-oriented languages, (3) event-based programming languages, and (4) domain-specific languages that are dependent on state transition of the system. Later behavioral semantics was introduced for state transition of the systems. However, they are still evolving.

15.4 FUTURE DEVELOPMENT OF PROGRAMMING LANGUAGES

Supercomputers currently are in the range of petascale computing (10^{15} operations per second), and the laptop computations are in gigascale computing (10^9 operations per second). The massive supercomputers of today are being built using clusters of computers. As the technology evolves, and today's laptops move to terascale computing and massive parallel supercomputers move to exascale computing, the need for concurrent multithreaded computation will be felt more. Number of spawned threads and their management will become a serious issue. However, as our experience shows, humans are not good in programming low-level programming constructs, as they cannot comprehend enhanced complexity caused by low-level instructions. Thus new high-level concurrency constructs that automatically handle low-level locks, and mapping of computations on processors have to be devised. Program and data segments will be mapped to graphs, and these graphs will be automatically mapped to different regions of clusters of processors.

For large-scale software development, object-oriented programming will be integrated with graph-based concurrent programming. However, as the supercomputers come to desktop and laptops, the languages will need to acquire interactive capability provided by event-based programming and web programming in addition to current effort to integrate object-oriented programming and concurrent programming. More and more languages will have a multiparadigm approach. We already see this direction in languages such as C#, Java, Ruby, and Scala, and to a limited extent on massive parallel computing languages, such as X10 and Chapel. Table 15.1 shows the characteristics of various programming paradigms, and Figure 15.1 illustrates a possible paradigm distribution from high-end computing to more interactive low-end commercial computing.

TABLE 15.1 Useful Characteristics of Popular Programming Paradigms

Paradigms	Characteristics
Procedure-driven imperative	Memory reuse, input driven
Class inheritance/object oriented	Code reusability, code-migration, large-scale software development
Concurrency	Efficient execution, multiple interacting subtasks
Event driven	Interaction, reactive systems
Web programming and multimedia	Better perception and resource sharing
Declarative languages	Better comprehensibility, free from low-level control
Agent based	Autonomous, intelligence, and reactive

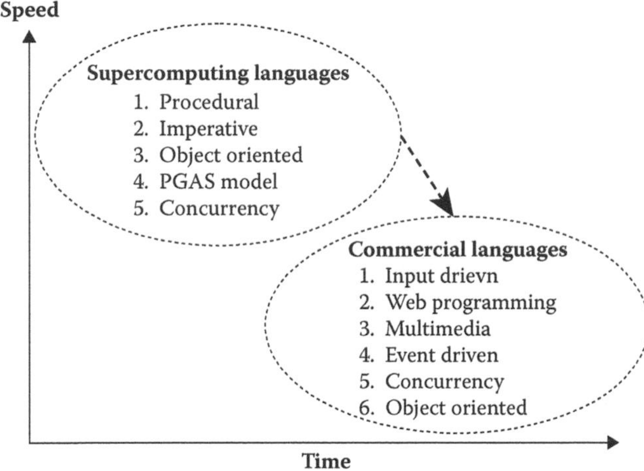

FIGURE 15.1 Future paradigm-integration with time.

At the supercomputing end, there is more need to develop large-scale software for input-driven computation. At the personal computing end, there is a need for interaction, web sharing of resources, and graphic visualization. Because parallel computers will be in both the supercomputing end and at the personal computing end in the future, concurrent programming will be a needed paradigm in languages. Future commercial languages will integrate all of the above programming paradigms with more emphasis on the integration of event driven, web programming, object-oriented, and concurrent programming paradigms. The languages will support both mutable and immutable objects, and will support a declarative programming paradigm mixed with traditional imperative programming constructs. The trend has already started with multiprogramming paradigms, such as in C#, F#, Ruby, and Scala.

Figure 15.1 shows that as the time progresses, supercomputing language constructs will be further augmented with event driven and web programming paradigms, and the supercomputing languages and construct will transform to commercial computing languages. The supercomputing speed will keep changing.

Already effort has started on the development of languages for "exascale" computing. The language designers envision a graph of tasks to be modeled in a language supported by PGAS. The language will support component reuse, and components will use XML-like component description at low levels to handle heterogeneity. It will support asynchronous iteration and multilevel parallelism.

One effort that has been tried for many years is the development of a more-algorithmic language, where low-level data-abstractions and constructs will be hidden from the programmer. Programmers will specify high-level constructs and data-abstractions that would be automatically translated to low-level constructs by using program translators that will use off-the-shelf libraries of software. As the software complexity increases, such high-level specification and the use of already developed libraries will progressively increase.

Another effort is to model a domain using state diagrams and to develop languages with instructions that perform a specific transition from one state to another. For example, we can have a language for automated automobile control or for automaton control. Already there are many domain-specific languages. One of the advantages of domain-specific language is that experts in the domain can program in this language.

FURTHER READING

Barnett, Mike, Rustan, K., Leino, M., and Schulte, Wolfram. "The Spec# programming system: An overview." *Lecture Notes in Computer Science*. Springer-Verlag, 3362. 2005. 49–69.

Deursen, Arie van and Klint, Paul. "Domain specific language design require feature specification." *Journal of Computing and Information Technology—CIT 10*, 1. 2002. 1–17.

Mernik, Marjan, Heering, Jan, and Sloan, Anthony M. "When and how to develop domain-specific languages." *ACM Computing Surveys*, 4(37). 2005. 316–344.

Roman, Gruia-Catalin, Pico, Gian P., and Murphy, Amy L. "Software engineering and mobility: A roadmap." In *The Future of Software Engineering*. Editor: A. Finkelstein. New York, NY: ACM Press. 2000. pp. 241–258.

Sheard, Tim. "Languages of the future." In *Proceedings of the ACM Conference on Object-Oriented Programming Systems, Languages and Applications - OOPSLA 2004,* Vancouver, BC, Canada. New York, NY: ACM Press. 2004. 116–119.

APPENDIX I

Supported Paradigms in Languages

C^T—concurrent thread of task based; **C^P**—data-parallelism; **C^D**—distributed;

C^S—synchronous concurrency; **E**—event based; **F**—functional; **I**—imperative;

L—logic; **M**—multimedia; **O^I**—object-oriented with class inheritance;

O^F—object oriented with flat objects; **V**—visual; **W**—web.

Note: Some of the languages have no clear standardized version. Some of the dates are approximate, based upon latest compiler development.

Language	Revision	Paradigms
Algol W	1996	**I**
ALICE	2000	**I, E, M, V**, educational multimedia language
Ada	2012	**I, C^T, O^I**
C	1999	**I, C^T** (library), **C^P** (variations)
C++	2011	**I, C^T** (library), **O^I, V** (variations)
C#	2010	**I, E, C^T, M, O^I, V, W** (with .NET)
Chapel	Evolving	**I, C^P, O^I** (for massive parallel computing)
Claire	2009	**I, F, L, O^I, W**
Clojure	2011	**F, C^T**, scripting language
COBOL	2002	**I, O^I**, business programming language
ECL^{i}PS^e	1997	**L**, constraint logic programming language
Emerald	1994	**I, C^D, O^F**, distributed computing language
Estrel	1991	**I, C^S**, synchronous hardware modeling language
F#	2005	**I, F, O^I, E, V, W**
FORTRAN	2008	**I, O^I, C^P** (variations such as HPF)

(Continued)

Language	Revision	Paradigms
Haskell	2010	**F, C^{T}**
Java	Continuous	**I, E, O^{I}, C^{T}, M, V, W,**
Javascript	1995	**I, E, O^{F}, W**, web scripting language
Lisp family	1994	**I** (limited), **F** (main), **O^{I}** (variations), **C** (limited)
Lua	2006	**I, F, O, C^{T}, W** (game-scripting language)
ML	1998	**I** (limited support), **F** (main)
Modula-3	1991	**I, O^{I}, C^{T}**
Oz/Mozart	2007	**I, F, L, O^{I}, C^{T}**
Perl	Various	**I, F** (limited), **O^{I}**, scripting language
PhP	2004	**I, O^{I}, W**, web scripting language
Prolog	Various	**L, O^{I}** (library), **V** (variations), **C^{T}** (variations)
Python	Continuous	**I, F, O^{I}**, regular language use for scripting
Ruby	2012	**I, F, O^{I}**, integrated multiparadigm
Scala/EScala	2012	**I, F, O^{I}, C^{T}**, integrated multiparadigm
SMIL	2008	**I, M, W** (for web computing)
VRML/X3D	1996/2005	**I, M, W**, 3D modeling languages
X10	Evolving	**I, C^{P}, O^{I}**, for massive parallel computing
XML/HTML based	Various	**M, V, W**, for web-based computing

APPENDIX II

Data Abstractions Summary

1. Mutable data entity
2. Immutable data entity
3. Independent explicit pointers/reference
4. Extra precision basic types
5. Strings
6. Arrays/indexed sequences
7. Hash table/maps/key-value table
8. Named tuples/record/struct
9. Ordinal type
10. Sets
11. Recursive data types/lists
12. Universal polymorphism/subtype
13. Objects
14. Class and inheritance (single or multiple inheritance)
15. Modules/package/namespace

▪—built-in/built-in library; **A**—aliases used as references; **Δ**—limited; **D**—dynamically typed; **L**—through library; **M**—media objects; **R**—reference instead of pointer; **ρ**—range, **S**—simulated by other data structure; **V**—variations of the language support; **X**—absent

Languages	1	2	3	4	5	6	7	8	9	10	11	12	13	14	15
ALICE	▪	X	X	X	▪	▪	X	X	X	X	▪	X	M	▪	X
Ada 2005	▪	Δ	▪	▪	▪	▪	▪	▪	▪	▪	▪	▪	▪	▪	▪
C	▪	X	▪	▪	▪	▪	X	▪	▪	S	▪	X	X	X	▪
C++	▪	X	▪	▪	▪	▪	▪	▪	▪	▪	▪	▪	▪	▪	▪
C#	▪	X	▪	Δ	▪	▪	▪	▪	▪	S	▪	▪	▪	▪	W
Chapel	▪	X	X	▪	▪	▪	▪	▪	▪	▪	X	▪	▪	▪	▪
Claire	▪	▪	X	X	▪	▪	X	▪	ρ	▪	▪	▪	▪	▪	▪
Clojure	L	▪	R	L	▪	▪	▪	▪	ρ	▪	▪	▪	▪	▪	▪
ECLiPSe	X	▪	▪	D	▪	▪	▪	▪	S	▪	▪	▪	▪	▪	▪
Emerald	▪	▪	X	X	▪	▪	X	▪	▪	▪	▪	▪	▪	X	▪
Estrel	▪	X	X	X	X	X	X	X	X	X	X	X	X	X	▪
F#	▪	▪	R	▪	▪	▪	▪	▪	▪	▪	▪	▪	▪	▪	▪
Fortran 2008	▪	X	▪	▪	▪	▪	X	X	▪	X	▪	▪	▪	▪	▪
Haskell	X	▪	X	▪	▪	▪	S	▪	▪	▪	▪	▪	▪	▪	▪
Java	▪	X	R	▪	▪	▪	▪	X	▪	▪	S	▪	▪	▪	▪
Javascript	▪	Δ	X	D	▪	▪	▪	X	S	X	X	X	▪	X	X
Lisp	▪	▪	X	D	▪	▪	▪	S	V	S	▪	▪	V	V	X
Lua	▪	▪	X	X	▪	▪	▪	X	S	S	▪	V	▪	S	▪
ML	▪	▪	▪	▪	▪	▪	▪	▪	▪	▪	▪	▪	X	X	▪
Modula-3	▪	X	▪	▪	▪	▪		▪	▪	▪	▪	▪	▪	▪	▪
Perl	▪	X	L	L	▪	▪	▪	S	▪	S	S	▪	▪	▪	▪
PhP	▪	X	X	X	▪	▪	▪	X	X	X	X	L	▪	▪	▪
Python	▪	▪	A	▪	▪	▪	▪	▪	▪	▪	▪	▪	▪	▪	▪
Ruby	▪	▪	R	D	▪	▪	▪	X	ρ	▪	ρ	▪	▪	▪	▪
Scala	▪	▪	X	▪	▪	▪	▪	▪	▪	▪	▪	▪	▪	▪	▪
SMIL	▪	X	X	X	▪	X	X	X	X	X	X	X	M	X	▪
X10	▪	▪	X	▪	▪	▪	X	▪	X	L	▪	▪	▪	▪	▪

APPENDIX III

Control Abstractions Summary

1. Destructive assignment
2. Multiple/parallel assignment
3. If-then-else
4. Case/switch statement
5. For loop/definite iteration
6. While-loop/indefinite loop
7. Do-while (repeat-until) loop
8. Iterators
9. Functions/subprograms
10. Guards
11. Exception handling
12. User-defined threads/tasks
13. Monitors/mutual exclusion using locks/protected objects/synchronized methods
14. Other concurrency constructs
15. Distributed computing

▪—built-in support; **C**—cluster-based computing; D—general loop and exit; **L**—interfaced with library; **L**^{T}—temporal loop; **M**—external message passing library like PVM or MPI; **N**—not part of current specification; **R**—simulated using tail-recursion; **S**—can be simulated using existing constructs; **T**—as a trait; **U**—until-loop; **V**—language variations support; **W**—web programming support; **X**—not supported in standards

	Control Abstractions														
Language	**1**	**2**	**3**	**4**	**5**	**6**	**7**	**8**	**9**	**10**	**11**	**12**	**13**	**14**	**15**
ALICE	▪	X	▪	X	▪	▪	X	X	▪	X	X	▪	X	▪	X
Ada	▪	▪	▪	▪	▪	▪	▪	▪	▪	▪	▪	▪	▪	▪	V
C	▪	X	▪	▪	▪	▪	▪	X	▪	X	X	L	L	V	M
C++	▪	▪	▪	▪	▪	▪	▪	▪	▪	X	▪	L	L	X	X
C#	▪	▪	▪	▪	▪	▪	▪	▪	▪	▪	▪	▪	▪	▪	▪
Chapel	▪	X	▪	▪	▪	▪	▪	▪	▪	X	N	▪	▪	▪	C
Claire	▪	X	▪	▪	▪	▪	U	▪	▪	X	▪	X	X	▪	X
Clojure	▪	▪	▪	▪	▪	▪	X	▪	▪	X	▪	▪	▪	▪	▪
ECL^{i}Ps^{e}	▪	▪	▪	X	▪	R	R	▪	▪	X	Δ	Δ	▪	X	▪
Emerald	▪	X	▪	X	▪	▪	D	X	▪	X	X	▪	▪	X	▪
Estrel	▪	X	▪	▪	▪	▪	▪	X	▪	X	▪	X	X	▪	X
F# 3.0	▪	X	▪	▪	▪	▪		▪	▪	X	▪	▪	▪	▪	W
Fortran 2008	▪	X	▪	▪	D	D	D	X	▪	X	Δ	V	V	▪	M
Haskell	X	X	▪	▪	▪	S	S	S	▪	▪	▪	▪	▪	▪	▪
Java	▪	Δ	▪	▪	▪	▪	▪	▪	▪	X	▪	▪	▪	X	▪
Javascript	▪	▪	▪	▪	▪	▪	▪	▪	▪	X	▪	X	X	X	W
Lisp	▪	▪	▪	▪	▪	▪	▪	▪	▪	X	▪	X	X	▪	X
Lua	▪	▪	▪	X	▪	▪	▪	▪	▪	X	▪	L	L	▪	W
ML	▪	▪	▪	▪	S	▪	X	▪	▪	V	▪	▪	▪	▪	▪
Modula-3	▪	X	▪	▪	▪	▪	▪	X	▪	X	▪	▪	▪	X	▪
Perl	▪	▪	▪	▪	▪	▪	▪	▪	▪	X	▪	▪	X	▪	V
PhP	▪	▪	▪	▪	▪	▪	▪	▪	▪	X	▪	X	X	X	W
Python	▪	▪	▪	▪	▪	▪	X	▪	▪	X	▪	▪	▪	X	▪
Ruby	▪	▪	▪	▪	▪	▪	U	▪	▪	X	▪	▪	▪	X	▪
Scala	▪	▪	▪	▪	▪	▪	X	T	▪	▪	▪	▪	▪	▪	W
SMIL	▪	X	X	X	▪	X	X	X	X	X	X	X	X	▪	W
X10	▪	X	▪	▪	▪	▪	▪	X	▪	L	▪	▪	▪	▪	C

APPENDIX IV

Websites for Languages

Language	Websites
Alice	http://www.alice.org/
Ada	http://www.sigada.org/education/ and http://www.ada-auth.org/arm.html
C	Multiple available on the Internet
C++	Multiple available on the Internet including Microsoft Visual Studio
C#	Multiple available on the Internet including Microsoft Visual Studio
Chapel	http://chapel.cray.com/
Claire	http://www.claire-language.com/
Clojure	http://clojure.org/
ECL^{i}PS^e	http://www.eclipseclp.org/
Emerald	http://www.emeraldprogramminglanguage.org
Esterel	http://www-sop.inria.fr/meije/esterel/esterel-eng.html
F#	http://research.microsoft.com/en-us/um/cambridge/projects/fsharp/ http://msdn.microsoft.com/en-us/library/dd233154.aspx
FORTRAN	http://j3-fortran.org/
Haskell	http://www.haskell.org
Java	http://www.java.com and http://www.oracle.com/us/technologies/java/overview/index.html
Javascript	http://msdn.microsoft.com/en-us/library/ie/d1et7k7c(v = vs.94).aspx
Lisp	Multiple including GNU Lisp site
Lua	http://www.lua.org
ML	http://www.smlnj.org
Modula-3	http://www.modula3.org
Perl	Multiple available on the Internet
PhP	http://www.php.net/
Prolog	http://www.sics.se/software; http://www.gprolog.org; http://www.swi-Prolog.org
Python	http://www.python.org/
Ruby	http://www.ruby-lang.org
Scala	http://www.scala-lang.org
SMIL	http://www.w3.org/AudioVideo/ and http://www.w3.org/TR/2008/REC-SMIL3-20081201/
X10	http://x10-lang.org/

APPENDIX V

Principle of Locality

A program executes within a small subset of the local environment during a small time interval. This currently active part of the local environment is called *locality*. The *principle of locality* states that a program will use the same locality in the near future. The locality changes with the control flow: (1) processes different parts of a large data structure, (2) moves to different parts of a large program, or (3) calls another subprogram.

Principle of locality is important for the efficient execution and better space utilization of the executing programs, because modern operating systems execute multiple processes—active part of the program at the same time—and allocate a limited amount of RAM to each active process. Owing to the space limitation in the RAM, only a small part of the program is brought into the active memory, and this active memory needs to be managed continuously to avoid inactive memory blocks of a large data structure. As the locality of execution changes, new parts of the data space are brought into allocated user space, causing overhead of bringing information from slower secondary storage devices such as hard disks.

APPENDIX VI

Virtual Memory and Page-Faults

A programmer writes a program in a logical space assuming that all the data space and code space are contiguous. However, in reality, at low level, program and data are stored in RAM and hard disks: the active part of a program is stored in RAM, while the remaining parts of the program and data files are stored in the hard disk. There are two memory spaces: the logically contiguous programmer's view of the memory space called *virtual memory*, and the interspersed physical memory space, where actual program and data are stored. Virtual memory is mapped into physical memory by the operating system. A memory block could be either variable-sized or fixed-sized, depending upon the operating system support. Variable-sized blocks are called *segments*, and have the size of a called subprogram. Fixed-sized blocks are called *pages*.

A page (or segment) is brought from secondary storage to RAM if the address being accessed by the RAM in virtual memory is missing from the RAM. The process of bringing missing pages from hard disks to the RAM is called *page-fault*, and has significant data transfer overhead. As the number of page-faults increases, the actual CPU utilization decreases. The process scheduler brings more processes to improve the CPU utilization. The execution of more processes causes more page-faults and even lesser CPU utilization. Eventually, the CPU stops doing any useful work. This phenomenon is called *thrashing*, and it must be avoided.

APPENDIX VII

Program Correctness and Completeness

A program is *correct* if the set of the generated solutions is a subset of all possible solutions available for the problem. A program is *complete* if the set of all possible solutions to the problem is included in the set of generated solutions. There is a possibility of incorrect solution in a program that cares only for the *completeness* of the program. Similarly, there is a possibility of a program being incomplete if the program tries only correct solutions. We have to look for both correctness and completeness of solutions when we write a program.

In real life, it is difficult to write a large software that is both correct and complete. Most of the large programs are neither correct nor complete. A software has many logical errors called *software bugs*. The number of these bugs increases significantly with the increase in the size of a software.

There are multiple ways to remove software bugs: (1) the programmer goes through the logic over and over again, (2) the programmer runs the program with different input data samples using debuggers to locate and fix the bugs every time the program does not return the desired output value, and (3) using automated program analysis tools to derive program correctness. Techniques 1 and 2 are quite common. However, technique 3 is uncommon for large software due to the involved computational complexity it requires.

APPENDIX VIII

Complexity of Algorithms

During the translation of higher-level constructs to lower-level intermediate code and during the execution of programs, the issue of program efficiency has to be addressed. Efficiency of programs is affected by many factors such as: (1) layout of data structures; (2) frequency of retrieval of data stored in hard disk; (3) the absence of efficient algorithms; and (4) data transfer overhead, especially when employing multiple processors to solve a complex task.

One way to study execution efficiency is to examine whether underlying algorithms work efficiently as the size of input data is increased. For example, in a linear complexity algorithm, execution time increases proportionately to the increase in the input data size. An algorithm has *quadratic complexity* if the time taken to execute a program increases by a factor of $O(m^2)$ if the size of input data is increased by a factor of m. In *cubic* algorithms, execution time increases by $O(m^3)$ when the size of input data is incremented by m times. Beyond cubic algorithms, algorithms do not scale well to large data size, and quickly become too slow to execute. Such algorithms, where the power coefficients are limited by a constant as the upper bound, are called *polynomial algorithms*, and in practice the power-coefficient-value > 4 are not considered good solutions. The classes of efficient algorithms are *constant complexity, log(log(N)) complexity, log(N) complexity,* and *linear complexity (O(N)),* since the rate of increase of execution time is less than or equal to increase in the input data size. Some of the examples of *log(N)* complexity are binary-search algorithms, and examples of *constant time* algorithms are hashing techniques.

An interesting class of algorithms are *exponential algorithms*, where the time increases exponentially with the increase in the input data size. As the input data size is increased, exponential algorithms quickly reach a level where execution time is too large to execute on any existing computer. If we do not know whether a given problem has a known polynomial complexity solution and a class of such NP problems can be transformed to this problem, then the problem is called *NP-complete* (nondeterministically polynomially complete). In the implementation of programming languages, we want to avoid

NP-complete problems, since their currently known implementations are inefficient. Currently, NP-complete algorithms use approximation, heuristics (intelligent guessing based on mathematical functions or known solutions in the past), and restricting the inputs and parameters to improve the execution efficiency of the algorithms.

The issue of execution-time efficiency is quite complex and cannot be just modeled by the study of algorithmic complexity. There are other issues to consider: (1) the overhead of fetching data from hard drives into RAM, (2) the use of the principle of locality, (3) bus congestions, and (4) the overhead of data transfers. These issues are discussed in a course of *Operating Systems*.

Additional References

Abrahams, David and Gurtvoy, Alesky, *C++ Template Metaprogramming: Concepts, Tools, and Techniques from Boost and Beyond*, Addison-Wesley/Pearson Education, Stoughton, MA, 2009.

ACM Computing Surveys: Special Issue on Programming Language Paradigms, 21(3), 1989.

Ada 2012 Language Reference Manual, ISO/IEC 8652:2007(E) edition 3, 2012, available at http://www.ada-auth.org/standards/ada12.html. Accessed on October 5, 2013.

Agesen, Ole and Hölzle, Urs, "Type Feedback vs. Concrete Type Inference: A Comparison of Optimizing Techniques for Object Oriented Languages," in *Proceedings of the Tenth ACM SIGPLAN Conference on Object-oriented Programming Systems, Languages, and Applications*, 1995, pp. 91–107.

Allen, Randy and Kennedy, Ken, *Optimizing Compilers for Modern Architectures: A Dependence Based Approach*, Morgan-Kaufman/Academic Press, San Francisco, CA, 2002.

American National Standard Institute, *Information Technology—Programming Languages—C, ANSI X3.9899*, 2011, available at http://www.iso.org. Accessed on October 10, 2013.

American National Standard Institute, *Information Technology—Programming Languages—LISP, ISO/IEC 9899*, 2011, http://webstore.ansi.org/RecordDetail.aspx?sku=ANSI+INCITS+226–1994+(R2004), last accessed on October 10, 2013.

Andrews, Gregory R., "Paradigms for process interaction in distributed programs," *ACM Computing Surveys*, 23(1), 1991, 49–90.

Andrews, Gregory R. and Schneider, Fred B., "Concepts and notations of concurrent programming," *ACM Computing Surveys*, 15(1), 1983, 3–43.

Appel, Andrew W., *Modern Compiler Implementation in Java*, 2nd edition, Cambridge University Press, Cambridge, UK, 2004.

Appel, Andrew W., "Semantics Directed Code Generation," in *Proceedings of the 12th POPL Conference*, New Orleans, LA, 1985, pp. 315–324.

Armstrong, Joe, "History of Erlang," in HOPL III: *Proceedings of the Third ACM SIGPLAN Conference on History of Programming Languages*, San Diego, CA, 2007, pp. 6-1–6-26.

Arnold, Ken, Gosling, James, and Homes, David, *The Java Programming Language*, 4th edition, Addison-Wesley Professional/Pearson, Stoughton, MA, 2005.

Atkins, Margaret S., "Mutual recursion in Algol 60 using restricted compilers," *Communications of the ACM*, 16(1), 1973, 47–48.

Atkinson, Malcolm P. and Buneman, O. Peter, "Types and persistence in database programming languages," *ACM Computing Surveys*, 19(2), 1987, 105–190.

Attiya, Haggit, Guerraoui, Rachid, Handler, Danny, and Kouznetsov, Peter, "Synchronizing without Locks is Inherently Expensive," in *Proceedings of the Twenty-Fifth Annual ACM Symposium on Principles of Distributed Computing*, Denver, CO, 2006, pp. 300–307.

Aycock, John, "A brief history of just-in-time," *ACM Computing Surveys*, 35(2), 2003, 97–113.

Bacon, David F. and Sweeney, Peter F., "Fast Static Analysis of C++ Virtual Function Calls," in *Proceedings of the Eleventh ACM SIGPLAN Conference on Object-oriented Programming, Systems, Languages and Applications*, Portland, OR, 1996, pp. 324–341.

Baden, Scott, "Berkeley FP User's Manual," in *Ultrix-32 Supplementary Documents*, Volume II (revised), Digital Equipment Corporation, Merrimack, New Haven, CT, 1984, pp. 2.359–2.391.

Banerjee, Utpal, "Dependence Analysis," *Volume 3 of Loop Transformations for Restructuring Compilers*, Kluwer Academic Publishers, Norwell, MA, 1997.

Bansal, Arvind K. and Uddin, Sharif, "Multimodal Triggers for Automated Filtering and Reactivity based on WWW Content," in *Proceedings of the IASTED International Conference of Internet and Multimedia Systems and Applications*, Las Vegas, NV, 2004, pp. 311-316.

Baxter, William and Bauer, Henry R. III, "The Program Dependence Graph and Vectorization," in *Proceedings of the 16th ACM SIGPLAN-SIGACT symposium on Principles of programming languages*, Austin, TX, 1989, pp. 1–11.

Bays, Carter, "A comparison of next-fit, first-fit and best-fit," *Communications of the ACM*, 20(3), 1977, 191–192.

Beech, David, "A structural view of PL/I," *ACM Computing Surveys*, 2(1), 1970, 33–64.

Bishop, Judy, "The effect of data abstraction on loop programming techniques," *IEEE Transactions of Software Engineering*, 16(4), 1990, pp. 389–402.

Black, Andrew P., Hutchinson, Norman C., Jul, Eric, and Levy Henry M., "The development of the Emerald programming language," in *HOPL III: Proceedings of the Third ACM SIGPLAN Conference on History of Programming Languages*, New York, NY, 2007, pp. 11-1–11-51.

Black, Andrew, Hutchinson, Norman, Jul, Eric, Levy, Henry and Carter, Larry, "Distribution and abstract types in Emerald," *IEEE Transactions on Software Engineering, SE*-13(1), 1987, 65–76.

Boehm, Hans-Jurgen and Adve, Sarita V., "Foundations of the C++ Concurrency Memory Model," in Proceedings of the SIGPLAN Conference on Programming Language Design and Implementation, Tucson, AZ, 2008, pp. 68–78.

Borning, Alan H. and Ingalls, Daniel H. H., "Multiple Inheritance in Smalltalk-80," in *Proceedings of the AAAI-82: The National Conference on Artificial Intelligence*, Pittsburgh, PA, 1982, pp. 234–237.

Brown, Deryck F. Moura, Harmono and Watt, David A., "ACTRESS: An Action Semantics Directed Compiler Generator," in *Proceedings of the 4th International Conference on Compiler Construction*, Paderborn, Germany, *LNCS 641*, Springer Verlag, Berlin Heidelberg, Germany, 1992, pp. 95–109.

Buckley, Gael N. and Silbershatz, Avi, "An effective implementation for the generalized input-output construct of CSP," *ACM Transactions of Programming Languages*, 5(2), 1983, 223–235.

Budd, Timothy A., *Multiparadigm Programming in Leda*, Addison-Wesley Longman/Pearson, Reading, MA, 1995.

Buhr, Peter A. and Harji, Ashif S., "Implicit-signal monitors," *ACM Transactions on Programming Languages and Systems*, 27(6), 2005, 1270–1343.

Calliss, Frank W., "A comparison of module constructs in programming languages," *ACM SIGPLAN Notices*, 26(1), 1991, 38–46.

Cardelli, Luca, "The Functional Abstract Machine," *Technical Report TR - 107*, AT&T Bell Labs, 1983.

Cardelli, Luca, "A language with distributed scope," *ACM Transaction of Computing Systems*, 8(1), 1995, 27–59.

Caromel, Denis and Henrio, Ludovic, *A Theory of Distributed Objects: Asynchrony-Mobility-Groups-Components*, Springer Verlag, Berlin Heidelberg, Germany, 2005.

Caseau, Yves and Laburthe, François, *Introduction to Claire Programming Language*, available at http://www.dcs.gla.ac.uk/~pat/cp4/claire/Claire_3.2.pdf. Accessed on October 5, 2013.

Chamberlain, Bradford L., Deitz, Steven J., and Navaroo, Angles, "User - defined Parallel Zippered Iterators in Chapel," in *Proceedings PGAS 2011: Fifth Conference on Partitioned Global Address Space Programming Models*, Galveston Island, TX, 2011, available at http://pgas11.rice.edu/papers/ChamberlainEtAl-Chapel-Iterators-PGAS11.pdf. Accessed on October 5, 2013.

Chandi, Kanianthra M. and Mishra, Jaydev, "The drinking philosopher's problem," *ACM Transactions on Programming Languages and Systems*, 6(4), 1984, 632–646.

Chen, Kai, Porter, Joseph, Sztipanovits, Janos, and Neema, Sandeep, "Compositional specification of behavioral semantics for domain-specific modeling languages," *International Journal of Semantic Computing*, 3(1), 2009, 31–56.

Church, Alonzo, "The Calculi of Lambda Conversion," *Annals of Mathematical Studies 6*, Princeton University Press, Princeton, NJ, 1941.

Clay, Sharon R. and Wilhelms, Jane, "Put: language-based interactive manipulation of objects," *IEEE Computer Graphics and Applications*, 2(16), 1996, 31–39.

Cohen, Jacques, "Garbage collection of the linked data structures," *ACM Computing Surveys*, 13(3), 1981, 341–367.

Cook, Robert P. and LeBlanc, Thoms J., "A symbol table abstraction to implement languages with explicit scope control," *IEEE Transactions of Software Engineering*, SE-9(1), 1983, 8–12.

Cooper, Eric C. and Morriset, J. Gregory, "Adding Threads to Standard ML," *Technical Report CMU-CS-90-186, Department of Computer Science, Carnegie Mellon University*, Pittsburg, PA, 1990.

Courtois, Pierre-Jacques, Heymans, F., and Parnas, David L., "Concurrent control with "readers" and "writers," *Communications of the ACM*, 14(10), 1971, 667–668.

Cunningham, H. Conrad and Church, James C., "Multiparadigm programming in SCALA, Tutorial Presentation," *Journal of Computing Sciences in Colleges*, 24(5), 2009, 99–100.

Dahl, Ole-Johan, Dijkstra, Edsger W., and Hoare, Charles A. R., *Structured Programming*, Academic Press, London, UK, 1972.

Dann, Wanda P., Cooper, Stephen, and Pausch, Randy, *Learning to Program with ALICE*, 3rd edition, Pearson/Prentice Hall, Boston, MA, 2012.

Danvy, Olivier and Filinski, Andrzej, "Abstracting Control," in *Proceedings of the 1990 ACM conference on LISP and functional programming*, Nice, France, 1990.

Demers, F. N. and Malenfant, J., "Reflection in Logic, Functional and Object-Oriented Programming: A Short Comparative Study," in *Proceedings of the IJCAI'95 Workshop on Reflection and Metalevel Architectures and their Applications in AI*, Montréal, Québec, Canada, 1995, pp. 29–38.

DeRemer, Franklin R., "Simple LR(K) grammars," *Communications of the ACM*, 14(7), 1971, 453–460.

Dershem, Herbert L. and Jipping, Michael J., *Programming Languages: Structures and Models*, 2nd edition, PWS Publishing Company, Boston, MA, 1995.

Dice, Dave, Shalev, Ori, and Shavit, Nir, "Transactional Locking II," in *Proceedings of the Twentieth International Conference on Distributed Computing*, 2006, Rhodes Island, Greece, pp. 194–208.

Dijkstra, Edsger W., "Guarded commands, nondeterminacy, and formal derivation of programs," *Communication of the ACM, 18(8)*, 1975, 453–457.

Dolby, Julian, "Automatic Inline Allocation of Objects," in *Proceedings of the SIGPLAN'97 Conference on Programming Language Design and Implementation*, Las Vegas, NV, 1997, pp. 7–17.

Emmerich, Wolfgang, Mascolo, Cecilia, and Finkelstein, Anthony, "Implementing Incremental Code Migration with XML," in *Proceedings of the 22nd International Conference on Software Engineering (ICSE2000)*, Limerick, Ireland, 2000, pp. 397–406.

Feuer, Alan R. and Gehani, Narain, editors, *Comparing and Assessing Programming Languages: Ada, C, Pascal*, Prentice Hall Software Series, Prentice Hall, Englewood Cliffs, NJ, 1984.

Finkel, Raphael A., *Advanced Programming Language Design*, Addison-Wesley, Menlo Park, CA, 1996.

Flanagan, David and Yukihiro Matsumoto, *The Ruby Programming Language*, O'Reilly Media, Sebastopol, CA, 2008.

Friedman, Daniel P., Wand, Mitchell, and Haynes, Christopher T., *Essentials of Programming Languages*, 3rd edition, MIT Press, Cambridge, MA, 2008.

Frigo, Matteo, Leiserson, Charles E., and Randall, Keith H., "The Implementation of the Cilk-5 Multithreaded Language," in *Proceedings of the International Conference on Language Design and Implementation*, 1998, Montreal, Canada, pp. 212–223.

Furr, Michael, Jong-hoon (David), Foster, Jeffrey S., and Hicks, Michael, "The Ruby Intermediate Language," in *Proceedings of the 5th Symposium on Dynamic languages*, Orlando, FL, 2009, pp. 89–98.

Gagnon, Etienne M. and Hendren, Laurie J., "SableVM: A Research Framework for the Efficient Execution of Java Bytecode," in *Proceedings of the Java Virtual Machine Research and Technology Symposium*, Monterey, CA, 2001, pp. 27–40.

Gasiunas, Vaidas, Satabin, Lucas, Mezini, Mira, Núñez, Angel, and Noyé, Jacques, "EScala: Modular Event - Driven Object Interactions in Scala," in *Proceedings of the Tenth International Conference on Aspect - Oriented Software Development*, Porto de Galinhas, Brazil, 2011, pp. 227–240.

Gehani Narain H. and Roome, W. D., *The Concurrent C Programming Language*, Silicon Press, 1989.

Geist, Al, Beguelin, Adam, Dongarra, Jack, Jiang, Weicheng, Mancheck, Robert, and Sunderam, Vidyalingam S., *PVM: Parallel Virtual Machine: A User's Guide and Tutorial for Networked Parallel Computing*, MIT Press, Cambridge, MA, 1994.

Gharachorloo, Kourosh, Adve, Sarita V., Gupta, Anoop, Hennessy, John L., and Hill, Mark D., "Programming for different memory consistency models," *Journal of Parallel and Distributed Computing*, 15(4), 1992, 399–407.

Ghezzi, Carlo and Jazayerri, Mehdi, *Programming Language Concepts*, 2nd edition, John Wiley & Sons, New York, NY, 1987.

Goldberg, Adele and Robson, David, *Smalltalk-80: The Language*, Addison-Wesley, Reading, MA, 1989.

Goldstein, Seth C., Schauser, Klaus E., and Culler, David E., "Lazy threads: implementing a fast parallel call," *Journal of Parallel Distributed Computing*, 37(1), 1996, 5–20.

Goodenough, John B., "Exception handling: issues and a proposed notation," *Communication of the ACM*, 18(12), 1975, 683–696.

Gosling, James, Joy, Bill, Steel, Guy, Bracha, Gilad, and Buckley, Alex, *The Java™ Language Specification, Java SE 7 Edition*, Oracle America Inc., Redwood City, CA, available at http://docs.oracle.com/javase/specs/jls/se7/jls7.pdf. Accessed on October 20, 2013.

Graham, Paul, *ANSI Common-Lisp*, Prentice Hall, Englewood, NJ, 1996.

Griswold, Ralph E., Poage, J. F., Polonsky, I. P., *The SNOBOL4 Programming Language*, 2nd edition, Prentice Hall, Englewood, NJ, 1971.

Grove, David, DeFouw, Greg, Dean, Jefffrey, and Chambers, Craig, "Call Graph Construction in Object-oriented Languages," in *Proceedings of the Twelfth ACM SIGPLAN Conference on Object-oriented Programming Systems, Languages and Applications*, Atlanta, GA, 1997, pp. 108–124.

Guercio, Angela and Bansal, Arvind K., "TANDEM—Transmitting Asynchronous Non Deterministic and Deterministic Events in Multimedia Systems over the Internet," in *Proceedings of the International Conference on Distributed Multimedia Systems*, San Francisco, CA, 2004, pp. 57–62.

Guercio, Angela and Bansal, Arvind K., "Towards a Formal Semantics for Distributed Multimedia Computing," *International Conference on Distributed Multimedia Systems*, Redwood City, CA, 2007, pp. 81–86.

Guercio, Angela, Bansal, Arvind K., and Arndt, Timothy, "Language constructs and synchronization in reactive multimedia systems," *International Transaction on Computers and Software Engineering*, 1(1), 2007, 52–59.

Gunter, Carl A., *Semantics of Programming Languages: Structures and Techniques*, The MIT Press, Cambridge, MA, 1992.

Gupta, Rajiv, "The Fuzzy Barrier: A mechanism for High Speed Synchronization of Processors," in *Proceedings of the Third ASPLOS*, Boston, MA, 1989, pp. 54–63.

Guttag, John V., "Abstract Data Types and the Development of Data Structures," *Communication of the ACM*, 20(6), 1977, 396–404.

Haller, Phillip and Odersky, Martin, "Scala actors: unifying thread-based and event-based programming," *Theoretical Computer Science*, 410, 2006, 202–220.

Hansen, Per Brinch, "Distributed processes: a concurrent programming concept," *Communications of the ACM*, 21(11), 1978, 934–941.

Hejsberg, Anders, Torgersen, Mats, Wiltamuth, Scott, and Golde, Peter, *C# Programming Language*, 4th edition, Addison-Wesley, Westford, MA, 2010.

Hennessey, John L., and Patterson, David A., *Computer Architecture: A Quantitative Approach*, 4th edition, Morgan Kaufmann, San Francisco, CA, 2007.

Herlihy, Maurice P., "Wait-free synchronization," *ACM Transaction of Programming Languages and Systems*, 13(1), 1991, 124–149.

Herlihy, Maurice, Luchangco, Victor, and Moir, Mark, "Flexible Framework for Implementing Software Transactional Memory," in *Proceedings 21th Annual ACM SIGPLAN Conference on Object-oriented Programming, Systems, Languages, and Applications*, Portland, OR, 2006, pp. 253–262.

Herlihy, Maurice P. and Moss, Elliot B., "Transactional Memory: Architectural support for Lock-free Data Structures," in *Proceedings of the 20th Annual International Symposium on Computer Architecture*, San Diego, CA, 1993, pp. 289–300.

Ho, Wilson W. and Olsson, Ronald A., "An approach to genuine dynamic linking," *Software Practice and Experience*, 21(4), 1991, 375–390.

Hoare, Charles A. R., "Hints on Programming Language Design," in *Essays in Computer Science*, edited by Cliff B. Jones, Prentice Hall, Upper Saddle River, NJ, 1989, pp. 193–216.

Hoare, Charles A. R., "Monitors: an operating system structuring concept," *Communications of the ACM*, 17(10), 1974, 549–557.

Hoare, Charles A. R., "Recursive data structures," *International Journal of Computer and Information Sciences*, 4, 1975, 105–132.

Hogen, Guido and Loogen, Rita, "Efficient Organization of Control Structures in Distributed Implementation," in *Distributed Implementation, Compiler Construction, LNCS 786*, Springer Verlag, Berlin, Germany, 1994, pp. 98–112.

Hogger, Christopher J., "*Essentials of Logic Programming*, Oxford Science Publications, Oxford, UK, 1991.

Hopcroft, John E., Motwani, Rajeev, and Ullman, Jeffrey D., *Introduction to Automata Theory, Languages, and Computation*, 3rd edition, Pearson/Addison-Wesley Longman Publishing Co., Inc. Boston, MA, 2008.

Horowitz, Ellis, *Programming Languages: A Grand Tour*, 3rd edition, Computer Software Engineering Series, Computer Science Press, Rockville, MD, 1987.

Horowitz, Susan, Reps, Thomas, and Binkley, David, "Interprocedural slicing using dependence graphs," *ACM Transactions of Programming Languages and Systems*, 12(1), 1990, 26–60.

Ierusalimschy, Roberto, de Figueiredo, Luiz Henrique, and Celes, Waldemar, 'The Evolution of Lua," in *Proceedings of the Third ACM SIGPLAN Conference on History of Programming Languages*, HOPL III, San Diego, CA, 2007, pp. 2-1–2-26.

Ingerman, Peter J., "Thunks: A way of compiling procedure statements with some comments on procedure declarations," *Communications of the ACM*, 4(1), 1961, 55–58.

ISO/IEC 9899:2011 Information technology—Programming languages—C," available on http://www.iso.org/iso/home/store/catalogue_ics/catalogue_detail_ics.htm?csnumber=57853. Accessed on October 10, 2013.

Jafffar, Joxan and Maher, Michael J., "Constraint logic programing: a survey," *Journal of Logic Programming*, 20, 1994, 503–581.

Jang, Myeong-Wuk, Ahmed, Amr, and Agha, Gul, "Efficient Agent Communication in Multiagent System," *LNCS 3390*, edited by R. Choren et al., Springer Verlag, Berlin Heidelberg, Germany, 2005, pp. 236–253.

Jones, Neil D., *Semantics Directed Compiler Generation, LNCS 94*, Springer Verlag, Berlin, Germany, 1980.

Jul, Eric, Levy, Henry M., Hutchinson, Norm, and Black, Andrew, "Fine-grained mobility in the emerald system," *ACM Transactions Computer System*, 6(1), 1988, 109–133.

Kay, Alan, "The Early History of Smalltalk," in *Proceedings ACM History of Programming Languages Conference II*, SIGPLAN Notices, 28(3), 1993, 67–95.

Keene, Sonya E., *Object-oriented Programming in Common Lisp: A Programmer's Guide to CLOS*, Addison-Wesley, Reading, MA, 1989.

Kennedy, Andrew and Syme, Don, "Design and Implementation of Generics for the .NET Common Language Runtime," in *Proceedings of the SIGPLAN Conference on Design and Implementation*, Snowbird, UT, 2001, pp. 1–12.

Kernighan, Brian W. and Ritchie, Dennis M., *The C Programming Language*, 2nd edition, Prentice Hall, Englewood Cliffs, NJ, 1988.

King, Peter, Schmitz, Patrick, and Thompson, Simon, "Behavioral Reactivity and Real Time Programming in XML: Functional Programming Meets SMIL Animation," in *Proceedings of ACM Document Engineering*, Milwaukee, WI, 2004, pp. 57–66.

Kishon, Amir S., Hudak, Paul, and Consel, Charles, "Monitoring Semantics: A Formal Framework for Specifying, Implementing and Reasoning about Execution Monitors," in *Proceedings of International Conference on Programming Language Design and Implementation, ACM Sigplan Notices*, 26(6), 1991, 338–352.

Knuth, Donald E., Morris, James H. and Pratt, Vaughan R., "Fast pattern matching in strings," *SIAM Journal of Computing*, 6(2), 1977, 323–350.

Kogan, Alex and Petrank, Erez, "A Methodology for Creating Fast Wait-Free Data Structures," in *Proceedings of the 17th ACM SIGPLAN Symposium on Principles and Practice of Parallel Programming*, New Orleans, LA, 2012, 141–150.

Kowalski, Robert, "The early history of logic programming," *Communications of the ACM*, 31(1), 1988, 38–43.

Lamport, Leslie, "Time, clocks, and the ordering of the events in a distributed system," Communications of the ACM, 21(7), 1978, 558–565.

Larus, James and Kozyrakis, Christos, "Transactional memory," *Communications of the ACM*, 51(7), 2008, 80–88.

Lerdorf, Rasmos and tatroe, Kevin, *Programming PhP*, O'Reilly & Associates, Sebastopol, CA, 2002.

Li, Kai and Hudak, Paul, "Memory coherence in shared virtual memory systems," *ACM Transactions on Computer Systems*, 7(4), 1989, 321–359.

Liskov, Barbara and Scheifler, R., "Guardians and actions: linguistic support for robust distributed systems," *ACM Transaction on Programming Language and Systems*, 5(3), 1983, 381–404.

Lomet, David B., "Making pointers safe in system programming languages," *IEEE Transactions of Software Engineering*, SE-11(1), 1985, 87–96.

Louden, Kenneth C., *Programming Languages: Principles and Practice*, 2nd edition, PWS-Kent, 2003.

Luck, Michael, Ashri, Ronald, and D'Inverno, Mark, *Agent Based Software Development*, Artech House, Norwood, MA, 2004.

Luckam, David C. and Polak, Wolfgang, "Ada exception handling: an axiomatic approach," *ACM Transactions on Programming Languages and Systems*, 2(2), 1980, 225–233.

Marlow, Simon, *Parallel and Concurrent Programming in Haskell*, 2012, available at http://community.haskell.org/~simonmar/par-tutorial.pdf. Accessed on October 20, 2013.

Mascolo, Cecilia, Picco, Gian P., and Roman, Gruia-Catalin, "A Fine-Grained Model for Code Mobility," in *Proceedings of 7th European Software Engineering Conference (ESEC/FSE 99)*, edited by O. Nierstrasz and M. Lemoine, Toulouse, France, 1999, pp. 39–56.

Meyer, Bertrand, *Touch of Class—Learning to Program Well with Objects and Contracts*, Springer-Verlag, New York, NY, 2009.

Miller, George A. and Johnson-Laird, Phillip N., *Language and Perception*, Harvard University Press, Cambridge, MA, 1976.

Milner, Robin, Tofte, Mads, Harper, Robert, and MacQueen, David, *The Definition of Standard ML–Revised*, MIT Press, Cambridge, MA, 1997.

Milojicic, Dejan S., Douglis, Fred, Paindaveine, Yves, Wheeler, Richard, and Zhou, Songnian, "Process migration," *ACM Computing Surveys*, 32(3), 2000, 241–299.

Mitchell, John C., *Concepts in Programming Languages*, Cambridge Press, Cambridge, UK, 2002.

Mohr, Eric, Kranz, David A., and Halstead, Robert H. Jr., "Lazy task creation: A technique for increasing the granularity of parallel programs," *IEEE Transactions on Parallel and Distributed Systems*, 2(3), 1991, 264–280.

Moss, Chris, *Prolog++: The Power of Object-oriented and Logic Programming*, Addison-Wesley, Wokingham, England, 1994.

Najork, Mark, "Obiq-3D Tutorial and Reference Manual," *Research Report 129, Digital Equipment Corporation*, Systems Research Center, Palo Alto, CA, 1994.

Naur, Peter, "Revised report on the algorithmic language algol 60," *Communication of the ACM*, 6(1), 1963, 1–13.

Neamtiu, Lulian, Foster, Jeffrey S., and Hicks, Michael, "Understanding source code evolution using abstract syntax tree matching," *ACM SIGSOFT Software Engineering Notes*, 30(4), 2005, 1–5.

Nyygard Kristen, and Dahl, Ole-Johan, "The Development of the Simula Languages," in *Proceedings of the History of Programing Languages*, ACM SIGPLAN NOTICES, 13(8), 1978, 439–493.

Odersky, Martin, Altherr, Philippe, Cremet, Vincent, Dragos, Iulian, Dubochet, Gilles, Emir, Burak, McDirmid, Sean, Micheloud, Stéphane, Mihaylov, Nikolay, Schinz, Michel et al., "An Overview of Scala Programming Language," *Technical Report LAMP-REPORT-2006-001, École Polytechnique Fédérale de Lausanne (EPFL)*, Lausanne, Switzerland, 2006, available at http://www.scala-lang.org/docu/files/ScalaOverview.pdf. Accessed on October 20, 2013.

Oppliger, Rolf, "Privacy Enhancing Technologies for the World Wide Web," *Computer communications*, 28, Elsevier, 2005, 1191–1197.

Pagan, Frank G., *A Practical Guide to ALGOL* 68, John Wiley & Sons, London, UK, 1976.

Pagan, Frank G., *Semantics of Programming Languages: A Panoramic Primer*, Prentice Hall, Englewood Cliff, NJ, 1981.

Parr, Terrence J. and Quong, Russell W., "ANTLR: A predicated-LL(K) parser generator," *Software Practice and Experience*, 25(7), 1995, 789–810.

Pausch, Randy, Burnette, Tommy, Capeheart, A.C., Conway, Matthew, Cosgrove, Dennis, DeLine, Rob, Durbin, Jim, Gossweiler, Rich, Koga, Shuichi, and White, Jeff, "Alice: rapid prototyping system for virtual reality," *IEEE Computer Graphics and Applications*, 15(3), 1995, 8–11.

Peyton Jones, Simon L., *Implementation of Functional Programming Languages*, Prentice Hall, Exeter, UK, 1987.

Peyton Jones, Simon L. and Singh, Satnam, "A Tutorial on Parallel and Concurrent Programming in Haskell," in *Proceedings of the 6th International Conference on Advanced Functional Programming*, Springer Verlag, Berlin Heidelberg, Germany, 2009, pp. 267–305.

Pierce, Benjamin C., *Types and Programming Languages*, MIT Press, Cambridge, MA, 2002.

Plainfossé, David and Shapiro, Marc, "A Survey of Distributed Garbage Collection Techniques," in *Proceedings of the International Workshop on Memory Management, LNCS* 986, Springer Verlag, 1995, London, UK, pp. 211–249.

Pohl, Ira, *C++ distilled: A Concise ANSI/ISO Reference and Style Guide*, Addison-Wesley, Reading, MA, 1996.

Pozrikidis, Constantine, *XML in Scientific Computing*, Chapman and Hall/CRC Press, Boca Raton, FL, 2012.

Pratt, Terrence W. and Zelkowitz, Marvin V., *Programming Languages: Design and Implementation*, 4th edition, Prentice Hall, Upper Saddle River, NJ, 2001.

Programming Languages—Ruby, IPA Ruby Standardization WG Draft, 2010, available at http://www.ipa.go.jp/files/000011432.pdf, Accessed on October 20, 2013.

Rajwar, Ravi, Herlihy, Maurice, and Lai, Konrad, "Virtualizing Transactional Memory," in *Proceedings of the 32nd International Symposium on Computer Architecture*, 2005, Madison, WI, pp. 494–505.

Ramesh, S., "A New Efficient Implementation of CSP with Output Guards," in *Proceedings of the Seventh International Conference on Distributed Computing Systems*, Berlin, Germany, 1987, pp. 266–273.

Rao, Anand S. and Georgeff, Michael, "BDI Agents: From Theory to Practice," in *Proceedings of the First International Conference on Multi-Agent Systems*, San Francisco, CA, 1995, pp. 312–319.

Reiser, Martin and Wirth, Niklaus, *Programming in Oberon—Steps Beyond Pascal and Modula*, Addison Wesley, Kent, UK, 1992.

Ritchie, Dennis M., "The development of C language," *ACM History of Programming Language Conference II, SIGPLAN Notices*, 28(3), 1993, 201–208.

Robinson, John A., "A machine-oriented logic based on the resolution principle," *Journal of the ACM*, 12(1), 1965, 23–41.

Rojemo, N., *Garbage Collection and Memory Efficiency in Lazy Functional Languages, Ph.D. thesis*, Department of Computing Science, Chalmers University, Gothenburg, Sweden, 1995.

Roy, Peter V., Haridi, Seif, Brand, Per, Smolka, Gert, Mehl, Michael, and Scheidhauer, Ralf, "Mobile objects in distributed Oz," *ACM Transactions on Programming Languages and Systems*, 19(5), 1997, 804–851.

Ruby, Sam, Thomas, Dave, and Hansson, David H., *Agile Web Development with Ruby 4*, 3rd edition, The Pragmmatic Bookshelf, Frisco, TX, 2013.

Rustan, K., Leino, M., and Nelson, Greg, "Data abstractions and information hiding," *ACM Transactions of Programming Languages and Systems*, 24(5), 2002, 491–533.

Rutishauser, Heinz, *Description of ALGOL 60*, Springer Verlag, New York, NY, 1967.

Sammet, Jean E., "The early history of COBOL," *ACM History of Programming Language Conference*, SIGPLAN Notices, 13(8), 1978, 121–161.

Schoinas, Ioannis, Falsafi, Babak, Lebeck, Alvin R., Reinhardt, Steven K., Larus, James R., and Wood, David A., "Fine-grain Access Control for Distributed Shared Memory," in *Proceedings of the Sixth International Conference on Architectural Support for Programming Languages and Operating Systems*, San Jose, CA, 1994, pp. 297–306.

Schwartz, Jacob T., Dewar, Robert B. K., Dubinsky, Ed, and Schonberg, Edmond, *Programming with Sets: An Introduction to SETL*, Springer Verlag, New York, NY, 1986.

Scott, Dana and Strachey, Christopher, "Towards a Mathematical Semantics for Computer Language," in *Proceedings of Symposium on Computers and Automata*, Polytechnic Institute of Brooklyn Press, 1971, Brooklyn, NY, pp. 19–46.

Scott, Michael, *Programming Language Pragmatics*, 3rd edition, Morgan Kaufman, Burlington, MA, 2009.

Sebesta, Robert W., *Concepts of Programming Languages*, 10th edition, Pearson/Addison Wesley, Upper Saddle River, NJ, 2011.

Sethi, Ravi, *Programming Languages; Concepts and Constructs*, 2nd edition, Addison-Wesley, Redwood City, CA, 1996.

Shapiro, Ehud Y., "The family of concurrent logic programming languages," *ACM Computing Surveys*, 21(3), 1989, 413–510.

Siek, Jeremy G. and Lumsdaine, Andrew, "A language for generic programming in the large," *Journal Science of Computer Programming*, 76(5), 2011, 423–465.

Sites, Richard L., "Algol W Reference Manual," *Technical Report STAN-CS-71-230*, Computer Science Department, Stanford University, Stanford, CA, 1972.

Skillicorn, David B. and Talia, Domenico, "Models and languages for parallel computation," *ACM Computing Surveys*, 30(2), 1998, 123–169.

Smith, Brian C., "Reflection and Semantics in a Procedural Language," *Technical Report 272*, Laboratory for Computer Science, MIT, Cambridge, Massachusetts, 1982.

Smith, Brian C., "Reflection and Semantics in Lisp," in *Proceedings of the 14th Annual ACM Symposium on Principles of Programming Languages*, 1984, Munich, Germany, pp. 23–35.

Soares, Luiz Fernando G., Rodrigues, Rogério Ferreira, Cerqueira, Renato, and Barbosa, Simone Diniz Junqueira, "Variable Handling in Time-Based XML Declarative Languages," in *Proceedings of the 2009 ACM Symposium on Applied Computing*, Honolulu, Hi, 2009, pp. 1821–1828.

Sommerville, Ian, *Software Engineering*, 9th edition, Pearson/Addison-Wesley Upper Saddle River, NJ, 2010.

Sperber, Michael, Dybvig R. Kent, Flatt, Matthew, Straaten, Anton V., Findler, Robby, and Matthews, Jacob, Revised report on algorithmic language scheme, *Journal of Functional Programming*, 19(S1), 2009, 1–301.

Srinivasan, Raj, "RPC: Remote Procedure Call Protocol Specification Version 2," *Internet Request for Comments 1831*, 1995, available at http://www.rfc-archive.org/getrfc.php?rfc=1831. Accessed on October 5, 2013.

Stansifer, Ryan, *The Study of Programming Languages*, Prentice Hall, Engelwood Cliff, NJ, 1995.

Steele, Guy L. Jr., *Common LISP—The Language*, 2nd edition, Digital Press, Waltham, MA, 1990.

Steele, Guy L. Jr. and Gabriel, R. P., "The evolution of LISP," *ACM History of Programing Languages Conference II, SIGPLAN Notices*, 28(3), 1993, 231–270.

Stevenson, Dorothy E., *Programming Language Fundamentals by Example*, Auerbach Publications, 2006.

Strachey, Chris, "Fundamental Concepts in Programming Languages," *Higher-Order and Symbolic Computation*, 13, 2000, 11–49.

Strom, Robert E., Bacon, David F., Goldberg, Arthur P., Lowry, Andy, Yellin, Daniel M., and Yemini, Shaula A., *Hermes: A Language for Distributed Computing*, Prentice Hall, Englewood Cliff, NJ, 1991.

Stroustrup, Bjarne, *The C++ Programming Language*, 3rd edition, Addison-Wesley, Reading, MA, 1997.

Sunderam, Vidyalingam S., "PVM: A framework for parallel distributed computing," *Concurrency—Practice and Experience*, 2(4), 1990, 315–339.

Syme, Don, Battocchi, Keith, Fisher, Jomo, Hale, Michael, Hu, Jack, Hoban, Luke, Liu, Tao, Lomov, Dmitry, Margetson, James, McNamara, Brian, Pamer, Joe, Orwick, Penny, Quirk, Daniel, Smith, Chris, Taveggia, Matteo, Malayeri, Donna, Chae, Wonseok, Matsveyeu, Uladzimir, and Atkinson, Lincoln, *The F# 3.0 Language Specification*, available at http://research.microsoft.com/en-us/um/cambridge/projects/fsharp/manual/spec.pdf. Accessed on October 20, 2013.

Tanenbaum, Andrew S., *Structured Computer Organization*, 5th edition, Pearson Prentice Hall, Englewood Cliff, NJ, 2005.

Taubenfeld, Gadi, *Synchronization Algorithms and Concurrent Programming*, Pearson/Prentice Hall, Upper Saddle River, NJ, 2006.

Tennet, Robert D., *Principles of Programming Languages*, Prentice Hall, Englewood Cliffs, NJ, 1981.

Thomas, Dave, Fowler, Chad, and Hunt, Andy, *Programming Ruby: The Pragmatic Programmers' Guide*, 2nd edition, Pragmatic Bookshelf, Frisco, TX, 2005.

Tilevich, Eli and Smaragdakis, Yannis, "Portable and Efficient Distributed Threads for Java," in *Proceedings of the ACM/IFIP/USENIX Middleware Conference*, Springer Verlag, New York, NY, 2004, pp. 478–492.

Tucker, Allen B. and Noonan, Robert E., *Programming Languages: Principles and Paradigms*, 2nd edition, McGraw Hill, New York, NY, 2007.

Turner, David A., "Miranda: A non-strict Functional Language with Polymorphic Types," in *Proceedings IFIP International Conference on Functional Programming Languages and Computer Architecture, LNCS 201*, Springer Verlag, Berlin, Germany, 1985, pp. 1–16.

W3C World Wide Web Consortium. *Synchronized Multimedia Integration Language-SMIL 3.0 Specification*, W3C Candidate Recommendation, 2008, available at http://www.w3.org/TR/SMIL3/. Accessed on October 5, 2013.

Wadler, Philip L., "The Essence of Functional Programming," in *Proceedings of the 19th Annual ACM Symposium on Principles of Programming Languages*, Albuquerque, NM, 1992, pp. 1–14.

Wampler, Dean and Payne, Alex, *Programming Scala*, O'Reilly Media, Sebastopol, CA, 2009.

Wang, Cheng, Wu, Youfeng, Borin, Edson, Hu, Shiliang, Liu, Wei, Sager, Dave, Ngai, Tin-fook, and Fang, Jesse, "Dynamic Parallelization of Single—Threaded Binary Programs using Speculative Slicing," in *Proceedings International Conference on Supercomputing*, Yorktown Heights, NY, 2009, pp. 158–168.

Watt, David A., *Programming Language Design Concept*, John Wiley & Sons, West Sussex, UK, 2004.

Weber, Adam B., *Modern Programming Languages: A Practical Introduction*, Franklin, Beedle & Associates Inc., WilsonVille, OR, 2003.

Wirth, Niklaus, *Algorithm + data Structure = Programs*, Prentice Hall, Englewood Cliffs, NJ, 1976.

Wirth, Niklaus, "Modula: A language for modular multi-programming," *Software Practice and Experience, 7*, 3–35.

Wirth, Niklaus, *Programming in Modula-2*, 3rd edition, Springer Verlag, Berlin Heidelberg, Germany, 1985.

Wolfe, Michael, *High Performance Compilers for Parallel Computing*, Addison Wesley, Boston, MA, 1996.

World Wide Web Consortium, *Extensible Style Sheet (XSL) Version 1.1*, 2006, available at www.w3.org/TR/xsl. Accessed on October 5, 2013.

World Wide Web Consortium, *Simple Object Access Protocol (SOAP) 1.1, 2000*, available at http://www.w3.org/TR/2000/NOTE-SOAP-20000508/. Accessed on October 5, 2013.

World Wide Web Consortium, *XML Current Status*, available at http://www.w3.org/XML/. Accessed on October 5, 2013.

World Wide Web Consortium, *XML Path Language (XPath), Version 2.0*, 2007, available at www.w3.org/TR/xpath20/. Accessed on October 20, 2013.

World Wide Web Consortium, *XSL Transformations (XSLT), Version 2.0*, 2007, available at www.w3.org/TR/xslt20/. Accessed on October 5, 2013.

Wu, Daniel, Agarwal, Divyakant, and Amar El Abdali, "StratOSphere: Mobile Processing of Distributed Objects in Java," in *Proceedings of the 4th annual ACM/IEEE international conference on Mobile computing and networking*, 1998, Dallas, TX, pp. 121–132.

Zhou, Neng-Fa, "Parameter passing and control stack management in prolog revisited," *ACM Transactions of Programming Languages and Systems*, 18(6), 1996, 752–779.

Zibin, Yoav and Gil, Joseph, "Two-Dimensional Bi-Directional Object Layout."in *Proceedings of the European Conference on Object-oriented Programming*, edited by L. Cardelli, LNCS 2743, Springer Verlag, Berlin Heidelberg, Germany, 2003, pp. 329–350.

Index

B

D

M

N

O

P

S

U

X

Y

Z

Milton Keynes UK
Ingram Content Group UK Ltd.
UKHW051928141024
449569UK00027B/1401